Handbook of Sociotechnical Systems

This handbook presents a bold rethinking of human systems integration (HSI) within sociotechnical systems (STSs), capturing the shift from traditional human–machine interaction to a dynamic, multi-agent systems approach. Drawing from various disciplines, including human factors, ergonomics, systems engineering, and artificial intelligence (AI), the handbook explores the integration of technology, organizations, and people in increasingly complex and digitalized environments. The evolution reflected in these pages aligns with broader societal trends, from the rise of digital engineering and virtual prototyping to the growing necessity for holistic, context-aware design in life-critical domains.

Fully updated, this new edition provides the tools and perspectives to shape resilient, intelligent, and human-compatible systems. As a comprehensive and reorganized reference spanning 14 sections and 45 chapters authored by leading experts from diverse fields, the handbook explores models, methods, and real-world applications that support the design, development, and management of complex STSs. Key topics include human–autonomy teaming, performance support, scenario-based design, organizational maturity, and the crucial role of tangibility in digital systems. For professionals focused on AI integration, safety assurance, or enhancing user experience across system lifecycles, the resource offers actionable insights and proven frameworks to meet the challenges of modern system complexity. Readers gain a thorough, helicopter view of HSI and STSs in various usable contexts.

Handbook of Sociotechnical Systems: A Human Systems Integration Approach, Second Edition, is an essential resource for engineers, designers, researchers, policymakers, practitioners, and educators who work with people, technology, and organizations. The content proves especially valuable for those involved in high-stakes, high-complexity environments such as aerospace, defense, healthcare, energy, and public systems.

Handbook of Sociotechnical Systems

A Human Systems Integration Approach

Second Edition

Edited by
Guy André Boy

CRC Press
Taylor & Francis Group
Boca Raton London New York

CRC Press is an imprint of the
Taylor & Francis Group, an **informa** business

Designed cover image: Shutterstock

Second edition published 2026
by CRC Press
2385 NW Executive Center Drive, Suite 320, Boca Raton, FL 33431

and by CRC Press
4 Park Square, Milton Park, Abingdon, Oxon, OX14 4RN

CRC Press is an imprint of Taylor & Francis Group, LLC

[First edition published by CRC Press 2011]

Library of Congress Cataloging-in-Publication Data
Names: Boy, Guy A., editor.
Title: Handbook of sociotechnical systems : a human systems integration approach /
edited by Guy André Boy.
Description: Second edition. | Boca Raton, FL : CRC Press, 2026. |
Includes bibliographical references and index.
Identifiers: LCCN 2025043781 | ISBN 9781041122289 (hardback) |
ISBN 9781041127550 (paperback) | ISBN 9781003666431 (ebook)
Subjects: LCSH: Sociotechnical systems. | Human-machine systems. |
Telecommunication–Social aspects. | Telematics.
Classification: LCC T59.7 .H36 2026
LC record available at https://lccn.loc.gov/2025043781

ISBN: 9781041122289 (hbk)
ISBN: 9781041127550 (pbk)
ISBN: 9781003666431 (ebk)

DOI: 10.1201/9781003666431

Typeset in Times
by Newgen Publishing UK

Contents

SECTION II Theoretical Approaches and Methodologies

SECTION III Modeling and Simulation Approaches

SECTION IV Human–Automation Interaction and Performance

SECTION V Emerging Technologies and Human Experience

About the Editor

Guy André Boy, PhD, is an Emeritus Professor who previously held the FlexTech Chair University Professorship at CentraleSupélec (Paris Saclay University) and ESTIA in France, while also earning fellowships with INCOSE, the Air & Space Academy, and the International Academy of Astronautics. Throughout his distinguished career, Dr. Boy served as University Professor and Dean at Florida Institute of Technology (FIT) in the USA, simultaneously working as IPA Chief Scientist at NASA Kennedy Space Center, following his leadership as President and Chief Scientist at the European Institute of Cognitive Sciences and Engineering, a research center jointly established by Airbus and Thales. His academic influence spans numerous prestigious institutions where he taught Human-Centered Design and human systems integration (HSI), including ISAE-SUPAERO, École Polytechnique, FIT, ESTIA, and CentraleSupélec, building upon his earlier experience as a group lead at both ONERA (the French Aerospace Institute) and NASA Ames Research Center. As a prolific scholar, Dr. Boy has authored more than 250 refereed articles and 15 books, while also making significant contributions through his leadership roles as Chair of the INCOSE HSI working group (2015–2024), ACM SIGCHI Co-Chair (1995–1999), and Chair of the IEA Aerospace technical committee (2008-2024), cementing his legacy as a pioneering figure in human-centered design and cognitive engineering.

Contributors

Audrey Abi Akle, PhD
Assistant professor in Design Engineering at ESTIA Institute of Technology, France, focusing on decision-making in complex systems design with human-centered approaches. Teaches design methods and leads the "Research Project" module. Holds a PhD from CentraleSupélec, an engineering degree from ESTIA, France, and an MSc from the University of Salford, UK.

Julien Ambrosino, PhD
Leads Lectra's Startup Studio and serves as Research Associate at ESTIA Institute of Technology, France. Focuses on creativity-driven innovation across fashion tech, AI, and sustainability. Earned a PhD in Innovation & Creativity Management, designs innovation workshops, and bridges applied research with industry to transform ideas into market-ready solutions.

Kahina Amokrane-Ferka, PhD
Senior Researcher in Human–Machine Interaction and system architect at IRT SystemX, France. Specializes in user-centered adaptive systems. Holds a PhD from the University of Technology of Compiegne. Her research interests include Human AI Teaming, task allocation, and User Experience.

Thierry Bellet, PhD
Senior Researcher at LESCOT, University Gustave Eiffel, France. Holds a PhD in Cognitive Ergonomics and a master's in artificial intelligence, Ergonomics, and Psychology. Research focuses on Human Driver Modeling (COSMODRIVE) and Cognitive Engineering for co-piloting systems and human-like automated vehicles.

Arnon Bentur, PhD
Professor Emeritus at Technion, Israel, with expertise in construction materials. Held positions as Vice President, Dean, and Director. Published numerous papers and ten books, served as RILEM President, and chaired multiple educational committees in Israel's higher education system.

Benjamin Berton
PhD candidate at Polytechnique Montréal, Canada, researching Cognitive Engineering and Human–Autonomy Teaming in commercial aviation, emphasizing cognitive modeling for evaluating function allocation schemes.

Jawahar Bhalla (JB)
Systems engineering leader with 30+ years of experience across defense, transport, and infrastructure sectors. Former President of SESA, Australia. Systems thinker, modeling and simulation expert. Fellow of Engineers Australia with qualifications in Aerospace Engineering and Computer Science. Currently pursuing a PhD at the University of Adelaide, researching autonomous systems frameworks.

Alexandra A. Boeing
Applied Organizational Research Leader at Future of Work Institute, Curtin University, Australia. Engages with industry and government to improve workplace practices while developing novel methodologies for evaluating human elements in complex engineering systems.

Jean-Charles Bornard, PhD
Research engineer at LESCOT Laboratory, Gustave Eiffel University, France, with a PhD in Cognitive Engineering. Focuses on cognitive modeling of human drivers through COSMODRIVE and developing Virtual Human-Centered Design platforms to guide the creation of human-like automated vehicles.

Élodie Bouzekri, PhD
Assistant Professor at the University of Brest, France, specializing in Human–Computer Interaction, focusing on interaction design and collaboration with autonomous systems. Chairs the IFIP TC 13 Working Group 6 on Human Work Interaction Design.

Guy André Boy, PhD
Emeritus Professor and former FlexTech Chair at CentraleSupélec & ESTIA Institute of Technology, France. Fellow of INCOSE, Air & Space Academy, and International Academy of Astronautics. Former NASA Chief Scientist, Dean at Florida Institute of Technology, USA, and author of 250+ articles and 15 books on Human-Centered Design and HSI.

Stélian Camara Dit Pinto, PhD
Research scientist with a PhD in industrial engineering from Paris-Saclay University, France. Research focused on the human systems integration paradigm for digital twins. He is a postdoctoral fellow at Long Island University, USA, designing human digital twins for medical decision-making.

Belinda S. Cham, PhD
Research fellow at Future of Work Institute with a background in Organizational Psychology, Curtin University, Australia. Worked with the Defense Science and Technology Group for 9 years, optimizing human performance. Focuses on designing work systems for sustained performance in high-risk environments.

Zhi Ching Chen
Bachelor's Student at Beihang University. Research focuses on intelligent software engineering, integrating model-driven and data-driven approaches. Developed tools for H-CPS system modeling.

Robyn Coman, PhD, CPE, AHPRA
Registered physiotherapist, certified professional ergonomist, and Senior Lecturer in OHS at the University of Wollongong. Researching physiotherapy and ergonomic interventions in health, aged care, and disability sectors at the Centre for Occupational, Public and Environmental Research in Safety and Health (COPERSH).

Kevin Crowston, PhD
Distinguished Professor of Information Science at Syracuse University. His research examines new ways of organizing enabled by information technology; theoretical characterizations of coordination problems and alternative methods for

managing them; and the design and empirical evaluation of systems that support people working together.

Brigitte Daniel Allegro
Specialist in Experience-Based Dependability and Systems Engineering in the nuclear and aeronautics sectors. Introduced systems engineering at Airbus, France, in the 1990s and created the ©SystemTree conceptual model. Active in academic and industrial fields promoting Systems Thinking.

Peter Denno, PhD
Computer scientist and industrial engineer at NIST's Engineering Laboratory, USA, researching smart manufacturing, supply chains, and human-centered AI. Previously worked in aerospace for ten years. Currently leads a project on human/machine teaming for manufacturing digital twins.

Alexandre Disdier, PhD
Post-doctoral researcher and assistant professor at ESTIA Institute of Technology, France. FlexTech chair member with a PhD in Complex Systems Engineering. Research early design stages in Systems Engineering, focusing on human systems integration approaches, including simulation and contextual modeling.

Marko Djapan, PhD
Associate Professor in Industrial Engineering at the University of Kragujevac, Serbia. Former researcher at Politecnico di Torino, Italy. Research interests include occupational safety and health, collaborative robotics, ergonomics, Industry 4.0/5.0, and workplace organization.

Aleksandar Djordjevic, PhD
Associate Professor of Information Engineering at the University of Kragujevac, Serbia, and Computer Integrated Manufacturing Center director. Research focuses on databases, information systems, web applications, Internet of Things, and their application in Quality 4.0.

Ondrej Doule, PhD
Expert in human–system performance improvement. Former NASA JSC MBSE specialist and director of Florida Institute of Technology's Human Spaceflight Laboratory, USA. Led R&D projects for the European Commission, NASA, and the FAA. Developed the first AIAA ANSI Spacecraft Ontology standard.

Philippe Doyon-Poulin, PhD
Associate Professor at Polytechnique Montreal, Canada, focusing on human factors in aviation, error prevention, and decision-making with automated systems. Authored over 25 aircraft certification reports on flight deck usability and pilot error.

Mica Endsley, PhD
Dr. Endsley is president of SA Technologies and a former Chief Scientist of the U.S. Air Force. She has also held the positions of Visiting Associate Professor at MIT in the Department of Aeronautics and Astronautics and Associate Professor of Industrial Engineering at Texas Tech University. Dr. Endsley is a recognized world leader in the design, development, and evaluation of

systems to support human situation awareness (SA) and decision-making, as well as methods for improving SA through the training of individuals and teams. She has published over 200 articles on situation awareness and the effects of artificial intelligence and automation on human performance. Dr. Endsley has a PhD in Industrial & Systems Engineering from the University of Southern California and is a fellow and former President of the Human Factors and Ergonomics Society.

Frank Ole Flemisch, Prof. Dr.-Ing.
Professor of human systems integration at RWTH Aachen University, Germany. Former aerospace engineer at NASA, research group leader at the German Aerospace Center. Coined terms "highly automated driving" and "cooperative automation." Currently, Fraunhofer FKIE's Principal Scientist-Human Systems – Evangelist, and Vice-Chair of NATO's Human Factors & Medicine Panel.

Juntao Gao, PhD
Associate Professor and Master's Supervisor at Northeast Petroleum University, China. Research focuses on software engineering, process mining, business process management, and big data analytics. CCF Software Engineering Committee member and Heilongjiang Provincial Backbone Young Teacher.

Nirit Gavish, PhD
Head of Industrial Engineering and Management at Braude College and Senior Lecturer/Teaching Associate at Technion, Israel. Research focuses on human factors, human–computer interaction, decision making, augmented/virtual reality, and cognitive training.

John Gill, PhD
INCOSE ESEP with four decades in defense technology acquisition. Former chief scientist at Kirtland AFB Advanced Weapons Lab, global systems project manager, Systems Engineering Director at BAE Systems, and Senior Director of Air Autonomy at Scientific Systems Company, USA.

Mark A. Griffin, PhD
John Curtin Distinguished Professor and Director of the Future of Work Institute at Curtin University, Australia. Global leader in health and safety research and expert in multilevel data modeling. Fellow of the US Society for Industrial/Organizational Psychology with Presidential recognition.

Gudela Grote, PhD
Professor of Work and Organizational Psychology at ETH Zürich, Fellow of the Society for Industrial and Organizational Psychology, and Past President of the European Association of Work and Organizational Psychology. She studies the relation between technology, organization, and work, and uncertainty regulation by individuals and organizations.

Holly A. H. Handley, PhD
Professor in Engineering Management and Systems Engineering at Old Dominion University, USA. Licensed Professional Engineer focusing on socio-technical systems. Author of "The Human Viewpoint for System Architectures" and co-developer of the Human Readiness Levels standard.

Katrina L. Hosszu
Applied Organizational Research Specialist at Future of Work Institute, Curtin University, Australia. Works at the intersection of human-systems integration and organizational psychology, collaborating with industry to design work systems supporting employee performance and well-being.

Gopal R. Jani
Information technology engineer, researcher, and entrepreneur pursuing PhD in Human-Centered Design. Two decades of experience leading telecom and 5G infrastructure projects. Heads BhumiTat Technologies, India, focusing on 5G services, artificial intelligence, cloud, blockchain, and Internet of Things innovations.

Karina Jorritsma, PhD
Professor and applied organizational psychology researcher, Curtin University, Australia, with 15+ years of experience bridging academic thinking with industry challenges across healthcare, mining, aviation, emergency services, and military sectors. Specializes in designing employee performance and well-being interventions.

Vivek Kant
Researcher in systems design, human factors, and sociotechnical systems. He heads Human Factors studios at IIT Kanpur, India. Holds Electrical Engineering, Cognitive Science, Psychology, and Systems Design Engineering degrees. Former Postdoctoral Fellow at NTU Singapore.

Osmo Kauppila, DSc
Degree program director and university lecturer of Industrial Engineering and Management at the University of Oulu, Finland. Adjunct professor at Luleå University of Technology. Research interests include HSEQ management, supply chain resilience, and Quality 4.0.

Grace A. L. Kennedy, CPEng, CSEP
Principal Consultant in Human Systems Engineering at Acmena Group. Over 16 years' experience in the integration of human factors and systems engineering. Specializing in the modelling of sociotechnical systems in the development and introduction of new technologies. Chair of INCOSE Human Systems Integration Working Group.

John Paul Kennedy, MPhys, PhD
Lecturer in STEM Education with Adelaide University Education Futures and a member of the Centre for Change and Complexity in Learning (C3L). Focused on industry-engaged research around data analytics for attitudinal change.

Kyoung-Yun Kim, PhD
Professor of Industrial and Systems Engineering at Wayne State University, USA, directing the Computational Intelligence and Design Informatics Laboratory. Research focuses on Design Science and Smart Manufacturing. Fellow of the Society for Design and Process Science and Editor-in-Chief of JIDPS.

Prosper Ameh Kwei-Narh, PhD
Senior Human and Organizational Factors scientist at the Institute for Energy Technology, Norway. PhD from BI Norwegian Business School. Research interests include shared team mental models, talent management, and team performance.

Paul Lambertson
Senior aerospace engineer with 20+ years of experience focusing on human–machine interfaces for flight decks. Led projects across all Boeing Commercial Aircraft, USA, involving human factors, ergonomics, and testing. Currently leads flight deck product development for simulation and strategy.

Danielle Lamoureux
Finance graduate from Bentley University with a master's in business Analytics from the University of Delaware, USA. Former Compliance Vice President at JP Morgan Chase, certified Anti-Money Laundering expert, and Combatting Terrorist Financing Specialist.

Jérémy Legardeur, PhD
Full professor at ESTIA Institute of Technology, University of Bordeaux, France. Founder of "The 24h of Innovation" event. Research focuses on creativity and sustainable innovation methods for socioecological transition. Chief scientist of the BALI Chair promoting textile circularity. Published over 150 papers.

Suvi Leinonen, MSc
Doctoral Researcher at the University of Oulu, Finland, with 15+ years of industry experience in supply chain management, leadership, and logistics. Research interests include supply chain resilience, organizational resilience, dynamic capabilities, and supplier relationship management.

Nicolas Lelouche
Computer engineer and research associate at NIST with the University of Maryland, USA. ECE Paris graduate. Former PREP researcher with Georgetown University. Interests include artificial intelligence for manufacturing, programming side projects, and human–computer interfaces.

Dimitri Masson, PhD
Assistant professor at ESTIA Institute of Technology, France, specializing in Human System Integration. Research focuses on systems supporting human cognitive activities. Co-chair of the INCOSE Human System Integration Working Group and INCOSE Technical Leadership Institute member. PhD from the University of Grenoble Alpes, France.

Christophe Merlo, PhD
Professor at ESTIA Institute of Technology, France, and current Dean. Former consulting engineer in CAD/CAM and PLM. Research focuses on cyber-physical systems design processes using human-centered approaches. Author of 190+ publications, including 20 international journals. PhD from the University of Bordeaux.

Michael E. Miller, PhD
Professor of Systems Engineering at Air Force Institute of Technology, USA, teaching human factors and

human-agent teaming. PhD from Virginia Tech with 20 years in product design and 15 years in academia. Author of 100+ patents, 50+ journal articles, and two books.

Chloé Morel
Human Factors and Ergonomics specialist with a master's from the University of Paris, France. Member of the ESTIA PEPSS platform since 2016. Evaluates innovative systems including AR/VR in industrial, military, and aeronautic fields. Works on projects involving human-centered design and human–machine performance.

Peter Mörtl, PhD
Leads human factors group at Virtual Vehicle Research Center, Austria, focusing on driver monitoring, cognitive modeling, and adaptive human–computer interaction. 25+ years of experience in human factors research across aviation and automotive engineering. Holds a doctorate from the University of Oklahoma.

Marie-Pierre Pacaux-Lemoine, PhD
Researcher at Université Polytechnique Hauts-de-France specializing in Human–System Integration and Human–Machine Cooperation. Research focuses on the dynamic adaptability of functions shared between humans and machines and has been published extensively on applications in transportation, manufacturing, and military domains.

Sharon K. Parker, PhD
Australian Research Council laureate fellow, Director of the Centre for Transformative Work Design, and John Curtin Distinguished Professor of Organizational Behavior at Curtin University. Her research focuses on job and work design, employee performance, proactive behavior, organizational change, work stress, and quasi-experimental designs.

Elina Parviainen
Process analyst at Meyer Turku Shipyard, Finland, with extensive experience in manufacturing processes. Focus areas include human system integration, ergonomics, and systems thinking – former president of Finnish and Nordic Ergonomics Societies and chair of IEA's Promotion Committee.

Véronique Pilnière, PhD
Associate Professor in Management Sciences and Ergonomist at ESTIA Institute of Technology, France, since 2004. Specializes in complexity, organizational change, and professional risk management. Practices action research and specialize in the interactive approach of Palo Alto as presented by the Gregory Bateson Institute.

Romain Pinquié, PhD
Professor of digitally mediated collaborative systems design at the University of Grenoble Alpes, France. Creates interactive three-dimensional systems modeling interfaces combining human–computer interaction, artificial intelligence, and CSCW – scientific coordinator of the MIMESIS joint research lab, developing immersive virtual environments for model-based systems engineering.

Michael Preutenborbeck
Industrial engineering researcher at RWTH Aachen University, Germany. Group lead of balanced human systems integration research and assistant to Principal Scientist at Fraunhofer FKIE. Research focuses on participatory design of sociotechnical systems.

Carolyn Psenka, PhD
Cultural anthropologist and research associate at Wayne State University's Industrial and Systems Engineering Department, USA. Works with Computational Intelligence and Design Informatics Laboratory since 2012. Research includes user experience with sustainable product development systems.

Kine Reegård, MPhil
Senior Human and Organizational Factors scientist at the Institute for Energy Technology, Norway. Specializes in human factors across the petroleum, nuclear, aviation, and healthcare sectors. Research focuses on how technologies impact work experiences.

Arto Reiman, DSc
Research team leader at the University of Oulu, Finland, focusing on workplace well-being and productivity. Has 20 years of experience in occupational safety, ergonomics, and human factors. Former president of the Finnish Ergonomics Society and active in international ergonomics networks.

Alexander Rudolph, PhD
Engineering manager at EUMETSAT with a PhD in robotics from Darmstadt University, Germany. Holds ESEP and SEZERT certifications. Background spans control theory, nuclear power, aircraft systems, and automated vehicles. He focuses on achieving system acceptance during transition to operational use.

Yuichi Saito, Dr.- Eng.
Assistant Professor, Institute of Systems and Information Engineering, University of Tsukuba, Japan. His research interests include smart collaboration between humans and machines, shared control, and vehicle dynamics. Currently the Co-Chair of the Technical Committee on Shared Control in the IEEE Systems, Man, and Cybernetics Society.

Parag Sanghani, PhD
Management expert with 25+ years of experience in infrastructure and education. Vice Chancellor of P P Savani University, India, with postdoctoral research in IT governance from Queensland University of Technology. Bridges technology, education, and leadership for societal development.

Marc Sango
Senior project manager in systems engineering at SNCF with a PhD from the University of Lille, France. Leads SNCF's System Engineering Working Group and co-leads the Engineering Environment Team within Europe's Rail program System Pillar.

Rachael Sexton
Mechanical engineer at NIST's Information Modeling and Testing Group, USA. Led Knowledge Extraction & Application Project and co-founded Technical Language Processing Community. Researches text analysis and network science for knowledge management in technical domains.

Lance Sherry, PhD
Professor at George Mason University and Director of the Center for Air Transportation Systems Research, USA. Pioneered data analytics and stochastic simulation for complex adaptive systems. Award-winning work in human–computer interaction and designing autonomous systems.

Jackelynne P. Silva-Martinez, PhD
NASA SpaceX Starbase Resident Office Lead for Human Landing System and Artemis Mission Integrator, USA. Previously served as Human Systems Engineering Lead for Commercial LEO Development Program and Flight Controller for ISS and founded NASA's Agile Community of Practice.

Jaemun Sim, PhD
Research Associate at Wayne State University, USA, since 2016. Published in leading decision science journals. Research interests include human–computer interaction, creativity science, and AI applications in business, traffic systems, healthcare, and manufacturing.

Anabela Simoes, PhD
Former Full Professor at Lisbon University and LUSOFONA University, Portugal. Past President of the Portuguese Ergonomics Society and International Ergonomics Association council member. Member of the HUMANIST Association since 2008, developing research through European-funded projects.

Caitlyn A. K. Singam
INCOSE-accredited systems engineer researching astrobiology and autonomous space systems. Program Director for INCOSE SySTEAM Initiative and Assistant Editor for Systems Engineering Body of Knowledge. NSF Fellow with degrees from the University of Maryland, USA.

Amandeep Singh, PhD
Postdoctoral Researcher at Polytechnique Montreal and Research Associate at University of Waterloo, Canada. She specializes in human factors and has 45+ peer-reviewed publications. Research interests include transportation systems, biomechanics, cognitive ergonomics, and predictive modeling.

Tero Sotamaa, MSc
Doctoral researcher at the University of Oulu, Finland, focusing on risk management and resilience in industrial companies. Research emphasizes technological capabilities in manufacturing SMEs. Has 15+ years of operational experience in automotive organizations, including BMW, Toyota, and Ford.

Jean-Jacques Speyer
Graduate engineer with extensive aviation experience, Vrije Universiteit Brussel (VUB, Belgium) and Massachusetts Institute of Technology (MIT, USA).

Instrumental in two-person crew certification at Airbus for multiple aircraft models. Expert in cockpit design, human factors, and flight operations at Airbus. Professor teaching aircraft certification and safety regulations at VUB.

Wei Tan, PhD
Faculty member at Civil Aviation University of China with a PhD from Florida Institute of Technology, USA. Holds senior ground instructor license and SMS auditor training. Research focuses on human factors in aviation, safety management, and human–machine interaction in civil aircraft.

Ari Teppo
Product Manager at Nokia, Finland, with 34+ years of experience in manufacturing and systems engineering. Bachelor in Automation. Led MBSE initiatives, developed architecture frameworks, and trained teams in SysML – extensive background in test engineering and quality management.

Michael C. Todd, PhD
Assistant Professor and Deputy Department Head at the USAF Academy in Computer Science. His research focuses on augmented intelligence systems, and he has 14 years of experience as a Special Agent and Computer Crimes Investigator.

Marcel Usai
Researcher at Fraunhofer FKIE with an electrical engineering background from RWTH Aachen University, Germany. PhD candidate focusing on interaction patterns for fluid human-machine cooperation. Previously worked at the Institute of Industrial Engineering and Ergonomics.

Leanne Treadwell, BEng
Chemical Engineer and Risk Management specialist with over 30 years' experience in practical application of OHS management systems as a safety manager, regulator, and consultant across industries ranging from heavy manufacturing, IT, transport, security, and aged care. She teaches Risk Management and OHS at the University of Wollongong.

Stéphane Valès
CEO of Ingenuity I/O, France, focusing on Systems of Systems with humans and artificial intelligence. Created Ingescape solution bridging MBSE, Human–Computer Interaction, and Human Factors. 20 years of experience deploying complex interactive systems across aerospace, transport, energy, and defense sectors. Contributed to publications in CSCW and interactive software engineering.

Yilong Yang, PhD
Associate Professor at Beihang University's School of Software, China. Research focuses on intelligent software engineering, integrating model-driven and data-driven approaches. Developed tools for HarmonyOS and automotive systems. Executive committee member for China Computer Federation and program committee member for international conferences, including MoDELs, ICSE, and INCOSE.

Bingjie Zeng
Master Student at Northeast Petroleum University. Research focuses on intelligent software engineering, with an emphasis on model-driven methods for artificial intelligence systems and Human-Cyber-Physical Systems (H-CPS).

Avigdor Zonnenshain, PhD
INCOSE Fellow and Senior Research Fellow at Gordon Center for Systems Engineering, Technion, Israel. Expert in quality, reliability, and systems engineering at RAFAEL and the Prime Minister's Office. Active in the Israel Society for Quality and past president of the INCOSE Israel .

Foreword

Dr. Azad M. Madni, NAE Gordon Prize and IEEE Simon Ramo Medal Recipient

AIAA Honorary Fellow, ASME Honorary Member, IEEE Life Fellow, INCOSE Life Fellow and recipient of Pioneer, Founders and Benefactor Awards, ACM Fellow, AAAS Fellow, IISE Fellow, AIMBE Fellow, SME Fellow, WAS Fellow.

University Professor of Astronautics, Aerospace and Mechanical Engineering, University of Southern California; CEO and Chief Scientist, Intelligent Systems Technology, Inc.

I am a member of the National Academy of Engineering, and have the designation of University Professor of Astronautics, Aerospace and Mechanical Engineering in the University of Southern California. I am the Executive Director of USC's flagship Systems Architecting and Engineering Program in which human systems integration (HSI) is taught. I have been investigating HSI approaches as part of my sponsored research in augmented intelligence and human performance modeling for more than four decades. Over that period, the HSI field has evolved from its classical human factors engineering roots to its current expanded version, which embraces concepts from cognitive psychology, computer science, social sciences, learning science, and artificial intelligence (AI) and machine learning (ML). These new perspectives have enriched the field by introducing new perspectives in HSI methodology and tools.

HSI has become increasingly more important with the growing complexity of sociotechnical systems (STSs) on the one hand, and ongoing advances in behavioral modeling, and AI/ML technologies on the other. HSI came into existence with the recognition that humans are an integral part of complex STSs and that human considerations cannot be an afterthought when architecting, engineering, and integrating STSs. HSI evolved from its human factors engineering roots to address a broader range of disciplines, including manpower, personnel, training, and system safety. The US military played a key role in its evolution through its initiatives such as US Army's MANPRINT and US Navy's HARDMAN, which emphasized incorporating human considerations into upfront systems engineering. NASA also contributed to HSI methodology and tools development through programs conducted by its human systems integration Division at Ames Research Center.

I have known Professor Guy André Boy, the editor of this volume, for almost two decades. An authority on human-centered design, he has published extensively on the subject over the past two decades. As chair of the INCOSE HSI working group for a decade, he helped bring HSI into the mainstream of Systems Engineering. In this volume, he has organized key HSI topics that include recent advances in each into fourteen standalone sections comprising forty-five chapters written by leading experts in the field. This structure makes it easy for readers to readily access and peruse specific HSI topics of their immediate interest. Recent advances in model-based HSI methods, human–autonomy teaming, human-in-the loop modeling and simulation, as well as key topics such as complexity, sustainability, AI, learning systems, ontological and transdisciplinary approaches, and digital engineering are specifically addressed to give the reader a holistic and comprehensive view of the subject.

This timely and cogent volume is well suited for a broad readership, including researchers, practitioners, instructors, and students. It adds to the existing body of knowledge in the field and discusses new topics from different perspectives that allow readers to rapidly familiarize themselves with recent advances in the field. It clarifies the connection between HSI and systems engineering, and discusses several new concepts such as frameworks, ontologies, and human-computer teaming from an HSI perspective.

This edited volume is a valuable addition to the HSI body of knowledge. It is written from the perspective of providing a comprehensive overview of this rapidly evolving field. Whether your interest is in integrating AI with humans, or safety assurance, or user experience enhancement, this book is a useful resource for you.

Preface

Technology is a useful servant but a dangerous master.

Christian Lous Lange

Humans are the center of any system. Even the most complex systems do not operate fully on their own, for their own purposes. They are an extension of human will – a means to achieve people's desired ends – and as such require effective guidance and direction, a blending of system capabilities with the constraints and variability of the world in which they exist. Technology cannot be considered to consist in a vacuum, but only as a part of the larger sociotechnical system in which it is imbedded.

New technologies, including automation, robotics, and artificial intelligence (AI), have rapidly changed the nature of people's relationship with technology. These systems have moved from interactions that are largely mechanical – focused on physical controls, perceptual readability of dials, and the ability to hear alarms – to a human–machine interaction that is primarily cognitive, emphasizing high levels of information exchange and joint interactions to achieve dynamic goals. Understandability and predictiveness of system behavior are now fundamental to people's ability to work within any complex technological system. Most recently, AI large language models (LLM) provide new forms of interaction using natural language that in some ways provides a simpler means of communication for people and technology, but is also fraught with many hidden challenges as people struggle to make accurate assessments of information veracity and reliability. Easy is not always the best when performance matters.

Against this backdrop, effective human–system integration (HSI) takes center stage in the development of useful technological systems. Today's technologies must be designed to allow people to be successful in their central role of directing system behavior to meet human goals. This primary design objective cannot be achieved as an afterthought. It requires a focused and systematic approach from system inception through to the design and testing of any technological system. The complex ways in which the full set of technologies, people, and processes interact and operate in context is a fundamental engineering consideration for system development.

The *Handbook of Sociotechnical Systems: A Human Systems Integration (HSI) approach* provides a set of tools, methods, and approaches to guide this endeavor. Bringing together classical fields of work including human factors and ergonomics, information technology, and systems engineering with relevant domain knowledge, HSI bridges the gap between the various fields of engineering and science to provide a holistic approach to the design of complex systems. HSI contributes a systematic and principled approach to creating human-centered systems that is essential for the development of new technologies and complex systems in today's world.

Mica R. Endsley

Section I

Foundations and Frameworks

1 Introduction

Guy André Boy

This handbook follows the *Handbook of Human–Machine Interaction: A Human-Centered Design Approach*, published in 2011. It could have been called *Handbook of Human–Machine Interaction: An Human Systems Integration (HSI)* approach. However, the emphasis on social issues is so crucial at the start of the 21st century that we have preferred to focus on Sociotechnical Systems (STSs) issues. Moreover, the systemic aspect of the evolution of human–machine interaction has become so important that it should feature prominently in the title of this handbook. Finally, we maintain the tradition of specifying the type of approach used in this contribution. We also wanted to highlight the emerging HSI discipline and the need for "systemic integration" in engineering today. Indeed, HSI is developing worldwide with (1) the creation of the first HSI international conference, held in Biarritz, France, in 2019, supported by INCOSE,[1] (2) the signing of the historical Memorandum of Understanding on HSI between INCOSE and IEA[2] at the worldwide level in 2020, and (3) the recent reinforcement of the *Human–Systems Integration* Committee of the US NASEM's BOHSI.[3]

The term "system" must be defined: "A system is a set of interdependent elements, coherently organized to accomplish something" (Meadows, 2008). Following this assertion, a system has three main components: elements, interactions, and a function or objective. However, within the INCOSE HSI Working Group, we agreed that a system is defined by a structure and a function (Boy, 2020; Boy & Kennedy, 2023). It can be a physical or a cognitive entity. It can be a natural (e.g., a human) or an artificial entity (e.g., a machine). A function can be a function of functions. A function is usually defined by a role, in a teleological sense, a context of validity, and a set of resources that are systems themselves. This definition makes clear that a system is a system of systems. In the same way, a structure is a structure of structures where functions can be allocated.

The system's behavior emerges from the association of the behaviors of its elements. The reason is that a system can adapt by reacting to internal interactions (i.e., interactions between subsystems) and external interactions (i.e., interactions with other systems). This is where the concept of emergent functions and structures enters play (Boy, 2023). By putting the system to work, we can discover emergent behaviors that induce emergent functions and structures that were not anticipated in the first

DOI: 10.1201/9781003666431-2

place. In addition, following Meadows's definition, the term STS denotes that this kind of system explicitly includes interdependent elements (i.e., people and machines considered as systems), interactions that induce a social organization of some sort, and functions.

This handbook concerns STSs using an HSI approach (Boy & Kennedy, 2023; Boy & Narkevicius, 2013). The STS concept refers to integrating technology, organizations, and people (the TOP model), where complexity and life-criticality are significant concerns. An STS comprises structures and functions integrated according to social, legal, and ethical rules and roles. Many authors have studied and produced methods and tools to treat them. For example, an aircraft, a nuclear power plant, or a governmental agency can be considered STSs, given their reliance on various kinds of technology, complex structures and functions, and different personnel working together in a system-of-systems way (e.g., top management, divisions, departments, suppliers, and individual experts). A sociotechnical system of systems (STSoS) is based on systems that operate and are typically managed as independently functioning (autonomous) entities with their own goals. However, they must collaborate with other subsystems to achieve the higher objectives of the overall STSoS.

This handbook is articulated around several approaches and perspectives from various backgrounds and disciplines, such as systems engineering, human factors and ergonomics (HFE), information and computer sciences, and different industrial sectors. These categories can be put orthogonally with others, such as philosophy, sociology, anthropology, engineering sciences (STEM: science, technology, engineering, and mathematics), and arts (i.e., introducing an "A" to STEM to make STEAM!). The shift from Industry 4.0 to Society 5.0, where human and societal elements are central in the design, development, manufacturing, certification, operations, maintenance, and dismantling of STSs, involves change management, resilience, safety, and many other issues related to their digitalization.

The significant evolution in HSI today is software accumulation in all parts of our lives. Consequently, engineering has evolved toward digital engineering. We inverted the way technology is made. Instead of starting with tangible physical things and integrating software (i.e., going from hardware to software), we start with software, virtual prototypes, and digital simulations, and make hardware through three-dimensional (3D) printing, for example (i.e., from software to hardware). This evolution makes the user interface a syndrome of the 20th century, designed and developed to compensate for engineering design flaws by not considering humans and organizations. Instead, the user interface is an element of the overall system, designed and developed from the start of the system's life cycle. The good news is that virtual prototypes of the overall system can be tested with humans in the loop very early during the design phase. However, we now need to consider the shift from virtual to tangible. This is why tangibility has become a contemporary issue in HSI.

Tangibility can be seen from two perspectives: physical and figurative (or cognitive and even socio-cognitive). Tangibility is related to grasping, touching, or sensing something. Physical tangibility refers to concrete objects, such as a pen, a car, or a nuclear power plant. Figurative tangibility deals with abstractions like an idea, an argument, or a conceptual model. In highly digital STSs, these two facets of tangibility must be accurately articulated. This relates to situation awareness at the individual level (e.g., a human interacting with a computer or a car) and the collective level (e.g., a system of systems where systems include humans and machines), where situation awareness must be shared among several human and machine agents.

Another concept that emerged from the evolution of STSs is flexibility. For a long time, we automated people by obliging them to follow operational procedures (i.e., automation of people) and monitor automation (automation of machines). Automation has undoubtedly improved our lives and, in some cases, created major issues. Indeed, if automation is well-defined in well-known expected situations, it can become a nightmare when an unexpected situation arises. In this case, people must solve problems alone (i.e., manually or by monitoring some technological automation), often without predefined procedures. There are many reasons why automation fails in unexpected situations. Still, one is crucial: the level of experience and expertise of the people involved, articulated with the degree of flexibility to correctly solve the problem. A significant issue here is risk-taking (i.e., what should we do when everything has been prepared correctly, and an unexpected event occurs?) Another cause of failure is that design engineers do not enough consider operational context. Operations procedures or automation are not context-free; they depend on context, and when, for some reason, context changes, they are not valid anymore. Decision-makers are not faced with problems independent of each other, but with dynamic situations that emerge from complex systems operations, generating interdependent issues that are difficult to grasp. Today's leaders manage increasingly bigger messes rather than problems (Ackoff, 1999). An important question is how to manage this complexity.

For a long time, HSI has focused on individual physical and cognitive problems and solutions. Today, major issues relate more to organizational, social, and even societal issues – enhanced connectivity forces social interaction among people through mediating machines and between people and smarter machines. Social networks have changed our ways of doing things, for example. This handbook will address these social problems and explore potential solutions related to competencies, roles, organizations, and technological developments. HSI is a broader topic than the individual disciplines it involves (e.g., systems engineering, human factors, and information technology). HSI could be seen as a systemic human–computer interaction (HCI) evolution that broadly combines human-centered design (HCD) and systems engineering (Boy & Narkevicius, 2013). For that matter, we will take the INCOSE HSI Primer Volume 1 (Boy & Kennedy, 2023) as an introductory reference for HSI.

It is interesting to notice that HSI can be seen from various perspectives depending on the background of the stakeholders (e.g., industrial engineering, HFE, human–computer interaction, artificial intelligence, systems engineering, and other fields). This is another way to consider HSI in context. For example, regulators and those responsible for certifying a new machine or STS need to assess different levels of maturity, safety, efficiency, sustainability, and comfort for people who will later be involved in operations. How do we define maturity in HSI terms? What do we mean by sustainability? How is safety defined? And so on.

Of the many points covered in this handbook, the current development of artificial intelligence (AI) in correlation with HSI and systems engineering (SE) is crucial.

For example, the concept of systems of systems in SE converges with that of multi-agent systems in AI. In addition, the digital twin concept is emerging as a conceptual tool for STS's design, development, and operations management. This handbook will address these perspectives, issues, and related solutions. Finally, analysis, modeling, design, simulation, and evaluation methods and tools will be presented to support STS design and management. Concrete examples in industry and everyday life will illustrate them.

The handbook is organized into 9 sections and 45 chapters, including the introduction and the conclusion:

Section I is dedicated to foundations and frameworks. It includes six chapters: An introduction to HSI of STSs by the editor; Avigdor Zonnenshain, John Gill, and Daniele Lamoureux describe HSI in the Global Environment; Jackelynne Silva-Martinez provides an account of agile and value-driven organizations; Caitlyn Singam develops an integrated model of STEAM education; Holly Handley proposes a systems architecture approach to designing human-centric systems; and Brigitte Daniel Allegro wanders in system thinking trees.

Section II is dedicated to theoretical approaches and methodologies. It includes five chapters: Gopal Jani and Parag Sangahani describe the India Story of HSI and its Cognitive Effects on Society; Vivek Kant proposes his viewpoint on human–machine interaction in sociotechnical systems (STSs), search for metatheoretical design guidance; Jérémy Legardeur and Julien Ambrosino propose a historical and genealogical account of creative approaches to hybridization for stimulating impact innovation; Michael Preutenborbeck and Frank Flemisch introduce us to human–systems dialectics and exploration, balancing system design thinking for the agile and participatory integration of STSs; Audrey Abi Akle proposes a design space exploration approach for the design of complex systems.

Section III is dedicated to modeling and simulation approaches. It includes five chapters: Thierry Bellet uncovers a virtual human-centered design method to support new developments in vehicle automation; Dimitri Masson and Guy André Boy give a comprehensive account of scenario-based systems integration; Marc Sango and Guy André Boy propose MBHSI4Rail, a model-based human system integration method for railway automation; Jawahar Bhalla presents an account on modeling and human-in-the-loop simulation; and Romain Pinquié proposes advancing human-centered computing in model-based systems architecting towards a Metaverse.

Section IV is dedicated to modeling and simulation approaches. It includes five chapters: Chloé Morel and Guy André Boy report on pilot-virtual assistant integration and issues of operational performance, trust, and collaboration; Lance Sherry proposes quantitative measures of performance of standard operating procedures (SOPs); Wei Tan provides an account of onboard information systems; Peter Moertl gives an account on human acceptance of automation and HSI; and Belinda Cham, Alexandra Boeing, Karina Jorritsma, and Mark Griffin present human performance in STSs: a time-based perspective.

Section V is dedicated to emerging technologies and human experience. It includes five chapters: Nirit Gavish emphasizes the Metaverse and its effect on human perception and behavior; Anabela Simoes addresses how to reach the goal of getting tangible technology for aging users; Michael E. Miller presents an application of a system modeling language extension to human–AI team design; Yilong Yang, Bingjie Zeng, ZhiChing Chen, and Juntao Gao extend problem frames for supporting human–cyber–physical system (PF-HCPS) collaboration; and Stéphane Valès proposes Ingescape, a hands-on solution for Model-Based HSI, from early design to post-deployment.

Section VI is dedicated to safety, security, and operational contexts. It includes five chapters: Tero Sotamaa, Ari Teppo, Arto Reiman, and Elina Parviainen present how to optimize overall system performance and human well-being in practice in high-technology manufacturing context; Alexander Rudolph presents HSI Considerations on Safety, Security, and Health; Jean-Jacques Speyer proposes to put the pieces in sync in the HSI of safety-management systems; Alexandre Disdier describes an example of what is meant by the operational context of military remote air traffic control centers; and Benjamin Berton, Amandeep Singh, and Philippe Doyon-Poulin report on human–autonomy teaming for future commercial aviation concepts of operations.

Section VII is dedicated to assessment, measurements, and management. It includes four chapters: Paul Lambertson addresses readiness levels assessments regarding technology (TRL), humans (HRL), and organizations (ORL), and other maturity levels; Grace Kennedy, Robyn Coman, Leanne Treadwell, and JohnPaul Kennedy develop design and measurement for sociotechnological change using OSCAR-TI; Christophe Merlo and Véronique Pilnière propose an account on change management; and Suvi Leinonen, Osmo Kauppila, Arto Reiman, Marko Djapan, and Aleksandar Djordjevic provide an account on dynamic capabilities for industrial supply chain resilience development.

Section VIII is dedicated to human–AI collaboration and autonomy. It includes five chapters: Frank Flemisch, Marie-Pierre Pacaux, Yuichi Saito, and Marcel Usai propose an account on human–systems cooperation and symbiosis in times of increasing autonomous abilities of technology; Gopal Jani and Parag Sanghani present an account on human–system integration and cognitive consonance for artificial intelligence; Kahina Amokrane-Ferka proposes a method for dynamic task allocation in human–autonomy teaming; Peter Denno, Carolyn Psenka, Nicholas Lelouch, Rachael Sexton, Jaemun Sim, and Kyoung-Yun Kim propose a design of engineering copilots for industrial systems and process engineering; and Gudela Grote proposes AI Compass: a design framework for human–AI integration.

Section IX is dedicated to future directions and applications. It includes five chapters: Avigdor Zonnenshain and Arnon Bentur develop the triptych engineering, society and environment from interaction to integration; Ondrej Doule proposes pragmatic tools for human–system integration; Stélian Camara Dit Pinto presents an HSI design of Digital Twin in the oil-and-gas domain; Kine Reegård provides an account of automation and sustainable working; and Conclusion and Perspectives, where Guy Andre Boy emphasizes the evolution of automation and digital in our industrial, institutional, and public environments during the last five decades, shifting from mechanical to cognitive to social issues, and at the exact times sifting from HFE to human–computer interaction (HCI) to human systems integration (HSI). Moreover, the concomitant evolution of artificial intelligence is building bridges between AI and systems engineering and, consequently, HSI.

This handbook is the result of an ongoing international collaboration among professors, research scientists, graduate students, and industry practitioners, aiming to produce a comprehensive and usable publication with a transdisciplinary, holistic approach. It is also intended for those who want to develop a solid understanding of HSI practice and/or who wish to teach both introductory and advanced graduate HSI courses. It specifically focuses on individuals who want to understand what a sociotechnical system is, the deficiencies in current practice that HSI overcomes, where and how it has been successfully applied, its benefits and payoffs, and how it is being deployed in different industries and across multiple applications.

NOTES

1 International Council on Systems Engineering.
2 International Ergonomics Association.
3 The *Board on Human–Systems Integration* (BOHSI), a standing board of the National Academies of Sciences, Engineering, and Medicine (NASEM), established in 1980, has a long history of providing stakeholders with objective, evidence-based advice concerning relationships between systems of individuals, organizations, technology, and the environment. BOHSI provides expert guidance on the research needed to expand the scientific and technical bases for designing technology that supports critical users' needs, while incorporating diversity, equity, and inclusion into the future of work.

ACKNOWLEDGMENTS

The completion of this handbook would not have been possible without the proactive efforts of individuals within various ecosystems, including the International Council on Systems Engineering (INCOSE), the International Ergonomics Association (IEA), and the Association for Computing Machinery (ACM). It is impossible to cite all names here, and they will surely recognize themselves. It took approximately two years to design, develop, write, and edit this handbook. Let us hope it will be helpful and inspiring for the entire STSs and HSI communities, whether you are a confirmed academic, an industry professional, or a student. The editor of this handbook would also like to thank CentraleSupélec Engineering School, Université Paris-Saclay, the ESTIA Institute of Technology, and their industrial partner companies, which have supported numerous projects on the HSI of STSs during his tenure as FlexTech Chair holder. Finally, many thanks to Becky Lane and James Hobbs from Taylor and Francis for their support and help in the final stages of producing this handbook.

REFERENCES

Ackoff, R. L. (1999). *Re-Creating the Corporation: A Design of Organizations for the 21st Century*. Oxford Academic, New York, NY. (online edition, 2023). Accessed from the Web on 10 May 2025, https://doi.org/10.1093/oso/9780195123876.001.0001

Boy, G. A. (2020). *Human Systems Integration: From Virtual to Tangible*. CRC Press–Taylor & Francis Group, USA. www.taylorfrancis.com/books/9780429351686

Boy, G. A. (2023). An epistemological approach to human systems integration. *Technology in Society Journal*. 74, 102298. https://doi.org/10.1016/j.techsoc.2023.102298

Boy, G. A. & Kennedy, G. Eds. (2023). *Human Systems Integration Primer Volume One*. INCOSE, San Diego, CA. ISBN 978-1-937076-12-2.

Boy, G. A. & Narkevicius, J. (2013). Unifying human-centered design and systems engineering for human systems integration. In *Complex Systems Design and Management*. Aiguier, M., Boulanger, F., Krob, D. & Marchal, C. (Eds.), Springer, UK. 2014. ISBN-13: 978-3-319-02811-8.

Meadows, D. H. (2008). *Thinking in Systems: A Primer*. Chelsea Green Publishing, Rizzoli, New York. ISBN-13: 978 1603580557.

2 Human Systems Integration in a Global Environment

John Gill, Avigdor Zonnenshain, and Danielle Lamoureux

INTRODUCTION

Humankind is the apex consumer of all resources on our planet. We have mastered using natural resources for our individual and collective benefit. Consequently, we must apply our diverse experiences and wisdom to use global resources responsibly. Systems engineers play a significant role in determining which resources are transformed and consumed to deliver the products and capabilities we provide. While we are predominantly focused on meeting the particular and well-documented needs of our product stakeholders, the authors assert that the needs of global stakeholders, those who may not buy or use the product, are equally important. Our perspective is that sociotechnical engineering (STE), performed as a complement to systems engineering, can achieve greater good for the global community while lowering resource consumption and waste produced over the product's life cycle (Gill, Zonnenshain, Lamoureux, 2021). We propose an STE framework to expose typically hidden costs and consequences to support decision-making and deliver better balanced solutions to all stakeholders – humankind and the planet. Our decisions as systems engineers lie within the joint domain, including global stakeholders, traditional system users, and developers. The framework is a comprehensive mechanism that includes other-than-engineering perspectives and requirements, such as environmental, educational, legal, moral, and other key considerations in the development process.

THE GENESIS OF THE STE FRAMEWORK CONCEPT

A system is often defined as the integrated collection of configuration items and other systems encompassed by the system boundary – a necessary and proper artifact of the system engineering process. This arbitrary boundary limits facts and concerns, which we must consider during the design process. It was also helpful during the early growth of systems engineering (SE) practice, processes, and procedures. As a discipline, we now have mature SE practices and procedures and are better equipped to expand our consideration of crucial factors related to the system lifecycle, for example, energy consumption and the creation of waste. Both factors are an immediate concern and a principal focus in our current design and product development process. However, they have little influence after a product has been deployed. While not in the domain

 DOI: 10.1201/9781003666431-3

of a currently practicing system engineer, the consequences of a product's intended and unintended use after deployment bear consideration. Such is the concern of all stakeholders, well beyond the small circle of product stakeholders (Gill, 2024). The proposed STE framework provides a venue to include these most critical concerns of all stakeholders. We refer to this superset of stakeholders as global stakeholders – those impacted by the deployed system, whether they use it or not. In our closed environment, we all experience the consequences of product development and deployment.

Consider the first law of thermodynamics; energy is neither created nor destroyed as work gets done, it is transformed. This law, coupled with the prior assertion that the traditional systems engineering boundary limits our consideration of the resources we consume and energy we transform over the product lifecycle, opens the door to our consideration of the ultimate costs borne by society. Innovative design, abundant energy, and robust manufacturing capabilities make almost anything possible – if we can dream it, we can do it. The STE framework was devised to expose the post-deployment costs we all bear and seeks to balance them against the needs and desires of the small subset of product stakeholders who can afford to acquire and use modern systems (Gill, Zonnenshain, Lamoureux, 2021). The STE framework is posed to give us insight into the question, "Are the capabilities delivered by our products over their lifecycle worthy of the resources being consumed and waste generated?" A distinctly different perspective than mere affordability.

THE SOCIOTECHNICAL ENGINEERING FRAMEWORK

The STE framework augments the SE framework and does not change how we currently perform our traditional processes. Proven SE processes are, in fact, the cornerstone of the STE framework. The STE framework includes global stakeholders in a better-balanced construct and incorporates four new elements: sociotechnical doctrine, a sociotechnical need statement, sociotechnical analysis, and sociotechnical validation (Gill, 2020). Sociotechnical doctrine is a belief statement regarding the need for and purpose of sociotechnical complements to the traditional SE process. A sociotechnical need statement describes the expectations and requirements of the global stakeholders. Sociotechnical analysis derives and communicates the consequences of system design decisions and societal usage patterns. Sociotechnical validation is the objective review of results obtained by implementing product design decisions from the global stakeholders' perspective. Figure 2.1 shows the STE framework as a natural complement to the SE framework.

While traditional SE spans the development of a user need statement to the eventual build, test, and deployment of a system that meets product stakeholder needs (Dori, et al, 2019), STE augments this well-defined and understood process by acknowledging the existence of global stakeholders whose needs must also be considered (Gill, et, al, 2021). The above distinction is evident if you accept that traditional SE is often procedurally bound by the determination of a system boundary, the admittedly arbitrary line that separates a specific system from other systems and the rest of the universe (Kennedy, Zonnenshain, Gill, Bhada, 2021). That boundary may be open or closed, but in either case, it poses the system as a subset of the universe in which it exists. The STE framework enables the holistic inclusion of the system's global

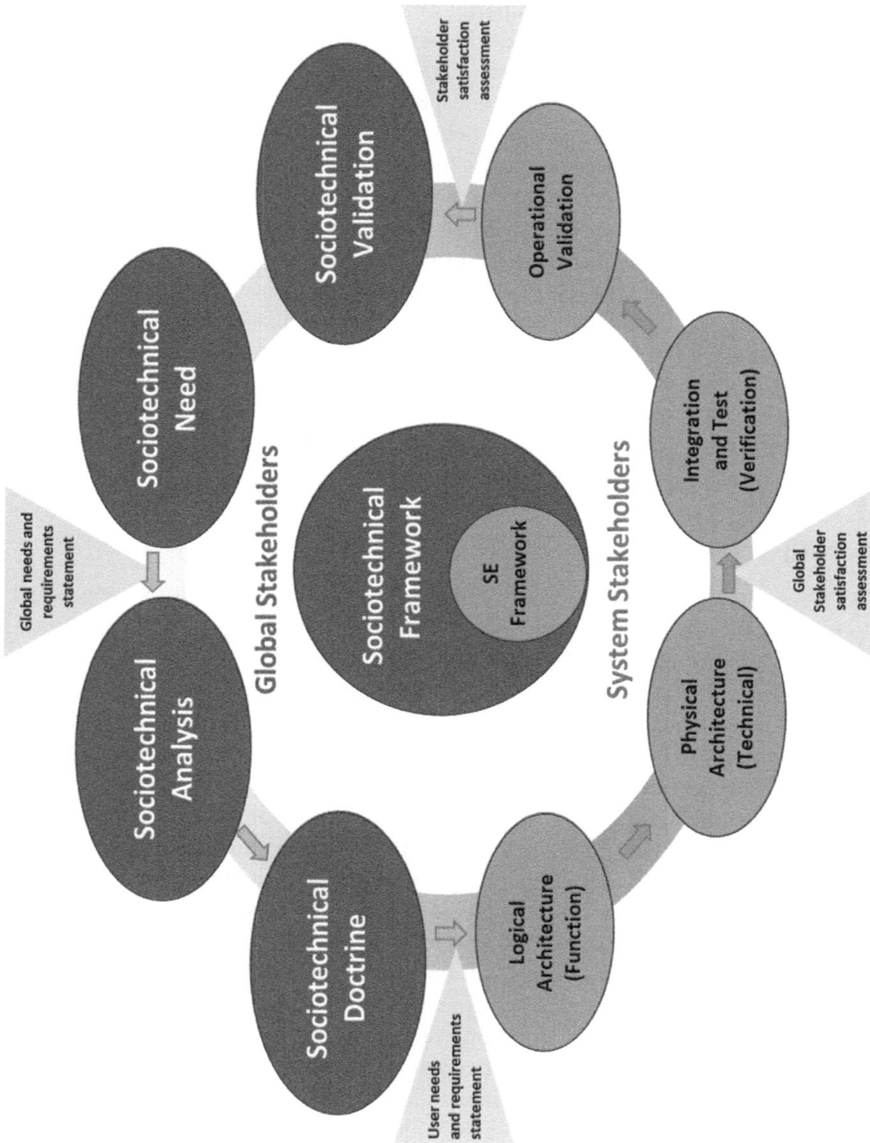

FIGURE 2.1 STE framework component diagram. STE framework elements are shaded green. SE, blue.

impacts (societal and natural costs), which have routinely been excluded from formal consideration by the traditional system boundary. STE invites designers and users to consider the global consequences of design and utilization decisions in all phases of a system's existence – from inception to disposal. Similarly, the STE framework is founded upon the reality that the Earth is a complex system with which humans interact. The Earth, its environment, and atmosphere serve as the outer boundary of consideration for STE and may be used as a practical boundary of consideration for STE thinking. In this context, the traditional SE product boundary defines a free body diagram for products and systems within the global environment. The STE domain exists beyond the conventional system boundary and the natural limits of that which we are willing to consider as our local/international environment.

THE BOUNDARYLESS NATURE OF STE

The notion of a boundaryless SE/STE framework lies in acknowledging that the traditional systems engineering product boundary is necessary but insufficient to produce systems that minimize unwanted emergent behaviors and outright detrimental side effects. The STE framework extends the boundary of consideration for systems engineers to selectively include system development and deployment aspects beyond the traditional boundary and is flexible/general enough to encompass all considerations that usually lie beyond that boundary. The STE framework makes us consider key elements that lie beyond the limits of the traditional design boundary and enables us to include the unavoidable consequences of the design choices we routinely make. It also places responsibility on the system user to consider whether the benefits gained from using the system outweigh the consequences of its use in a proposed activity. Accordingly, the STE framework proposes that product developers consider the longer-term global consequences of product deployment as part of the design process, from conception to disposal.

As with our SE process, the STE elements are intended to distill meaningful, quantitative, and actionable information from the options considered in the early stages of developmental efforts. The framework promotes these goals by incorporating an objective feedback function, providing a means to improve our product designs iteratively. It opens a door for including global stakeholder-generated alternatives, promoting creativity and innovation. Figure 2.2 depicts these aspects of the STE framework. The STE framework is a unique and novel augmentation to the SE framework. Consider that the SE framework is highly optimized to meet requirements within time and budget (Honour, 2009–2016). Acquisition customers determine the criteria that distinguish the best options from the bad.

Often, there is little to no consideration of the consequences of a design decision beyond the project stakeholders' purview. The STE framework augments the SE Framework and provides a means for global stakeholders to assess and express the consequences of design choices made by an acquisition team and product stakeholders. The STE framework offers insight into whether the products and services desired and used by a relatively small number of product stakeholders are worth the cost and burden borne by the population superset, the global stakeholders.

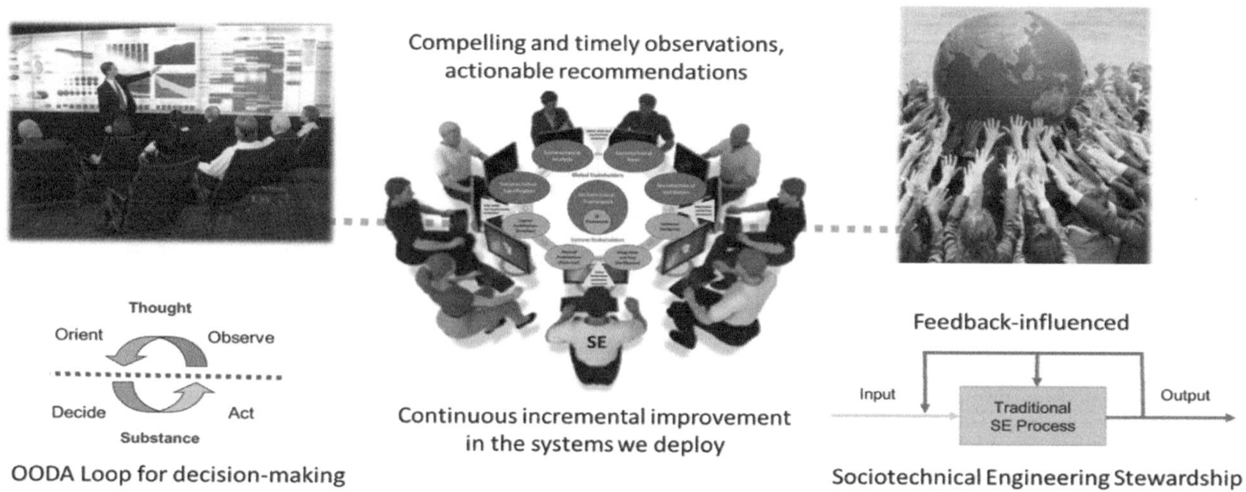

FIGURE 2.2 Scalable design enhancement informed/inspired by global stakeholder experience.

The integrated frameworks establish a global feedback loop for those who bear the ultimate costs of a product deployment and may have no input regarding its initial creation or use. The framework exposes insights that promote better design and implementation choices over time. It reveals the consequences of design choices made and implemented in a way that fosters better decision-making in future design iterations. It is intended to provide a vital external (non-product stakeholder) influence over the design process, which is currently weak or nonexistent. An integrated SE/STE framework will give global stakeholders a fair voice in the design process by providing a venue and process for them to voice their perspectives. The STE feedback loop is intended to enable product improvements from all corners of our system deployment universe that may easily, or with additional investment, be incorporated into pre-planned product improvement (P^3I) cycles. These improvements may arise from cross-disciplinary initiatives and innovations regarding materials selection, product manufacture, maintenance, and disposal processes. They are equally valuable whether they improve the product or reduce its negative impact on us. This novel feedback loop sets the stage for the iterative approach to decision-making and problem solving embedded in the time-honored observe, orient, decide, act (OODA) loop developed by Col John Boyd (Boyd, 1985). Vetted by decades of successful problem-solving in stressful circumstances, the OODA loop is an essential element of the STE framework. A structured feedback loop enables global stakeholders to exert influence over evolving designs.

POSING THE STE FRAMEWORK FROM A PRACTICAL PERSPECTIVE

Frameworks are valuable from a visionary standpoint. Making the vision credible by tracing a viable path to realization is equally essential. Figure 2.3 depicts typical STE team members and their areas of interest.

The inclusive nature of the STE framework is reflected in the areas of interest of participating team members. The framework's strength derives from the diverse backgrounds of those participating in STE-related exploration and problem-solving activities. Team members and constituent agencies supporting a STE initiative do not need to be fluent and proficient in all framework aspects. However, common imperatives and behaviors align and unify them.

STE TEAM MEMBER IMPERATIVES

Act with Global Integrity in Mind: Expose, quantify, understand, and communicate the impact of human technology and engineering activities in society and our natural world – objective clarity and consistency.

Promote Coherent, Focused Initiative: Align and organize scalable efforts to identify, specify, and resolve issues that inhibit development and sustainability. Lead/follow with vigor.

Improve Incrementally and Continuously: Deliver the greatest societal good from among viable options and apply effort to the highest priority activities as indicated by Pareto analyses. Curtail the spread of detrimental emergent

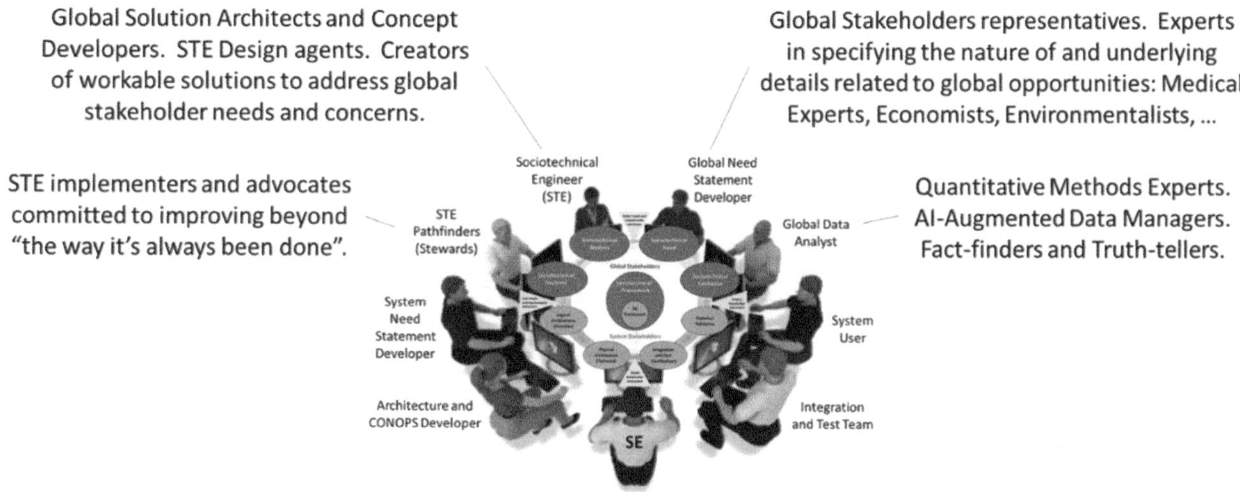

FIGURE 2.3 STE Team Members and potential areas of interest.

behaviors with the same intensity, intellect, and dignity with which you promote positive impacts.

STE proponents shall…

Expose the Objective Truth: Dare to identify key aspects of the problem in its natural context in pursuit of the root cause(s) of the issue. Seek to adequately understand relationships and dynamics before proposing a solution.

Craft Accurate Problem Statements: See beyond symptoms and understand a situation well enough to know what conditions, triggers, and influences precipitate them. Use quantitative measurement to the greatest reasonable extent.

Implement Sound, Principle-Based Solution Development Practices: Use design thinking – think broadly, then narrowly, and repeat until a viable, actionable solution results. Integrate the experiences and perspectives of global stakeholders.

Determine Which Actions will Make a Difference: Use a Pareto-like analysis to determine where limited resources may be applied most effectively. Prioritize prevention over prescription. Seek to maximize return by leveraging others' investments. Internalize lessons learned.

Alter the Trajectory of the Situation: Small, persistent actions may lead to a bigger result than a sweeping change. Deeply rooted problems take time to resolve. Accept that immediate improvement is not typical. Work to prevent issues instead of continuing to fix them.

Take Necessary Actions: Remain committed and attentive to issues you are passionate about resolving. Use limited energy wisely, conservatively, and promptly to preclude future problems and resolve current ones.

Monitor Outcomes: Assess the utility of your actions. Optimize their effectiveness using continuous improvement techniques. Promote working solutions, curtail fruitless efforts.

Become Influencers: Leverage the capabilities of social media to help others see the power of sociotechnical responsible decisions, actions, and behaviors. Promote global awareness.

A BETTER-BALANCED, AFFORDABLE SYSTEM ENGINEERING SOLUTION

Balance is an oft-used term in system engineering and project management. As implemented, it is an arbitrary concept considering its project-dependent, fluid definition and tailored application. INCOSE has adopted the position that "A system is affordable to the degree that the system performance, cost and schedule constraints are balanced over the system life, while mission needs are satisfied in concert with strategic investment and organizational needs." (INCOSE, 2024) One may claim to deliver a solution that is balanced within the confines of the product boundary. Yet, the solution fails the "balanced" metric at a global level. Defining the STE boundary of consideration to include global stakeholders offers a more complete perspective

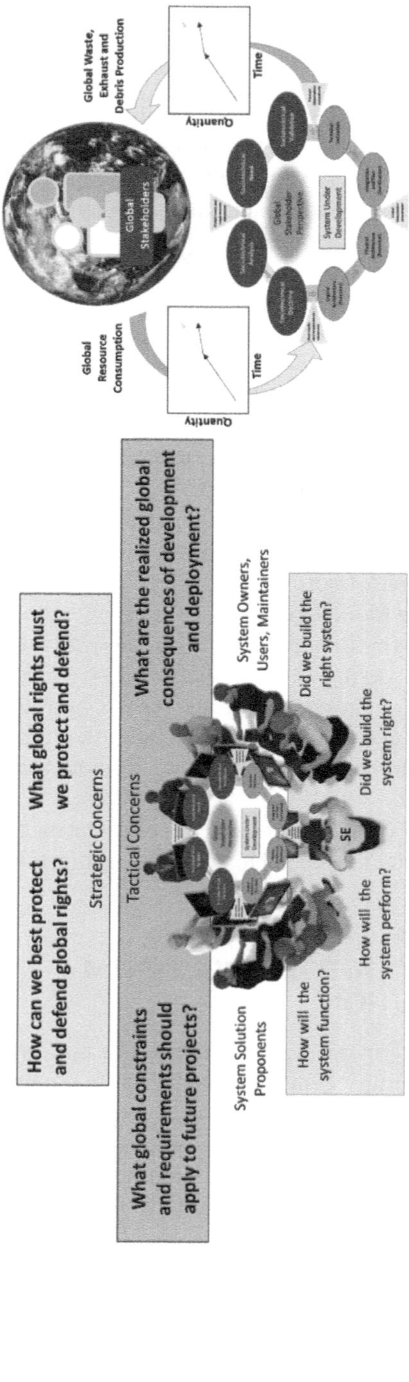

FIGURE 2.4 The STE framework is structured to promote comprehensive and pervasive balance.

regarding a balanced solution by incorporating typically excluded terms, perspectives, and stakeholders. Likewise, a balanced STE construct that relies more heavily on physical properties than cost-, schedule-, and performance-related parameters may better substantiate balance claims. Toward this end, we propose a qualitative balanced STE definition. A system solution is balanced if, from the global stakeholder perspective, there is or will be no growth of known-negative consequences implicit in the development and use of a system or product compared to those currently planned or deployed. In absolute terms, the goal of no harmful accumulation is unattainable in our current engineering endeavors. Every human-developed system consumes resources and generates by-products, waste, and debris over its life cycle. A key aspect of the STE framework is to help developers and users see this reality and focus their efforts on remediating and preventing negative consequences. Continuous improvement is a key attribute of the STE framework.

Awareness, acknowledgment, and ownership of potential good (or reduction/elimination of known bad) in our cyclic development processes often lead to a better outcome. Heuristically, focusing on relevant information and value prioritization leads to better decision-making. Pursuing global balance implies comparing the needs of a limited number of product users to the consequences imposed on global stakeholders. This attribute of the STE framework is evident in the two technical performance measure charts embedded in Figure 2.4 on the right. The left inset depicts resource consumption, and the right depicts waste, exhaust, and debris production over time. The STE framework can help global stakeholders exert their influence by exposing and normalizing acceptable measures of goodness regarding the utility of a given product, system, or capability and balancing them against their associated total cost, resource consumption, and consequence generation. While goodness values are typically generated to promote and advertise a system in its infancy, accurate values obtained after the system is put into service are often different, i.e., "That was the vision, here's the reality." The STE framework leverages the clarity of user experience to counterbalance our profit-driven development, deployment, and use of modern systems.

PRACTICAL APPLICATION OF THE STE FRAMEWORK

The STE framework offers a method for collaborating specialists to view a grand challenge holistically and its constituent parts, and organize and prioritize efforts to ameliorate emergent problems, using bottled/canned beverages as an example. Figure 2.5 depicts significant steps in manufacturing, delivering, and disposing of the beverage product for each constituent product element – bottle, beverage, labeling/packaging, and transportation.

It also suggests a format for tabulating each process step's resource consumption and waste generation. Within a holistic construct, stakeholders determine where waste and debris are created in the delivery and use of a product, focusing their attention on high-priority problems. A "Measure of User Benefit" is also at the framework's core, establishing a given product's relative utility or usage. For example, a clean water bottle may mean life to someone caught in a natural disaster that destroys local

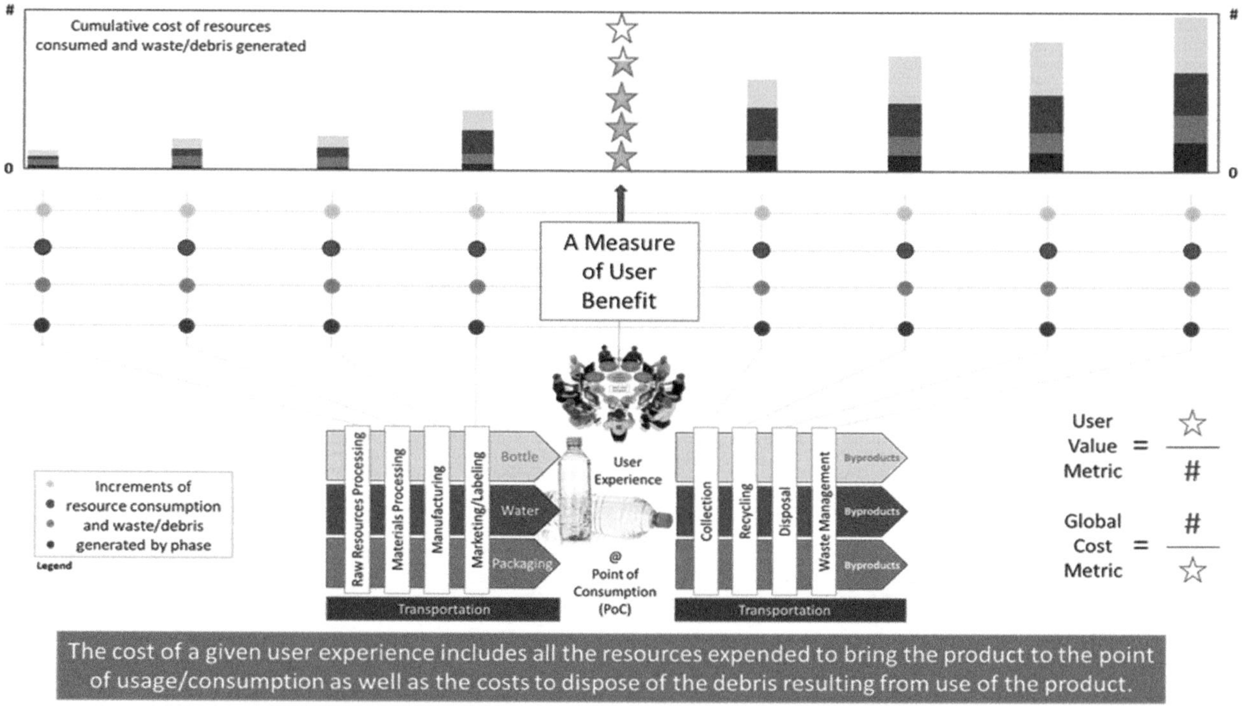

FIGURE 2.5 Notional Consumption/Waste Aggregation and Global Value Metric Determination.

infrastructure, whereas pouring that bottle onto your lawn has no obvious utility. The resources consumed and waste produced for each bottle may be similar, but the benefits differ. Each increment of resource consumption and waste/debris generated by phase (intersection dots on Figure 2.5) is individually profitable for the product or service provider, yet may ultimately result in a capability delivery that renders no utility (emptying a bottle onto your lawn). While the example is an individual bottle of water, this scenario plays out on a grand scale when considering the global deployment of a product or capability.

The framework offers a metric (Figure 2.5, lower right) intended to expose either the relative value of a given product/capability or its relative global cost. The metric provides a basis for comparing various circumstances and alternatives to prioritize where preventive and corrective measures may be applied and do good.

The metric has two variables: user benefit (☆) and global cost (#). User benefit, albeit arbitrary, is valuable primarily for relative comparison of alternatives. The user must determine how they will assess the value of a given product/capability/service to evaluate options. Cost, a tangible value, can be assessed at the individual instance level, the global level, and anywhere between. It is a helpful term because it causes viewers to consider invisible or hidden costs that are typically discounted; for example, a product/capability/service may be had for an identical cost ($) via two paths. However, one of the paths requires the use of toxic chemicals that impose long-term detrimental health implications. That path becomes a higher cost (or lower value) path owing to the inclusion of the global costs as required by the STE construct. Figure 2.6 depicts the increasing cost of a cup of coffee versus the marginal user benefit associated with its delivery method. Is robot-delivered coffee worth more to the recipient than getting coffee through other means?

While the STE framework is primarily offered as a means for us to incorporate the diverse perspectives of global stakeholders, it also exposes clear and ample information for product users to consider the natural consequences of their use of a given product or capability. The question is the same for both – are we receiving enough utility from using a given product to warrant the associated expenditure of resources?

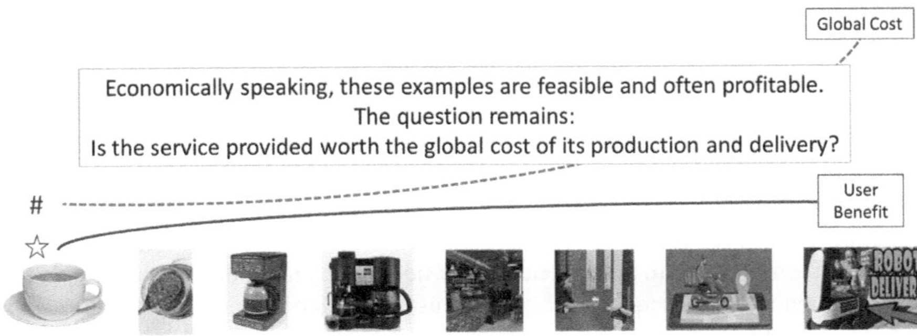

FIGURE 2.6 The global cost of a cup of coffee is increasing, With a marginal increase in user benefits.

CONCLUSION

"Sociotechnical" refers to the interrelatedness of social and technical aspects for individuals, organizations, and society. STE is an extension of the SE practice that incorporates societal perspectives to inform the product development process. The STE framework proposed herein promotes collaboration between engineers and societal stakeholders at key points in the design process. It respects the SE Product boundary as currently implemented and incorporates the natural environment beyond the arbitrary boundary within which SE products are developed, deployed, and used. It is formulated to preserve required product performance and improve user experience while reducing the system's global resource consumption and waste production throughout its lifecycle. The practical and inclusive framework enables nontraditional product stakeholders to exert fair and reasonable influence during design. In a practical sense, the STE framework implements a global stakeholder feedback loop wherein lessons learned by deploying and using systems inform design protocols that maintain performance expectations while inhibiting the propagation of design practices and decisions known to render negative global consequences. The framework is a scalable, closed-loop augmentation of the SE process and includes design considerations typically discounted by current practices. It integrates the perspective of global stakeholders in the SE process, accounting for resources consumed, waste produced, and other byproducts of system usage.

In its current implementation, the iterative but predominantly linear SE process is primarily driven by those who can afford to acquire and use the capabilities we deliver. Whether and how the capabilities are used is often determined by economics. The proposed STE framework and balance proposition extend our consideration to include the physical impact of a given capability on global stakeholders and the natural environment. The intent is to provide a venue for exploring an inevitable aspect of our discipline: should we develop and deploy capabilities simply because we can and/or are funded to do so? Specifically, the human systems integration (HSI) discipline is integral to helping us, as systems engineers, promote a more equitable, balanced, and global role in designing and integrating our products and capabilities into society under the moniker of STE. This paper offers a framework for accomplishing this goal and role. The primary questions to be addressed are "Is this product, system or capability worth the precious resources which will be consumed to develop and deploy it?" and, once deployed, "Are we using this product, system or capability in a responsible manner that merits the resource consumption as well as the natural byproducts and waste associated with its use and disposal?" The STE framework organizes and aligns our collective thinking concerning these essential questions. HSI is a key aspect of system engineering and extends beyond the ease and effectiveness with which we might use a system. The ultimate question is: Can we all live with the consequences of the systems we create and deploy? The framework offers a seat at the table and a timely voice to all stakeholders, whether they use a deployed capability or only experience the consequences of others' use. In collaboration with a diverse set of global stakeholders, our discipline is well positioned to take up this opportunity and meet its challenges using tools and techniques we have honed over prior decades in this proposed natural extension of the traditional SE process.

REFERENCES

Boyd, J. (1985). Observe, Orient, Decide, Act (OODA) Decision-Making Loop.

Dori, D., Sillitto, H., Griego, R., McKinney, D., Arnold, E., Godfrey, P., Martin, J., Jackson, S., & Krob, D. (2019). System Definition, System Worldviews, and Systemness Characteristics. *IEEE Systems Journal*, 14(2):1538–1548. Multiple other Sources: INCOSE, NASA, IEEE, DoD Acquisition.

Gill, J. (2020). *The Keynote Address was Delivered by Dr John Gill in November at the Gordon Center for Systems Engineering*, Neaman Institute for National Policy Research, Technion, Haifa, Israel.

Gill, J. (2024). *Professional Management and Systems Engineering Cohort. Keynote Address.* August, Georgia Institute of Technology, USA.

Gill, J., Zonnenshain, A., & Lamoureux, D. (2021). *Opportunities and Challenges of Sociotechnical Systems Engineering*. July, INCOSE International Symposium Proceedings, INCOSE, San Diego, CA, USA.

Gill, J., Zonnenshain, A., & Lamoureux, D. (2021). Opportunities and Challenges of Sociotechnical Systems Engineering. INCOSE International Symposium Proceedings, 31(1), pp. 453–467. DOI: 10.1002/j.2334-5837.2021.00847.x

Gill, J., Zonnenshain, A., & Lamoureux, D. (2021). Opportunities and Challenges of Sociotechnical Systems Engineering. Social/Sociotechnical and Economic Systems, Information Management. Wiley Digital Library. DOI: 10.1002/j.2334-5837.2021.00847.x

Honour, E. (2009–2016). Systems Engineering for Effective System Development. Multiple Training Engagements before Becoming the INCOSE President.

INCOSE Website, 2024 query. 2025 Web Query of the Most Important Attributes Required for Effective Decision-Making. AI-Generated Response.

Kennedy, G., Zonnenshain, A., Gill, J., & Bhada, S. (2021). Exploring the Synergies between Social Systems and Human Systems Integration. Panel at INCOSE HSI2021 Conference. Social/Sociotechnical and Economic Systems, Information Management. WileyDigital Library. DOI: 10.1002/j.2334-5837.2021.00847.x

3 Agile and Value-Driven Organizations

Jackelynne P. Silva-Martinez

AGILITY AND VALUE-DRIVEN ORGANIZATIONS

Heisterkamp (2019) defined agility as the "ability of an organization to adapt to fast-changing environments and achieve better outcomes." Agility may not be able to increase the predictability of value, but frequent customer feedback can increase the probability of value. Patton (2014) stated that the focus on Agile moved from producing output to producing outcomes. The scarcity of outcomes from Agile teams was identified as a gap in the Agile literature, as many studies mainly showed the excellent outputs from Agile teams. It was essential to determine the outcomes that can allow organizations to measure the progress of their Agile transformations. Therefore, a study of National Aeronautics and Space Administration (NASA) Agile teams permitted the identification of various outcomes from these teams and Agile leadership characteristics used as these teams increased value to their customers. Figure 3.1 summarizes the Agile Leadership Model (Silva-Martinez, 2023a, 2023b). An organization's design and management should follow a value-driven framework that focuses on increasing value for its end users in a sustainable manner. The early implementation of human systems integration (HSI) in developing teams entails establishing and maintaining communication and connection with the end users, which allows teams to increase their value-creation capabilities. This model provides organizations with a platform to base their Agile transformation activities on, planning for their workforce recruiting, development, and retention efforts as they create value-driven organizations.

Becoming agile involves continuous learning and improvement, allowing organizations to create sustainable value for the end user. There are misconceptions that the organization is already agile due to switching to an Agile method. For example, an innovative energy solutions company transformed into an Agile company, with smaller teams, training, and roles following the Agile Scrum methodology. Their goal was to have more efficient teams that allowed them to increase velocity, which is common in Agile transformations. Velocity, a team output, measures the work a team can do during a given cycle that would lead to sustained impacts in the market (Nir, 2018). However, this company suffered a loss in productivity as it

DOI: 10.1201/9781003666431-4

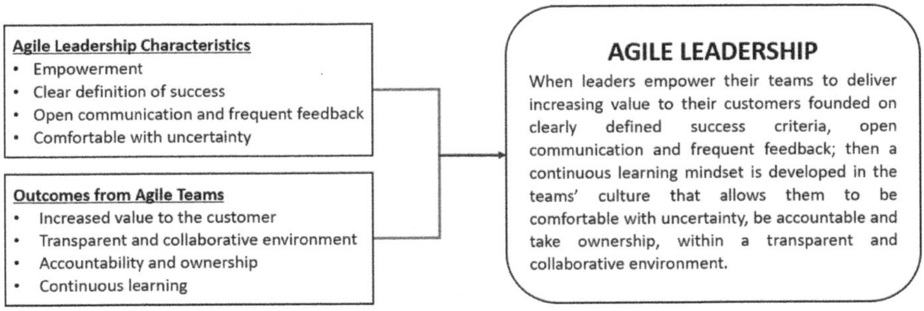

FIGURE 3.1 Agile Leadership Model, (From Silva-Martinez (2023a).

struggled to adjust to new roles and tried to determine interdependencies with other teams. This case study showed that when Agile transformation focuses on team velocity without understanding why and how their actions increase value to the end user, it can cause instability, and organizations cannot maintain performance. The benefits of creating value should outweigh the effort and/or cost. The following are examples of tools that help measure value.

Agile emerged from software development when a group of 17 independent thinkers about software development met to find ways to improve their processes. As a result of the meeting, instead of processes, they developed four values and twelve principles that allow a project to "do the best for the customer and deliver something timely and tangible and as promised" (Beck et al., 2001). Many industries widely use Agile, and organizations invest in Agile transformations. Examples are in human resources, which hires, trains, and develops employees, and in complex projects such as jet fighters (Cappelli & Tavis, 2018; Furuhjelm et al., 2015).

Agile enables organizations to adapt successfully to disruptive environments. Schwaber and Beedle (2001) sustained that Agile delivers results in complicated and complex environments with high uncertainty and disagreements on requirements. The aerospace company Saab Defense is a demonstration of that, where Agile was used in a complex aircraft system. By focusing on more minor releases, they were able to collect frequent feedback on the design and integration of their products (Furuhjelm et al., 2015). Sharifi and Zhang (2001) sustained that the growing turbulence in the business environment calls for an Agile transformation for organizations to maintain a competitive advantage. Organizations experience different needs and challenges, so their Agile implementation may require different approaches. Organizations will need a strategic plan that allows them to incorporate Agile transformations with set priorities and specific development plans for their employees.

TEAMS

NASA's Agile teams study participants suggested that Agile works because of the dynamics and interactions that occur among team members. This section discusses team development, self-organization, motivation, and feedback.

Team Development

Tuckman (1965) proposed a widely known four-stage model for team development: form, storm, norm, and perform. This was followed by Tuckman and Jensen (1977), who added one more stage to the adjournment model. During formation, members try to determine their roles, rules, and procedures. Then, in the storming stage, conflicts start to bubble up among team members and tasks, even resisting or rebelling against getting those tasks done. The group then moves into the norming stage, where behavior norms are set, commitment builds, and members find ways to work together. Then, proficiency in working together is reached following the set rules and procedures they developed in the previous stages. The last stage, not part of the original Tuckman's model, is related to the termination process. A team's dissolution can occur during organizational changes (Tuckman & Jensen, 1977).

Something to note here about Agile practices is that the set of people who start in the forming stage is still considered a group. While people use the terms group and team interchangeably, in Agile, a group is just that, several individuals; a team, instead, is several individuals working together to achieve a common goal. In a group, each member has individual accountability, whereas in a team, members work collaboratively with a shared accountability (International Consortium for Agile (ICAgile), n.d.). A team is the basic unit of performance, which combines multiple skills and experiences from all its members (Kozlowski & Ilgen, 2006). Opposite to structured organizational groupings, teams are more flexible and can refocus their performance objectives rather quickly because they are committed to delivering tangible performance results, which in Agile, those results are the value provided to the customer. Other researchers who also studied how to help groups become functional teams following set processes to achieve their tasks include Fisher (1970), Gersick (1988), Poole and Roth (1989), McGrath (1991), etc. For some, Tuckman's approach seemed to be a very sequential process for team development. For example, McGrath (1991) developed the time, interaction, and performance model that showed the team development process in a multi-functional manner. This model showed that team members can simultaneously engage in four functional modes: inception, problem solving, conflict resolution, and execution.

Similarly, Rickards and Moger (2000) found that teams are aware of a barrier during the storming stage in an unconscious team development process. However, some teams reported that the barrier is reduced when there are conscious efforts to address team members' roles and responsibilities. The other barrier that can be found in Tuckman's team development model is during the performing stage, as it is not always clear what steps are needed to exceed performance norms. Therefore, Rickards and Moger (2000) proposed a combination of creative leadership and team performance, where there can be complicated contingencies with expected diversity of teams addressed with seven team factors: platform of understanding, shared vision, favorable climate, resilience, idea owners, network activators, and learning from experience. Many of these are characteristics of leadership theories discussed earlier.

Self-Organized Teams

Self-organization "emerges from the collective action of the agents using local information to determine how they move from one state to another over time" (Walker & Dooley, 1999). In today's complex systems, decision-making must be done faster with well-informed risk-taking (there is always some uncertainty). To accommodate that, a team that self-organizes generates a rule that can start randomly, and then adopts a plan or action that generates a configuration that works for that specific team. Holland (1995) called this process a complex adaptive systems theory. Self-organization is beneficial in the capacity of complex systems for maintaining control while also enabling the decentralization of authority (Walker & Dooley, 1999).

Following Tuckman and Jensen's (1977),team development stage model, Agile teams form and reform as a new project, and iteration or increment starts; agile teams' beauty is that the members can move between stages. Some say Agile teams usually stay in the forming and storming stages. Moreover, while conflicts during the storming stage can sound like showstoppers, Agile teams can use those as stepping stones to develop better working methods. Srivastava and Jain (2017) performed a qualitative study using open-ended surveys and semi-structured interviews with 18 Scrum Masters. They determined their questions by using leadership functions typical of Scrum teams and leadership processes that come from self-organized and self-managed teams. Of notice was that 15 out of the 18 participants had certifications in both Project Management Professionals and as Scrum Masters. Based on the Scrum Masters' responses, the researchers proposed leadership approaches and behaviors: making decisions, facilitating and being open and transparent, communicating and committing to tasks, people-centricity, and maintaining cohesion. Silva and Guerrini (2018) extended self-organization to innovation networks and the characteristics that rely on complex systems theory. This manifests in a team adapting to the environment and making any needed changes within the team's system structure.

Team Motivation and Team Performance

Team motivation is correlated to team performance (Peterson, 2007). Martens (2013) conducted a quantitative study with college basketball teams and found a significant correlation between satisfied high task-motivated teams and achieving high performance. This result was in contrast to satisfied but moderate to low affiliation motivated teams, and achieving low performance. Similarly, Melo et al. (2013) performed case studies with three companies that allowed them to understand some factors impacting Agile team productivity. They used retrospective documentation, interviews, and direct observation of each company's work session. This resulted in cause-and-effect relationships between team productivity, team design choices, staff turnover, and team coordination.

Team Motivation

Human motivation has been studied by many researchers and explained through various theories, including Maslow's Need Hierarchy Theory (1943), McClelland's Need Theory (1975), Campbell et al.'s Expectancy Theory (1970), among many others. Of note from these theories is that motivation and behavior are different, as motivation is just one determinant of behavior. In addition to motivation, behavior can be influenced by aspects such as culture and situation (Maslow, 1943). Leaders can develop team culture through participative leadership to provide greater ownership of tasks and decision-making authority (Peterson, 2007). Once the team has matured in a motivational environment, they tend to develop improvements in processes and operations, which is a characteristic of self-organized Agile teams (Hoda et al., 2012).

High-Performing Teams

Many organizations focus on how they can successfully build a high-performance team. According to Natemeyer (2011), this can be addressed through team building. He defines team building as "the process of assessing current team effectiveness, and developing and implementing an action plan to improve" (p. 24). Natemeyer further presented eight steps in such team building process: (1) recognition that team building is desirable; (2) assessment of the teams' effectiveness; (3) discussion and analysis of the assessment results; (4) clarification of the teams' vision, mission, and values; (5) identifying the teams' key goals; (6) development of the team action plan; (7) implementation of the team action plan; and (8) action plan progress review and revision. Although motivation is internal to each individual, leaders and organizations can create an environment conducive to high performance (Scholtes, 1998). Empowerment is also a key ingredient to build self-organized, performing teams, with four components: (1) capable resources, (2) accurate information, (3) team member authority, and (4) accountability to complete the work (Peterson, 2007). The interview responses from the NASA Agile teams study showed that feedback was a key characteristic for high-performing teams.

Importance of Feedback

Feedback is a primary driver for Agile teams to go and update what is needed, implement changes based on customer feedback, and return an improved product in a shorter time (ICAgile, n.d.). Although some people may be thirsty to know how they are doing, some may not feel open to receiving feedback. Kouzes and Posner (2017) described a good way to approach giving feedback by asking people if they want it. The receivers can be less defensive, as it was their choice to receive feedback. This applies to both seeking and providing feedback.

Recognizing Contributions: Offering positive reinforcement is a form of feedback that strengthens a behavior and can occur more often by presenting something positive. This can be done through recognitions in front of the group, encouraging words given in person, personalized letters, and other genuine recognitions. Rewards do not need to be tangible; leaders can provide intrinsic rewards. Some people may feel appreciated and know their work is recognized by getting a higher responsibility in a project, a challenging assignment, or something similar. Litzky et al. (2006) stated that positive reinforcement can also be ineffective if not used appropriately, such as predictable recognitions, or standardized letters that may be impersonal, or uniquely monetary—showing team members that leaders believe in them is a powerful force that creates greater performance (Kouzes & Posner, 2017). Agile practitioners use recognition tools such as kudos cards to show appreciation for team members' contributions, all done regularly.

Celebrating Accomplishments: Kouzes and Posner (2017) emphasized celebrating different milestones with the group. People need encouragement to function at their best. Even when a project can be long term, it is important to celebrate small steps. This encourages people to continue working, to be happy for coworkers' accomplishments, and to aspire to get where a colleague is instead of feeling jealous. Goleman et al. (2013) considered this social support, where getting to know others personally allows them to be more empathetic. Knowing that team members and the leader care about one's well-being enhances productivity, as they know they are not alone in their efforts. "People have a deep desire to make a difference" (Kouzes & Posner, 2017, p. 111). Sometimes, in technical projects that take 10–15 years to see tangible results, as is common in the aerospace industry, employees may not feel like they are making an impact. This can worsen if, after investing 5 years in the project, it gets cut due to funding or a new administration with different priorities. Leaders of these technical teams can find ways to provide their employees with signs for how their work impacts the overall mission. This could be done through lunch and learn presentations by the team members, where they can share what they are working on, via stretch assignments where results can be seen in a shorter period, and through team celebrations after milestones. Team members need to feel appreciated, and celebrating small victories helps them to stay motivated in what they do. In Agile practices, activities are done to recognize and celebrate such accomplishments after delivering an increment of the product or service. This is so ingrained in Agile that those activities are called celebrations in most Agile methods.

ORGANIZATIONAL CHANGE

Agile transformations involve organizational changes. Kotter (2012) explained that organizational change is a transformation process that requires resilience and preparation to be successful. He identified eight main errors that organizations make when trying to make a change: (1) allowing too much complacency, (2) failing to create a sufficiently powerful guiding coalition, (3) underestimating the power of vision, (4) under-communicating the vision, (5) permitting obstacles to block the new vision, (6) failing to create short-term wins, (7) declaring victory too soon, and (8) neglecting to anchor changes firmly in the corporate culture (Kotter, 2012). By definition, these errors are obstacles to moving a vision forward.

Complacency: Complacency is one of the significant obstacles to organizational change, as it can come up at the beginning of the change process, in the middle, or when the change is already in progress. Kotter (2012) described nine sources of complacency: (1) absence of a significant and visible crisis, (2) too many visible resources, (3) low overall performance standards, (4) too much happy talk from senior management, (5) human nature with its denial, (6) kill the messenger of bad news, (7) lack of sufficient performance feedback from external sources, (8) internal measurement systems that focus on the wrong performance indexes, and (9) organizational structures that focus employees on narrow functional goals. Morse and Babcock (2007) explained that organizations must constantly update their processes to be more efficient according to internal and external sources. As new tools and methods are developed, and new personnel knowledgeable of recent academic findings are hired, organizations should learn from state-of-the-art research and technology that can be applied to their workplaces. Kouzes and Posner (2017) stated that "exemplary leaders know that they must transform how things are done. They actively rely on outsiders to seek innovative ideas beyond the boundaries of familiar experience."

Resistance: Resistance to change is an emotional or behavioral response to real or imagined threats to an established work routine (Kreitner & Kinicki, 2008, p. 546). Organizational change can be hard for employees because when they have mastered how to do their daily tasks, such a change produces some disruption. The reactions can vary based on people's backgrounds, what they were exposed to while growing up, or a similar change they experienced. Marshak (2006) identified some reasons people may resist change: surprise and fear of the unknown, fear of failure, individual's predisposition toward change, peer pressure, loss of status or job security, non-reinforcing reward systems, poor timing, among others. Kotter (2012) explained how structures and supervisors were obstacles to the change process. This can be interpreted as two types of resistance sources: the employees' characteristics and employees' responses to obstacles that the organization has failed to address. Marshak (2016) focused more on the personal characteristics affecting employees' predisposition to change, while Kotter (2012) focused more on organizations failing to communicate the rationale of the change and their lack of preparation for employees to embrace change.

Commitment to Change: Beer et al. (2011) presented commitment as the first of six steps to effective change: mobilizing commitment to change through joint diagnosis of business problems. It is important to seek help to define the problem from those executing the changes to gain their early buy-in and support. A broad review of the current processes can start with this. In addition, Herscovitch and Meyer (2002) found that employees' commitment to change positively predicted behavioral support for a change initiative.

Show the Benefits: Part of communicating the reasoning for a change is explaining to employees how the disruption will improve their day-to-day activities. Resistance will be less if participants see the change as reducing rather than increasing their present burdens (Burke et al., 2009).

Resilience: The willingness of people to be open to change is positively related to job satisfaction (Wanberg & Banas, 2000). Providing an environment where new ideas are valued is part of the cultural change needed for resilience to be achieved. Conner (2006) posits that cultural change goes along with resilience as part of the change structure, and it can be achieved through participative management. This is when leaders empower people to participate by developing the idea and the decision process and executing the change themselves (Conner, 2006). Participative management can be a significant aid in this area, as it shows the trust leaders must have in their employees to be accountable for their contributions to the overall mission.

Tolerance for Risk: Everything has some level of uncertainty. Organizations must provide room for failure as part of the employees' learning process. Risk-averse companies may not stay competitive if they are not searching outwardly for innovative processes and opportunities (Kouzes & Posner, 2017).

Uncertainty

Getting to explore space and being able to learn the origins of our planet is exciting. This task is very uncertain as well. Everything—from requirements development, design, assembly, test, launch, and operations—contributes to the success of a human spaceflight mission. Any issues at any stage can cause catastrophic accidents, where lives can be lost. One may remember Apollo 1 with three lives lost in 1967, the Space Shuttle Challenger disaster in 1986, and the Space Shuttle Columbia disaster in 2003, with seven lives lost in each of them (NASA, 2013; Sparrow, 2007). In addition, the NASA Safety and Mission Assurance Flight Safety Office developed an interactive tool with a list of reported significant incidents and close calls in human spaceflight of various missions that occurred at different flight phases (NASA, 2017).

Space is hard, and every phase is full of unknowns. Many assumptions are made at the start of a space mission because of the unknown environments to which either humans or instruments or equipment will be exposed, especially if there are no precursor missions that would provide some background information. This is the case of a mission to Mars, where the goal is to be able to send humans to the red planet. Before that, we first need to understand the environment, how it will affect the human body, and if humans can live and work there safely. Hence, orbiting satellites and robotic missions have been sent to study the red planet to prepare us for future crewed missions. The latest rovers sent were Curiosity in 2012, Insight in 2018, and Perseverance in 2020, which have helped better understand Mars by enhancing the search for minerals, water, and other environmental aspects (NASA, 2022). The latter, the Perseverance mission, flew during an unprecedented pandemic, where teams worldwide supporting the mission had to quickly adapt to new ways of testing, planning, and executing the mission from their homes.

Findings from precursor missions help scientists minimize uncertainty. They know that not everything can be predicted and that there will still be unknowns; yet, missions would have to be adjusted according to the new findings. It is similar to organizational change, where change is a dynamic process. Both the organization leaders and the scientists searching for the universe need to know that "the future would be made up of constantly shifting variables, and they were therefore going to be routinely surprised at the actual turn of events" (Conner, 2006). In other words, leaders should be prepared to be surprised. This was demonstrated in NASA's Agile team study, where one of the identified Agile leadership characteristics was being comfortable with uncertainty.

Global Strategic Leadership

The dynamics of how organizations operate have changed in the past century. This involves changing how leaders approach their everyday processes within their organizations. Things seem to go faster, which can be attributed to technology, and processes now need to be more agile while keeping in mind that we have people, with their capabilities and limitations, making this happen (Silva-Martinez, 2016). A main global strategic leadership element should be the increasing focus on the human aspect as part of the system of systems, represented by the organization using a HSI approach. Organizations could not function without team members or employees; hence, they deserve the highest attention to keep a project running and to make it productive and successful.

Today's global economy requires fast decision-making, being open to uncertainty, and quickly adapting to change. This starts from the recruitment process. McEntire and Greene-Shortridge (2011) explained how human resources experts are looking into different strategies to recruit and select high-level innovative professionals who can operate globally. One of those strategies is targeted succession planning, where organizations hire and train top-talent employees to fill leadership roles during organizational growth. This is followed by the retention process, where leaders help their employees develop and grow as professionals. A study looking at social and cultural factors that influence motivation showed that "one cannot look at the personal life and the workplace life of the employees separately, the two existences (one at work and one outside work) of the same individual are not in silos" (Choudhary et al., 2013). It is important to recognize that employees' motivations and values are generally shaped by their culture.

Cultural Intelligence: Cultural intelligence can be defined as the ability of a person to work in a culturally diverse society efficiently, and this person can quickly adapt to new cultural situations (Ang et al., 2007). "The potential for superior productivity of culturally diverse teams is high; they possess a breadth of resources, insights, perspectives, and experiences that facilitate the creation of new and better ideas" (Adler & Gundersen, 2008). Cultural intelligence increases the benefits of emotional intelligence for building trust, maintaining strong relationships, and establishing motivated and effective teams (Silva-Martinez, 2023c). Global leaders must understand the influences of culture on work performance and how tightly interdependent they are. Global leaders recognize cultural differences in communication and do not impose their approaches (Giri, 2006). Leaders should be aware of how these different means of communication work and employees' preferences for using them. When they have a message to communicate to the entire organization, they will know how to reach out to their employees more efficiently. Ms. Lisa Cox, trainer, coach, and consultant with Corporate Education Group, discusses how to lead successful virtual teams. She states that "organizations around the globe have come to recognize the benefits of using technology to replace face-to-face interactions in the name of improved productivity, lower costs and larger labor pools from which to draw talent" (Cox, 2019). In light of the pandemic, electronic communication has become necessary, bringing challenges and opportunities. However, it was recognized that people may be afraid of trying new computer tools that may facilitate Agile implementation.

Constant Learning: Kouzes and Posner (2017) sustained that all humans have leadership qualities and can become effective leaders, but must be constant learners to improve those skills. This process starts through self-exploration, getting to know one's strengths, weaknesses, values, and goals. Leaders are the stewards of God's creation and are responsible for developing their teams so that they can also be leaders. It is important to have a vision, which should result from everyone involved in the team. Team members should be encouraged to look outward for new ideas for improvement, reflecting on the past and present to make a better future. Living at the edge of the unknown and uncertainty may not sound pleasant for some people, and it is not easy. It is not taught in school and is not usually taught or discussed at work. Conner (2006) makes a distinction between two groups of people who may take this continuous change differently: "While type-D individuals tend to become immobilized and react with fear, denial, or complacency, type-O people recognize the discomfort as a signal to activate their coping mechanisms and adapt to the shifting circumstances." The aerospace industry, like many other industries, is a fluctuating curve of ups and downs, based on the priorities of each administration. For decades, programs have been started during one administration and canceled during the next one. Examples are Constellation, Space Shuttle, Asteroid Redirect, among other missions. Agencies and private companies supporting those missions ramp up with their designs, testing, hiring, and training personnel, only to learn that after a few years, they must stop the work for ongoing projects and lay off people. The latter also creates a knowledge gap as people move to other industries. Some space programs may last longer than the duration of an administration. That is the case of the International Space Station (ISS), which has been orbiting Earth for over 20 years with continuous human presence aboard the station (NASA, 2020).

Based on these examples, one can say that the aerospace industry is used to this continuous change. However, that does not prevent these changes from affecting people and their families when they lose jobs and need to find places where their skills can be utilized and valued. This is also a significant change for young professionals who come to the industry to make a difference and find that another program has just been canceled. Things have been changing again over the past few years, with the Space Policy Directive 1 to return to the Moon sustainably by 2024 (NASA, 2019), and then moved to 2025 due to budget, litigations from contractor companies, and the pandemic; one can assume this may continue to change. While going back to the Moon is exciting because of the engineering challenge for this generation to step up and have a continuous presence on the Moon, there is still some uncertainty about what the future will look like with new administrations. As leaders, it is wise to "view change as an unfolding continuum and demonstrate a high tolerance for its ambiguity" (Conner, 2006).

CONCLUSION

Global leaders need to constantly be learning, which is also one of the characteristics highlighted in the journey of agilists. This includes new or improved ways of communication, operations of foreign markets, expansion of their understanding of cultures, understanding of employees' motivations, and embracing of socio-cultural and socio-technical influences in work performance. To be effective, global leaders need to use an integrated leadership style to incorporate different strategic elements according to different situations. Finally, integrating the human element as part of the global leadership strategy should always be a priority: human systems integration. Agile leadership has the human element as a priority, which grants empowerment and motivation of team members to perform at their fullest potential.

REFERENCES

Adler, N., & Gundersen, A. (2008). *International dimensions of organizational behavior* (5th ed.). Mason, OH: South-Western Cengage.

Ang, S., Van Dyne, L., Koh, C., Ng, K., Templer, K., Tay, C., & Chandrasekar, N. (2007). Cultural intelligence: Its measurement and effects on cultural judgment and decision making, cultural adaptation, and task performance. *Management and Organization Review*, 3(3), 335–371. https://doi.org/10.1111/j.1740-8784.2007.00082.x

Beck, K., Beedle, M., van Bennekum, A., Cockburn, A., Cunnigham, W., Fowler, M., Greening, J., Higsmith, J., Hunt, A., Jeffries, R., Kern, J., Marick, B., Martin, R., Mellor, S., Schwaber, K., Sutherland, J., & Thomas, D. (2001). *Manifesto for agile software development*. https://agilemanifesto.org/

Beer, M., Eisenstat, R., & Spector, B. (2011). Change through persuasion. In *HBR's 10 must-reads on change management*. Boston, MA: Harvard Business Review.

Burke, W., Lake, D., & Paine, J. (Eds.) (2009). *Organization change: A comprehensive reader*. Hoboken, NJ: Jossey-Bass.

Bushe, G.R. & Marshak, R.J. (2016). The Dialogic Mindset: Leading Emergent Change in a Complex World. *Organization Development Journal*, 34(1).

Campbell, J. P., Dunette, M. C., Lawler, E. E. & Weick, K. E. (1970). *Managerial behavior performance and effectiveness*. New York: McGraw-Hill.

Cappelli, P., & Tavis, A. (2018). HR goes agile. *Harvard Business Review*, 96(2), 46–52. https://hbr.org/2018/03/hr-goes-agile

Choudhary, S., Choudhary, A., & Joshi, S. (2013). Social and cultural factors influence motivation. *Journal of Business and Management*, 14(6), 6–14.

Conner, D. R. (2006). *Managing at the speed of change*. New York, NY: Villard Books.

Cox, L. (2019). *How to lead successful virtual teams*. Corporate Education Group.

Fisher, B. A. (1970). Decision emergence: Phases in group decision-making. *Communication Monographs*, 37, 53–66.

Furuhjelm, J., Segertoft, J., Justice, J., & Sutherland, J. J. (2015). Owning the sky with agile means building a jet fighter faster, cheaper, and better with Scrum. Semantic Scholar. www.scruminc.com/wp-content/uploads/2015/09/Release-version_Owning-the-Sky-with-Agile.pdf

Gersick, C. J. (1988). Team and transition in work teams: Toward a new model of group development. *Academy of Management Journal*, 31, 9–41.

Giri, V. (2006). Culture and communication style. *Journal Review of Communication*, 6(1–2), 124–130. https://doi.org/10.1080/15358590600763391

Goleman, D., Boyatzis, R., & McKee, A. (2013). *Primal leadership: Unleashing the power of emotional intelligence*. Watertown, MA: Harvard Business Review.

Heisterkamp, T. (2019). Application of organizational agility to enhance the predictability of costs and increase the probability of creating value. *Journal of Creating Value*, 5(2), 164–175. https://doi.org/10.1177/2394964319885520

Herscovitch, L., & Meyer, J. (2002). Commitment to organizational change: Extension of a three-component model. *Journal of Applied Psychology*, 474–487.

Hoda, R., Noble, J., & Marshall, S. (2012). Developing a grounded theory to explain the practices of self-organizing agile teams. *Empirical Software Engineering*, 17(6), 609–639. https://doi.org/10.1007/s10664-011-9161-0

Holland, J. H. (1995). *Hidden order*. Reading, MA: Addison-Wesley.

International Consortium for Agile (ICAgile). (n.d.). International Consortium for Agile. www.icagile.com/

Kotter, J. (2012). *Leading change*. Boston, MA: Harvard Business School Press.

Kouzes, J., & Posner, B. (2017). *The leadership challenge* (6th ed.). Hoboken, NJ: John Jossey-Bass.

Kozlowski, S., & Ilgen, D. (2006). Enhancing the effectiveness of work groups and teams. *Journal of Psychological Science in the Public Interest*, 7(3), 77–124.

Kreitner, R., & Kinicki, A. (2008). *Organizational behavior* (8th ed.). New York, NY: The McGraw-Hill Companies, Inc.

Litzky, B., Eddleston, K., & Kiddler, D. (2006). The good, the bad, and the misguided: How managers inadvertently encourage deviant behaviors. *Academy of Management Perspectives*, 20(1), 91–103.

Marshak, R. (2006). *Covert processes at work*. San Francisco, CA: Berrett-Koehler Publishers.

Martens, R. (2013). Influence of participation motivation on success and satisfaction in team performance. *Research Quarterly, American Association for Health, Physical Education and Recreation*, 41(4), 510–518. https://doi.org/10.1080/10671188.1970.10615009

Maslow, A. H. (1943). A theory of human motivation. *Psychological Review*, 50, 394–395.

McClelland, D. C. (1975). *Power: The inner experience*. New York: Irvington.

McEntire, L., & Greene-Shortridge, T. (2011). Recruiting and selecting leaders for innovation: How to find the right leader. *Journal Advances in Developing Human Resources*.

McGrath, J. E. (1991). Time, interaction, and performance (TIP): A theory of groups. *Small Group Research*, 22, 147–174.

Melo, C. O., Cruzes, D. S., Kon, F., & Conradi, R. (2013). Interpretative case studies on agile team productivity and management. *Information Software Technology*, 55(2), 412–427. https://doi.org/10.1016/j.infsof.2012.09.004

Morse, L., & Babcock, D. (2007). *Managing engineering and technology* (4th ed.). Upper Saddle River, NJ: Pearson Prentice Hall.

Natemeyer, W. E., & Hersey, P. (2011). *Classics of organizational behavior* (4th ed.). Long Grove, IL: Waveland Press, Inc.

National Aeronautics and Space Administration (NASA) (2013). *Through a new lens*. NASA Safety Center: System Failure Case Study. https://sma.nasa.gov/docs/default-source/safety-messages/safetymessage-2013-04-01-safetyculturefivefactormodel.pdf?sfvrsn=28af1ef8_6

National Aeronautics and Space Administration (NASA) (2017). *Significant incidents and close calls in human spaceflight interactive edition*. Johnson Space Center Safety and Mission Assurance Flight Safety Office. Technical Report.

National Aeronautics and Space Administration (NASA) (2019). Forward to the moon: NASA's strategic plan for human exploration. www.nasa.gov/wp-content/uploads/2019/05/america_to_the_moon_2024_artemis_20190523.pdf?emrc=647a4c

National Aeronautics and Space Administration (NASA) (2020). 20 breakthroughs from 20 years of science aboard the International Space Station. www.nasa.gov/mission_pages/station/research/news/iss-20-years-20-breakthroughs

National Aeronautics and Space Administration (NASA) (2022). *Mars exploration program*. NASA Science. https://mars.nasa.gov/

Nir, M. (2018). Identify metrics that matter. In Michael Nir (Ed.), *The pragmatist's guide to corporate lean strategy* (pp. 43–50). Apress. https://doi.org/10.1007/978-1-4842-3537-9_5

Patton, J. (2014). *User story mapping*. Sebastopol, CA: O'Reilly Media, Inc.

Peterson, T. (2007). Motivation: How to increase project team performance. *Project Management Journal*, 38(4), 60–69. https://doi.org/10.1002/pmj.20019

Poole, M. S., & Roth, J. (1989). Decision development in small groups IV: A typology of group decision paths. *Human Communication Research*, 15(3), 323–356.

Rickards, T., & Moger, S. (2000). Creative leadership processes in project team development: An alternative to Tuckman's stage model. *British Journal of Management*, 11, 273–283.

Scholtes, P. R. (1998). *The leader's handbook: Making things happen, getting things done*. New York: McGraw-Hill.

Schwaber, K., & Beedle, M. A. (2001). *Agile software development with Scrum*. New York: Pearson Education.

Sharifi, H., & Zhang, Z. (2001). Agile manufacturing in practice: Application of a methodology. *International Journal of Operations and Production Management*, 21(5–6), 772–779. https://doi.org/10.1108/01443570110390462

Silva, A. L., & Guerrini, F. M. (2018). Self-organized innovation networks from the perspective of complex systems: A comprehensive conceptual review. *Journal of Organizational Change Management*, 31(5), 962–983. https://doi.org/10.1108/JOCM-10-2016-0210

Silva-Martinez, J. (2016). Human systems integration: Process to help minimize human errors, a systems engineering perspective for human space exploration missions. *Journal REACH—Reviews in Human Space Exploration*, 2, 8–23. https://doi.org/10.1016/j.reach.2016.11.003

Silva-Martinez, J. (2023a). Conceptualization of agile leadership characteristics and outcomes from NASA agile teams as a path to the development of an agile leadership theory. *Journal of Creating Value*. https://doi.org/10.1177/23949643231202894

Silva-Martinez, J. (2023b). Practical implications to becoming agile organizations: NASA case study. *Acta Astronautica*. https://doi.org/10.1016/j.actaastro.2023.11.044

Silva-Martinez, J. (2023c). *Diversity and inclusion in the aerospace industry and the role of leaders*. Las Vegas, NV: AIAA ASCEND.

Sparrow, G. (2007). *Spaceflight: The complete story from sputnik to shuttle—and beyond*. New York, NY: DK Publishing.

Srivastava, P., & Jain, S. (2017). A leadership framework for distributed self-organized teams. *Team Performance Management*, 23(5/6), 293–314. https://doi.org/10.1108/TPM-06-2016-0033

Tuckman, B. W. (1965). Development sequence in small groups. *Psychological Bulletin*, 63(6), 384–399.

Tuckman, B. W., & Jensen, M. C. (1977). Stages of small group development revisited. *Group and Organizational Studies*, 2, 419–427.

Walker, C. C., & Dooley, K. J. (1999). The stability of self-organized rule-following work teams. *Computational & Mathematical Organization Theory*, 5(1), 5–30.

Wanberg, C., & Banas, J. (2000). Predictors and outcomes of openness to changes in a reorganizing workplace. *Journal of Applied Psychology*, 85(1), 132–142.

4 Developing an Integrated Model of STEAM Education

Caitlyn A. K. Singam

INTRODUCTION

Conventionally, academic knowledge is portrayed as being essentially dichotomous, with subject-matter domains being generally classifiable under either the sciences or the humanities (Bod & Kursell, 2015; Bouterse & Karstens, 2015). In the education sector proper, this is usually discussed in terms of the STEM/HASS divide, with science, technology, engineering, and mathematics (STEM) and humanities, arts, and social sciences (HASS)[1] being named as the composite sectors of the sciences and humanities, respectively. Although STEM and HASS fields are each typically further separated into individual fields and subfields, often to the point of being re-agglomerated under different names such as MINT, STEMM, SHAPE, and HACD,[2] the fundamental separation of their corresponding representations of the sciences and humanities remains.

Beyond being a simple academic matter (in either sense of the phrase), the separation of science and humanities fields as different areas of academic knowledge has seeped into education not just as a semi-arbitrary means of convenient categorization but as a significant facet of educational systems. Even in cases where they are taught in parallel, such as during pre-collegiate education, STEM and HASS subjects are generally taught independently from one another with limited or infrequent cross-disciplinary curricular integration (Zhan and Niu, 2023), and with little centralization of content resources across the STEM/HASS divide (Klein and Newell, 2002). At the university level and beyond, the divide between STEM and HASS fields is only made more concrete by degree programs which typically result in STEM and HASS specialists being separated into completely different classes (American Academy of Arts & Sciences, 2014), if not completely different faculties and buildings (Chittle et al., 2023, p. 10), throughout their academic careers. The effect has been an enforced segregation between STEM and HASS fields and their associated knowledge sets, analytical approaches, and skill sets, at the cost of including interdisciplinary and cross-disciplinary skills on either side of the STEM/HASS divide (Singam, 2022b).

Under this divided view, STEM and HASS fields are often seen as not only as having different objects of study – the observable physical world, in the case of STEM, and sociocultural phenomena/aspects of human endeavor in the case of HASS – but as representing fundamentally different schools of thought with irreconcilably different

DOI: 10.1201/9781003666431-5

cultures. Proponents of splitting the sciences and humanities fields assert that the sciences and humanities are representative of entirely different worldviews and approaches (Wierzbicka, 2011; Zemlinskaia et al., 2022) that cause them to "produce knowledge" according to fundamentally different practices (Mazzolini, 2005, p. 201). Given the lack of a consistently definable point of demarcation between STEM and HASS fields, the necessity for separation between STEM and HASS fields has become increasingly ambiguous. The exact definitions of "the sciences" and "the humanities" as categorical terms have changed substantially over the centuries (Bod & Kursell, 2015), and remain a point of observable inconsistency today (Banerjee et al., 2024). Thus, the *assumption* of a demarcation between the sciences and the humanities has remained a point of insistence over the years, despite the exact location and nature of that divide being contentious (Bouterse & Karstens, 2015). Nonetheless, existing educational programs and most formal academic institutions are constructed such that STEM and HASS subjects are still disaggregated into separate, typically divergent, academic tracks.

Although uni-disciplinary tracks are effective for specialized training in single, narrow areas of study, such programs heavily limit the degree to which students can develop their abilities across multiple research areas or acquire familiarity with problem-solving in complex, multi-disciplinary contexts. As the international community becomes increasingly cognizant of the role of cross-disciplinary, interdisciplinary, and trans-disciplinary competency development[3] in promoting what the United Nations (UN) terms "global citizenship" (United Nations, 2023), there has been an increasing trend away from dichotomous STEM/HASS-separated education and toward a combined science, technology, engineering, arts/humanities, and mathematics (STEAM) approach to education. In countries like the United States, Australia, and Japan, this has prompted local and national governments to create STEAM policies, issue STEAM-centric education grants, and foster STEAM-focused collaborative initiatives to encourage and support the adoption of STEAM principles in classroom education (Dell'Erba, 2019; MacDonald et al., 2019; Matsuura & Nakamura, 2021).

The central idea behind STEAM as an educational philosophy is reflected in the STEAM acronym: placing the arts and humanities (the "A" in STEAM) on equal footing with the sciences and STEM, and intertwining STEM and HASS education as part of a more integrated approach to competency development (see Figure 4.1).

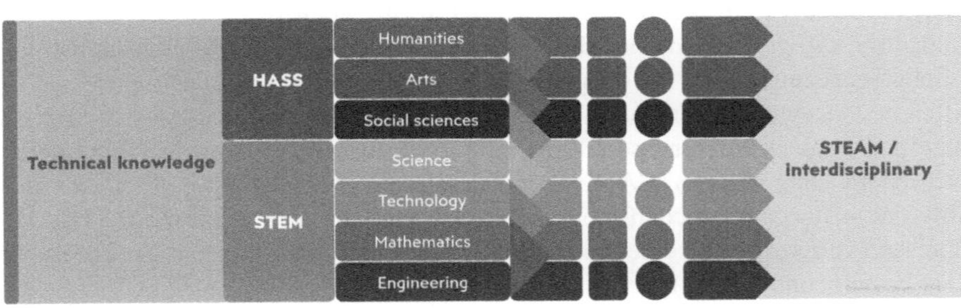

FIGURE 4.1 Graphic showing the relationship between STEAM, STEM, HASS, and their component disciplines/fields. (Graphic created by the author.)

Compared to specialty-specific education, STEAM aims to foster generalized interdisciplinary and transdisciplinary competencies along with more specialized skills and knowledge required to develop cross-disciplinary and multi-disciplinary competencies. This integrated vision of education has garnered burgeoning attention as employers continue to voice mounting frustration with a lack of employable job candidates able to demonstrate general, domain-transferable skills in the face of changing industries and technological landscapes (Bessen, 2014; Cunningham & Villaseñor, 2016; Taylor, 2005), even in specialist industry fields such as the biosciences and engineering (Cramer & Hamilton, 2017; Groeneveld et al., 2021). Integrated STEAM education has thus been seen as a potential means of bringing educational programs into better alignment with regional and global labor markets that are trending toward being "over- suppl[ied]" (Ulicna & Royale, 2015, p. 29) with uni-disciplinary specialists in both the sciences and humanities alike (Bennett, 2009; Charette, 2013; Itoh, 2014; Siekmann & Korbel, 2016). More broadly, STEAM education also offers promise as a more effective means of fostering socially responsible global citizens who have an understanding and appreciation of the value brought by specializations and perspectives other than their own, along with the interdisciplinary problem-solving skills needed to translate that understanding into personal and professional success in the global community.

THE SYSTEMS-BASED APPROACH TO STEAM EDUCATION

Despite consensus on the beneficial nature of STEAM education and the value of uniting (rather than separating) educational endeavors across the existing STEM/HASS divide, there has been a somewhat ironic *lack* of unity in envisioning how STEAM education should achieve its vision of integrating the sciences and the humanities.

Existing perspectives on STEAM education reflect a wide range of conceptualized approaches to STEAM education, ranging from those envisioning STEAM as a transformative curricular overhaul of existing educational programs to those that see STEAM as merely entailing arts integration (i.e., the injection of visual arts-based projects into existing STEM courses) and little else. Meta-analyses (Leavy et al., 2023; Mejias et al., 2021; Perignat & Katz-Buonincontro, 2019) of publications and policies regarding STEAM education have noted that such inconsistencies in conceptualizing STEAM education are widespread, to the point of making it challenging to identify a single unified, integrated concept of what the STEAM educational philosophy is meant to look like in practice. In many instances, inconsistencies across different conceptualizations of STEAM education have been noted to extend beyond mere pedagogical differences and reflect more deep-seated issues, including debate over the meaning of the STEAM acronym itself and what the "A" in STEAM – possibly one of the most recognizable aspects of the campaign to promote STEAM – represents in terms of included disciplines from the arts and humanities (Perignat & Katz-Buonincontro, 2019).

The documented inconsistencies in how various educational publications, policies, and recommended practices conceptualize STEAM education cannot simply be attributed to a lack of stakeholder familiarity with the overall STEAM. They cannot be dismissed as issues attributable to STEAM still being nascent. STEAM is neither a particularly novel concept, nor an obscure one: the acronym "STEAM" itself, about the interdisciplinary educational concept for bridging STEM and HASS fields, has been in continuous use in scholarly publications since at least 2006 (Marín- Marín et al., 2021), and has been garnering increasing international attention ever since. Figure 4.2 shows interest and awareness of STEAM as an educational philosophy are arguably on par with established areas of educational focus (e.g., the humanities), suggesting that some other factor contributes to the observed inconsistencies in conceptualizing STEAM education.

Strikingly, a common thread underlying STEAM stakeholders' varying conceptualizations of STEAM education is the different use cases that stakeholders have for STEAM education as a system. In other words, there are discernible differences among STEAM stakeholders regarding how and why those groups interact with educational systems. These produce notable differences in how those stakeholders envision the overall mission and objectives of implementing STEAM education. STEM specialists and industry stakeholders, for instance, primarily interact with the education system to develop the STEM technical workforce, and accordingly approach conceptualization of STEAM education as a means of better meeting industry workforce needs and addressing economic trends. This results in many STEM-oriented discussions of interdisciplinary STEAM education conceptualizing STEAM more as a narrowly scoped methodological tool to help educate STEM specialists (e.g., via arts integration), where arts and humanities competencies are sometimes seen as secondary "add-ons" to conventional STEM education (Mejias et al., 2021). In contrast, stakeholders seeking to foster idea exchange across existing academic bounds separating the sciences and humanities tend to favor conceptualizing STEAM education as a broader-scoped, pedagogical construct for enabling bi-directional enrichment of STEM and humanities coursework (e.g., through cross-disciplinary curricular alignment).

These variable interpretations or conceptualizations of the STEAM educational philosophy reflect a close bond between technical facets of STEAM education (e.g., methodological approaches or curricular content choices) and social elements of the educational landscape (e.g., stakeholder interests and socio-economic factors), where minor differences in the interfaces between those elements can yield remarkably different conceptualizations of the overall STEAM educational system. This poses an issue for an integrative effort such as implementing STEAM education, where continuous and cohesive integration of different segments of the STEAM educational concept is needed to enable effective interdisciplinary learning meaningfully (Mejias et al., 2021).

Socio-Technical Analysis and STEAM Education

STEAM education is nonetheless conventionally approached from a purely technical angle, with most discussions of STEAM education either focusing exclusively on the technical elements (e.g., pedagogy, teaching methodologies, and curricular content/

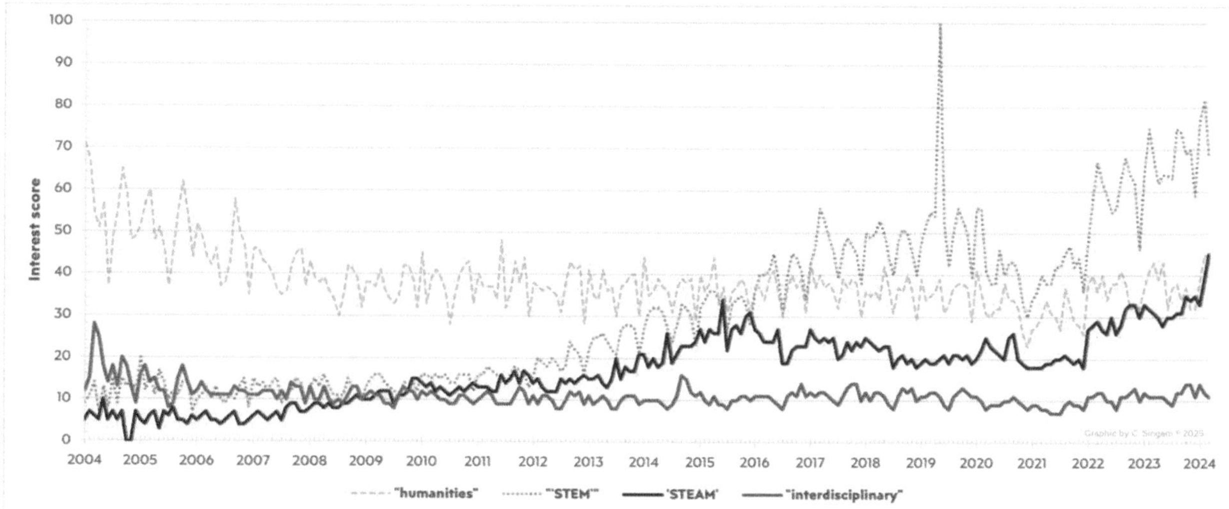

FIGURE 4.2 Monthly search term interest for the terms "humanities," "STEM," "STEAM," and "interdisciplinary" in Google's "Education" category, from 2004 to 2024. The search interest index (*y*-axis) is expressed as provided by Google Trends (quantity of searches, expressed on a 0–100 scale relative to the maximum value on the chart). (Graphic generated by the author; data from Google Trends, 2024.)

structure) with minimal consideration of the social and human elements involved therewith (e.g., student access to educational opportunities, or the societal and non-technical functions of educational systems). This produces a rather one-dimensional perspective of a system that needs to serve and smoothly reconcile the constraints from various stakeholder groups and use-cases. A more effective means of working toward an integrated model of STEAM education is to use the interdisciplinary toolkits offered by systems engineering (SE) and systems thinking (ST).

One can consider construction of a consistent, integrated model of STEAM education as an SE exercise in (1) reconciling the primary stakeholder needs of all the STEAM stakeholder groups (including stakeholders from the sciences and humanities alike), (2) translating those needs into desired functions for STEAM education, and (3) mapping that information into a conceptual systems architecture model aligned with the structure of existing educational infrastructures and contexts. SE and ST methodologies, such as stakeholder analysis, requirements flowdown, and system architecture design processes, offer a means of achieving those tasks toward an end goal of integrating social and technical elements of STEAM education into a cohesive and consistent system concept.

From a systems perspective, many emergent difficulties with consistent conceptualization of STEAM education are partially due to insufficient consideration of the human-system integration (HSI) elements and interactions relevant to educational systems. The relationships between stakeholder groups, their interests, and other social factors (e.g., access to education) applicable to the humans who interact with educational systems directly impact the desired and implemented technical structure and function of STEAM education. They thus must be necessarily considered as part of any meaningful effort toward an integrated and non-fragmented model of STEAM education. To that end, socio-technical systems (STS) analysis offers a promising means of reconciling interfaces between social and technical elements of complex systems with HSI elements. As part of the interdisciplinary SE/ST toolkit, STS analyses are well-suited to handling systems with multidisciplinary and interdisciplinary elements like STEAM education; in fact, SE and ST methodologies like STS are some of the interdisciplinary competencies that STEAM education is meant to help highlight (Singam, 2022b). One of the significant advantages of using STS analyses is that, unlike more conventional approaches to system analyses, which usually operate with a narrower scope and use element-by-element review processes, STS approaches allow for analyses to be conducted on weakly-bounded systems or across multiple systems when needed (Winby & Mohrman, 2018). This is attributable to the interface-centric methodology that is part and parcel of STS analyses, which encourages the conceptual outlining of system boundaries relative to their least nebulous points (i.e., their points of interaction with other entities or systems), making it easier to identify otherwise hidden interfaces. STS methods have also encountered newfound popularity as a means of handling highly dynamic systems in informatics contexts (Walker et al., 2008), hence making STS analysis a quality choice for analyzing an extensive, difficult-to-define system (or system concept) of information dissemination, like STEAM education.

Socio-Technical Elements of Education

Indeed, one can make the case that STEAM education is a prime candidate for evaluation from socio-technical perspectives due to its significant role in fulfilling both technical and social needs for STEAM stakeholders and broader society. Education is, after all, the primary means through which individuals acquire the unique knowledge, skills, and experiences needed to succeed as working professionals in their fields of choice (UNICEF, 2019), as well as how individuals gain the interdisciplinary and transdisciplinary skills needed to be self-sufficient and responsible citizens in broader society (United Nations, 2023).

From a technical perspective, education plays a critical role in shaping the overall character and composition of the technical workforce,[4] whose members shape evolving standards of technical knowledge and practice incorporated into future educational programs' dissemination (Singam, 2022a). Employers are widely reliant on education as a means of developing job candidates with the requisite job-specific knowledge and skills to fulfil evolving workplace needs (Buchanan et al., 2020), often to the extent of using educational qualifications as a semi-standardized shorthand for expressing specialty-specific knowledge/skill baselines required for various job positions (Chittister & Haimes, 2011). STEM fields (i.e., the sciences) have an outsized dependence on educational programs due to their need for substantial technical knowledge and skill proficiency even in entry-level positions (Singam, 2022a). For individuals pursuing careers in such fields, educational experiences tend to be highly differentiated in order to meet the distinct, specialized needs of each technical discipline and its associated professionals; as a result, participating in the educational route specific to one specialty can often lead to one losing the opportunity to participate in, or even being entirely unaware of, the educational experiences particular to another discipline (Singam, 2022b). Education often ends up being a career-long companion to most professional workforce members, chaperoning them from their formative years of generalized classroom education through their specialized training and job-specific experiential learning later on in their careers (Guterres, 2023).

Even outside the workplace, education has been deemed a fundamental necessity, ensuring individuals are equipped with the skills to be responsible global citizens and effective societal participants (UNESCO, 2013; United Nations, 2023). Education is well documented as a major driver of beneficial social change in several different small-scale and large-scale socio-economic metrics and observable phenomena (Table 4.1), and plays a significant role in shaping the societal environment in which individuals live, work, and interact. Indeed, education's societal significance is such that it has been formally recognized as a universal human right by the UN member states (Universal Declaration of Human Rights, 1948, art. 26), placing it on par in importance with rights such as those to life and liberty, which were similarly recognized as fundamental human rights by the same declaration (Universal Declaration of Human Rights, 1948, art. 3). Free and universal access to basic education has since been established as a legal requirement under international treaty (International Covenant on Economic, Social and Cultural Rights, 1966, art. 13), making attainment of some level of formal education a more common experience worldwide than accessing the internet even as of 2020.[5]

TABLE 4.1
List of Small-Scale (i.e., Individual Level, or per Person) or Large-Scale (i.e., Community-Wide or Regional) Socio-economic Phenomena Associated with or Positively Influenced by Educational Quality and/or Availability Increases

Factor	Scale	Citation
Public health	Large-scale	Hahn and Truman (2015)
Economic growth	Large-scale	Aghion et al. (2009), Hanushek and Woessmann (2008)
Technological innovation	Large-scale	Biasi et al. (2022)
Crime reduction	Large-scale	Lochner (2011)
Civic engagement	Large-scale	Center for Global Development (2006)
Social equity	Large-scale	United Nations (2023)
Sustainable development	Large-scale	UNESCO (2014)
Financial independence	Small-scale	Xiao et al. (2014)
Informed decision-making	Small-scale	Cutler and Lleras-Muney (2006)
Personal autonomy	Small-scale	United Nations (2023)

WORKING TOWARD AN STS FRAMEWORK FOR STEAM EDUCATION

It is thus evident that education is a system that is much more deeply embedded in its socio-economic and socio-technical environment than is typically suggested in most analyses of the STEAM educational space. Accordingly, a meaningful approach to conceptualizing STEAM education requires incorporating both the social and technical elements applicable to educational systems, which play essential roles in defining the operational constraints, requirements, and use-cases for STEAM education. Figure 4.3 presents a basic outline of some of the key socio-technical relationships that impact STEAM education and indicates the existence of these elements as complementary and interfacing pairs that shape the socio-technical landscape of the educational world.

In light of the significant social and technical elements applicable to STEAM education, STS analytical approaches for STEAM education have not become a widespread phenomenon. STEAM education tends to be analyzed segmentally instead, focusing on the socio-technical dyad's technical or social side (i.e., what would be represented by only half of Figure 4.3). This is further exacerbated by STEAM analyses being heavily skewed toward exploring technical facets of STEAM education (such as curricula and instructional methods) in the sciences context. Industry-driven workforce needs applicable to secondary and/or collegiate education (Marín-Marín et al., 2021), often with minimal consideration of aspects relevant to the humanities or earlier-stage education.

The STEAM community will nonetheless need to overcome the hurdle of developing a holistic understanding of the educational world if it desires to see its vision of an integrated STEAM educational system properly realized. In addition to considering how the various subject-matter domains under the STEAM umbrella need

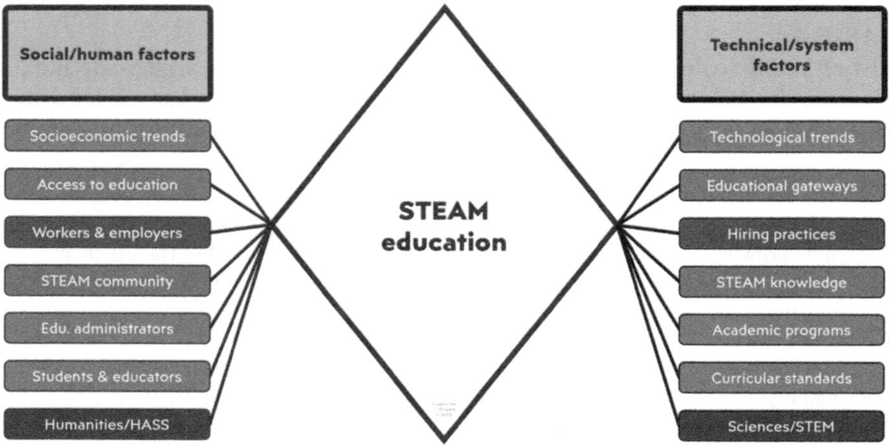

FIGURE 4.3 Graphic showing socio-technical relationships of relevance to STEAM education. Each pair of similarly colored boxes represents a socio-technical pair, with the "social" element in the left-side column and the corresponding "technical" element in the right-side column (Graphic created by the author.)

to be combined and integrated in order to promote the kind of interdisciplinary, adaptable competencies needed to prepare students for today's workforce, the STEAM community also needs to have better awareness of how the academic, professional, and socio-economic contexts surround STEAM interact with each other to potentially impact STEAM implementation (Figure 4.3). Prior approaches to analyzing STEAM through smaller-scope analyses have already set the stage for understanding many of the factors at play in terms of ensuring effective development of a unified, consistent concept model of STEAM education; it is now a matter of ensuring that these insights are coordinated into a cohesive concept.

CONCLUSIONS

STEAM, both from an educational and an occupational lens, has a demonstrated capacity to change the landscapes of the academic and professional world. Through interdisciplinary and multidisciplinary development of a unified STEAM concept model, it is now a matter of making that change a reality.

One of the great appeals of STEAM is that it provides a glimpse of a vision for a more unified educational world: a blend of specialist and generalist, scientific and artistic, educational and occupational, theoretical and practical – all brought together under the STEAM umbrella. It is undoubtedly an alluring vision, made all the more so by the fact that it is demonstrably practical (through case studies, pilot programs, etc.) in a way that most wide-reaching concepts are not, but comes with the caveat of requiring a high level of coordination across several different gaps and divides that are often poorly visible yet highly influential. Achieving the vision of truly integrated STEAM education then requires the ability to cohesively visualize the overall educational landscape holistically, which is rarely done, if only due to its sheer size and complexity. However, examining the socio-technical factors contributing to this landscape can yield insight into how the segmented and fractionated elements of the STEAM community can be more effectively interconnected into a unified framework.

Rather than approaching STEAM education in terms of a miscellany of disparate and often inconsistent concept-models, the socio-technical approach to STEAM education instead construes those different conceptualizations of the STEAM philosophy as differently constrained "views" of a complex, multi-dimensional system architecture that includes both social and technical dimensions. While contemplating the conceptual architecture of STEAM education based on a single dimension alone (e.g., the technical perspective only) or based only on a single subsection of the overall system architecture (e.g., only from a higher-education standpoint) is unlikely to facilitate effective development of the complete system concept, the intersectional view offered by socio-technical analysis methods can provide a substantially different angle from which to appraise the interactions and intersections between various elements of the larger system. In the case of a system for which there are a variety of disparate stakeholder groups – ranging from students and educators to industry professionals and broader society – understanding the interconnections and relationships among these stakeholders and their interests can be highly beneficial in constructing a cohesive system architectural concept that reconciles the varying needs, use-cases, and constraints applicable to various aspects of STEAM education. STEAM stakeholders already have a shared foundational philosophy upon which to build an integrated concept of STEAM education, along with many of the conceptual "puzzle pieces" or architectural views needed to start assembly of that concept. The additional perspective offered by socio-technical dimensions helps delineate significant interaction lines between STEAM-related social and technical elements (Figure 4.3). STS analytical approaches, combined with holistic SE/ST system architecture design methods, offer the tools needed to assemble the necessary puzzle pieces to construct a meaningful, integrated, and implementable concept of STEAM education.

NOTES

1 HASS appears to be the most common acronym for the humanities with international scholarly use at the time of writing; however, unlike STEM, it is not a widely standardized term.
2 See, e.g., de Vries (2016), NASEM (2018), Jones et al. (2020), and Root-Bernstein (2018), for instances of each respective acronym being used in published academic discourse.
3 Cross-disciplinary competencies involve skill in appropriately applying the specialized perspectives and methodologies of one discipline in the context of another (e.g., applying physics to a biology problem). Interdisciplinary competencies involve the integration of knowledge/skill elements from multiple different disciplines to achieve a common goal (e.g., creating a bio-electronic sensor). Trans-disciplinary competencies are areas of generalized understanding or skill (e.g., critical thinking or creativity) that are essentially universal and not specific to a given discipline or academic field.
4 The term "technical workforce" is used in this chapter to refer to the multi-disciplinary body of practitioners and professionals who utilize STEAM subject-matter knowledge and skills as part of their work.
5 In 2020, 86.3% of the world's population was estimated to have had some level of formal education (Samir et al., 2023); comparatively, the International Telecommunications Union

(ITU) estimates that approximately 60% of the world's population was estimated to have used the internet that year (ITU, 2021).

REFERENCES

Aghion, P., Boustan, L., Hoxby, C., & Vandenbussche, J. (2009). The causal impact of education on economic growth: Evidence from the US. *Brookings Papers on Economic Activity*, *1*(1), 1–73.

American Academy of Arts & Sciences. (2014, August 18). Enclosed in a College Major? Variations in Course-taking among the Fields. American Academy of Arts & Sciences Data Forum. www.amacad.org/news/enclosed-college-major-variations-course-taking-among-fields

Banerjee, P., Graham, L., & Given, G. (2024). A systematic literature review identifies inconsistencies in subjects' inclusion in research reports on STEM workforce skills in the UK. *Cogent Education*, *11*(1), 2288736.

Bennett, D. (2009). Academy and the real world: Developing realistic notions of career in the performing arts. *Arts and Humanities in Higher Education*, *8*(3), 309–327.

Bessen, J. (2014). Employers aren't just whining–the "skills gap" is real. *Harvard Business Review*, *25*. https://scholarship.law.bu.edu/shorter_works/186

Biasi, B., Deming, D., Moser, P., & Dillon, E. W. (2022). Education and Innovation. In M. J. Andrews, A. Chatterji, J. Learner, & S. Stern (Eds.), *The Role of Innovation and Entrepreneurship in Economic Growth* (pp. 537–558). University of Chicago Press.

Bod, R., & Kursell, J. (2015). Introduction: The humanities and the sciences. *Isis*, *106*(2), 337–340.

Bouterse, J., & Karstens, B. (2015). A diversity of divisions: Tracing the history of the demarcation between the sciences and the humanities. *Isis*, *106*(2), 341–352.

Buchanan, J., Allais, S., Anderson, M., Calvo, R. A., Peter, S., & Pietsch, T. (2020). *The Futures of Work: What Education Can and Can't Do* (Programme and Meeting Document ED-2020/FoE-BP/20; Futures of Education). UNESCO.

Center for Global Development. (2006). *Education and the Developing World: Why Is Education Essential for Development?* (Rich World, Poor World: A Guide to Global Development). Center for Global Development.

Charette, R. N. (2013). The STEM crisis is a myth. *IEEE Spectrum*, *50*(9), 44–59.

Chittister, C., & Haimes, Y. Y. (2011). Harmonizing ABET accreditation and the certification of systems engineers. *Systems Engineering*, *14*(3), 341–348.

Chittle, L., Kustra, E., and Houser, C. (2023). A Qualitative Examination of Science Faculty Members' Perceptions of Interdisciplinary Curriculum Development and Refinement. *Canadian Journal for the Scholarship of Teaching and Learning*, 14(2), 1–17. https://ojs.lib.uwo.ca/index.php/cjsotl_rcacea/article/view/11122

Cramer, J. M., & Hamilton, P. T. (2017). An internship may not be enough: Enhancing bioscience industry job readiness through practicum experiences. *Journal of Microbiology & Biology Education*, *18*(1). 1–5. https://doi.org/10.1128/jmbe.v18i1.1248

Cunningham, W. V., & Villaseñor, P. (2016). Employer voices, employer demands, and implications for public skills development policy: Connecting the labor and education sectors. *The World Bank Research Observer*, *31*(1), 102–134.

Cutler, D. M., & Lleras-Muney, A. (2006). *Education and Health: Evaluating Theories and Evidence* (Working Paper 12352). National Bureau of Economic Research.

de Vries, M. J. (2016). The "I" in MINT: A Tale of Two Translations. In *Proceedings of the 11th Workshop in Primary and Secondary Computing Education* (pp. 1–4). Association

for Computing Machinery, New York, NY, United States. https://doi.org/10.1145/2978249.2978250
Dell'Erba, M. (2019). *Preparing Students for Learning, Work, and Life through STEAM Education. Policy Brief.* Education Commission of the States.
Google. (2024). *Google trends* [CSV]. Google Trends.
Groeneveld, W., Vennekens, J., & Aerts, K. (2021). Identifying non-technical skill gaps in software engineering education: What experts expect but students don't learn. *ACM Transactions on Computing Education*, *22*(1), 1:1–1:21.
Guterres, A. (2023). *Our Common Agenda Policy Brief 10: Transforming Education* (Policy Brief 23–14197; Our Common Agenda). United Nations.
Hahn, R. A., & Truman, B. I. (2015). Education improves public health and promotes health equity. *International Journal of Health Services, 45*(4), 657–678.
Hanushek, E. A., & Woessmann, L. (2008). The role of cognitive skills in economic development. *Journal of Economic Literature*, *46*(3), 607–668.
International Covenant on Economic, Social and Cultural Rights (United Nations [UN]) 993 UNTS 3, CTS 1976/46, S Exec Doc D, 95-2 (1978), GAOR 21st Session Supp 16, 49, UN Doc A/6316, UN Doc A/RES/2200(XXI)A, Annex, UNTS Reg No I-14531.
Itoh, A. (2014). Japan's period of high economic growth and science and technology education: The role of higher education institutions. *Japan Labor Review*, *11*(3). 35–57.
ITU. (2021). *Measuring Digital Development: Facts and Figures 2021*. ITU.
Jones, M. M., Abrams, D., & Lahiri, A. (2020). Shape the future: How the social sciences, humanities, and the arts can SHAPE a positive, post-pandemic future for people, economies, and environments. *Journal of the British Academy*, *8*, 167–266.
Klein, J. T., and Newell, W. H. (2002). Strategies for Using Interdisciplinary Resources Across K-16. *Issues in Interdisciplinary Studies*, 20, 139–160.
Leavy, A., Dick, L., Meletiou-Mavrotheris, M., Paparistodemou, E., & Stylianou, E. (2023). The prevalence and use of emerging technologies in STEAM education: A systematic review of the literature. *Journal of Computer Assisted Learning*, *39*(4), 1061–1082.
Lochner, L. (2011). *Non-Production Benefits of Education: Crime, Health, and Good Citizenship*. Working Paper No. 16722. National Bureau of Economic Research. *Wittgenstein Center (WIC) Populationand Human Capital Projection (WIC3.002 (Beta))*.
MacDonald, A., Hunter, J., Wise, K., & Fraser, S. (2019). STEM and STEAM and the spaces between: An overview of education agendas pertaining to 'disciplinarity' across three Australian states. *Journal of Research in STEM Education*, *5*(1). 75–92.
Marín-Marín, J.-A., Moreno-Guerrero, A.-J., Dúo-Terrón, P., & López-Belmonte, J. (2021). STEAM in education: A bibliometric analysis of performance and co-words in Web of Science. *International Journal of STEM Education*, *8*(1), 41.
Matsuura, T., & Nakamura, D. (2021). Trends in STEM/STEAM education and students' perceptions in Japan. *Asia-Pacific Science Education*, *7*(1), 7–33.
Mazzolini, E. (2005). Humanities and science. In S. P. Restivo (Ed.), *Science, Technology, and Society: An Encyclopedia* (pp. 201–206). Oxford University Press.
Mejias, S., Thompson, N., Sedas, R. M., Rosin, M., Soep, E., Peppler, K., Roche, J., Wong, J., Hurley, M., Bell, P., & Bevan, B. (2021). The trouble with STEAM and why we use it anyway. *Science Education*, *105*(2), 209–231.
NASEM. (2018). *The Integration of the Humanities and Arts with Sciences, Engineering, and Medicine in Higher Education: Branches from the Same Tree* (D. Skorton & A. Bear, Eds.). National Academies Press.
Perignat, E., & Katz-Buonincontro, J. (2019). STEAM in practice and research: An integrative literature review. *Thinking Skills and Creativity*, *31*, 31–43.
Root-Bernstein, R. (2018). STEMM education should get "HACD." *Science*, *361*(6397), 22–23.

Samir, K. C., Moradhvaj, Potancokova, M., Adhikari, S., Yildiz, D., Mamolo, M., Sobotka, T., Zeman, K., Abel, G., Goujon, A. & Lutz, W. (2023). *Wittgenstein Center (WIC) Population and Human Capital Projection (WIC3.002 (Beta))* [dataset]. Zenodo.

Siekmann, G., & Korbel, P. (2016). *Defining "STEM" Skills: Review and Synthesis of the Literature. Support Document 1.* National Centre for Vocational Education Research.

Singam, C. A. K. (2022a). A Critical Analysis of the Systems Engineering Leadership Pipeline: Closing the Gender Gap. In A. F. Squires, M. J. Wheaton, & H. J. Feli (Eds.), *Emerging Trends in Systems Engineering Leadership: Practical Research from Women Leaders* (pp. 195–236). Springer International Publishing.

Singam, C. A. K. (2022b). A vision for universal and standardized access to systems competency education. *INSIGHT*, *25*(3), 30–34.

Taylor, A. (2005). What employers look for: The skills debate and the fit with youth perceptions. *Journal of Education and Work*, *18*(2), 201–218.

Ulicna, D., & Royale, R. (2015). *Does the EU Need More STEM Graduates?* Report (120/01). Danish Technological Institute.

UNESCO. (2013). *Right to Education* (ED-2013/SANS COTE; Programme and Meeting Document ED/ADG/2013/11; Education Sector Technical Notes). UNESCO.

UNESCO. (2014). *UNESCO Roadmap for Implementing the Global Action Programme on Education for Sustainable Development* (ED-2014/WS/34). p. 38).

UNICEF. (2019). *Global Framework on Transferable Skills.* www.unicef.org/media/64751/file/Global-framework-on-transferable-skills-2019.pdf

United Nations. (2023). *The Sustainable Development Goals Report 2023.* https://unstats.un.org/sdgs/report/2023/

Universal Declaration of Human Rights, G.A. Res. 217A (III), U.N. Doc. A/810 (Dec. 10, 1948)

Walker, G. H., Stanton, N. A., Salmon, P. M., & Jenkins, D. P. (2008). A review of sociotechnical systems theory: A classic concept for new command and control paradigms. *Theoretical Issues in Ergonomics Science*, *9*(6), 479–499.

Wierzbicka, A. (2011). Defining 'the humanities.' *Culture & Psychology*, *17*(1), 31–46.

Winby, S., & Mohrman, S. A. (2018). Digital sociotechnical system design. *The Journal of Applied Behavioral Science*, *54*(4), 399–423.

Xiao, J. J., Chatterjee, S., & Kim, J. (2014). Factors associated with the financial independence of young adults. *International Journal of Consumer Studies*, *38*(4), 394–403.

Zemlinskaia, T., Fersman, N., Kobchikov, V., & Holden, T. L. (2022). Formation of an Integral Worldview of Humanities and Technical Specialists. In Z. Anikina (Ed.), *Integration of Engineering Education and the Humanities: Global Intercultural Perspectives* (pp. 469–477). Springer International Publishing.

Zhan, Z., and Niu, S. (2023). Subject integration and theme evolution of STEM education in K-12 and higher education research. Humanities and Social Sciences Communications, 10(1), 1–13. www.nature.com/articles/s41599-023-02303-8#citeas

5 A Systems Architecture Approach to Designing Human-Centric Systems

Holly A. H. Handley

INTRODUCTION

The International Council on Systems Engineering (INCOSE) defines Systems Engineering (SE) as an interdisciplinary approach and means to enable the realization of successful systems (INCOSE, 2011). Systems engineers integrate other engineering disciplines through a structured system development process from initial concept to system deployment. Systems that include both human and technical components are called sociotechnical systems (STSs). Development of these systems must consider aspects of both the human (social) and technical subsystems to ensure system success. While many systems engineers understand that the human operator and maintainer are part of the system, they often lack the guidance to fully specify and incorporate human capabilities into the system design (INCOSE, 2011). In response, human systems engineers have developed tools that consider the human component during system development.

Human system engineering (HSE) focuses on identifying and improving methods to integrate human concerns into the conceptualization and design of systems (DOD, 1988). HSE encourages the early understanding of human roles and responsibilities, along with limitations and constraints that may impact system design (Handley, 2021). Incorporating HSE can reduce system risk by including the human component into the system design, ensuring that systems will not require expensive "train-arounds" or late-stage fixes to address issues of ineffective usability (INCOSE, 2009). The human component is considered critical to developing successful STSs during the SE process.

HSE tools are initiated at the architecting stage of system development. System architecting is the step in the SE process between the system concept and the detailed design. The resulting architecture description provides the baseline system structure and the roadmap for the design work. The system architecture stage leverages an architecture framework with viewpoints and models for describing the different perspectives of the envisioned system. However, most architecture frameworks lack a mechanism to explicitly capture human-centric data necessary to ensure the effectiveness of human-operated systems. To address this shortcoming, the Human Viewpoint was created to provide visibility on the human component as part of the system architecture (Handley & Smillie, 2008). The eight Human Views give a complete

DOI: 10.1201/9781003666431-6

perspective on the human roles, activities, and information flows required by a complex system. The Human Viewpoint organizes information for a comprehensive representation of human capabilities and provides a fully integrated set of products that can be used to inform and influence system design (NATO, 2010).

Likewise, the Human Readiness Levels (HRLs) are used to evaluate the maturity of the human considerations included in the developing system (ANSI/HFES, 2021). Designed as an adjunct to Technology Readiness Levels (TRLs), the HRLs are a measure of the readiness of the technology for use by human operators and maintainers. The nine-level scale provides a mechanism to assess the degree to which human human-focused requirements have been incorporated into design decisions (See, 2021). While the HRL tracks the readiness of a technology for human usability throughout the system lifecycle, the first four levels are assessed during the system architecting stage Each HRL has a series of supporting questions determining if the necessary human system requirements have been addressed at that level. These questions can be mapped to the individual components of the Human Viewpoint models to provide the data to support HRL decisions (Handley et al., 2023).

Both the Human Viewpoint and the Human Readiness Levels were developed to support system engineering of traditional STSs. However, as systems transform to include virtual team members, autonomous systems, and adaptive technologies, the next progression in Human System Engineering is to apply the HSE tools to STSs that include Artificial Intelligence (AI) components. AI-based systems may change over time, operate autonomously, and make decisions without human intervention; it may be difficult to fully predict how AI will behave in certain situations. The examples in this chapter explore using the HSE tools to better "design in" the human into future systems.

SYSTEMS ARCHITECTING

System architecting is refining a system's initial operational concept into a baseline representation of the envisioned system. System architecting aims to describe the logical, behavioral, and physical aspects of the system and its high-level structure (INCOSE, 2015). The architecting process is iterative; as more information is available, analyses on different alternatives are performed, and design decisions are made. The result of this stage is a viable description or "blueprint" for the subsequent detailed system design stage based on the arrangement of the system elements described in the system architecture.

Architecture frameworks provide the conventions, principles, and practices for describing system architectures (Wasson, 2016). A framework includes different viewpoints that describe the system from different areas of concern. For example, the Department of Defense Architecture Framework (DoDAF) defines eight viewpoints that break the system description into specific areas of interest: all, capabilities, data and information, operational, project, services, standards, and systems (DoD CIO, 2010).

The models within these viewpoints describe different system entities and characteristics important to that view, with some architecture elements providing linkages between views to ensure consistency. HSE augments the system architecting process to influence these early design efforts and inform the evolving system concept and definition of requirements with human aspects (INCOSE, 2015). The following description of the high-level steps of the system architecting process identifies the opportunities to include human-focused data in the resulting architecture description.

Architecture Definition

Architecture definition activities focus on determining the intended use of the architecture and what capabilities the proposed system will provide. It also identifies the system inputs and outputs and the constraints placed on the system by external factors (DoD CIO, 2010). Here, the operation concept, which will guide the architecture development of the system, can be augmented with a representation of the expected human roles regarding operational demands and system components. At this stage, the human contribution to the system operation and important human characteristics, capabilities, and limitations that may impact the system design are identified.

Architecture Development

During the architecture development stage, data on the developing system is collected and organized guided by an architecture framework; the architecture framework provides the repository of data that can then be used to render different views or models of the system (Handley & Smillie, 2010). System architecting activities include decomposing the operational requirements, describing the initial physical structure, and allocating functions to system nodes. The HSE effort describes the human-specific tasks supporting overall operational and system activities. It identifies the human roles, knowledge, skills, and ability requirements for specific tasks or technology. It defines the required information flows or interactions with the system and the expected human performance metrics.

Architecture Analysis

At different iterations of architecture development, areas of particular interest for stakeholders can be further investigated using the data collected in the architecture framework (INCOSE, 2015). The results can induce revisions to the architecture elements and structure. Analysis of human-focused data includes applying sets of constraints that may alter the human performance of specific roles or tasks. Simulation tools are used to predict dynamic aspects of human behavior, and the outcomes can inform on the feasibility of design decisions and forecast overall system performance (Handley & Smillie, 2010).

Architecture Description

The architecture description documents the results of the architecting activities. It provides a formal representation of the system, organized in a way that supports reasoning about the structure and behavior of the system (DoD CIO, 2010). Extending the system description with the Human Viewpoint provides a framework to collect and organize human-focused system data, resulting in a more complete set of system data. Likewise, utilizing the HRLs throughout the system architecting process provides human-focused evaluation criteria that progress with the system development. These HSE tools ensure that humans are adequately engineered in the system.

HSE TOOLS

The Human Viewpoint provides an architecture framework and models to augment the system architecture process. The HRLs identify how human considerations have been incorporated into architectural design decisions. Together, these tools support integration of the human user or operator into the SE effort, which is critical to the design of successful systems.

The Human Viewpoint

The Human Viewpoint provides a framework to capture the human-focused data generated through the system architecture process. The Human Viewpoint organizes the information about how humans interact with the system to model the interactions of human tasks with system functions and resources. The different views provide models that capture information on human capabilities, constraints, tasks, roles, networks, training, and metrics, providing a more complete representation of the total system (NATO, 2010). This ensures that the impacts of human capabilities and limitations on system performance are appropriately represented. By capturing the necessary human-focused data in the Human Viewpoint and integrating this view with the rest of the architecture framework, the improved framework provides a complete description of the system data. The set of eight views or models is defined as follows: (NATO, 2010).

- *Concept (HV-A):* This model aims to facilitate understanding of the human component of operational demands and other system components. It serves as the single reference point to depict how humans may impact performance and how the system design and operational context will impact humans.
- *Constraints (HV-B):* Expressing the capabilities and limitations of humans in the system is important to identify risks that may otherwise not be considered. The constraints are a repository for different parameters that may impact the human system and its relationship to the technical system.
- *Task (HV-C):* The task model describes the human-specific activities assigned to humans in a system; the term *task* refers to a piece of work that can be assigned to a single person.
- *Roles (HV-D):* A role represents a job function defining specific behavior within the context of an organization. This model defines the roles of humans in interacting with the system. A role may include additional information regarding the authority and responsibility conferred to the user in the role and the competencies required to do the job.

- *Human Network (HV-E):* The human network model captures the human-to-human communication patterns. These interactions and information exchanges inform task processes that achieve system functions.
- *Training (HV-F):* The training model categorizes the training required to prepare personnel to meet the job requirements and describes the required knowledge, skills, abilities, or other qualifications required for a specific role or task.
- *Metrics (HV-G):* The metrics model provides a repository for human-related performance criteria. It provides quantifiable performance metrics and assessment targets; it may include task standards and performance measures used to evaluate the ability of personnel to perform assigned tasks adequately.
- *Dynamics (HV-H):* Human dynamics support simulations of the human system by including data from other human view models; the design decisions recorded in the static models can be assessed through a dynamic evaluation of the human system's performance. Tradeoff analyses can also be conducted to determine the impact of system parameters on human performance metrics.

The Human Viewpoint can relate system engineering data to human system integration (HSI) activities, thus bridging the system engineering and HSI communities (HFIDTC, 2008). HSI is the interdisciplinary technical process for integrating multiple human considerations into SE practice (DOA, 2015). The principal goal of HSI is to ensure a safe and effective relationship between the human and the system (DAG, 2013). The Human Viewpoint considers human roles early in the design and provides the foundation for HSI requirements definition (Craft, 2020).

Human Readiness Levels

The HRL provides a scale to determine the extent to which human considerations have been considered in the design of systems. The HRL technical standard, *HFES/ANSI 400-2021 Human Readiness Level Scale in the System Development Process*, describes each of the nine HRLs along with the questions that should be answered affirmatively to ensure that the system design has met the requirements at that level (ANSI/HFES, 2021). The HRL scale aims to provide program management with a metric that indicates the system's state concerning the integration of humans and technology (Handley, 2023a). The HRLs utilize data from existing processes and tools to provide evidence to justify a rating and to support recommendations for transition to the next level. The Human Viewpoint represents one tool that can assess the HRL level by providing the data to answer the questions about how human-focused requirements are incorporated into design decisions (Handley et al., 2023). The nine levels of the HRL scale are categorized by the phase of system development that they evaluate and are described below (ANSI/HFES, 2021):

Basic Research and Development: Scientific research, analysis, and preliminary development are conducted. This phase culminates in a validated proof of concept that addresses human needs, capabilities, limitations, and characteristics.

- *HRL 1:* Basic principles for human characteristics, performance, and behavior observed and reported.

- *HRL 2:* Human-centered concepts, applications, and guidelines are defined.
- *HRL 3:* Human-centered requirements for human performance and human-technology interactions are established.

Technology Demonstrations: The technology is demonstrated at increasing levels of fidelity, first in the laboratory and later in relevant environments. This phase concludes with the demonstration of a representative system in a high-fidelity simulation or actual environment, followed by an evaluation of human systems designs provided by representative users.

- *HRL 4:* Completed modeling, part-task testing, and trade studies of human systems design concepts and applications.
- *HRL 5:* Human-centered evaluation of prototypes in mission-relevant part-task simulations completed to inform design.
- *HRL 6:* Human systems design is fully matured and demonstrated in a relevant high-fidelity, simulated, or actual environment.

Full-Scale Testing, Production, and Deployment: Final testing, verification, validation, and qualification occur, with human performance evaluations based on representative users. This phase concludes with operational use of the system and continued systematic monitoring of human system performance.

- *HRL 7:* Human systems design is thoroughly tested and verified in an operational environment with system hardware, software, and representative users.
- *HRL 8:* Human systems design thoroughly tested, verified, and approved in mission operations, using completed system hardware and software and representative users.
- *HRL 9:* System successfully used in operations across the operational envelope with systematic monitoring of human system performance.

By basing the HRL level determination on the data from the system architecting stage and existing HSI processes, the HRLs support the detection and mitigation of human systems issues early in the design to enhance usability and minimize human error in fielded systems (Savage-Knepshield et al., 2021). The Human Viewpoint data directly support assessing HRL levels 1—4. HRLs have the potential to minimize the cost of design changes through early identification of human issues and reduce human error in fielded systems by tracking the mitigation of identified issues through subsequent HRL assessments (Salazar et al., 2021).

HSE WITH AI SYSTEMS

A STS refers to the human–technology partnership for systems that depend on user interactions. These systems may have a variety of types of human interfaces, such as wearable systems that sense user inputs, remote controlled systems that may not be co-located with the user, systems that contain the human, such as self-driving cars, systems composed of teams of humans, such as a space shuttle launch, as well as more traditional types of human interface systems, such as a laptop computer or cell phone (Handley & Tolk, 2021). However, virtual team members, autonomous systems, and adaptive technologies are changing the nature of STSs, often migrating the definition of human roles from an operator to a technology supervisor (Handley, 2019a). In many cases, the interactions between people and technology represent an opportunity for AI applications to free up human workers from mundane tasks, such as data gathering, and support humans on complex tasks that require predictions or insights.

As AI is integrated into human-centered systems, the importance of understanding how human roles and tasks interact with the technology increases. This section provides two examples of how human SE tools can be adapted to AI-driven contexts. The first example illustrates the Human Viewpoint's application to virtual team members' design, an extension of the "teams of humans" STSs. The Human Viewpoint was designed to be adaptable to different applications, with the views defined as general categories of data that can be further specified based on the type of system (Handley & Smillie, 2008). This flexibility will extend the views to include STSs comprising a range of human and non-human agents (Handley, 2023b).

The second example illustrates the use of the HRLs during the development of a Machine Learning (ML) algorithm, a system component accessible through a traditional human system interface like a computer. The HRLs include specifications for intelligent technology where the human component is the authority rather than the user (See, 2018). Researchers have highlighted that integrating algorithms into human workflow remains a substantial challenge; developers may not fully consider the impact of AI on human processes (Asan & Choudhury, 2021).

Human Viewpoint Example: Virtual Team Member Design

The Human Viewpoint is a method for considering the human operator in systems design and development. As part of the architecting process, roles and tasks are often reallocated and balanced between humans and machines (Handley, 2019b). With AI-assisted systems, the architecture description may include human roles and non-human agents acting as "virtual team members." In this context, an agent perceives its environment, takes actions autonomously to achieve goals, and may improve its performance with learning (Miller et al., 2020).

The Human Viewpoint can be expanded to better capture the increasing use of non-human agents assigned as "roles" in the system architecture description (Handley, 2023b). For example, new military vehicle concepts include options for autonomous systems (Seffers, 2020). However, the level of involvement of the human crew with these systems may vary; in some cases, a human operator only performs supervisory tasks, in other cases, humans may be required to regain direct system control. The Human Viewpoint can be used to adjust the human role strategy, i.e., determine which decisions and activities systems are required to be performed by humans and which can be allocated to virtual team members under different operational conditions.

The following paragraphs describe the Human Viewpoint models extended to include the virtual team members of the military vehicle. For this example, there are two virtual team members: an autoloader (to replace the manual loading of ammunition in the gun) and an unmanned ground vehicle (UGV). There are four human operators: the commander, the driver, the UGV operator, and the gunner. Note that the previous ammunition loader role has been redesigned to be that of the UGV operator. However, this role must also take corrective action to restore the operation of the autoloader if it is in an error state. When the UGV operator is attending to the autoloader, the UGV must be able to operate unsupervised while the autoloader is brought back online.

The concept (HV-A) is intended to help the system architect visualize the human component of the STSs – it can be extended to include representations of different agent types, separate from their technology representation. For example, virtual team members are often represented by a grayed-out human figure or a robotic figure in a concept diagram. In this case, the autoloader is included in the concept diagram as an avatar soldier ready to load the gun, and the UAV is an avatar soldier ready to control the asset.

The constraints (HV-B) are used as a repository for different human capabilities and limitations that should be considered in the system's design. In this case, constraints that apply to the non-human agents can be added, such as environmental or operating conditions that may cause the technology to fail. The interaction of humans and agents may also induce a set of limitations, i.e., there may be limiting conditions that impact the ability of different human roles to retake control of the automation. In this case, each human role is evaluated on the following attributes: "qualified" indicates that a crew member has the knowledge, skills, and abilities required, providing redundancy to both human and virtual team member tasks; "available" indicates that the degree of workload the crew member is managing while performing tasks had excess capacity to accept responsibility for additional tasks; and "information access" indicates that the role has situation awareness to make a timely response to reassigned tasks.

The task (HV-C) model describes the human-specific activities assigned to humans in a system. With virtual team members included in the system description, this model must also include the agent-assigned tasks that may be reassigned to a human operator, if necessary, as well as any additional tasks, such as error checking or supervision, that will be required of the human agent. Thus, the task sets may change and need to be rebalanced. In this case, the military vehicle task sets are categorized by different modes of operation: In normal mode, the UAV controller tasks are focused on the maintaining situation awareness with the autonomous vehicle. When the autoloader (virtual team member) enters an error state, the UAV controller retakes control of the autoloader and performs error correction. The UAV operates autonomously during this state until the loader is returned online.

The roles (HV-D) are defined for the humans interacting with the system; a role represents a job function defining specific behavior within the context of the system. In the case of non-human agents, the role concept is still applicable – agents are assigned specific duties and responsibilities. For this example, the military vehicle crew is categorized into four human roles and two virtual team members. The four human roles include the commander responsible for the vehicle and crew. This driver operates the vehicle, the gunner operates the weapon, and the UAV controller supervises the autoloader and the autonomous system. The two virtual team members are the autoloader, which switches ammunition types and loads shells, and the uncrewed ground vehicle, which autonomously performs detection and tracking.

The human network (HV-E) captures the interactions of the humans in the system; it focuses on information exchanges that are required to complete task processes. This network can include a mix of human and non-human agents – the required information flows will determine the type of interactions required among agents. For example, the network model captures the backup schema of task assignments when the crew must switch to a different mode of operation. When the UGV controller (human) takes back control from the autoloader (virtual team member), the gunner supervises the UGV while it operates autonomously. The backup schema informs the crew design to ensure that all tasks have a responsible human in case the virtual team members cannot perform their assigned tasks.

The training (HV-F) model details the job requirements and the required knowledge, skills, abilities, or other qualifications required for a role or task. For the human agents, the training view now includes training for new tasks resulting from changes in the non-human agent's allocations. For example, the backup schema for the military vehicle requires cross-training of the crew for the changes in task responsibility required by the different modes of operation. The UGV operator has the most task cross-training requirements – if virtual team members go "offline," this role is required to perform these tasks until the non-human agent resumes its responsibilities.

The metrics (HV-G) model provides a data repository for human-related performance criteria. It provides objective measures used to evaluate the ability of personnel to complete assigned tasks. With systems that include virtual agents, the metrics must also account for the performance of the non-human agents that impact the human roles. For example, the autoloader must maintain acceptable performance; otherwise, it impacts the performance of the UGV operator, as this role brings the system online, as well as the gunner role, which must then supervise the autonomous vehicle. The metrics provide the criteria to assess whether the reconfigured team is performing adequately to meet the operational demands.

The dynamics (HV-H) supports simulations of the human system using data from other human views; the design decisions recorded in the static models can be assessed through a dynamic evaluation of the human system performance. This dynamic model captures aspects of the human system that may change over time or result from triggering events. For example, a mode transition graph defines the state-space of allowable transitions between the human crew and virtual team member assignments in response to planned and unplanned events. The transition graph identifies the necessary stages for the changes. It illustrates the alternative roles and automation configurations that may require a dynamic reallocation of tasks among both the human and virtual crew members. The optimum allocation of tasks may change over different scenarios, with the roles restructured as tasks are dynamically rebalanced between humans and automation based on availability and loading. The dynamics model can be used to better understand the transfer of control and redistribution of tasks across all agent types.

Human Readiness Example: ML Models

Including ML models in systems provides the ability to analyze data, automate tasks, and enhance decision-making capabilities; these models can process large volumes of data to make predictions and offer solutions. ML models utilize pattern recognition and learning algorithms to improve system functionality and efficiency. Embedding an ML model into a system is a multi-step process that includes assessing system compatibility, preparing data, training models, deploying them within the system, and continuously monitoring performance for optimization. Sociotechnical system evaluation, in this case, involves evaluating the system's fit with the human operator and monitoring the ML model to ensure a continuous fit with its system and operator. ML models also require an assessment of "social" aspects of their use of data, such as trust, transparency, and ethics. The HRLs can evaluate effective human use across the entire lifecycle of the AI-assisted system.

The work order prioritization example illustrates the applicability of the HRLs to an AI-assisted system. In this example, different maintenance employees, including engineers, technical staff, and managers, have access to many system failure records. This makes the maintenance work order prioritization process complex and time-consuming, as it depends on the humans' interpretation of the maintenance records, which is influenced by varying levels of experience and knowledge in the prioritization process. Incorporating an ML model into the process of maintenance prioritization can reduce the subjectivity caused by different human interpretations. This automation of the prioritization process strives to reduce subjectivity while minimizing human workload.

The work order prioritization example describes how the HRLs can assist in systematically incorporating and evaluating ML models into traditional systems. The following paragraphs describe how each level of the HRLs can be extended to an AI-assisted system. Additionally, the exit criterion for each level is described in two parts – the requirement for the model or data (technical) and the human (social) considerations.

HRL 1 ensures that basic principles for human characteristics, performance, and behavior are observed and reported. For AI-assisted systems, the focus of HRL 1 is on the data and definition of initial human-centric requirements. The exit criteria for HRL 1 include a thorough understanding and characterization of both the data needs for the ML model and the essential human behaviors, capabilities, and limitations that are relevant to the developing concept or proposed applications, which must be demonstrated. For the work order prioritization example, the operational effectiveness of the model is tied to the qualitative and quantitative attributes of the data it processes; in instances where data information is not adequately robust, the model may not yield satisfactory outcomes. Thus, identifying the types of maintenance records and failure narratives, which are the inputs to the ML model, must be completed (technically).

An initial assessment is conducted, focusing on the tasks and roles of maintenance professionals, such as engineers and managers, who will engage with the system. This preliminary exploration may involve deciphering their interpretative constraints related to maintenance records and their adeptness in manipulating AI-supported tools (social).

HRL 2 guarantees that human-centered concepts, applications, and guidelines are defined. For AI-assisted systems, the focus of HRL 2 is on data processing and human-centered design guidelines. The exit criteria for HRL 2 include completion of data acquisition, labeling, cleaning, text processing, and feature extraction, as well as establishment of key human-centered design principles, standards, and guidance for human interaction with the technology. For the work order prioritization example, the procurement and labeling of maintenance logs, followed by text manipulation and feature extraction, are executed to condition the data for model training (technical). The appropriate design guidelines, which will represent the interaction of the ML model with maintenance personnel, are identified. This might include a conceptual user interface that conforms to ergonomic standards and satisfactorily conveys ML outputs to the users (social).

HRL 3 requires establishing human-centered requirements to support human performance and human–technology interactions. Regarding AI-assisted systems, HRL 3 focuses on model selection, design, and human-centered requirements. The exit criteria for HRL 3 include, first, a suitable ML model must be determined and the accompanying design of experiment created, and second, the requisite human-centric analyses must be conducted, resulting in the identification of human-centric requirements and key performance parameters integrated into overarching system requirements. For the work order prioritization example, multiple ML models are identified, and experiments are structured for these models to identify the optimal model along with its hyperparameters (technical). The requirements that govern how the outcomes will be communicated to maintenance personnel are also identified. This might involve defining the modalities of notifications or alerts that will be utilized to inform staff regarding the urgency of various maintenance tasks (social).

HRL 4 requires the completion of modeling, part-task testing, and trade studies of human systems design concepts and applications. Regarding AI-assisted systems, HRL 4 focuses on model training, tuning, and part-task human testing. The existing criteria for HRL 4 include, first, adequate training and calibration of the ML model must be completed to satisfy predetermined performance metrics, and second, the assessment and characterization of human interactions and performance utilizing analytical instruments, modeling methodologies, and partial task testing with swift prototypes must be evaluated. For the work order prioritization example, the ML model, applied to the processed maintenance records, undergoes training and fine-tuning to achieve optimal performance (technical). Then, partial task testing with maintenance personnel, utilizing a preliminary prototype of the model, was completed. This might encompass simulated scenarios where staff prioritize maintenance tasks based on these scores (social).

HRL 5 assesses the human-centered evaluation of prototypes in mission-relevant part-task simulations to inform design. Regarding AI-assisted systems, HRL 5 focuses on the preliminary model validation/testing and mission-relevant prototype testing. The exit criteria for HRL 5 include, first, that the initial validation and examination of the ML model within a simulated or regulated environment should be completed, and second, that an assessment of the human interactions and performance should be completed while interfacing with the evolving system prototype. For the work order prioritization example, the ML model is validated within a simulated environment, utilizing a subset of maintenance records to ensure that the model's outputs are congruent with expert evaluations (technical). Additionally, mission-pertinent partial task simulations are executed, involving maintenance personnel and utilizing a high-fidelity prototype. This might encompass simulated scenarios where staff prioritize maintenance tasks based on outputs generated by the validated model (social).

HRL 6 evaluates whether the human systems design is fully matured and demonstrated in a relevant high-fidelity, simulated, or actual environment. In the case of AI-assisted systems, the focus of HRL 6 is on system integration, verification, and human-centric maturity. The exit criteria for HRL 6 include, first, successful integration of the ML model into the prevailing system, and confirmation of its functionality must be completed. Second, evaluating human interactions within the context of high-fidelity simulated or actual environments, utilizing a functional and realistic prototype, representative users, and a comprehensive array of usage scenarios and tasks must be completed. For the work order prioritization example, the ML model is integrated into the existing maintenance work order system, followed by verifying its functionality.

This ensures that the outcomes derived from the ML model accurately influence the system's prioritization algorithms (technical). Then, high-fidelity simulations or real-world tests with maintenance personnel, utilizing the integrated system, are executed. This might span a spectrum of scenarios, from standard maintenance tasks to emergencies, to evaluate the efficiency of the integrated model in aiding work order prioritization (social).

HRL 7 evaluates that the human systems design is thoroughly tested and verified in an operational environment with system hardware and software, including representative users. In the case of AI-assisted systems, the focus of HRL 7 is on operational testing and human-centric validation. The exit criteria for HRL 7 include, first, extensive testing of the ML model within an actual operational environment, which must be completed to ascertain adherence to all performance and safety criteria. Second, the assessment and characterization of human interactions and performance within an operational environment must be completed, employing the final development system, representative users, and a comprehensive spectrum of usage scenarios and tasks. For the prioritization of work order example, the ML model should be tested within a live operational environment, utilizing authentic maintenance records and work orders. Performance metrics, such as accuracy, precision, recall, and safety metrics like false positives and negatives, should be evaluated (technical). Additionally, operational tests should be conducted with maintenance personnel, utilizing the final integrated system within a live environment. This might span a variety of scenarios, from standard maintenance tasks to emergencies, to evaluate the efficiency of the integrated model in assisting work order prioritization (social).

HRL 8 evaluates that the human systems design is thoroughly tested, verified, and approved in mission operations, using completed system hardware and software and representative users. In the case of AI-assisted systems, the focus of HRL 8 is still on continuous monitoring and improvement. The exit criteria for HRL 8 include ongoing monitoring and periodic updating of the ML model, which must be completed to ensure it meets performance and safety criteria. Second, the evaluation and characterization of human interactions and performance with the actual system in mission operations during the full range of usage scenarios and tasks completed by representative users must be completed. Monitoring protocols should be established to regularly assess the ML model's performance and prioritize work order examples. This could include automated alerts for performance degradation and mechanisms for model retraining (technical). Additionally, extensive testing of the integrated system in mission operations should be completed, involving representative users and a full range of scenarios. This could include emergencies to assess how effectively the integrated ML model aids in work order prioritization (social).

HRL 9 evaluates that the system is successfully used in operations across the operational envelope with systematic monitoring of human system performance. In the case of AI-assisted systems, this HRL is focused on the ethical, societal, and operational monitoring and evaluation. The exit criterion for this HRL is twofold – the first focuses on the system itself and the ongoing monitoring, testing, and evaluation of the fielded system to ensure it supports the mission as intended. The second criterion is focused on the impact of using AI through assessing and mitigating ethical and societal implications, including data privacy, fairness, and transparency. For the prioritization of work order example, the ML model would undergo ongoing human systems monitoring to ensure that the intended levels of operational performance are achieved. This could include periodic evaluations involving human systems experts to assess the effectiveness of system outcomes.

Additionally, these evaluations would ensure that the ML model does not introduce bias in the maintenance work order prioritization process. This could involve periodic audits of the model's decision-making process and outcomes (technical). The ethical considerations include a thorough review of data usage, ensuring that data privacy is maintained, and that the model's decisions do not unfairly favor any group. This could involve a detailed analysis of the model's predictions and decision-making processes to ensure transparency and fairness in all operations (social).

CONCLUSION

Total system design and performance evaluation in complex STSs cannot be fully realized without accounting for the human component of the system. Comprehensive integration of human constraints and limitations into the SE effort is critical to designing successful systems. HSE provides methods for including the human in the system architecture stage. Tools such as the Human Viewpoint and the HRLs provide the means to identify and assess the characteristics and relationships between humans and systems in a traditional systems context.

The line between human roles and technology becomes blurred as systems transform to include AI-enabled technologies. Sociotechnical systems can now include virtual team members, autonomous systems, and adaptive technologies. This chapter provided examples to describe how both the Human Viewpoint and the HRLs can be adapted to systems that include new technologies, such as autonomous systems that act as virtual team members and adaptive technologies, i.e., ML models, which may act and learn in ways that are not transparent to the user. As the examples show, these HSE tools can continue to assist in the design and improve the resulting STSs. The human views and the HRLs enable successful HSE efforts for AI-assisted systems.

REFERENCES

ANSI/HFES (2021). *Human Readiness Level Scale in the System Development Process*. American National Standards Institute/Human Factors and Ergonomics Society. Retrieved from: my.hfes.org/online-store/publications

Asan, O. & Choudhury, A. (2021). Research Trends in Artificial Intelligence Applications in Human Factors Health Care: Mapping Review. *JMIR Human Factors*, 8(2) e28236. https://doi.org10.2196/28236

Craft, R. (2020). *The Place of Human Readiness Levels in HSI Management*. Report Sandia National Laboratories, Albuquerque, NM.

DAG (2013). Defense Acquisition Guidebook Chapter 6. Retrieved from: www.dote.osd.mil/Portals/97/docs/TEMPGuide/DefenseAcquisitionGuidebook.pdf

DOA (2015). *Soldier-Materiel Systems Human Systems Integration in the System Acquisition Process*. Department of the Army Regulation 602–2, Washington, DC.

DOD (1988). *Manpower, Personnel, Training, and Safety (MPTS) in the Defense System Acquisition Process*. DoD Directive 5000.53, Washington, DC.

DoD CIO (2010). The DoDAF Architecture Framework Version 2.02. U.S. DoD Deputy Chief Information Officer. Retrieved from: http://dodcio.defense.gov/Portals/0/Documents/DODAF/DoDAF_v2–02_web.pdf

Handley, H. (2019a). *Human System Engineering: The Human Viewpoint for System Architectures*. Springer, Cham, Switzerland: Springer Nature Switzerland AG.

Handley, H. (2019b). A Cluster Approach to Task Assignment. *Proceedings of the 2019 Annual IEEE International Systems Conference (SYSCON)*, Orlando, FL.

Handley, H. (2021). A Retrospective of the Human Viewpoint and Its Applicability to Future Systems. *Systems Engineering*, 25(1), 91–103. https://doi.org/10.1002/sys.21603

Handley, H., See, J. & Savage-Knepshield, P. (2023). Human Readiness Levels and Human Views as Tools for User-Centered Design. [Manuscript submitted for publication].

Handley, H. (2023a). Human System Integration Framework (HSIF) Activities to Support Human Readiness Levels (HRLs). *Ergonomics in Design: The Quarterly of Human Factors Applications*, 33(1), 44–51. Retrieved from: https://journals.sagepub.com/doi/full/10.1177/10648046231152050

Handley, H. (2023b). A Human View Dynamics Model for Regaining and Relinquishing Control. *Proceedings of the 2023 Annual IEEE International Systems Conference (SYSCON)*, Vancouver, British Columbia, Canada.

Handley, H. & Smillie, R. (2008). Architecture Framework Human View: The NATO Approach. *Systems Engineering*, 11(2), 156–164.

Handley, H. & Smillie, R. (2010). Human View Dynamics – The NATO Approach. *Systems Engineering*, 13(1), 72–79.

Handley, H. & Tolk, A. (2021). *A Framework for Human System Engineering Applications and Case Studies*. Wiley-IEEE Press, Hoboken, New Jersey, USA.

HFIDTC (2008). *The Human View Handbook for MODAF (1st issue)*. Human Factors Integration Defence Technology Centre. Retrieved from: www.dau.edu/acquipedia-article/dod-architecture-framework-dodaf

INCOSE (2009). *Systems Engineering Handbook. Appendix M – Human Systems Integration*. International Council of Systems Engineering, San Diego, CA.

INCOSE (2011). *Systems Engineering Handbook: A Guide for System Life Cycle Processes and Activities* (H. Cecilia, Ed.) (3.2 ed.). San Diego, CA.

INCOSE (2015). *Systems Engineering Handbook: A Guide for Systems Life Cycle Processes and Activities* (4th ed.). Wiley, Hoboken, NJ.

Miller, M., McGuirl, J., Schneider, M. & Ford, T. (2020). Systems Modeling Language Extension to Support the Modeling of Human-Agent Teams. *Systems Engineering*, 23, 519–533. https://doi.org/10.1002/sys.21546

NATO (2010). *Human Systems Integration for Network Centric Warfare (Intégration des systèmes humains dans les opérations réseaux centrées)*, RTO Technical Report, TR-HFM-155. North Atlantic Treaty Organization, Brussels, Belgium, February 2010.

Salazar, G., See, J. E., Handley, H. A. & Craft, R. (2021). Understanding Human Readiness Levels. *Proceedings of the Human Factors and Ergonomics Society Annual Meeting*, 64(1), 1765–1769. http://doi.org/10.1177/1071181320641427

Savage-Knepshield, P., Hernandez, C. & Sines, S. (2021). Exploring the Synergy between Human Systems Integration and Human Readiness Levels: A Retrospective Analysis. *Ergonomics in Design: The Quarterly of Human Factors Applications,* 29(4), 16–24.

See, J. (2018). Incorporating Human Readiness Levels at Sandia National Laboratories. *Journal of Human Performance in Extreme Environments*, 14(1), 1. DOI: 10.7771/2327-2937.1085

See, J. (2021). Human Readiness Levels Explained. *Ergonomics in Design: The Quarterly of Human Factors Applications,* 29(4), 5–10.

Seffers, G. (2020). Army Seeks Open Architecture For Optionally Manned Fighting Vehicle. AFCEA. Retrieved from: www.afcea.org/content/army-seeks-open-architecture-optionally-manned-fighting-vehicle.

Wasson, C. (2016). *System Engineering Analysis, Design, and Development: Concepts, Principles, and Practices*. Wiley, Hoboken, NJ. ISBN: 9781118442265.

6 Wandering in Systems Thinking Trees

Brigitte Daniel Allegro

INTRODUCTION

During the two-day debates about Governing the Anthropocene: Cyber-systemics, systemic inquiry, and institutional reform? In Hanover in 2015, Ray Ison opened the first debate on "An investigation about framing and re-framing choices? Reframe how we think and act? How is governance (or governing) understood? How do we frame what is to be governed?"

We brainstormed the topic from many angles, drilling into issues and opportunities. What was particularly interesting was the diversity of approaches, clearly influenced by the mix of perspectives brought by the participation of philosophers, social and natural scientists, political theorists, psychologists, economists, doctors, environmentalists, and a very few engineers. These inquiries highlighted the different cyber systemic lineages and communities in our conversation and the significant fragmentation of the cyber systemic field. However, the Systems Tree resonated and provided a common conceptual framework for engaging the discussion about the Anthropocene between the participants around our table.

The original idea for the Systems Tree started in 2012 with an analysis of these two words, Systems and Thinking, as a basis for teaching Systems Thinking in the Defense Industry in France in 2013.

QUESTIONING CONCEPTS AND PRINCIPLES FOR SYSTEMS THINKING TEACHING

In a tweet format, #SystemThinking is to observe an overall situation to underdand it (Daniel Allegro, 2013). Could the teaching be based on guidelines, procedures, or principles, or could it be based on conceptual foundations for providing a framework? It is important to understand what Concepts and Principles mean before we explore the Systems Tree. *A concept* is a mental representation that is neither true nor false but is useful, for example, the concept of color.

 DOI: 10.1201/9781003666431-7

A Concept Has a Semantic Value

The correlation between language and thought is a significant philosophical debate (Panaccio, 2011). As far as we are concerned, we face this issue when we try to translate a concept from one language into another.

For example, there exist concepts in English that have no equivalence in French, such as "pattern," "togetherness," or "wholeness." If we try to define the concept in a specific context, we lose the potential and mental representation of the concept in English.

Let us consider the concept expressed by the word "pattern" in English. We need different French words "*motif*" (in the context of music), "*patron*" (a template in the context of fashion design) or "*modèle*" (a model in the context of design), "*schéma*" (a graphic in the context of drawing), "*ébauche*" (a sketch in the context of fine art or architecture). When "pattern" is used with a notion of evolution in English, we would refer to "tendance" in French. Six different French words to translate the word "pattern."

The Semantic Dimension of a Concept is Essential for Its Own Identity

That is why we must be aware of the importance of this semantic dimension in multicultural environments. It is often experienced when sharing concepts across cultures that there is a change in semantic value, which leads to misunderstanding. If we know this, we can take additional steps to capture the essence and identity of concepts.

How to Share a Concept with Somebody

If you consider a concept a mental representation, as we stated previously, you can use different media to share it with others, not only the media of language and words. You can also sketch or draw, you can use materials, you can use music, you can use your body to display feelings, such as facial mimicry as clowns do, through dance, or many other ways. It is how artists usually express some concepts in their art. Concepts can be used to form complete thoughts when combined with other concepts.

Principles

Let us have a look at "principles" (Wikipedia, 2023). A principle is a fundamental truth or proposition that is the foundation for a system of beliefs or behavior or a chain of reasoning. That is a guide for behavior or evaluation. It can be desirable or an inevitable consequence of something, such as the laws observed in nature or how a system is constructed. For example, if you consider these three pigments, red, blue, and yellow, a principle is that if you mix yellow with blue, you get a green color. As a conclusion for questioning Concepts and Principles, the Systems Tree aims to provide a framework integrating a set of concepts of soft skills (attitudes represented by the branches of the tree) with a set of Systems' concepts issued from cybernetics, represented by the tree's roots. It does not provide Principles for Systems.

Nevertheless, at this stage, here is one candidate principle that most system practitioners may agree on: Systems can form bonds with other systems to form new systems with properties distinctly different from the sum of the component systems, with beneficial or detrimental emergent behavior.

According to Gall (1977) in his book "Systemantics: How Systems Work and Especially How They Fail," many principles of systems are highlighted. Here are my top three:

- Complex systems exhibit unexpected behaviors (as the perceiver perceives them as complex).
- Systems tend to expand to fill the known universe (as our knowledge expands, so does our perception).
- New systems generate new problems (or new systems generate new systems – due to relational theory).

THE SYSTEMS TREES CONCEPTS

Genesis of the Systems Trees

The *Systems Tree* has been designed as a mind map and inspired by a growing fern, an organic living item. The different sketches evolved during 2012. Its branches and roots are fractal by choice, and it is expected that over time, the tree can adapt and grow from its main branches and its primary roots. In French, we would speak of "arborescence." This concept includes the concept of the "tree" and the concept of the "potential of growing" (teleology).

Nevertheless, we can use the framework of the *Systems Tree* to transform data into knowledge, "*to encode*" and formalize, "*to decode*" a situation, and then "*encode.*" Let us take an example. Data a human being selects can be transformed into information when contextualized (the "*Context*" root). This information may be used for a dedicated "*Purpose*," let us say, to design an artifact. The human being can aggregate this information with other information to define its "*Togetherness,*" to describe the "*Behavior*" of the intended artifact, its "*Structure,*" its "*Relationships*" with external systems, and its "*Structures.*" From this example, we can easily perceive the "encoding" of data through the "roots." The human being performs the "decoding" when *perceiving* a situation, *detecting the influences* of the context, *understanding the system* (the new artifact to design), and maybe *reassessing* the preliminary design to find other alternatives.

Wandering in the Branches of the *Systems Tree*

Systems Thinkers free themselves from the "three unities" rule of classical theatrical drama: time, place, and action. They dive into past events and look into the future to understand a present-day situation, considering the variety of outcomes and consequences.

Attitudes, the Branches of the *Systems Tree*

How do you recognize Systems Thinkers? They approach a subject in its wholeness and accept different perspectives of diverse stakeholders in a situation. At any time, they can play the role of any other actor. The attitudes reveal the preparation of the mind and the body of human beings who intend to understand a situation or act. They reflect their knowledge, know-how, and experience, as well as their emotional reactions and cognitive positions toward a situation.

At this point, there is a tribute to the Waters Foundation Systems Thinking Institute (2006), which realized that many educators might not have available resources to attend in-person workshops. The Waters Foundation began creating free, online learning modules. One of them was called "Habits of a Systems Thinker." The Waters Foundation created a set of cards, at that time. This point was taken into account as part of the state of the art to frame the branches of Systems Tree and the attitudes of a Systems Thinker in 2012.

A Systems Thinker Perceives. Like shepherds on a mountain or sailors on the ocean, Systems Thinkers open their eyes. They scan and scrutinize their environment and can detect tiny changes.

- *A Systems Thinker Delimits:* They enlarge their range of vision to delimit a system's boundaries and the exchanges with its environment. They define and characterize the outside and the inside of the system. How they figure out the boundaries determines how they embrace the system and observe its behavior. The system and the environment interact. One can consider that the environment sets the context for the system, including its dynamic relationships. This establishes a tension that shapes the behavior of the system.
- *A Systems Thinker Sees the Big Picture:* Systems Thinkers notice events, observe situations, and detect changes. Even if all of them seem independent, unique, and unpredictable, they consider the world a system where elements may be influenced by one another: events emerge. Seeing the big picture means that they can picture all kinds of influences and can correlate internal, detailed interactions, embedded within the system, to external observable events. They link the facts they observe to better understand a situation.
- *A Systems Thinker Changes the Viewpoint:* Systems Thinkers select meaningful scales for the subject, a large or a small mesh, and they change their point of observation. What they discover depends on the viewpoint they take. By changing the viewpoint, the angle of sight, and the point of focus of the observation, this attitude improves their understanding of the system from the inside and the outside (thinking both inside and outside of the box). By changing the settings of their perspective, they escape the standard frames of reference (established wisdom), move beyond their comfort zone, and their familiar environment. They get out of deadlocks and liberate their creative acuity. They can put themselves

in someone else's shoes, or even in the position of organizations, animals, plants, or objects. Changing perspective means likewise moving from the "here and now" to tomorrow, in the future, with short-term and long-term perspectives, considering past events at the same time. It also means changing the time scale and seeing across the whole lifespan. Systems Thinkers change the settings of the metronome beat and can figure out simultaneously situations which evolve slowly and quickly.

- *A Systems Thinker Discerns:* Systems Thinkers zoom in and out, change the focus of their observations, and adjust this new focus to detect and explore system details. Zooming out and in, again and again, to correct the early focus; iterating and observing the behaviors so that connections are formed to understand the system better. This is a way to start to recognize patterns. They expand the view to encompass other elements that interact with or within the system. Details are considered at the same time to reveal the system. All aspects of the situation will progressively come out of these observations and lead to a constant re-evaluation.

A Systems Thinker Detects the Influences: As a Systems Thinker, all of your senses will merge. You will gain awareness as your understanding grows.

- *A Systems Thinker Is Attentive To:* Systems Thinkers are attentive to facts and to people. They notice how events occur, how people react, and how the environment behaves. They listen to the others. They concentrate on what people mean when saying one thing or another and try not to give a personal interpretation. With their intention to understand, they "decode" information to gain knowledge.
- *A Systems Thinker Looks for Influential Factors:* Systems Thinkers assume that nothing is ever totally isolated or independent. They look around, progressing in a larger environment. They also look at other environments, situations, and events, thus avoiding the potential trap of forgetting multiple interactions. They make connections between the elements they observe. They identify the essential elements. They look for links, relationships, and interactions between the elements as observed during a period. Do they influence immediately or not? Are there influential factors? They observe, make assumptions, experiment, and test. They modify the context and notice how elements change. They observe and assess the results and change the actions as necessary, gaining insight progressively. They correlate their observations and identify the influential factors. They seek to bring value.
- *A Systems Thinker Anticipates the Effects*. Systems Thinkers always question themselves: "If I do this action, what will happen then?" or "Does the effect impact the initial act?" They look for influences between facts. They distinguish an effect and a cause. They consider today's events as a result of the past. They think about possibilities for the future based on current events. They search for low, minor, or hidden signals in the present

day as indicators of future happenings and as latent potential for change in future direction. They propel their quest in the short and long term at the same time. They can highlight circular causality in their observations.

- *A Systems Thinker Identifies Possible Leverage:* Facing a situation, Systems Thinkers look for possible causes beyond the apparent effects. They try to find essential elements that have leverage actions. Among these, they consider the fulcrum and the lever to stimulate significant effects for small effort. A minor change can have a significant impact on the system's behavior. A local change somewhere can influence the entire system. Sometimes we can see the reverse situation where we must spend a considerable amount of effort to effect just a small change, for instance, within organizations that are overly risky or bureaucratic.
- *A Systems Thinker Seeks Implicit Needs:* All too often, people think in terms of already known solutions rather than what might be possible. They do not dream enough. Doing so, they limit and constrain themselves, and even more, they limit the other people who could be solicited to answer their real questions. They try to capture real needs, the ones that are not formulated, which are called implicit needs. The needs of some people may be perceived as constraints for others. Needs and constraints polarize the system.

A Systems Thinker Seeks to Understand the System: Systems Thinkers seek to understand the system by developing different mental models built from their perspectives. Indeed, while seeking to understand the system's complexity, they use existing knowledge or develop new concepts, mental models, and principles to build a helpful representation of the situation. The more diverse the perspectives, the more valuable this will become, giving a better chance to capture the many facets of the system, which might not be evident at first glance. There are two primary key points: first, the choice of viewpoints, and second, the integration of these viewpoints into their mental model.

- *A Systems Thinker Pictures the System:* Systems Thinkers consider the structure of the system. What are the elements of the system? They distinguish the constituting elements and their relative position in the system structure. They consider the topology of the system. Systems Thinkers consider how the elements interact within the system and how they contribute to the system's behavior. They picture the system and visualize its functioning in their mind.
- *A Systems Thinker Identifies Stocks and Flows:* Systems Thinkers are aware of delays and inertia in a system. They explore the stocks (quantities) and flows to understand the system's fluidity. They detect and identify stocks accumulated by the system somewhere. They identify inflows and outflows during a period. They can detect how the flows influence the states of the stocks in the system, during a period, whether it is a growing stock, a decreasing stock, or a stable one. They think about how sharing or retention of information affects a decision.

- *A Systems Thinker Observes Changes along the Life Cycle.* Systems Thinkers carefully watch how elements evolve, move, and constantly change over time. They discern in these findings the emergence of trends and behaviors of the system. They observe the capacity of self-adaptation of some systems and the collapse of others. Systems Thinkers propel their minds into the system's future and imagine what might be. What will the system become when in operation, while being maintained, or retired?
- *A Systems Thinker Imagines Scenarios:* Systems Thinkers will script scenarios to better understand interactions between the system and its environment. They play the roles of different stakeholders throughout the system life cycle. They imagine the system at different stages of its life. For an engineered system, for instance, they imagine what influence this idea might have on potential stakeholders, in a development stage, in a production stage, in operation, or the retirement stage. By describing scenarios, they can share a vision between themselves and with different stakeholders with the appropriate wording.

A Systems Thinker Reassesses: Ptolemy was a keen observer. He linked the facts, measured phenomena, and perceived the cycles of planets. His theory was based on a geocentric vision of the world. "The Earth is fixed at the center of the universe. The other objects turn around the earth thanks to a strength which leads to a circular orbital movement." However, he rejected the observations he could not explain! Copernicus looks like a Systems Thinker. He reconsidered Ptolemy's theory. He tried to understand the unexplained observations of Ptolemy, and he conceived a new theory based on a heliocentric world vision. The Earth turns around its axis in a day and turns around the sun in a year. The planets turn around the sun." He developed a new paradigm by exploring the apparent paradox and changing the previous perspective.

- *A Systems Thinker Reconsiders the Mental Models:* Systems Thinkers consider their intuition when exploring a situation. People may act and formulate an opinion based on intuition. However, although intuition is often a source of innovation and creativity, Systems Thinkers know that it might also limit their perception. Putting aside our beliefs and certainties and questioning ourselves is fruitful. They raise doubts about their convictions and beliefs. They try to find and consider how they limit their perception of the present situation and their perception of the future. They go beyond their initial intuition.
- *A Systems Thinker Accepts the Uncertainty of a Situation.* Systemic situations may appear with paradox or contradictions. Analyzing these gray areas is quite complex and uncomfortable for many people, especially those with a fixed mindset. When Systems Thinkers take another viewpoint when looking at the situation, they may find out that they are a part of the problem because of their existing mental models. This is interesting, is not? In these situations, changing the viewpoint could be sufficient to realize that there is no problem (it was just the attitude!) (Watzlawick et al.,

1974). Systems Thinkers accept the uncertainty of a problematic situation. They respect and manage the tensions generated by the uncertainty. They do not try to solve them straight away. They would rather try to understand the tensions as these might reveal new insight into potential solutions. They adopt iterative reasoning. They accept that the gray area exists and avoid the urge to make immediate assumptions. They take time to discern the links between elements that affect the situation and consider the wholeness over time.

- *A Systems Thinker Looks for Alternatives and Other Options.* Systems Thinkers look for alternatives. They make assumptions and evaluate other options' feasibility in different environments. They explore new ways, new approaches, and new means. They do not restrict their imagination. That is how artists behave while creating. They can find shortcuts to solve a situation (Berthoz, 2009, 2013).

Concepts, the Roots of the Systems Tree

At this point, there is a tribute to Rosnay (1975), a key pioneer in France on systemic analysis who wrote "Le macroscope" combining the words "microscope" and "telescope" which was a seminal work for systemic analysis on diverse topics such as organism, biology, cells, and also economy, ecology, and cities. When we architect and design systems, the Systems Tree roots are essential features of the systems that we have to master to gain a complete understanding of a situation.

Togetherness: According to Boardman and Sauser (2013), the essence of a system is togetherness, the drawing together of the various parts and relationships they form to produce a new whole. Energy is to biology as togetherness is to systems. Togetherness is revealed when autonomous agents commit to a collective vision, when willingness to belong constitutes sacrificing individual objectives for a common mission, and when envisioning something superior that no individual could ever achieve. This feature represents a compound of different concepts.

- *Integrity:* Integrity encompasses the energy that keeps the system in its wholeness. The system's behavior will change if essential elements are added or taken away. Removing the heart from the body changes the way the body works!
- *Boundary:* The boundary allows an observer to mark a difference between the entity perceived as a system and its interactions with the environment, through boundaries. It is a key point to define the System of Interest of a human being and the viewpoints mentioned previously in the attitudes (Checkland & Poulet, 2006).
- *Identity:* The identity represents the uniqueness of an individual or a group. This concept is at a crossroads of sociology and psychology and is used too in philosophy, biology, and geography. It is helpful for Systems Thinkers to recognize identity as it *serves multiple functions, acting as a "self-regulatory structure" that provides meaning, direction, and a sense of self-control.*

Structure: The structure displays the cohesion of the elements of the system. It refers to the Component and Composition concepts. It shows how components are arranged and interconnected.

- *Component:* Basic element.
- *Composition:* The composition highlights the relationships between the components.
- *Hierarchy Structure:* The system is decomposed into components according to levels (two or more). It is often used to develop products (PBS, CBS, etc.), organize human activities (WBS, OBS), and establish a chain of command (who reports to whom).
- *Network Structure:* The network structure highlights links, flows, and interactions between the components. It is often used to capture properties of physical systems (electricity, water, gas, telecommunication, internet) while providing relationships between individuals or groups.
- *Holarchy Structure:* Each holon is influenced by and influences larger wholes. Bidirectional influences: information flows between smaller and larger systems. Examples of holarchy: (letter–word–sentence–paragraph–chapter–book) or (molecule–cell–tissue–organ–organism).
- *Relationships:* They represent how things are connected. These include **Dependencies**, where components rely on the other for existence.
- *Influence:* This is when components can alter others, and interaction is when the relationship is bidirectional between the components.

Behavior: The system's behavior relates to the effects produced by its components in operation. It reveals the dynamic cohesion of the system. An open system interacts with its environment to maintain its balance (homeostasis). The behavior can be featured by:

- **States**: stable, dynamic, convergence, divergence
- **Patterns** observed in the dynamics of behavior (movement along time from one state to another), system archetypes
- **Cycles** of repetition versus life-changing occurrences can be observed throughout the system's existence.

Purpose: It represents the perceived goal, the reason, and the aim for the system's existence. A human use imposes extrinsic purpose, while intrinsic purpose is irrespective of human use or opinion. Having a look at teleology opens our mindset. Teleology (from Greek telos, for end or purpose) is a philosophical attempt to describe things in terms of their apparent purpose, directive principle, or goal. *According to Aristotle, an acorn aims to become a fully grown oak tree.*

- **Drivers** represent attraction factors that motivate the system despite stress or pressure.

- **Functions** performed toward a perceived outcome are characterized by transforming Inputs into Outputs, including data, information, materials, or energy.

Context: Human beings perceive a context and form the setting for observers. It considers the **Stakeholders' Perspectives or needs**. Different observers in different positions (both physically and mentally) perceive and understand the properties and characteristics of the system in different ways. The **emergence** of new situations or events relies on context, as the creation of these properties and characteristics involves the active participation of the observers. The emergence is a predictable or unpredictable behavior occurring due to a system in operation. (Sartenaer, 2018). **Living organisms** develop capabilities to perceive and cope with nature.

The Shadow of the Systems Tree

In Jungian psychology, the shadow or "shadow aspect" may refer to an unconscious aspect of the personality that the conscious ego does not identify with. *According to Jung (1875–1961), everyone carries a shadow, and the less it is embodied in the individual's conscious life, the blacker and denser it is.*

When illustrating the *Systems Tree* during training, looking at it from a negative perspective might be helpful. In opposition to the attitudes of the Systems Thinker, Traps and shadowy concepts identify blocking points or sticking points in thinking in different situations. Recognizing these blocking points when problems are endless is a reminder to change one's attitude and think differently. Awareness of these detrimental attitudes and the adverse effects is a first step to changing something in our mindset if we want to handle the situation more successfully.

Likewise, when considering the shadow of the *System Tree*, this might reveal that the System of Interest or the enabling systems are not sufficiently explored for understanding a situation.

THE INTENTIONALITY

While teaching Systems Thinking in 2013, at the end of the five-day introductory tutorial, each student was expected to share with the group how to change their behavior, mindset, and ways of working with their colleagues personally. We met two months later and shared our experiences.

It was a means to identify, through these actions, how they were moving from an intended action (from an interior and subjective point of view) to an effective act (the result of the action in their daily job). The "intention for acting" was tightly connected to the "act domain," "the doing."

At that point of the *Wandering in the Systems Trees*, a stimulating position is to consider the features related to the *thinking of Systems Thinkers* (Daniel Allegro, 2013). For instance, when one discerns interactions of elements, they develop their capabilities to perceive a situation.

They show skills to step back, to focus in, and to focus out, and to change the mindset with which they are seeing things. These are considered qualities of *Lateral Thinking* in the context of this paper. When somebody seeks implicit needs or is attentive to expectations of a customer, or can accept, for some time, the uncertainty of a situation, this person shows listening, empathy, and adaptability skills. These are considered qualities of *Flexible Thinking*. When somebody sees the big picture, imagines scenarios, pictures the system, reconsiders mental models, and looks for other options to solve a problem, this person shows skills of integrating and synthesizing, developing initiatives, and being creative.

The Potential of Intentionality for Systems Thinking

Tutorial scenarios were drawn across a broader context than Engineering. They extended to artistic field such as dancers performing a choreography (how they think about their posture and their attitude and how they prepare their mind and their body), the artistic field such as art (how artists think about their piece of work, how they conceptualize their ideas, how they visualize their intended action before their acting, modeling the material, should it be a splash of painting on a canvas, a line to draw on a paper, a rough cut in the matter, etc.)

Intentionality Underpins the Attitudes of Systems Thinkers

We saw previously that attitudes reflect the preparation of the mind and the body to perform a gesture or an act, for instance, as a gymnast or a dancer. This attitude is therefore linked to an intention to act. Our mental model considers the action as an emergent process ("*processus en devenir*" in French, which means a potential to grow). The intentionality is therefore the property of the activity of a human being as a whole, who perceives, thinks, and acts in an environment under specific circumstances.

There is no coincidence between the intended action and the practical action. The intended action has a quality of generalization, while the effective act is unique and has the quality of singularity and spontaneity. So, the act is considered by the agent via a specific description that reports back the agent's intention at that time. The general concept of intentionality can acquire the quality of specificity only when the circumstances have realized the meaning of the intention. Interestingly, the action's meaning can be defined before, during, or after the act. It is linked to the action itself being a component of it.

Furthermore, there is a difference between *reflex* (an action without conscious thought) and *intended action*. If we can report on an intended action, this capability highlights skills acquired in a specific know-how and practice. We can act because we are aware of our potential to do it. Considering the agent's viewpoint, it gives a broader and more complete understanding of the action as he/she integrate their declared intentions with what is observed. Nevertheless, one should be aware of the agent's subjectivity, as the story reported afterwards may have been transformed to make it more attractive or to reflect the change of mindset for any reason.

CONCLUSION

As experience feedback, the *Systems Tree & Anti-tree* frameworks improve effectiveness in understanding the individual intended actions, the related justifications given afterwards, the revealed acts, the viewpoint of each team member, and the gap between intentionality and acting (Figure 6.1). It contributes to increasing awareness

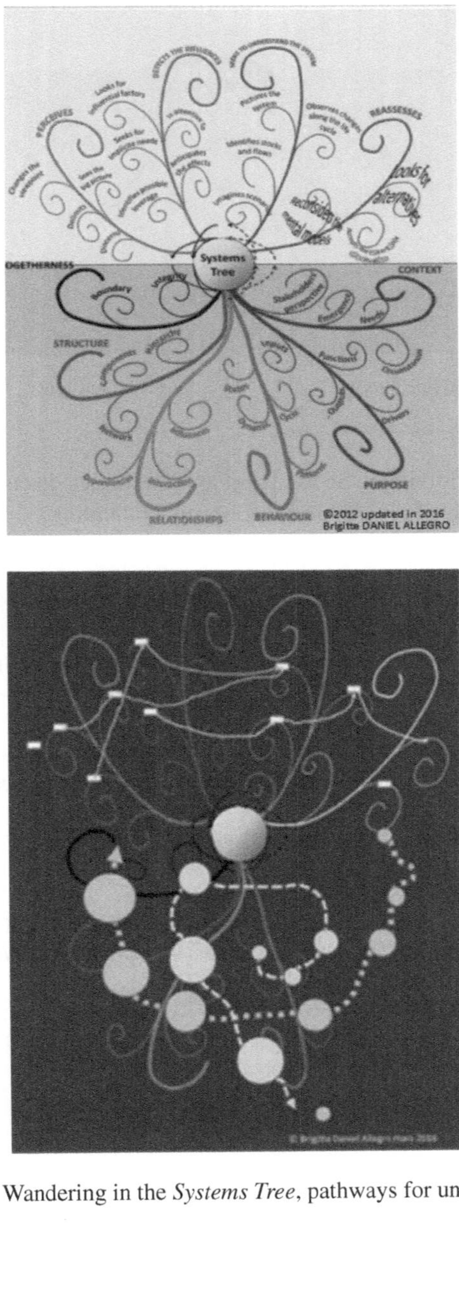

FIGURE 6.1 (A–C) Wandering in the *Systems Tree*, pathways for understanding systems.

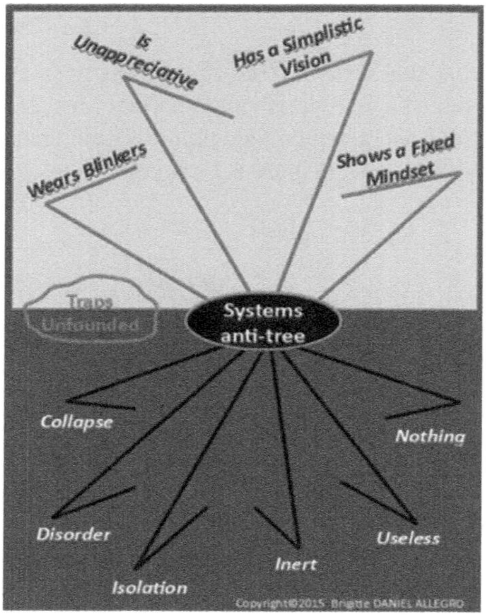

FIGURE 6.1 (A–C) (Continued)

of the behaviors of individuals in the team and facilitates discussions. Your attitudes and hard skills are engaged in your quest for understanding.

REFERENCES

Berthoz, A. (2009). *La simplexité.* Odile Jacob.

Berthoz, A. (2013). *La vicariance Le cerveau créateur de mondes.* Odile Jacob.

Boardman, J. & Sauser, B. (2013). *Systemic Thinking. Building Maps for Worlds of Systems.* Wiley.

Checkland, P. & Poulet, J. (2006). *Learning for Action.* Wiley.

Daniel Allegro, B. (2013). *Think Systems and Act Systems While Using Systems Engineering.* ENSTA Paris.

Gall, J. (1977). *Systemantics: How Systems Work and Especially How They Fail.* Quadrangle/The New York Times Book Co.

Ison, R. (2015). *Cyber-Systemic Possibilities? Systemic Inquiry.* The Open University, Proceedings of the 59th Annual Meeting of the International Society for the Systems Sciences, Vol. 1 No. 1, https://journals.isss.org/index.php/proceedings59th/issue/view/15.

Panaccio, C. (2011). *Qu'est-ce qu'un concept?* J.VRIN.

Rosnay, J. (1975). *Le macroscope. Vers une vision globale.* Editions du Seuil.

Sartenaer, O. (2018). *Qu'est-ce que l'émergence?* J.VRIN.

Section II

Theoretical Approaches and Methodologies

7 Human System Integration and Its Cognitive Effects on Society

The Indian Story

Gopal R. Jani and Parag Sanghani

INTRODUCTION: FRAMING HUMAN SYSTEM INTEGRATION IN A COGNITIVE CONTEXT

Human system integration (HSI) encompasses a multidisciplinary approach that draws on cognitive psychology, ergonomics, artificial intelligence (AI), and social sciences to ensure systems are designed with humans at the center (Boy, 2023). India presents a unique context due to its sociocultural diversity and varying degrees of digital literacy. Consequently, HSI plays a pivotal role in bridging technological innovation with the cognitive realities of everyday users. This chapter examines how cognitively informed design principles have shaped India's digital transformation, emphasizing the unified payments interface (UPI) as a successful societal adaptation to HSI (Dev et al., 2024).

We address the key questions: Why is HSI a relevant framework for analyzing Indian technology adoption over more traditional cognitive human factors or innovation management? How do HSI-based interventions create cognitive ease, reduce decision fatigue, and foster trust (Calikli & Bener, 2014)? Moreover, most importantly, what are the cognitive outcomes of such systemic integrations?

UNDERSTANDING THEORETICAL FOUNDATIONS: HSI AND COGNITIVE FRAMEWORKS

While traditional human–computer interaction (HCI) and innovation management explore the usability and diffusion of innovations, HSI emphasizes system-wide integration of human cognitive needs. In this context, Cognitive Load Theory (Sweller, 1988), Usability Heuristics, and Confirmation Bias (Calikli & Bener, 2014) become critical lenses. For instance, UPI minimizes extraneous cognitive load via intuitive design and builds trust using real-time feedback (Figure 7.1) and biometric security (Sweller, 1988; Mayer & Moreno, 2003).

DOI: 10.1201/9781003666431-9

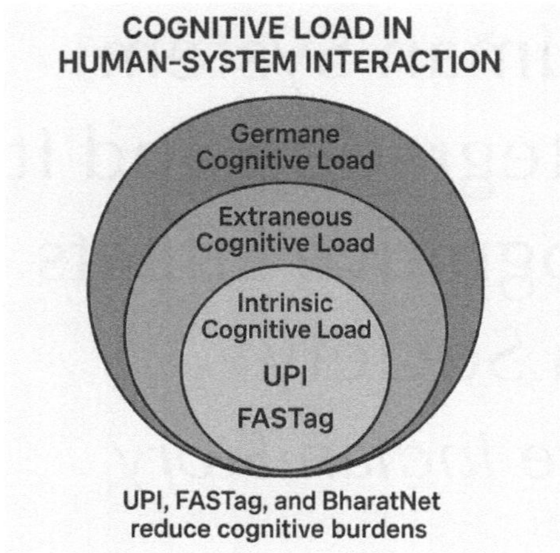

FIGURE 7.1 Real-time feedback reduces cognitive friction.

These features are not merely design affordances; they actively shape cognitive behaviors and reduce resistance to adoption (Nanda, 2023). HSI, therefore, is a more suitable approach for societies like India, where technological, cultural, and educational variables intersect (Hanham, 2023).

CASE STUDY

The UPI Revolution: Cognitive Design and Inclusivity

UPI employs cognitive principles to ensure accessibility across India's linguistically and educationally diverse population. Using biometric authentication, multilingual support, and minimalistic interface design reduces user anxiety and fosters trust (Kalyuga et al., 2003). Unlike password-driven systems, biometric cues appeal to universal human identifiers, eliminating the need for high literacy levels (Dev et al., 2024).

Real-Time Feedback and Trust Formation. Cognitive bias, specifically confirmation bias, is cleverly utilized via real-time payment feedback, reinforcing user belief in the system's reliability (Calikli & Bener, 2014). This reassures users and reduces cognitive dissonance in transactions. Decision fatigue is minimized as UPI's interface avoids overwhelming the user with choices (Mayer & Moreno, 2003).

The UPI transaction flow (Figure 7.2) shows that the cognitive journey from initiation to confirmation minimizes decision fatigue. One of the intriguing cognitive aspects of UPI is its integration of biometric authentication. This feature speaks to India's diversity by transcending language barriers and literacy concerns. UPI taps into the universally recognized aspect of human identity by utilizing fingerprints or facial recognition, minimizing the cognitive burden of remembering complex passwords or PINs. The reassurance of biometric security fosters user confidence,

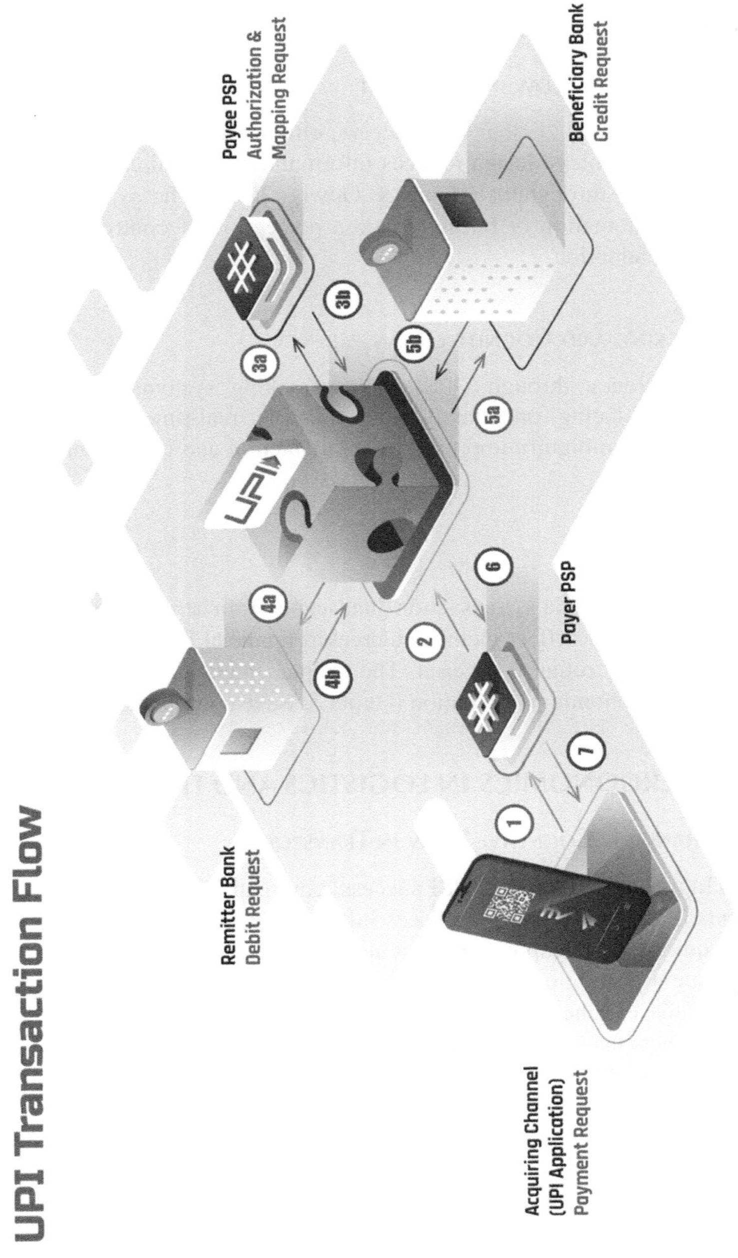

FIGURE 7.2 UPI transaction flow and cognitive touchpoints.

contributing to the widespread acceptance of UPI across demographics (Drummond & Fischhoff, 2017).

COGNITIVE ERGONOMICS IN LOGISTICS AND TRAVEL – FASTTAG

Enhancing Cognitive Flow in Transport

India's FastTag system is another HSI success, eliminating the cognitive burden of toll payments. Its seamless design reduces information load, minimizes user stress, and ensures quicker throughput (Heinz & Opwis, 2013). The system's feedback mechanism and automation of billing create a perception of control, thus encouraging widespread adoption.

Real-Time Tracking and Cognitive Transparency

Cognitive transparency through GPS-enabled tracking systems allows users to monitor goods, reducing uncertainty and enabling real-time decision-making. Reducing cognitive ambiguity improves user satisfaction and operational efficiency (Disdier et al., 2023).

Societal Impacts

UPI has transformed India's transaction landscape from street vendors to urban professionals. The ripple effect includes increased financial literacy, entrepreneurial ventures, and reduced reliance on cash. The chapter maps how UPI is a cognitive enabler for socio-economic participation (Gupta & Balasubramaniam, 2024).

COGNITIVE ERGONOMICS IN LOGISTICS AND TRAVEL

FastTag: Enhancing Cognitive Flow in Transport

India's FastTag system is another HSI success, eliminating the cognitive burden of toll payments. Its seamless design reduces information load, minimizes user stress, and ensures quicker throughput. The system's feedback mechanism and automation of billing create a perception of control, thus encouraging widespread adoption. As in Figure 7.3, the exponential increase in toll transactions completed via FastTag across India reflects the success of human-centered system design in promoting user adoption.

Real-Time Tracking and Cognitive Transparency

Cognitive transparency through GPS-enabled tracking systems allows users to monitor goods, reducing uncertainty and enabling real-time decision-making. The reduction in cognitive ambiguity improves user satisfaction and operational efficiency.

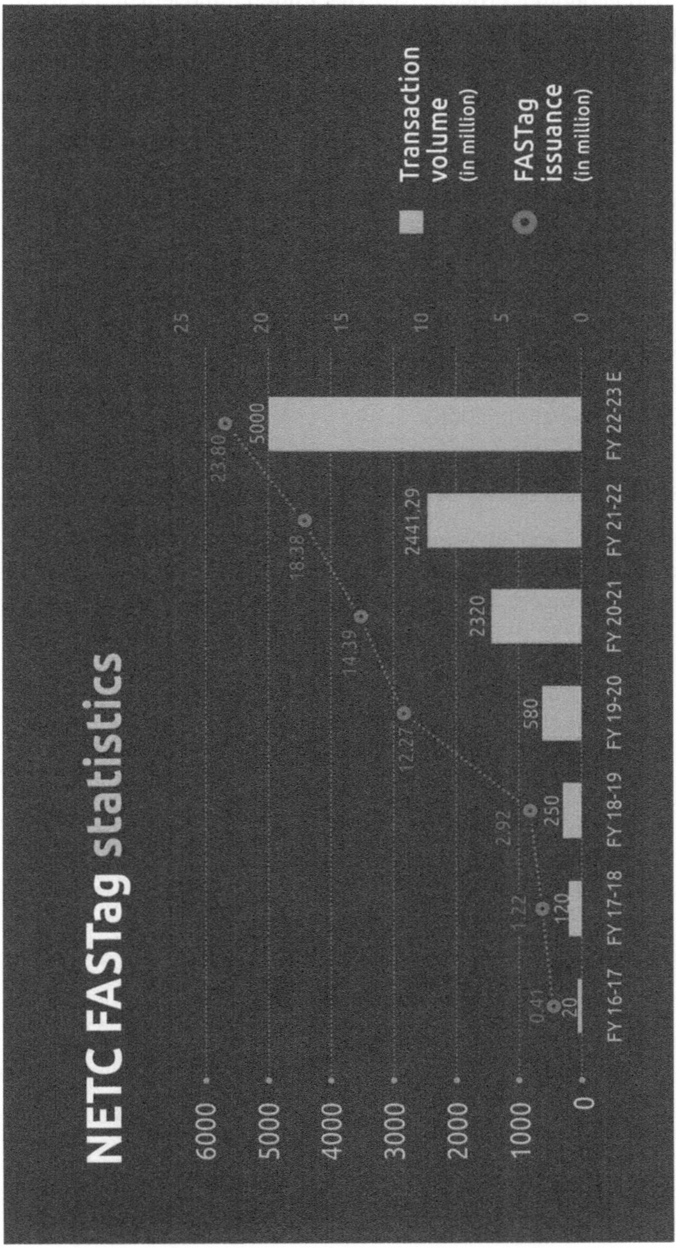

FIGURE 7.3 Growth in FastTag transaction volume from 2016 to 2024.

Impact on Consumption Patterns and Economic Growth

The cognitive impact of FastTag extends beyond logistics, influencing consumption patterns and economic growth. By expediting the movement of goods, FastTag contributes to the availability of products, impacting buying behaviors and consumption. This impact resonates through various sectors, from perishable goods to e-commerce, as it enables faster delivery, encourages greater consumption, and drives economic growth.

INNOVATION AND COGNITIVE SHIFTS IN INDIAN SOCIETY

Bharat Net: Fostering Digital Inclusivity

The BharatNet project catalyzes cognitive empowerment by providing high-speed internet to rural India, as shown in regional coverage data (Figure 7.4) (Bharat Broadband Network Limited [BBNL], 2024). It enables access to education, digital services, and economic opportunities, aligning HSI with social cognition and equity (Kalyuga, 2007; Kester et al., 2006).

Adani's Sustainability Drive: Shaping Environmental Consciousness

Adani's green logistics and data-driven sustainability efforts show how HSI can steer collective cognitive awareness. Real-time dashboards, environmental indicators, and transparent metrics help shift public behavior toward eco-consciousness (Picard, 1997).

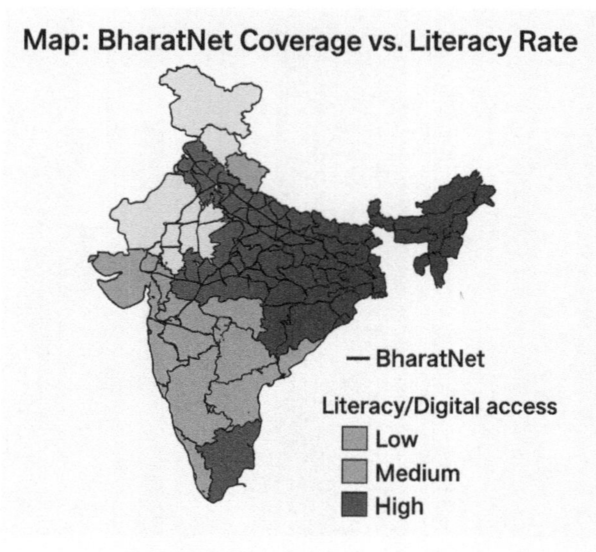

FIGURE 7.4 BharatNet coverage and regional digital inclusion indicators.

CHALLENGES AND RISK LANDSCAPES IN HSI

Cognitive Diversity and Ethical Design

Designing for India requires navigating high cognitive variability: cultural context, language, and educational background demand adaptive interfaces. Universal design principles and iterative feedback loops must be integrated into all systems (Boy, 2023).

Ethical Considerations: Integrating human systems and technology raises ethical dilemmas. Data privacy, preventing algorithmic bias, and safeguarding against unintended consequences are pivotal challenges. The cognitive impact of these ethical concerns extends beyond user experience, shaping perceptions of trust and system reliability.

Fraud Detection and Prevention

The "Anatomy of Fraud 2023" report outlines that 55% of India's digital frauds involve UPI (Bureau & Praxis, 2023). HSI can reduce fraud risk by embedding cognitive friction at high-risk points, using behavior-based anomaly detection, and AI-driven fraud alerts. These interventions align security with cognitive heuristics.

In the contemporary landscape of digital transactions, the battle against fraud has assumed paramount significance. A recent report has shed light on the intricate web of fraud detection and prevention challenges, particularly within digital payments. This revelation becomes even more crucial considering the escalating popularity of the UPI, which has become the focal point of over half of the reported digital payment frauds in India. UPI systems increasingly rely on behavior-based fraud detection to reduce user vulnerability, as shown in recent trend analyses (Figure 7.5). The insights gleaned from "The Anatomy of Fraud 2023," a collaborative effort between digital solutions platform Bureau and consultancy firm Praxis, underscore the complexity of these challenges and their potential impact on India's digital economy.

Identity-Related Frauds and Regulatory Safeguards

Identity-related frauds emerge as a prevalent concern across diverse industries. The report emphasizes that in 2021, 65% of all financial services sector fraud was account-related. These encompassed both account takeover and fraudulent account creation. Similarly, the e-commerce industry faced its share of challenges, with 54% of the reported frauds being account-related (Drummond & Fischhoff, 2017).

The report acknowledges the proactive stance taken by the Indian government in curbing these fraudulent activities. Initiatives such as stringent know-your-customer (KYC) norms and regulations tailored to real-money gaming sectors serve as essential layers of protection for consumers. These regulatory measures are pivotal in fortifying the digital payment ecosystem and safeguarding users against identity-related fraud.

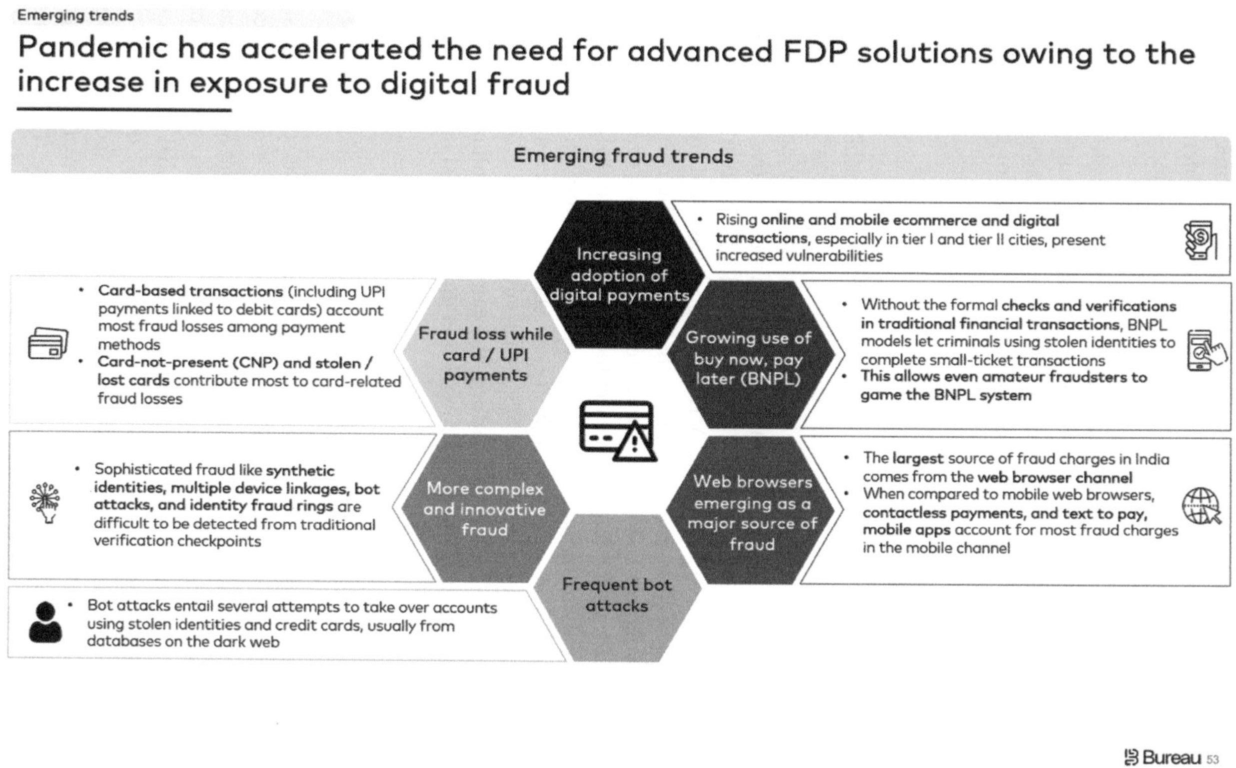

FIGURE 7.5 Fraud detection loop in UPI systems based on user behavior analytics.

OPPORTUNITIES AND THE PATH FORWARD

HSI offers a vision for inclusive, adaptive, and ethical technology design. Future systems should incorporate.

- Personalized cognitive modeling
- Voice-based interaction for illiterate users
- Real-time emotional analytics for adaptive UX
- Human-centered automation balancing control and trust

As India moves toward Industry 5.0, HSI will bridge cognitive needs and systemic functionality (Boy, 2023; Hanham, 2023).

Future of Human System Integration in India

The evolution of technology has ushered in a new era where the harmonious interaction between humans and complex systems is paramount. HSI is a pivotal approach that enhances user experiences and shapes societal norms and behaviors. As India hurtles toward digital transformation, the future of HSI within its diverse and dynamic landscape is ripe with possibilities. This article delves into the potential trajectories that the future of HSI may take in India, examining its implications, challenges, and transformative impacts (Sweller, 1988; Kalyuga, 2007).

CONCLUSION: TOWARD A COGNITIVELY ALIGNED SOCIETY

The Indian experience illustrates how HSIs can drive large-scale behavioral and societal transformation when informed by cognitive frameworks. From UPI and FastTag to BharatNet and sustainable logistics, the cognitive impacts are measurable and enduring. By framing design around human cognition, India leaps over technical limitations and fosters a cognitively inclusive digital future (Boy, 2023; Hanham, 2023).

The path forward lies in embracing HSI as a strategic framework, integrating it into policy, education, and industry to build systems that genuinely resonate with human complexity.

Conclusion Hypothesis: HSI in India's technological advancements can lead to a more inclusive, efficient, and cognitively aligned society, transforming industries and user experiences. As HSI continues to guide India's technological evolution, it promises to bridge gaps, enhance user interactions, reshape behaviors, and foster societal evolution. The challenges and ethical considerations inherent in this journey will be instrumental in ensuring the path ahead is innovative and ethically responsible. The future of HSI in India is a journey of potential, challenges, and ethical considerations, guided by the imperative of aligning technology with human cognitive capabilities.

REFERENCES

Boy, G. A. (2023). Model-based human systems integration. In A. M. Madni & N. Augustine (Eds.), *Handbook of model-based systems engineering* (pp. 471–499). Springer. www.researchgate.net/publication/372639063_Model-Based_Human_Systems_Integration

Bureau & Praxis. (2023). *The anatomy of a fraud report*. Bureau & Praxis. www.praxisga.com/reports-and-publications/technology/anatomy-of-fraud-report-2023

Calikli, G. (2014). Empirical analysis of factors affecting confirmation bias levels of software engineers. *Software Quality Journal*, 22(3), 387–405. www.researchgate.net/publication/303689485_Empirical_analysis_of_factors_affecting_confirmation_bias_levels_of_software_engineers

Dev, H., Gupta, R., Dharmavaram, S., & Kumar, D. (2024). From cash to cashless: UPI's impact on spending behavior among Indian users and prototyping financially responsible interfaces. arXiv preprint arXiv:2401.09937. https://arxiv.org/abs/2401.09937arXiv

Disdier, A., Masson, D., Brethomé, T., Jankovic, M., & Boy, G. A. (2023). Toward a human systems integration approach to designing and operating a remote and virtual air traffic control center. *Proceedings of the International Conference on Engineering Design (ICED 2023), Bordeaux, France*. Design Society, Cambridge University Press. www.cambridge.org/core/journals/proceedings-of-the-design-society/article/toward-a-human-systems-integration-approach-to-the-design-and-operation-of-a-remote-and-virtual-air-traffic-control-center/990443E85D16BF38974E6D57FFB65A52

Drummond, C., & Fischhoff, B. (2017). Development and validation of the scientific reasoning scale. *Journal of Behavioral Decision Making*, 30(1), 26–38. https://doi.org/10.1002/bdm.1906

Gupta, H., & Balasubramaniam, V. S. (2024). Behavioral economics in UI/UX: Reducing cognitive load for sustainable consumer choices. *International Journal of Research in Modern Engineering & Emerging Technology (IJRMEET)*, 12(12), 1–10. https://ijrmeet.org/behavioral-economics-in-ui-ux-reducing-cognitive-load-for-sustainable-consumer-choices/IJRMEET

Hanham, J. (2023). Integrating cognitive load theory with other theories, within and beyond educational psychology. *British Journal of Educational Psychology*, 93(1), 239–250. https://doi.org/10.1111/bjep.12612

Heinz, S., & Opwis, K. (2013). Cognitive load in eCommerce applications: Measurement and effects on user satisfaction. *Proceedings of the 15th International Conference on Human-Computer Interaction*, 1–10. Advances in Human-Computer Interaction. www.academia.edu/5745025/Cognitive_Load_in_eCommerce_Applications_Measurement_and_Effects_on_User_SatisfactionAcademia

Kalyuga, S. (2007). Expertise reversal effect and its implications for learner-tailored instruction. *Educational Psychology Review*, 19(4), 509–539.

Kalyuga, S., Ayres, P., Chandler, P., & Sweller, J. (2003). The expertise reversal effect. *Educational Psychologist*, 38(1), 23–31. www.researchgate.net/publication/48829036_The_Expertise_Reversal_Effect

Kester, L., Kirschner, P. A., & van Merriënboer, J. J. G. (2006). Just-in-time information presentation: Improving learning and troubleshooting skills. *Contemporary Educational Psychology*, 31(2), 167–185. www.sciencedirect.com/science/article/abs/pii/S0747563201000115

Mayer, R. E., & Moreno, R. (2003). Nine ways to reduce cognitive load in multimedia learning. *Educational Psychologist*, 38(1), 43–52. https://psycnet.apa.org/record/2003-04803-006

Nanda, N. (2023). Cognitive load and screen design: Creating interfaces that reduce mental burden. *International Journal of Human Computer Interaction and Data Mining*, 6(2), 1–10.

Picard, R. W. (1997). *Affective Computing*. MIT Press. adrjournalshouse.com

Sweller, J. (1988). Cognitive load during problem solving: Effects on learning. *Cognitive Science*, 12(2), 257–285. https://doi.org/10.1207/s15516709cog1202_4

8 Human–Machine Interaction in Sociotechnical Systems

Search for Meta-Theoretical Design Guidance

Vivek Kant

INTRODUCTION

With the growth of organizational and informational interdependencies and other novel interactions, modern-day interconnected sociotechnical systems (STSs) are rapidly developing. A prominent direction for human–machine interactions (HMIs) in complex systems was developed at the Risø Laboratories in Denmark under the novel area of cognitive systems engineering. This ecological interface design (Vicente & Rasmussen, 1992) approach has undergone multiple variations and refinements in the past several decades. It derives from the insights of Jens Rasmussen, an engineer, and J.J. Gibson, a psychologist (Bennett & Flach, 2019; Burns, 2013; Burns & Hajdukiewicz, 2004; Vicente, 2002; Kant & Sudakaran, 2021).

While the foundations of HMI design from this direction have been explicated by several researchers (e.g., Bennett & Flach, 2011; Burns & Hajdukiewicz, 2004), a broader question is how to develop meta-theoretical directions that can serve as the guidance of HMI in complex systems. The development of such meta-theoretical directions is essential as they provide directions for HMI for the future. Thus, while the technology may change, these meta-theoretical design principles will provide directions and insight for HMI designers for future applications and enable newer methods for analysis and synthesis in the future.

The chapter is divided into six sections in the quest for these meta-theoretical design directions for HMIs. Second section delineates the problem of information in HMI in complex STSs. Third and fourth sections address the information for perception, action, and cognition from a psychological and systems design perspective. The chapter concludes with a convergence of the psychological and systems design basis for the HMI design and eight insights for design guidance for HMI design for complex STSs (Section 5 and 6).

DOI: 10.1201/9781003666431-10

PROBLEM OF HMI INFORMATION IN COMPLEX SOCIOTECHNICAL SYSTEMS

The notion of information has moved to the forefront of everyday lives worldwide. This move has partly been due to the explosion of media and communication technologies and a convergence of academic disciplines such as information technology, informatics, computer sciences, and other related technological sciences. In particular, the value of information has been developing in greater detail in the field of human–computer interaction and interaction design, which aims toward designing interactions between people and technologies.

This field of study has turned toward phenomenology to develop the philosophical foundations of interaction. In this approach, Heideggerian thought and Merleau-Ponty's approach have been rather prominent in discussing the foundational basis of interaction. In addition, with the growth of the past phenomenological philosophy of technology, the human–technology relationship has received further impetus regarding its conceptual foundations. While the insights from these various disciplines support one conception of human technology interactions, they fall short when these interactions are part of STSs (Figure 8.1). In short, human–technology interactions in STSs require the phenomenological viewpoint of human–technology relationships and an engineering philosophy viewpoint on the functioning of complex systems.

Information challenges in HMI in complex STSs relate to two main issues. First, the nature of STSs requires a particular set of principles that is different from the design of the interaction of products. Second, due to the interaction being shaped by myriad elements of the STSs (regulations, laws, teams, individuals, etc.), the HMIs in these systems are systemic. Third, because of the first challenge, the HMI design for STSs requires considering two different kinds of information, which must be taken together to devise HMI. These two kinds of information result from various sources and must be coherently comprehended for a complete characterization of the HMI design.

Thus, this paper highlights that human–technology interaction in STSs requires a convergence of these two kinds of information for a complete characterization (Figure 8.2):

1. Information for perception, action, and cognition (e.g., real-time visual displays).
2. Information for joint optimization and systems reliability (e.g., optimization data for system stability).

These two viewpoints are necessary because of the particular characteristics of humans in complex STSs. First, the operator is part of the overall system's functioning. At the same time, the operator is not a passive entity. The human is an active entity that acts on the world around it. In other words, the technological system is the person's operational environment. This duality and mutuality between the human and the machine provide a significant basis for the design of HMI: *The human is a system's component, and the system is the human's environment* (Rasmussen, 1976, 1986; Rasmussen et al., 1994).

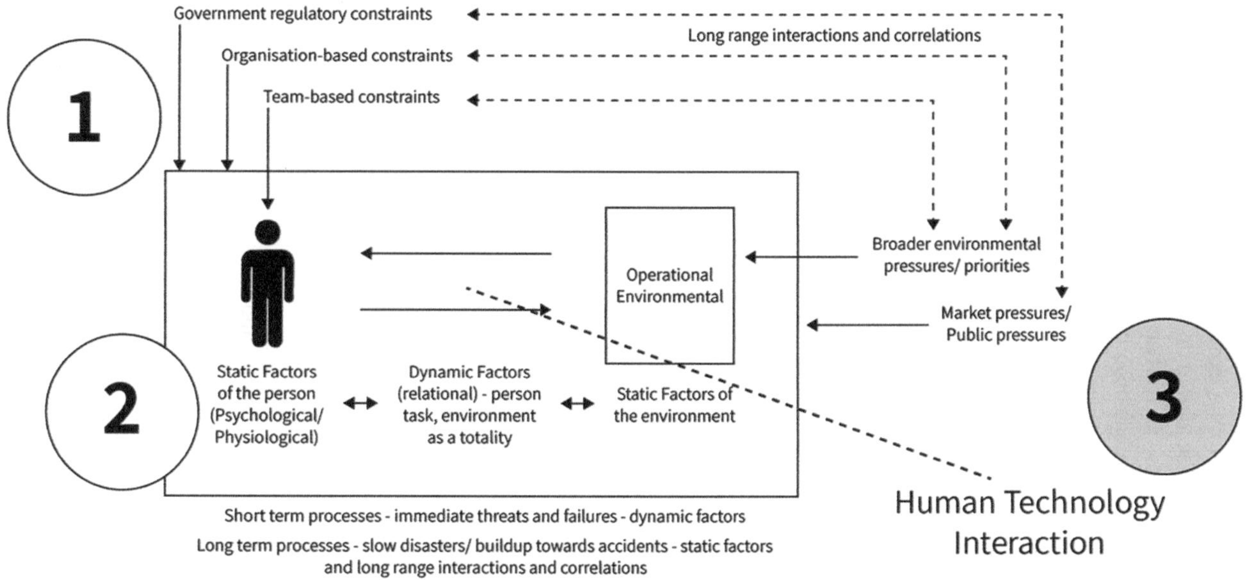

FIGURE 8.1 A sociotechnical system: (1) sociotechnical systemic factors that affect behavior; (2) factors involved in the human–operational environment coupling; (3) human–technology interaction, which involves the design of the HMI and incorporates concepts of affordances and reciprocity (Kyriakidis et al., 2018).

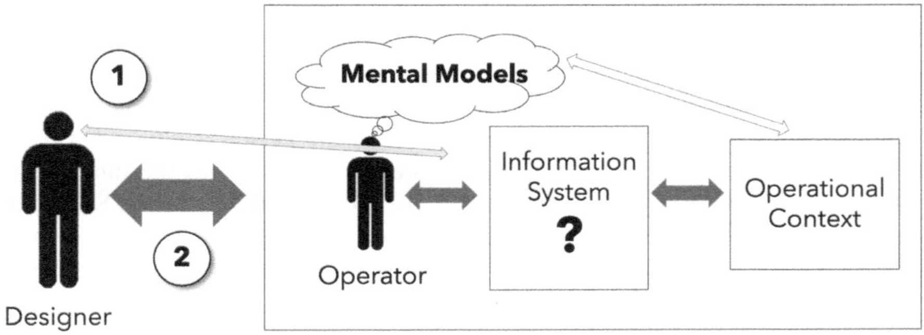

FIGURE 8.2 (1) Pathway to support the design for information for perception, action, and cognition (intentional stance). (2) Pathway to support information for joint optimization and systems reliability (design stance).

This insight leads us to characterize two sets of information flows (Figure 8.2). The first requires the operator's perspective, such that the information flow is between the operator and the environment and back from a phenomenological perspective. The basis of ongoing sense experience is crucial to emphasize, as the everyday lived experience of operators is based on the understanding of technology and the ongoing transactions in the realm of the here and now. This means that the operators act not only in terms of a rational understanding but also depend on the limits of memory and perception, considering everyday fatigue.

In contrast, the operator is also a system component. Thus, a common language is needed to characterize machines and humans. As such, it requires a discussion of both psychology and information design for perception, action, and cognition. Engineers often employ a common language, such as control theory, to characterize the human-in-the-loop scenarios where information engineering and sensor fusion techniques are used to optimize both the human and the machine jointly (Hollnagel & Woods, 2005). Thus, to successfully design for humans, the person and the environment must be taken together, along with the case for successfully designing for people. In the following sections, the background of information for perception, action, and cognition has been explored, both from a psychological and informational perspective, as the basis of HMI design in complex systems.

INFORMATION FOR PERCEPTION, ACTION, AND COGNITION – THE PSYCHOLOGICAL PERSPECTIVE

This section briefly introduces the concepts of embodiment, embeddedness, and situatedness based on the insights from the work of the psychologist J.J. Gibson. The idea of situatedness entails understanding, knowing, and acting based on circumstances in which people find themselves. As circumstances change, new opportunities arise; consequently, people understand the changing circumstances and formulate potential lines of action. However, the term "situatedness" does not imply that knowing and acting are merely reactive. Rather, situatedness implies that rational planning and foresight would not be enough to characterize human knowing and acting completely. Therefore, local contingencies must be addressed along with sensible planning to understand human knowing and acting comprehensively.

In other words, how does behavior emerge in situations and shape the situations themselves? Along with situatedness, the idea that the person is embedded in the environment and the person–environment system should serve as the fundamental unit of analysis has been briefly discussed in the section outlining the STSs. A third aspect to be noted is the issue of embodiment. Embodiment theory came into vogue with the demise of standard cognitive psychology (information processing psychology). With the growth of embodied cognitive science (Clark, 2011, 2015), there was a growing need for understanding the role of the body in the mind. Embodied cognitive science emphasizes the primacy of the role of the body in the process of knowing and acting.

Gibson's (1966, 1979) approach was to understand perception and action as ecological concepts; i.e., the perception and action of animals could only be understood about their environments. Therefore, ecological psychology emphasizes this relationship; in other words, environmental psychology presents a functional–relational perspective on understanding behavior, emphasizing reciprocity between the entities of the ecosystem. Regarding ecological psychology, the animal is immersed in its lifeworld, constantly knowing and acting. Knowing and acting (perception and action) are two sides of the same coin. It is an ongoing process in which the animal is constantly involved. The animal in its environment requires further clarification. For Gibson, the mind was an embodied and embedded concept. Avoiding the traditional mind–body dualism, Gibson recognized the mind as emerging out of the activities of the embodied animal in its environment.

Gibson provided the concept of "affordances" as a cornerstone of ecological psychology. Affordances are value-laden properties of environmental entities present for the good or ill of the animal. Further, affordances are not static but change over time. They belong not simply to the object but also to the animal and must be understood from the animal's perspective. An ecosystem can hold multiple animals at multiple spatial and temporal scales. It is also to be noted that, as a direct realist, Gibson did not envision affordances as a static property of the object. An object can have multiple affordances based on the animals interacting with it and the context. Further, the affordances are not psychical qualities changing on the whim of the observer; these are realistic properties of the environment.

This brief discussion of Gibson's approach raises a few essential aspects for the upcoming framework. First, the workers and their environments are to be considered as reciprocal entities, together as a whole. Second, perception and action (knowing and acting) are ongoing processes in which the person is constantly involved. Third, the same environment can hold multiple people and social roles (Costall, 1995; Lombardo, 1987; Heft, 2001). Due to their backgrounds and varied experiences, these people will interact with their environments differently. The environment holds different meanings and values for other individuals. These individuals occupy various niches in the environment. These niches involve the consideration of meanings and values that have evolved and changed due to the person's interaction with the environment.

A key idea discussed in this section was that Gibson provided an ecological approach to studying knowing and acting, emphasizing reciprocity. In contrast to the traditional psychology theories, Gibson promulgated an approach where the mind, specifically human knowing and acting, is treated as an embodied and embedded concept. This ecological reality sustains knowing and acting. In the next section, the insights from Gibson's approach have been used to highlight salient aspects of the informational design for HMIs in STSs.

INFORMATION FOR PERCEPTION, ACTION, AND COGNITION – THE SYSTEMS DESIGN PERSPECTIVE

Throughout his career, Jens Rasmussen proposed a framework for addressing systems design. This framework considers the dimension of function–structure linkage, correct functioning, and accounting for human activity in technical contexts. Rasmussen's primary goal was to support the operator's traditional ways of knowing and acting in the technical context of nuclear power plants. The problem that Rasmussen approached was the reliability of nuclear power systems, i.e., how to ensure that the nuclear power system was functioning correctly for the specified conditions and duration of time.

In studying the problem, he found that the human element was crucial to system reliability. Further, due to the influx of digital computers as a medium for controlling the power plant functioning, the challenge for Rasmussen and the Risø group was to develop concepts that would aid in systems design for supporting operator activity (Kant, 2017). For this reason, Rasmussen and his research group conducted several field studies on operators' and electronic troubleshooters' activities. These studies emphasized that the operators' traditional ways of knowing and acting must be supported. This traditional knowing consisted of a unique notion of "process-feel." The operator approached any situation based on a generic, intuitive understanding gained from experience; thus, having a generalized "feel" for the situation. This intuitive understanding was also found in electronic troubleshooters during the fault-finding and repair of electronic equipment.

To account for the traditional "process-feel," Rasmussen formulated a model of the human as a system's component and part of the overall system. In addressing operator decision-making, he found that experts behaved differently from novices; specifically, experts made intuitive leaps during the reasoning process. Rasmussen incorporated new knowledge structures, such as the decision ladder, to address the behavior of novices and experts. Further, he stressed the need to address the operators' strategies in conducting their work. Rasmussen found that operators are attuned to routine situations; however, non-routine situations, even though rare, put extensive demands on the operator. For operators to successfully handle routine and non-routine situations, the designers need to address the operators' cognitive activity. Rasmussen reasoned that the designer had to share an interpretative understanding of the operator's activities to address the operator's knowing and acting in situations. In other words, the designer needs to understand the situation from the operator's point of view and actions.

Another key insight was that operators use multiple ways of understanding the scenario as the event unfolds. They often shift between various construals to cope with the situation. Therefore, to support operators' knowing and acting, these multiple constructs, or mental models, must be brought together at various levels of abstraction. Further, as these abstractions were part of a technical system, the integrity and correct functioning had to be accommodated.

Like the categories used for representing the system in human-centric terms, Rasmussen also addressed human performance in multiple categories. According to Rasmussen, any activity requires understanding motor skills, rules governing behavior, and knowledge categories (SRK taxonomy) that enable the person to cope with the situation. Thus, the performance of the individual is best viewed in light of the person's goals and expectations. Rasmussen regards humans as goal-oriented systems. Being goal-oriented, people have their reasons for acting in a particular manner. Rasmussen emphasizes that this view of reasoned activity needs to be addressed in terms of generic categories rather than specific instances for systems design. This formulation of human activity in generic categories led Rasmussen to address its linkages with perceptual meaning and, more broadly, symbolism. Following the philosophers Cassirer and Whitehead, Rasmussen recognized the human as a symbol-manipulating animal. Rasmussen linked the generic categories of performance with perceptual content. In doing this, Rasmussen used Whitehead's distinction between signals, signs, and symbols, corresponding with skills, rules, and knowledge. Thus, Rasmussen linked meaning-making along with generic categories of human performance.

In a book on cognitive engineering in 1994, Rasmussen, along with his colleagues (Rasmussen et al., 1994), distilled many of the main ideas of the past decades of research into a comprehensive framework for the engineering analysis, design, and evaluation of advanced information systems. In the 1994 book, the ideas presented initially in Rasmussen's 1986 book were extended to STSs. Specifically, Rasmussen et al. (1994, p. xi) *"seek an optimal combination of a conceptual basis with a practical and realistic sense for 'what it is really like out there'. "* The authors adopt a functional perspective and focus on qualitative models to address modern work systems. They mention that at a systemic level, work systems act as self-organizing adaptive systems and are dynamic. Thus, Rasmussen et al. (1994) adopt a technical–rational perspective toward systems modeling (further developed in Rasmussen, 1997; see De Bruijn & Herder, 2009 for the difference between technical–rational and actor-based approaches).

In their approach, they stress the modeling of purposes and associated constraints. These systems have goals and constraints implicitly reflected in the practice and customs inherent in the operator's work activity. The authors acknowledge the inherent variability in human behavior. Human behavior has a history based on the decisions and choices made. These choices have been made in the presence of other alternatives. Even though the alternatives were not selected, these potential choices remain hidden in the work activity. From the design perspective, these potential choices are meaningful as they provide insight into the possibilities of activities and challenges that shape the behavior of the work system as a whole. The key insights derived from Rasmussen's approach can be used to model the human as part of the system through qualitative insights and structured representations of goals and activities.

CONVERGENCE OF THE PSYCHOLOGICAL AND DESIGN BASIS FOR INFORMATION FOR HMI

In designing for people involved in their everyday activities, the design engineer has to understand the conduct of people involved in their work activity via an interpretive approach. This focus on human conduct in situations eschews a division based on mental and physical, social and biological, knowledge and activity, among other widespread dualisms. Work involves the body as much as the mind, and the behavior of the individuals is best understood in terms of a social endeavor. In short, there is a necessity for a holistic understanding of the work. With technological leaps in material science, information-processing devices are decreasing in size; they are becoming ubiquitous and embedded in work environments. These technological devices become "invisible" to those involved in these "smart" environments. Thus, this requires a holistic embodied mode of interaction for systems that consider people as embedded in their work environments. Therefore, the designer should treat the notion of knowing and acting as intertwined, meaningful constructs for human–technology interaction design that have to be understood from the point of view of the people involved (essentially, an interpretive approach).

The designer shares a dual relationship with the subject matter under consideration. First, the workers are treated as people, i.e., ethical and moral creatures, who are involved in their everyday work activities. In such an approach, the designer shares an interpretative understanding of the worker's perspective – the designer adopts an intentional stance (Dennett, 1989). The intentional stance provides emphasis on actor-based approaches to systems engineering. In other words, it supports perception, action, and cognitive information.

Along with actor-based approaches, the system can be addressed in terms of an engineering-based approach in which the people of the system are treated as parts of a STSs. In this engineering-based approach, the designer ensures the system functions correctly. Therefore, the designer adopts a design stance toward the subject matter (Dennett, 1989). The design stance supports the information for "joint optimization" of the human and the technology while considering the reliability of the overall system (Hollnagel & Woods, 2005).

In adopting the intentional stance, the designer gives primacy to the actor-based approaches to systems modeling, whereas, in adopting a design stance, the designer stresses a technical–rational perspective in systems modeling. Both these stances are crucial for systems design, and the designer balances these perspectives to ensure proper safety and health for the people involved, as well as the correct functioning of the overall system (Figure 8.2 shows the designer and two stances).

A primary requirement for the design framework to work successfully is to understand the nature of ongoing activity. A few salient aspects need to be highlighted. First, without being reduced to mechanisms, operators form a part of the overall system, and the system serves as their environment. In other words, people need to be addressed as embodied agents embedded in their sociotechnical milieu. This relation between the person and the system puts a theoretical demand on the designer (i.e., the designer should treat the design of the technology as a systemic concept). Focusing on one part, the operator or the environment, will render the design, at best, meaningless and, at worst, hazardous. Thus, the designer's aim should be to design with a focus on the entirety of the situation that unfolds (i.e., considering both the human and the surrounding system).

Second, when people are immersed in their everyday work environments, they act meaningfully toward the entities in their surroundings. These meanings are not always preordained but emerge as the events unfold. Thus, people use an interpretive process to modify their meanings for the objects of their environments. As Rasmussen has shown, supporting people's everyday reasoning is essential for routine and non-routine situations. For designers, the essential message is to design interfaces that support these people's everyday reasoning in technological work settings.

Third, as everyday reasoning is functional, it focuses on what is going on and how activities are accomplished rather than why things are the way they are. This emphasis on the "what" and "how" is brought about strongly in Rasmussen's view of errors. Rasmussen points out that errors were after-the-fact labels attributed to the operators. From the operator's perspective, the person chose the particular outcome because it seemed meaningful. Further, from the study of error reports, Rasmussen found that in most cases, the operators knew what to do but, in many instances, did not know how to accomplish the task. Rasmussen identifies that everyday activity is laced with ambiguities, and humans must make choices to circumvent obstacles and accomplish their goals. For people to succeed in their outcomes, the designers should provide the users, operators, and other related people with meaningful information to avoid mental traps and support the variability in their activity. Thus, the third aspect brought into relief is the need for the designers to support the notion of "what" and "how" of human conduct to provide meaningful choices to the people involved.

Fourth, in supporting the functional reasoning of the people, designers need appropriate concepts to represent human conduct. Such concepts should reflect the person's functional reasoning and account for the system's correct functioning. Rasmussen used a qualitative approach focusing on generic categories rather than details and quantitative variables. Such a categorical qualitative mode of analysis relates more easily to the operator's reasoning and is amenable to systems design. As Rasmussen indicates, formal language understanding focuses on variables and causal modes of representation. In case of failures in the system, making changes in formal representations becomes cumbersome due to the presence of a large number of variables and their updating to exemplify the changes. In contrast, a natural language description of the system allows for accounting for the person's perspective and provides a less cumbersome method for accounting for failures. Thus, the fourth aspect brought into light is using a functional description to characterize the system.

The fifth aspect of being illuminated was the requirement for treating the human as a symbol-manipulating creature. Humans are creative improvisers who not only adapt to aspects of their environments but also often adopt and appropriate things for their purposes. These creative symbol manipulators use an interpretative process to negotiate meanings and values in their environment. Therefore, any representation of the work context should support multiple levels of interpretative abstractions. Such interpretive flexibility is required in dealing with entities because meaning is a relational concept. As discussed by J.J. Gibson, affordances, objects, and their meanings can only be discussed concerning the knowing agent. Rasmussen also deals with interpretive understanding while employing Whitehead's classification of signals, signs, and symbols corresponding to the objects in the environment and human performance. Following Whitehead, Rasmussen mentions that the object's meaning depends on the context in which the agent perceives it. In other words, it depends on the intentions and expectations of the perceiver.

The objects of the environment, thus, have meanings and are valuable to the actors based on the situations. The meaning and value of the object may change as the event unfolds. The important insight for the designer is that to design robust systems, flexibility in interpretation and abstraction should be provided. From a designer's perspective, when the people and the associated technological environment are treated *in toto*, the designer also requires an interpretative understanding to address the system at multiple levels of abstraction. This will allow the designer flexibility in addressing all the aspects of the system.

The previous two sections highlighted J.J. Gibson's approach toward HMI design for STSs. While researchers have developed both of their thoughts, their use for deriving meta-theoretical needs of HMI design has not been significantly addressed. Therefore, based on the above discussions, the following insights need to be highlighted for HMI design in STSs:

1. *Design for Embodied, Embedded, and Situated Behavior:* This will ensure that the operator's presence is considered in a context and that she/he recognize her/his limitations and capabilities while involved in activities. This will include designing for body-based considerations and constraints of the surroundings. This will ensure the operator is involved in co-creating and unfolding the situation.
2. *Design for Human-Environmental Mutuality:* This will ensure that the operator and environment are addressed coherently and together. For example, the design of control rooms should ensure that the line of sight, reach of various physical parts, and other environmental features are considered along with the operator in terms of their co-constitution and mutuality of functioning.
3. *Design for the Duality of Humans as a System's Component and System as the Human's Environment:* This will ensure that the human's capabilities from a phenomenological viewpoint have to be considered. In contrast, the design of the overall system is being considered for joint optimization.
4. *Design to Enable Information for Perception, Action, and Cognition, Along with Information for Joint Optimization and Systems Reliability:* The key idea is to ensure that the information design to support the operator from a phenomenological viewpoint from displays is such that interacting with it does

not interfere with the overall functioning of the system. Both viewpoints of information should support the operator and ensure overall system reliability.

5. *Design for Humans While Recognizing That They Are Active, Meaning-Makers, and Creative Improvisers in Their Everyday Life Course:* This will ensure that the information design is such that it actively supports sense-making for operators. The information design should be such that the operator is supported during fault finding and actively understands the unfolding situation.
6. *Design for HMI as an Ecological Concept While Paying Specific Attention to the Actor's Niche:* This supports the view that the HMI is not limited to the here and now but requires considerations as an ecosystemic construct. Thus, the operator's niche in a control room must be supplemented with information from the whole ecosystem.
7. *Design for Balancing Adaptation and Creative Appropriation (at Times of Vital Necessity) Needs to be Balanced:* The essential challenge here is to recognize that operators adapt to their circumstances, which may result in mistakes and problems. In addition, individuals have their idiosyncrasies, which allow them to assess and act advantageously in particular situations. A proper design should strike a balance between accounting for the possibilities of creative appropriation and the ensuing adaptation of operators in their work environments.
8. *Design for Flexibility and Interpretive Understanding:* The key idea, in this case, is to design HMI and information in such a manner as to ensure flexibility in use as well as an interpretive understanding of the information to enable rapid decision-making.

CONCLUSION AND FUTURE DIRECTIONS

As the informational interdependencies of various parts of modern-day systems are increasing, there is a growing need to move away from design for a particular application or domain toward more meta-theoretical principles required for design. This current chapter introduced a few of them, deriving from insights presented by Jens Rasmussen and J.J. Gibson. This complementary approach was taken to support both the psychological and design-oriented approaches to support both the psychological and design-oriented approaches of information for perception, action, and cognition in HMI design. Further, in the future, these principles can be extended in terms of other challenges that are important for HMI design, such as managing cognitive overload, safeguarding ergonomic usability, and balancing automation with human oversight, as well as cognitive transparency, autonomy, and data privacy. With the growth of new technologies such as artificial intelligence and machine learning, there is a change in how people will interact with future technologies. These meta-theoretical design principles can play a significant role in designing human interaction with these technologies. In the future, the design directions presented in this chapter, regardless of the form of the technology involved in that particular STSs, can join others to serve as meta-theoretical directions for HMI design in STSs.

REFERENCES

Bennett, K. B., & Flach, J. (2011). *Display and interface design: Subtle science, exact art.* Boca Raton, FL: CRC Press.

Bennett, K. B., & Flach, J. (2019). Ecological interface design: Thirty-plus years of refinement, progress, and potential. *Human Factors*, *61*(4), 513–525. http://doi.org/10.1177/0018720819835990

Burns, C. M. (2013). Ecological interfaces. In J. D. Lee & A. Kirlik (Eds.), *The Oxford handbook of cognitive engineering*. New York, NY: Oxford University Press. http://doi.org/10.1093/oxfordhb/9780199757183.013.0039

Burns, C. M., & Hajdukiewicz, J. R. (2004). *Ecological interface design.* Boca Raton, FL: CRC Press.

Clark, A. (2011). *Supersizing the mind: Embodiment, action, and cognitive extension.* Oxford; New York: Oxford University Press.

Clark, A. (2015). *Surfing uncertainty: Prediction, action, and the embodied mind.* Oxford; New York: Oxford University Press.

Costall, A. (1995). Socializing affordances. *Theory & Psychology*, *5*(4), 467–481.

De Bruijn, H., & Herder, P. M. (2009). System and actor perspectives on sociotechnical systems. *IEEE Transactions on systems, man, and cybernetics-part A: Systems and Humans*, 39(5), 981–992. DOI: 10.1109/TSMCA.2009.2025452

Dennett, D. C. (1989). *The intentional stance*. Cambridge, MA: MIT Press.

Gibson, J. J. (1966). *The senses considered as perceptual systems.* Westport, CT: Greenwood Press.

Gibson, J. J. (1979). *The ecological approach to visual perception.* Boston: Houghton Miffin.

Heft, H. (2001). *Ecological psychology in context: James Gibson, Roger Barker, and the legacy of William James's radical empiricism.* Mahwah, NJ: L. Erlbaum.

Hollnagel, E., & Woods, D. D. (2005). *Joint cognitive systems: Foundations of cognitive systems engineering*. Boca Raton, FL: CRC Press.

Kant, V. (2017). Supporting the human life-raft in confronting the juggernaut of technology: Jens Rasmussen, 1961–1986. *Applied Ergonomics*, *59*, 570–580.

Kant, V., & Sudakaran, J. S. (2021). Extending the ecological interface design process – integrated EID. *Human Factors and Ergonomics in Manufacturing & Service Industries*, *32*(1), 102–124. http://doi.org/10.1002/hfm.20933

Kyriakidis, M., Kant, V., Amir, S., & Dang, V. N. (2018). Understanding human performance in sociotechnical systems – steps towards a generic framework. *Safety Science*, *107*, 202–215.

Lombardo, T. J. (1987). *The reciprocity of perceiver and environment: The evolution of James J. Gibson's ecological psychology*. Mahwah, NJ: L. Erlbaum.

Rasmussen, J. (1976). Outlines of a hybrid model of the process plant operator. In T. B. Sheridan & G. Johannsen (Eds.), *Monitoring behavior and supervisory control* (pp. 371–383). Boston, MA: Springer US.

Rasmussen, J. (1986). *Information processing and human–machine interaction: An approach to cognitive engineering*. New York: North-Holland.

Rasmussen, J. (1997). Risk management in a dynamic society: a modelling problem. *Safety science*, *27*(2–3), 183–213.

Rasmussen, J., Pejtersen, A. M., Goodstein, L. P. (1994). *Cognitive systems engineering*. New York: Wiley.

Vicente, K. J. (2002). Ecological interface design: Progress and challenges. *Human Factors*, *44*(1), 62–78. http://doi.org/10.1518/0018720024494829

Vicente, K. J., & Rasmussen, J. (1992). Ecological interface design: Theoretical foundations. *IEEE Transactions on Systems, Man, and Cybernetics*, *22*(4), 589–606. http://doi.org/10.1109/21.156574

9 From the History and Genealogy of Creative Approaches to Hybridization for Stimulating Impactful Innovation

Jérémy Legardeur and Julien Ambrosino

INTRODUCTION

In this chapter, we will show how several creative methods and tools approaches have been developed since 1960, and we will try to elaborate on the genealogy among them and highlight the need to hybridize them:

- Frame collective creative and innovation processes like the double diamond, Design Thinking approach, stage-gate process, CPS (Creative Problem-Solving) approach, etc.
- Redefine the STSs and their ecosystem where different methods/tools could be helpful to "redefine the box" like SWOT matrix, strategic canvas, nine windows of TRIZ.
- Facilitate the people and humans by stimulating the teams, such as six thinking hats, brainstorming rules, SCAMPER method, mind mapping techniques.
- Define innovative solutions and roadmaps by "out of the box" thinking that involves intentionally bringing new concepts and knowledge (C-K theory) to the system or finding new markets (Blue Ocean Strategy, Lean Startup)
- Solve innovative problems by "in the box" thinking within a defined or closed system using existing resources (rather than introducing new ones), seeking to remove obstacles or conflicts between the problem and the solution with methods like TRIZ, or the Advanced Systemic Inventive Thinking, ASIT. CONDITIONS TO ENCOURAGE THE EMERGENCE AND DIFFUSION OF INNOVATIONS

During our various industrial field studies (Legardeur, 2001) concerning Socio-Technical Systems (STSs), we noticed that innovation situations very quickly led to

DOI: 10.1201/9781003666431-11

a certain amount of disruption in the traditional design process and required existing systems to be called into question to accommodate the complex and unstable nature of these situations.

In this context, we were able to identify and characterize three main conditions for innovation:

- First, there can be no innovation with constant organization, i.e., without a certain amount of organizational change and the players' networking. When new ideas are introduced in the early stages of a project, the challenge of innovative design is above all to manage a certain tension that arises between a system for qualifying these new ideas, led by the promoters of relatively innovative proposals, and a system for de-qualifying them, led by the promoters of more routine solutions. So our first condition, which characterizes the dissemination of innovations in their upstream phases, concerns the need to rethink the collective and the various resources differently. We assert that the dynamics of innovation depend on the need to identify, mobilize, and develop new networks of players. These networks, which are the result of processes of interest and enrolment in the sense of the Actor-Network Theory ANT (Latour, 1987) that are often not very formalized, offer stakeholders appropriate structures for encouraging the exchange of ideas and information. The network structure allows ideas, information, and knowledge to be shared flexibly, less prescriptively, and contractually than in traditional company circuits.
- Second, there can be no innovation without constant knowledge and the creation and sharing of new knowledge. Introducing new ideas leads most players to re-examine existing practices and knowledge, and therefore implies the development of new knowledge within the network. We are therefore in a situation where knowledge will be constantly reexamined during action by the arrival of new ideas and, as a result, will contribute to the construction of new knowledge. The challenge is, therefore, to organize the emergence and confrontation of knowledge when new ideas emerge. New ideas are often characterized by a lack of knowledge on the part of stakeholders about certain aspects (technical, economic, etc.), which explains the emergence of tensions that can seriously slow down or stop innovative projects. In this context, designing to incorporate a new idea involves developing new knowledge, which needs to be updated and discussed within the network. It requires real co-development with the players who can provide the missing expertise.
- Third, there can be no innovation without a rethink of evaluation tools and systems. In the case of more routine projects, the network of stakeholders involved and the main criteria used are established and stabilized relatively quickly. On the other hand, in the case of innovative projects, we have seen that this map is very unstable with the gradual arrival of new players, bringing new evaluation criteria at different project stages. The confrontation between the different points of view implies largely informal exchange processes that require new supports and new means of instrumentation to encourage the exchange and evaluation of new ideas.

In this way, the process of adopting or rejecting a new concept is part of a complex socio-technical process that is not solely a technical matter, and which leads to a reopening of the design process and its perimeter: a reopening with the arrival of new practices, new players and new knowledge, a reopening with the arrival of new criteria and new means of evaluation to compare new proposals within the design team.

METHODS AND TOOLS TO STIMULATE CREATIVITY

During an innovative design process, the stakeholders mobilize various methods and tools to help them in their creative activities. According to some authors (Cross et al., 1992; Cross, 1993), they can be grouped into two main categories: creativity methods and rationalization methods. Creativity methods are frequently associated with innovation because they crystallize both a significant specificity and a necessity of the innovation process: the stimulation for the emergence of new ideas and the exploration of new spaces.

In the literature, creativity methods are often described as a starting point for innovation, seeing them as a means of divergence, without being too specific about where they place them in the design process. Some authors define the framework more precisely for using creativity methods in certain phases of a new product design process (Dorst & Cross, 2001). However, historically speaking, if we look at the genesis of creativity techniques, we can see that they were not developed for industrial innovation. Indeed, after the Second World War, Guilford (1959) distinguished between convergent intelligence, as measured by IQ, and divergent intelligence, which he made synonymous with creativity. In the Anglo-Saxon world of the '60s and '70s, the selection of students or the recruitment process for certain positions included a series of creativity tests, designed to measure the divergent intelligence of candidates. It was to provide pedagogical support to satisfy these tests that the first methods were devised, notably by psychologists such as Paul Torrance: these are exercises in the mind, personal or collective, aimed at maximum proliferation.

The literature on creativity methods is increasingly important, with some authors proposing a classification of 172 methods (Ngassa et al., 2003) twenty years ago. Consequently, our aim is not to present an exhaustive state of the art but rather to explain the main principles of these group creativity techniques. We propose to refine our state of the art on this question of creativity, which seems "commonly" inseparable from innovation, but also a promising way of varying the identity of an object. We will analyze different creativity methods, including "collective animation" and the TRIZ and ASIT methods, which focus on finding technical solutions to a given problem.

However, what is creativity? According to many authors, creativity is an individual's ability to create, produce new and feasible ideas, and combine and reorganize elements (Osborn, 1953). Creativity is an intellectual process that aims to provoke as many associations as possible to arrive at a new synthesis and arrangement from which new concepts, unexpected strategies, and innovations will emerge (Roy, 1978). However, this process is challenging to grasp, with the great inventors unable to explain what happened to them when they came up with their invention. The complexity of creativity can be illustrated by the different disciplines that study it: psychology, sociology, management, engineering sciences, philosophy, etc. (Amabile, 1996), but also by the introspection of the human brain that seems necessary. Many authors (Wallas, 1926; Koestler, 1965) have tried to model what goes on in a person's head, from when they are confronted with a problem to when they find a creative solution. Koestler proposes a three-phase model of the creative process:

- A logical phase in which the problem is formulated, data is gathered, and an initial search is made for solutions.
- An intuitive phase that involves distancing oneself from the problem, maturing, and decanting before enlightenment.
- A critical phase in which the discovery or solution is examined, checked, and finalized.

By superimposing this creative process model with that of Wallas, we can see that the intuitive phase can be broken down into two sub-phases: Incubation and Illumination. Incubation aims to bypass "the mental auxiliary that prevents us from overthinking, called fixation" (Horowitz, 1999). Illumination represents the famous "flash of insight" synonymous with the emergence of creative solutions. Creativity methods and techniques are mainly aimed at harnessing this intuitive phase, which characterizes the production of ideas.

From an operational point of view, these idea-generation techniques are generally carried out in groups, as the mechanisms of linguistics, particularly the construction of metaphors, are favored in the context of a group of individuals. The classic approach is to have a multi-disciplinary group of six to eight willing and motivated participants and a facilitator who uses a range of facilitation techniques, such as the six hats (de Bono, 1973) and games, to reduce the distance between participants.

Once the problem has been identified, which must be sufficient to avoid making enormous creative efforts in vain to solve poorly identified problems there is generally an initial, divergent phase of idea generation, followed by a second phase consisting of evaluating and sorting the ideas: a "delay between the generation of ideas and the sorting out" is recommended to allow the necessary distance so as not to "kill off" specific ideas too quickly "out of a protective reflex against change." The creative process itself, the production of ideas, begins with an initial phase, corresponding to Wallas' "illumination," where everything can be said without censorship or self- censorship, followed by a second phase where the numerous ideas produced are refined, classified, and grouped without eliminating any of them or depriving them of their originality and where combinations of ideas can be made. In the end, the ideas are formalized in the form of an idea sheet, but it should be noted that there is generally no mechanism to address how to develop them further when this is outlined.

Regarding these group creativity methods, one thing stands out: the sheer number of existing techniques. There are hundreds of them, each based on a particular feature or cognitive mechanism, and there is very little information about their effectiveness and use in companies. The question of choosing one or more techniques for a creativity session does not seem to be very rational ("Which one or ones should

I choose?"), Moreover, we can therefore think that the effectiveness of a collective creativity session depends mainly on the facilitator, on his or her experience of the techniques to be used and on his or her skills in leading the group, but also on how the session is conducted from a formal point of view.

Our second observation is that the legitimacy of creativity methods themselves, from the point of view of their formalism, has been called into question by specific authors. For example, creativity techniques are presented as recipes, like cooking, without any framework for theoretical reflection that could provide any rigor.

HISTORY OF THE EVOLUTION OF CREATIVITY METHODS

By exposing them to a variety of problem-solving approaches, people's creative thinking improves, and the process becomes more agile, particularly in complex sociotechnical environments (Amabile, 1997). Despite this observation, during the workshops in which we participated, the approaches of many practitioners tended to converge towards using a single method deployed on distinct case studies.

Many scientific studies seek to compare them to one another according to a series of criteria, which are not necessarily harmonized (Boshear & Albrecht, 1977; Davies, 2011; Eppler, 2006; McFadzean, 1998a, 1998b, 1999, 2001; Shah et al., 2000). Given this abundant literature, which encourages the use of different methods by considering each of them as competing elements, we do not seek to introduce a new frame of reference for comparison between methods.

The proposal of a theory of the evolution of creativity methods is a postulate, like any scientific theory, which we attempt to explain here. Developing a theory of the evolution of creativity methods, tools, and techniques would offer several advantages.

Practitioners and participants will be able to facilitate workshops more effectively. By providing them with more tools, they can draw on complementary methods, tools, and techniques better adapted to their contexts. In this way, they can revisit and adapt some of them to make them more effective.

For researchers, understanding, following, and even anticipating the approaches taken by the creators of these methods is crucial.

Through a series of examples designed to support this point of view, we introduce the three evolutionary strategies of creativity methods by analogy with the laws of evolutionary theory (Darwin, 1859): mutation, selection, and hybridization. We then introduce a representation highlighting the links of kinship between the methods to observe the need for the methods to evolve.

Finally, the last logic of hybridization, explicitly introduced by the proposals of Legardeur (2009), has particularly focused our attention. To this end, we will describe a novel example of combining two creativity tools.

MUTATION AND SELECTION: TWO FUNDAMENTAL STRATEGIES IN THE EVOLUTION OF METHODS

Many of the existing creativity methods result from developments that have optimized the efficiency of the approaches in given contexts. We postulate that these fundamental developments can follow two paths.

The first strategy consists of improving an existing method by modifying its approach to better adapt it to the case under study. This is known as a mutation.

The second strategy consists of selecting one or another approach that corresponds better to the case under study. This is known as selection.

To illustrate these proposals, several examples of mutation and selection are presented below.

Successive Examples of Mutations

Let us take the example of the first form of mutation of creativity methods. In addition to setting out and relying on the four rules of creativity, the Osborn (1953, p. 271) and his pupil Clark (1962, p. 76) both drew on their experience of animation to develop other practices, such as the use of elementary mathematical operations which consist of adding parts, deleting others, etc. These operations are designed to see through the creative process. These operations aim to see the problem from another angle. With this principle of animation introduced, a few years later, the SCAMPER method was born and formalized (Eberle, 1971). Part of brainstorming as a creativity method had thus been transformed into a new method.

Another example illustrates the phenomenon of mutation of methods linked to the early work of Alex Osborn. He quotes John Dewey, a psychologist and philosopher, who states the importance of observing new and old facts in the creative process (Osborn, 1953, p. 130). Without a formal link being established, this logic of observing past elements to provide new avenues for imagining the future is taken up again, notably in the famous TRIZ theory (Theory of Inventive Problem Solving) (Altshuller & Williams, 1984). Its creator, the Russian Genrich Altshuller, uses patent databases to set out physical principles for imagining new inventions. In TRIZ, this process of observing past elements has even led to a dedicated graphical representation tool: the "9 screens" (or "9 windows"), where the entire left-hand side of the tool represents the past states of the systems under consideration. However, TRIZ's systematic approach allows other concepts to be encompassed (such as evolutionary trends) and thus constitutes a new mutation.

Numerous methods have thus mutated or derived from TRIZ, seeking to simplify the theory. For example, the work by Genadi Filkivski and then theorized by Roni Horowitz is part of this dynamic. The SIT method was developed as part of a doctoral thesis (Horowitz, 1999) and was, in turn, simplified once again to create ASIT, a new method consisting of using a series of elementary operators, such as deletion, division, etc., to imagine a series of new solutions. These successive mutations have enabled the creation of new adaptations of ASIT for practitioners like EcoASIT (Tyl et al., 2009, 2010, 2013).

Other methods and techniques could be cited as examples of mutation, such as the Synectics method (Gordon, 1961), which, in the case of the personal analogy mechanism, seems to have mutated. Personal analogy consists of putting oneself in someone else's shoes and empathizing with them. It could be linked to the persona method developed by Jung et al. (1964) and then taken by many Design Thinking practitioners. This mechanism of personal analogy is effective; in groups constrained in this way, it enables them to produce more ideas than groups in conventional conditions.

Examples of Selections

We want to illustrate the comparison between the methods mentioned above: ASIT and SCAMPER, and then between brainwriting and brainstorming.

About the first two, the practitioner may find strong similarities between the operators of SCAMPER and ASIT and thus face the dilemma of choosing between one or the other of the methods. Roni Horowitz responds openly by stating that the ASIT method is more cognitively demanding for the user. To justify this choice, he relies on a significant difference based on the fundamental principle of ASIT, which is to work in a closed world. In other words, the new product imagined must not include any object not present in the immediate environment of the original product. In this way, the practitioner can voluntarily choose between one or the other methods according to his or her needs. ASIT can thus be used to think "inside the box" without using any means external to the immediate environment. Conversely, SCAMPER can be used to think "outside the box" by allowing more freedom thanks to a less constrained cognitive framework that may tend to limit creativity.

As for the next two, formulating proposals in writing so that they can then be shared with the rest of the group certainly has its roots in the writings of Preyer (1884) on the mental development of children. It was only later that the brainwriting technique came to replace brainstorming in idea-generation sessions. Some studies (Madsen & Finger, 1978) have argued that it is superior to brainstorming, even outperforming it by a factor of 4. There are many ways of using paper as a medium for generating ideas. Some of these methods and techniques involve writing down proposals individually before giving them to a nearby participant, such as the 6-3-5 method (Rohrbach, 1969), or another participant at random, such as brainwriting pooling (Geschka et al., 1976). In addition to the action of ideation, other derived methods seek to reach a collective consensus, which is obtained by the successive integration of individual ideas, such as the Successive Integration of Problem Elements method (Warfield et al., 1975), or taking into account the selection stage, such as the Nominal Group Technique (NGT) (Taylor et al., 1958; Taylor & Faust, 1952; Van De Ven & Delbecq, 1974). We should also mention methods based on writing on paper, allowing the solution to be drawn rather than described, such as the 5-1-4 G method, renamed C-sketch (collaborative sketching) (Shah et al., 2001).

Hybridization of Methods

The last hybridization strategy we are introducing does not consider the different methods, tools, and techniques of creativity as competing entities, but, on the contrary, to be brought into synergy. This strategy goes beyond that of mutation because it implies the use of at least two methods, and it goes beyond that of selection as it does not simply consist of choosing one of them according to this or that criterion.

In one of the first modern creativity methods (if not the first), the brainstorming approach (Osborn, 1953) insists on the flexibility of the facilitation process through numerous practices to stimulate creativity among its participants. These go far beyond today's four rules commonly used in group creativity sessions. His recommendations on the various stages of the session, from preparation to feedback, including facilitation techniques, are numerous and form the basis of many more modern approaches.

This logic, defended by Alex Osborn, is not limited to considering creativity methods as unitary elements to be chosen for a given case study. However, the wealth of work carried out over more than half a century implies retaining several to deal more effectively with the same case study. This is the hybridization of methods (Legardeur, 2009). Our analysis of the methods used by these authors indicates that their complementarities provide fertile ground for their hybridizations, just like DNA hybridization.

We argue here that hybridization of methods can occur in two distinct ways. Either simultaneously, i.e., the methods are used simultaneously to enhance the creative process, or sequentially, i.e., the methods are used one after the other, on condition that the results of the first are exploited by the second.

In the case of simultaneous hybridization, we hybridized brainwriting and the discovery matrix simultaneously (Ambrosino et al., 2016). In our practice of idea generation with participants, given its performance in groups, because it limits procedural blocking mechanisms, we use brainwriting in different experimental protocols to solicit participants to generate ideas individually. What is special is that we use it to generate ideas for a single problem and around several sub-problems to better structure the overall problem. To increase the level of cognitive constraint, these problems are confronted with ways of responding to them in a discovery matrix (Moles, 1954).

GENETIC HERITAGE OF SOME CREATIVITY METHODS

Figure 9.1 shows the evolution of some of the creativity methods that have particularly interested us in this work. Each of the three strategies mentioned above is exemplified, and their modeling could enable us to imagine new methods.

On the one hand, the mutation strategy could multiply many existing methods with strong development potential.

On the other hand, selection and hybridization strategies could then be based more on historical elements. In other words, having already identified that a given method may have already been selected in each context, it would seem more straightforward to make the same choice if that context arises again. The analysis can be more refined about the hybridization strategy: to multiply approaches and create highly original hybridized methods, it seems wiser to hybridize methods with more distant kinships.

The family tree representation highlights stages of change to visualize the evolution of methods dynamically. It also illustrates:

- A clearer vision of the changes caused or undergone, answering questions such as: what are the links between approaches, and why have new approaches had to be adapted?

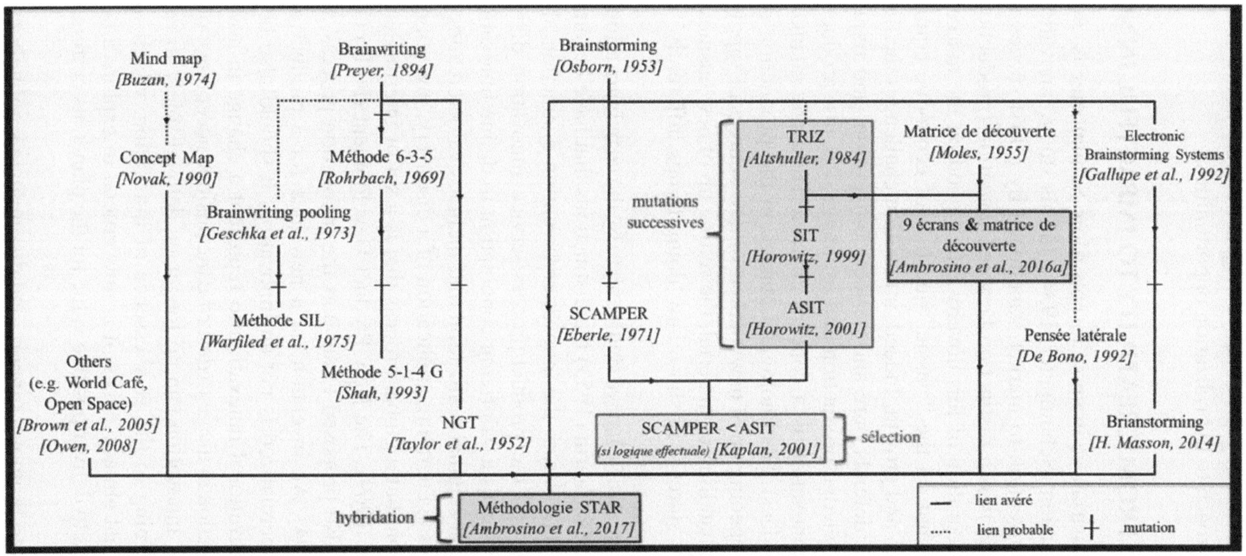

FIGURE 9.1 Genetics of some creativity methods (Ambrosino, 2018).

- A more unified and objective representation, to be able to make a better quality selection between methods by integrating the history of the criteria for these mutations,
- A distribution tracing the distance of genetic links would enable a more systematic approach to hybridization opportunities.

CONCLUSION: FROM CREATIVITY TO IMPACTFUL INNOVATION

Historically, the Austrian economist Joseph Aloïs Schumpeter is credited with the first study of innovation (Schumpeter, 1934). In this study, innovation is defined as the passage from invention to market, sanctioned by a first successful commercial transaction. There is thus a distinction between invention (production of a novelty) and innovation (integration of this novelty into a good on the market). In this way, the author focuses not on the "classic" question of economic equilibrium, but on economic dynamics and growth. Indeed, innovation challenges the theory of general equilibrium, which defends pure and perfect competition, and requires us to consider forms of knowledge production and exchange, not just products and services. For Schumpeter, innovation is a concept characterizing any change in the supply of products or services: "Technological change in the production of commodities already in use, the opening up of new markets or new sources of supply, Taylorisation of work, improved handling of material, the setting up of new business organizations such as department stores – in short, any 'doing things differently' in the realm of economic life

all these are instances of what we shall refer to by the term Innovation" (Schumpeter, 1939). We owe to MacLaurin (1953) the distinctions that have become classic today for characterizing innovation: the innovation/invention distinction, the incremental/radical distinction, and the product/process/service innovation distinctions. Later, Knight (1967) introduces a valorizing connotation to define innovation as a process whose outcome is an original realization that includes value-creating attributes. It is the original, progress-generating application of a discovery, invention, or concept. It is not a single action, but a global process that is a set of interrelated sub-processes (Myers & Marquis, 1969). The New Product Development (NPD) (Wheelwright & Clark, 1992) was also proposed to support the innovation process, like the " stage gate" (Cooper, 1994). An invention, or a new idea for a product or process, that has not been commercialized is not an innovation. It should also be noted that we can speak of organizational innovation to represent a change in how a company is organized, or innovation in usage to describe the positive acceptance of new practices.

However, today, innovation can no longer be defined solely in terms of its economic value. Returning to the very essence of innovation, which must constantly seek to improve what already exists, it is now crucial to explore a new approach to innovation that aims to deliver positive value for the planet, its current occupants, and future generations.

The aim here is to promote, support, and disseminate resolutely creative and disruptive ideas, to bring about exceptional innovations that can now be described as "neutral or positive impact," i.e., capable of reducing or even reversing environmental and social impacts.

These "impact innovations" will henceforth be defined as "new services, products, or processes that will find economically viable markets to ensure the sustainability of their uses, while neutralizing and making positive social and environmental impacts." To achieve this, we need to mobilize and combine all the creative minds in the world, especially those who aspire to bring about this new paradigm. That is why training new generations to hybridize all forms of creativity to invent and overturn today's models is so important. Tomorrow, the current and next generations must, more than ever, become experts in collective intelligence, which means learning to work in the unknown, the undefined, to detect weak signals, to take an interest in singularities, to work outside one's comfort zone in exploring areas of "non-art" and their uncertainties.

REFERENCES

Altshuller, G. & Williams, A. (1984). Creativity as an exact science: The theory of solving inventive problems. *Journal of Development Economics*, 91(1), 64–76.

Amabile, T. M. (1996). *Creativity in Context*. Westview Press, Harvard.

Amabile, T. M. (1997). Motivating creativity in organizations: Doing what you love and loving what you do. *California Management Review*, 40(1), 39–58. https://doi.org/10.2307/41165921

Ambrosino, J. (2018). *Apports de l'hybridation de méthodes de créativité pour l'émergence de projets collaboratifs d'innovation dans les pôles et clusters: proposition de la méthodologie de facilitation d'ateliers STAR et de l'outil de brainstorming électronique IdeaValuation*. Thesis. University of Bordeaux, France.

Ambrosino, J., Legardeur, J. & Lattes, P. (2016). An example of hybridization between the "discovering matrix" tools during the ideation phases of interclustering projects. In *The Design Conference Proceedings. 14th International Design Conference*, Faculty of Mechanical Engineering and Naval Architecture, University of Zagreb; The Design Society, Glasgow, Dubrovnik, Croatia. 897–906.

Boshear, W. C. & Albrecht, K. (1977). *Understanding People: Models and Concepts*. University Associates. ISBN 0883901153.

Clark, C. C. (1962). *Brainstorming*. Dunod, Paris.

Cooper, R. G. (1994). Third generation new product processes. *Journal of Product Innovation Management*, 11(1), 3–14. https://doi.org/10.1111/1540-5885.1110003

Cross, N. (1993). Science and design methodology. *Research in Engineering Design*, 5, 63–69.

Cross, N., Dorst, K. & Roozenburg, N. (1992). *Research in Design Thinking*. Delft University Press, NL.

Darwin, C. (1859). *On the Origin of Species by Means of Natural Selection*. John Murray, London.

Davies, M. (2011). Concept mapping, mind mapping, and argument mapping: What are the differences, and do they matter? *Higher Education*, 62(3), 279–301. https://doi.org/10.1007/s10734-010-9387-6

De Bono, E. (1973). *Lateral Thinking: Creativity Step by Step*. Harper & Row.

Dorst K., Cross N. (2001). Creativity in the design process. Co-evaluation of problem solution, Design Studies, 22(5), 425–437.

Eberle, R. F. (1971). *Scamper: Games for Imagination Development*. D.O.K. Publishers. Retrieved from https://books.google.fr/books?id=fPmQKQAACAAJ

Eppler, M. J. (2006). A comparison between concept maps, mind maps, conceptual diagrams, and visual metaphors as complementary tools for knowledge construction and sharing. *Information Visualization*, 5(3), 202–210. https://doi.org/10.1057/palgrave.ivs.9500131

Geschka, H., Schaude, G. R. & Schlicksupp, H. (1976). Modern techniques for solving problems. *International Studies of Management & Organization*, 6(4), 45–63. https://doi.org/10.1080/00208825.1976.11656211

Gordon, W. J. J. (1961). *Synectics: The Development of Creative Capacity*. Harper & Brothers Publishers. ISBN-13: 978-0060324308.

Guilford, J. P. (1959). Traits of creativity. In Anderson, H. H., Ed., *Creativity and Its Cultivation*. Harper & Row, New York, pp. 142–161.

Horowitz, R. (1999). *Creative problem solving in engineering design*. Thesis. Tel-Aviv University, Israel.

Jung, C. G., Cahen, R. & Cahen, R. (1964). *Dialectique du moi et de l'inconscient*. Gallimard, France.

Knight, K. (1967). A descriptive model of the intra-firm innovation process. *The Journal of Business*, 40(4), 478–496.

Koestler, A. (1965). *Le cri d'Archimède*. Calman-Lévy.

Latour, B. (1987). *Science in Action: How to Follow Scientists and Engineers through Society*. Open University Press, Milton Keynes.

Legardeur, J. (2001). *Méthodes et outils pour l'innovation produit/process*. PhD Thesis. INP Grenoble.

Legardeur, J. (2009). *Le management des idées en conception innovante: pour une hybridation des outils d'aide aux développements créatifs*. HDR, Bordeaux University, France.

Maclaurin, W. R. (1953). The sequence from invention to innovation and its relation to economic growth. *The Quarterly Journal of Economics*, 67(1), 97–111. https://doi.org/10.2307/1884150

Madsen, D. B. & Finger, J. R. (1978). Comparison of a written feedback procedure, group brainstorming, and individual brainstorming. *Journal of Applied Psychology*, 63(1), 120.

McFadzean, E. (1998a). Enhancing creative thinking within organizations. *Management Decision*,36(5), 309–315. https://doi.org/10.1108/00251749810220513

McFadzean, E. (1998b). The creativity continuum: Towards a classification of creative problem-solving techniques. *Creativity & Innovation Management*, 7(3), 131–139. https://doi.org/10.1111/1467-8691.00101

McFadzean, E. (1999). Creativity in MS/OR: Choosing the appropriate technique. *Interfaces*, 29(5), 110–122. https://doi.org/10.1287/inte.29.5.110

McFadzean, E. (2001). Critical factors for enhancing creativity. *Strategic Change*, 10(5), 267–283. https://doi.org/10.1002/jsc.542

Moles, A. (1954). *La création scientifique (Kister)*. René Kister editor, Genève, p. 237.

Myers, S. & Marquis, D. G. (1969). *Successful Industrial Innovation*. National Science Foundation, Washington.

Ngassa, A., Bigand, M. & Yim, P. (2003). A new approach for the generation of innovative concepts for product design. Proceedings of ICED 03, *Stockholm, Août*.

Osborn, A. (1953). *Applied Imagination: Principles and Procedures of Creative Thinking*. Scribner, New York.

Preyer, W. (1884). *Éléments de physiologie générale*. Alcan.

Rohrbach, B. (1969). Kreativ nach Regen-Methode 635, eine neue Technik zum Losung von Problemen. *Absatz-Wirtschaft*, 12, 73–75.

Roy, M. (1978). Processus de la Créativité. *Canadian Journal of Education*, 3(1), 61–74. https://doi.org/10.2307/1494512

Schumpeter, J. A. (1934). *The Theory of Economic Development*. Harvard University Press, Cambridge, MA.

Schumpeter, J. A. (1939). *Business Cycles: A Theoretical, Historical and Statistical Analysis of the Capitalist Process*. New York, Toronto, London McGraw-Hill Book Company. P. 461.

Shah, J. J., Kulkarni, S. V. & Vargas-Hernandez, N. (2000). Evaluation of idea generation methods for conceptual design: Effectiveness metrics and design of experiments. *Journal of Mechanical Design*, 122(4), 377. https://doi.org/10.1115/1.1315592

Shah, J. J., Vargas-Hernandez, N. O. E., Summers, J. D. & Kulkarni, S. (2001). Collaborative sketching (C-sketch) – an idea generation technique for engineering design. *The Journal of Creative Behavior*, 35(3), 168–198. https://doi.org/10.1002/j.2162-6057.2001.tb01045.x

Taylor, D. W., Berry, P. C. & Block, C. H. (1958). Does group participation when using brainstorming facilitate or inhibit creative thinking? *Administrative Science Quarterly*, 3(1), 23. https://doi.org/10.2307/2390603

Taylor, D. W. & Faust, W. L. (1952). Twenty questions: Efficiency in problem solving as a function of the group size. *Journal of Experimental Psychology*, 44(5), 360–368. https://doi.org/10.1037/h0054376

Torrance, E. P. (1966). *Torrance Tests of Creative Thinking: Norms-Technical Manual. Research Edition. Verbal Tests, Forms A and B. Figural Tests, Forms A and B*. Personnel Press. Retrieved from https://books.google.fr/books?id=_4dUYAAACAAJ

Tyl, B., Legardeur, J., Millet, D., Falchi, A. & Ranvier, B. (2009). A new approach in the development of a creative method to stimulate sustainable innovation, *International Conference on Green and Sustainable Innovation, December 2–4, Chiang Rai*, pp. 754–759.

Tyl, B., Legardeur, J., Millet, D. & Vallet, F. (2010). *Stimulate creative ideas generation for eco-innovation: An experimentation to compare eco-design and creativity tools*. IDMME-Virtual Concept 2010, Bordeaux, France.

Tyl, B., Legardeur, J., Millet, D. & Vallet, F. (2013). Adaptation of the creativity tool ASIT to support eco-ideation phases. *Sustainable Intelligent Manufacturing International Conference, June, Lisbon, Portugal*. CRC Press, pp. 437–443.

Van De Ven, A. H. & Delbecq, A. L. (1974). Effectiveness of and decision nominal, processes' Delphi. *The Academy of Management Journal*, 17(4), 605–621.

Wallas, G. (1926). *The art of thought*. Harcourt Brace, New York.

Warfield, J. N., Geschka, H. & Hamilton, R. (1975). *Methods of Idea Management, Approaches to Problem Solving, No. 4*. Academy for Contemporary Problems, Columbus, August.

Wheelwright, S. C. & Clark, K. (1992). *Revolutionizing Product Development. Revolutionizing Product Development: Quantum Leaps in Speed, Efficiency, and Quality*. The Free Press. ISBN: 0029055156, 9780029055151.

10 Human–Systems Dialectics and Exploration

Balancing System Design Thinking for the Agile and Participatory Integration of Sociotechnical Systems

Michael Preutenborbeck and Frank Ole Flemisch

INTRODUCTION: WHY DIALECTICS AND EXPLORATION FOR HUMAN–SYSTEMS INTEGRATION?

In theory and practice, Human–Systems Integration (HSI) is a science, art, and craft to integrate humans, technology, organization, and the environment into sound sociotechnical systems, which is under a constant tension between the pushes of scientific and technological innovation, and the pull of individual, organizational, or societal requirements. An increasing acceleration in organizations and societies, described, e.g., by Hollnagel and Woods (2005) or Rosa (2015), is being answered by HSI with a growing demand for agility in the design, engineering, and testing of sociotechnical systems (STSs).

In practice, an example of this increasing agility is the introduction of teleoperated cars into traffic systems or artificial intelligence (AI) into office software and thereby into organizational systems. Another example from the defense domain is the agile introduction of drones in the battlefield, as observed in the wars of Nagorno-Karabakh, Ukraine, and the war in Israel and Palestine.

How do we make HSI, in theory and practice, fit for these pushes of innovation and pulls of demands, and be agile enough for the needs of the 21st century?

HSI addresses all steps of the systems life cycle. However, it can have a remarkable impact when applied initially, i.e., in the ideation and concept exploration phase. On the other hand, the knowledge about the impact is still low in the early stages of the system's life cycle (Collingridge, 1982). This Collingridge dilemma can be systematically addressed by dialectically balancing a system's ideation and system exploration between a more constructive and critical perspective (Figure 10.1).

Increasing speed and agility in the interplay of theory and practice does not mean that theory becomes less important; the contrary is true: When practice pushes even

DOI: 10.1201/9781003666431-12

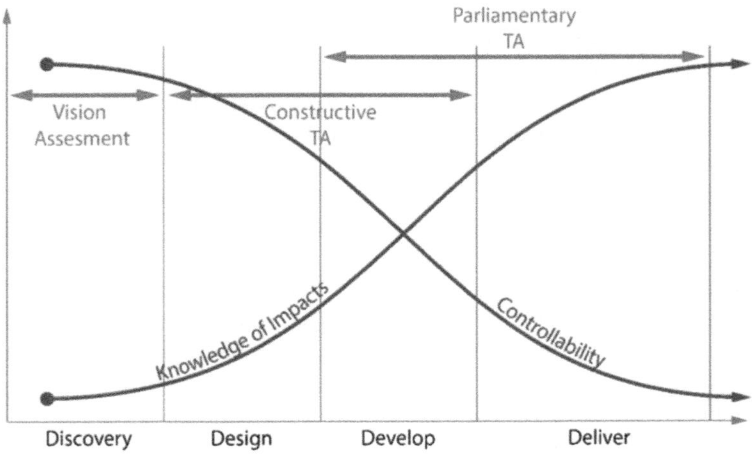

FIGURE 10.1 Collingridge dilemma for new socio-technical systems and its connections to vision assessment, constructive technology assessment, and parliamentary technology assessment (Flemisch et al., 2022).

more and faster toward innovative systems, we need even more or better theory to counterbalance this tension field. With increasing speed and agility and an increasing number of stakeholders in an increasingly interconnected world with often contradicting opinions and divergent tension fields, the art and craft of HSI must be complemented with a science of balancing and integrating those tension fields efficiently into well-balanced systems. We see the philosophical stream of Dialectics *(Greek: διαλεκτική, dialektikḗ)* as a promising base to develop a Human Systems Dialectics, from which an innovative Human Systems Exploration (HSE) and with that a more balanced HSI is more likely.

WHY AND HOW PHILOSOPHY CAN HELP US TO DO A BETTER HSI: TOWARD A HUMAN–SYSTEMS DIALECTICS

What is special about HSI, compared to e.g., material science, mechanical engineering, psychology, organizational, or environmental science? We are convinced that this is the necessity and ability to bridge, balance, and integrate a variety of even contradicting perspectives into a meaningful whole, from which we can derive meaningful and concrete action. Besides system theory, which can support us with this special challenge?

Applied science, like engineering or HSI, has always profited from an interplay of philosophy, natural science, and liberal arts. This interplay is constantly threatened by the centrifugal forces of specialized science, in which specialized disciplines try to establish their social systems in the sense of (Luhmann, 1984), by declaring right from wrong thinking and methods, and with that differentiating them from other disciplines. On the other hand, HSI already has a long history of working against these centrifugal forces. Still, with the increasing acceleration of technological advances, HSI will need even stronger allies in theory.

We found a strong theoretical ally in a philosophical stream called Dialectics. The word dialectics roots back into Greek διαλεκτική, dialektikḗ, e.g., by Greek philosophers like Heraklit and Platon, and means literally "the art of dialogue." Dialectic thinking has been intensely described by Hegel (1845), traces of this can be found already in Kant (1781), and later in Fichte and Schelling, Marx and Engels, and Horkheimer and Adorno (2004). Dialectic approaches can also be found, e.g., (Reckwitz, 2021).

At the heart of the concept lies the challenge of handling two or more contradicting statements about the world. According to Hegel, several ways of handling these contradictory statements were later called the thesis and antithesis. The traditional way is to clarify what is true or false and abandon the false statement (e.g., Popper, 1968). In many cases, a more sophisticated way is possible: Thesis and anti-thesis can be brought together into a new statement, "synthesis" about the world, which is much more powerful. Hegel uses the German word "aufheben," which can have several meanings: (1) to cancel, (2) to contain and preserve, and (3) to lift. In the synthesis, all three meanings of "aufheben" can come together: The synthesis cancels the opposition between the thesis and the anti-thesis. Both thesis and anti-thesis are contained in the synthesis, which is "higher" because it is more powerful in describing the world. Hegel was convinced that the dialectic principle is one of the most important principles for progress in philosophy and science.

Two hundred years after Hegel, we interpret Hegel's dialectic principles as shown in Figure 10.2: Starting from spanning or diverging into two different statements about the world T1 and T2, these can be balanced and synthesized into a synthesis T1T2. It can also be extended toward other statements, e.g., T3, and balancing and synthesizing iteratively applied operations. It is important to state that operations like diverging, balancing, synthesizing, and extending are cognitive and/or social operations of one or several agents, like humans or an AI, leading to different cognitive states. It is also important to note that dialectic thinking starts with expanding from a unified or unclear view into divergent thinking, before it can use the tension to lift towards a synthesis.

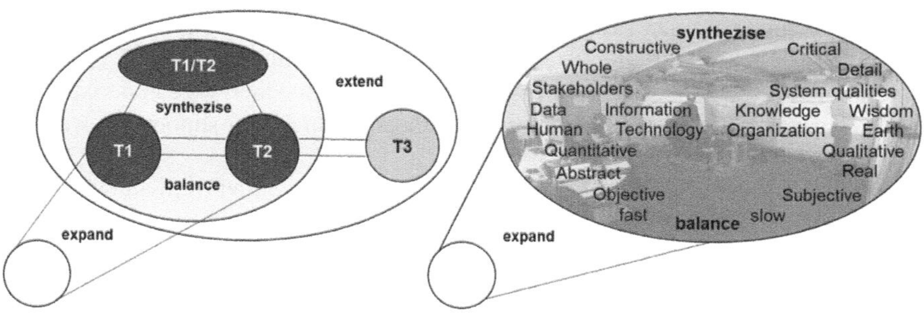

FIGURE 10.2 Left: Basic principles and operations of dialectic thinking and acting (extended from Hegel, 1845). Right: Examples for tension fields in Human–Systems Integration.

In everyday language, "tension field" describes a psychological tension between people, organizations, or more generally between systems and/or subsystems. To connect this to theory, we see a proper connection between dialectics and Kurt Levine's field theory. We understand tension fields as psychological fields in the sense of Kurt Levine (1951), who describes psychological forces (or tensions), e.g., between people and the environment as driving forces for thinking and acting (Lewin, 1951). These tensions can be caused by different perceptions and statements of different cognitive agents about the world, which can be seen as thesis and antithesis in the most essential form. Dialectics is therefore about handling tension fields in a way that leads to "aufheben," to an uplifting of the solutions, to containing different ideas, and finally cancelling the tension (Figure 10.2).

Applied to HSI, the following opposing theses create tension fields and can describe fundamental challenges of the art, craft, and science of HSI:

- Different perspectives on **humans and technology** were the initial motivation for HSI. Early calls for user-centered design (Norman, 1986; Cooley, 1989) evolved into integrative approaches balancing views on humans, technologies, organizations, and environments (Booher, 2003; Flemisch, 2003; Flemisch et al., 2008).
- **Constructive and critical** refer to contrasting mindsets: solution-oriented vs. problem-oriented thinking. HSI benefits from combining both in a dialectic process where better solutions evolve through addressing their limitations.
- *Stakeholders and system qualities* reflect conflicting priorities. Different stakeholders emphasize different system qualities, requiring HSI to find balanced compromises.
- **Data, information, knowledge, sense, and wisdom** describe a layered view of information processing. Turning data into meaningful insight is central to human and organizational decision-making (Elliot, 1934; Weick et al., 2005).
- Qualitative vs. quantitative perspectives have long been debated. It is now widely accepted that combining both (mixed methods) leads to more robust results in complex systems (Reichardt & Rallis, 1994).
- Objective vs. subjective perspectives highlight the interplay between stakeholder viewpoints and attempts to approximate objective truths. While constructivist approaches (Förster) question objectivity, pragmatic integration is possible (Molter, 2017).
- Fast vs. slow refers to the growing gap between human and technological reaction speeds. Technology-driven acceleration (Kahneman, 2013; Rosa, 2015) can cause harmful side effects such as flash crashes or autonomous conflict escalation.

In summary, human systems dialectics can structure the complex challenge of HSI into tension fields with opposing or diverging poles, which can be balanced and synthesized into solutions. Expanding Levins' field theory metaphorically from psychological systems to socio-technical systems, these poles or dimensions can form complex fields or spaces which contain a vast number of options, impossible to determine, e.g., with experiments alone, but necessary and possible to explore and navigate to at least some reasonable, well-balanced solutions.

BALANCING TENSION FIELDS AND NAVIGATING THE SPACES FROM GOOD AND BAD IDEAS TO SOUND, HUMANE SYSTEMS: HSE

HSE is a participatory design methodology that attempts to resolve tension fields and conflicts inherent in the design process under the leitmotif of balance. As described above, Hegel's dialectical concept of tension fields forms the methodological foundation. HSE offers a framework for analyzing and designing socio-technical systems that integrate humans, technology, organizations, and the environment. The interplay between the use, design, and value spaces is at the heart of HSE. The use space focuses on the context of system usage, capturing the interactions between users, other stakeholders, and technology. Solution concepts are generated, expanded, and refined in the design space. The value space integrates stakeholder values and goals to evaluate system qualities systematically. Based on Flemisch et al. (2022), a definition of these fundamental terms is provided:

- *Use Space:* The term *use space* is a spatial metaphor that describes the context of use, use cases, scenarios, and stakeholders of a socio-technical system. It systematically analyzes how a system can be used in varying contexts. The use space aims to develop an understanding of the interactions between users and technologies to optimally align the system design with the actual needs and requirements of the user in varying use situations. Use space correlates to what Guy Boy calls context (Boy, 2020). Boy describes context as a multi-layered construct characterized by various dimensions such as time, situation, behavior, perspective, relationships between actors, discourse, and dialogue. He divides context into verbal, social, and historical context. He emphasizes the relevance of extensional and intentional meanings of words and the necessity of integrating historical contexts when defining use situations.
- *Design Space:* The term *design space* is a spatial metaphor that describes the variety of direct design dimensions of a system, e.g., colors, shapes, or interaction design, as well as indirect dimensions like user education and training. The design space is used to systematically analyze various design options to develop creative solutions. The design space helps to structure the complexity of design decisions and provides an overview of all relevant design dimensions.
- *Value Space:* V*alue space* is a spatial metaphor that describes integrating the values, goals, and needs of all relevant stakeholders within a socio-technical system. It serves as a foundation for systematically capturing, analyzing, and evaluating system qualities and requirements from a human or technological perspective. The value space allows for identifying, understanding, prioritizing, and balancing manifold and potentially conflicting values and requirements.

The use, design, and value space intersect to form a cohesive system, where design considerations in the design space meet user needs in the use space and are assessed through a value-based lens. This intersection fosters a participatory design process, accommodating the complexity of STSs. By iteratively defining and evaluating these spaces, HSE facilitates the design and development of innovative human–machine systems, ensuring their relevance and value in both local and global contexts. HSE closes the gap between creative design methods, such as design science research and design thinking, and development that is more conventional approaches from systems and human factors engineering.

The description of use, design, and value space integrates concepts from systems engineering, e.g., systems thinking as part of the problem-solving cycle, for instance, Züst (2004), Lindemann (2009), Haberfellner et al. (2019), with the orchestration of problem and solution spaces from design thinking, e.g., Vom Brocke et al. (2020). HSE thus addresses a classic challenge, that the problem and solution space are initially two separate areas that designers and engineers need to connect to create practical solutions (Gerritsen, 2019). Navigating the problem space, which involves understanding the situation and identifying what makes it problematic, can often be more challenging than exploring the solution space. Gerritsen suggests that both spaces are integral to the design and can be bridged by exploring STSs from two perspectives. The first perspective involves moving from problem to solution, which includes evaluating existing solutions and imagining better alternatives. The second perspective involves moving from the solution to the problem, which focuses on identifying innovative combinations of people, technology, organization, and environment to address problematic situations proactively.

HSE's '*exploration*' aspect metaphorically represents the journey through known and unknown territory in system design, similar to historical explorations such as those by Lewis and Clark of the western USA, or the exploration of space (NASA, 2013). According to Flemisch et al. (2022), the HSE metaphor assumes that an exploration typically involves:

- a start and an end (and can therefore be organized as a project or part of projects),
- a higher openness and non-determinism compared to deterministic procedures (e.g., compared to an experiment),
- a higher level of uncertainty and risk compared to well-known procedures,
- a higher chance to discover something new, direction, mandate, and/or some restrictions or boundaries.

Figure 10.3 illustrates a meta-model of HSE, which provides a multi-layered approach to understanding explorative system design. From general and abstract theories at the top to concrete and individual applications at the bottom, the model encapsulates the progress from general concepts on the left to detailed, real-world implementations on the right.

At the center is the local human–machine system, embedded in a system of systems and the global environment. The use and design space is the solution space

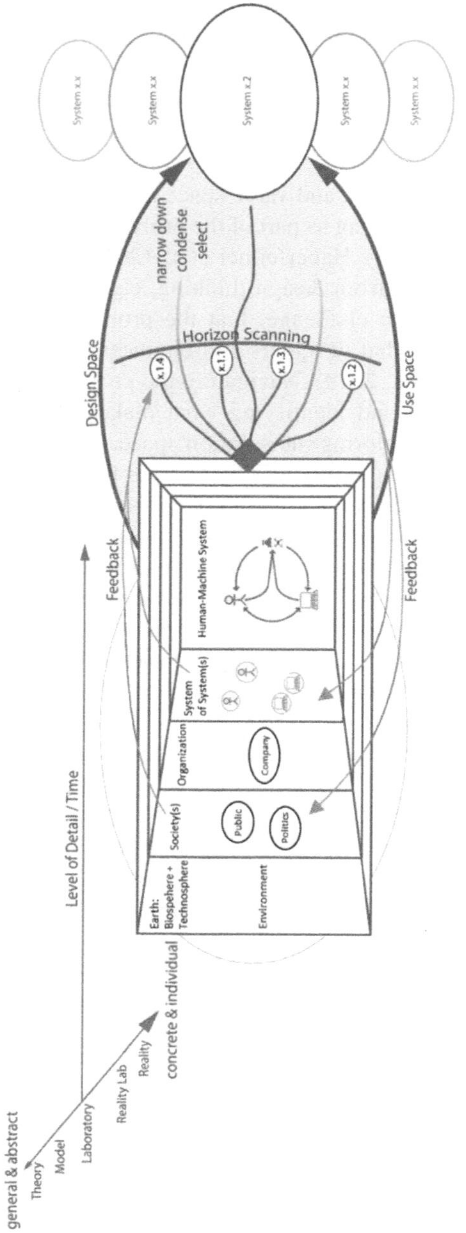

FIGURE 10.3 Holistic model of human–systems ideation and exploration. (Based on Flemisch et al., 2022; Flemisch et al., in press.)

in which ideas are opened, expanded, narrowed down, condensed, and finally selected in an explorative process. The explorative nature of the model is represented by red design paths, where promising paths serve as the basis for further iterations (System x.2), and less promising paths are discarded. This evaluation is part of the value space and allows feedback information to be fed into the system, enabling continuous adaptation and improvement. The meta-model shows the dynamics between theory and practice and provides a structured approach to analyzing and designing systems in different contexts and at different levels of complexity. New system concepts can be designed in a participatory process by superposing the use, design, and value space and involving all relevant stakeholders. This approach addresses the high complexity of socio-technical systems and systems of systems characterized by numerous variables and diverging stakeholder interests. It counters this complexity by integrating stakeholders into the design and development process, thus fostering creativity and stakeholder-driven innovation.

HSE: PROCESS MODEL AND METHODS

HSE is a paradigm, a set of methods, and most of all, a set of processes. Process models are structured approaches that organize the sequence of processes (Versteegen, 2002). The description of process models has its scientific origins in software development but is now used in many other domains, especially in project management. They specify the activities and tasks to be carried out, the sequence in which they are to be carried out, and the products (artifacts) to be produced in this process (Versteegen, 2002). Fundamental process models from literature are, e.g.:

- Waterfall model, e.g., Royce (1970),
- V-model, e.g., Boehm (1979), and
- Spiral model, e.g., Boehm (1988).

Following the argumentation that HSE can also be organized as a project, a process model is presented below, which intends to organize the phases and activities of HSE. As shown in Figure 10.4, the exploration process can be structured into several consecutive phases. This model is a combination of the double diamond model (British Design Council, 2004) and the innovation turbine (Flemisch et al., 2022), integrating older concepts of a spiral and iterative development process (e.g., Boehm, 1979, 1988). As an extension of the British Design Council's double diamond model, the underlying metaphorical inspiration is a Mach turbine (Flemisch et al., 2022). The thrust begins with a small initial spark, followed by expansion and subsequent compression, with peak pressures at the Mach points (diamonds). The process then develops into a more extensive cycle of expansion and compression, leading to increasing thrust and propulsion. According to the double diamond, the first step expands possibilities and the search for new solutions (divergent activities).

In contrast, the second step narrows the space to a manageable selection (convergent activities). Divergent and convergent phases can follow each other indefinitely and have different foci. Figure 10.4 shows an example with three different foci: an ideation, a structuring, and a prototyping phase.

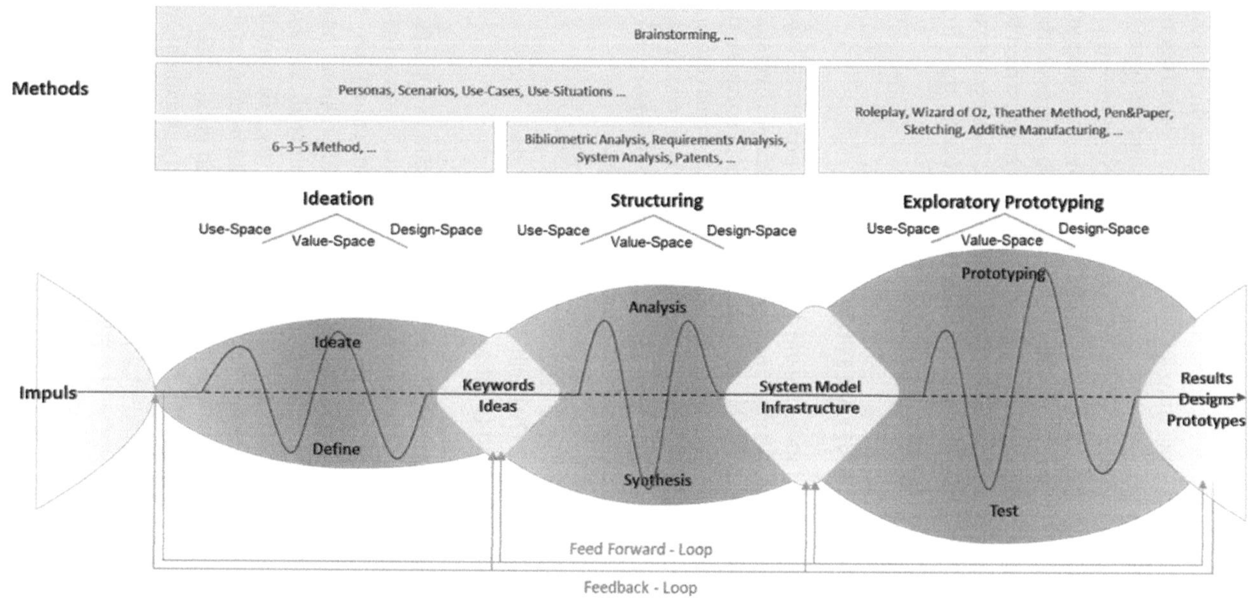

FIGURE 10.4 Exploration and innovation of a turbine for balancing tension fields from initial ideas to results, designs, and prototypes. Here is an example with three phases. (Based on (Flemisch et al., 2022; Preutenborbeck et al., 2024.)

The impulse is the initial spark for an HSE process and can encompass various aspects. According to systems engineering approaches (e.g., Haberfellner et al., 2019), system development or optimization begins with defining and analyzing the problem to be solved. HSE may start with a problem or an idea for which the problem has not yet been identified. It broadens the scope by including solutions from other domains, such as nature and biology. An early example is the H-metaphor, which transferred concepts like "Loose Reins" and "Tight Reins" from equestrian interaction to highly automated vehicles (Flemisch et al., 2003).

The ideation phase includes activities from the initial impulse to developing keywords, ideas, and concepts. Often more intuitive than rational, it emphasizes qualitative over quantitative aspects. Its goal is to develop a preliminary understanding of the use, design, and value space without prematurely structuring the problem. System qualities and early design patterns can be identified through use cases, scenarios, and stakeholder perspectives. HSE employs a broad method portfolio from systems and human factors engineering and design, including brainstorming and stakeholder workshops. For a comprehensive overview of exploratory ideation methods, see Flemisch et al. (2022).

The structuring phase consolidates understanding of the use, design, and value space. Bibliometric analyses can reveal current approaches and knowledge gaps, helping identify use cases, design patterns, values, and system qualities. Empirical methods, such as interviews or observations, further capture stakeholder needs. Once this structure is defined, a system analysis can explore interactions between humans and co-systems within a socio-technical system of systems. These findings deepen understanding of systemic complexity.

The exploratory prototyping phase emphasizes stakeholder-driven design and co-creation (Preutenborbeck et al., 2024). Stakeholders explore the system context using infrastructure from the structuring cycle. In exploroscopes—spaces specialized for exploration—they creatively modify prototypes, try different elements, and co-design system concepts. Requirements are expressed explicitly or implicitly through interaction. This reveals users' mental models, informing later development stages. A central tool here is the exploration map (e.g., mind-net map), which summarizes the use, design, and value space, documents design decisions, and supports transparent development.

The HSE turbine illustrates this process, which aims at generating concrete results, designs, and prototypes for further development. Activities balance constructive and creative design with critical evaluation, e.g., through experiments or field studies. Such validation ensures that proposed solutions are innovative but also practical and applicable in real-world contexts.

EXAMPLE: HUMAN–SYSTEM DIALECTICS AND HSE IN THE "INTERNET OF PRODUCTION"

The methodological foundation of HSE dates to early work on automated driving, e.g., Schieben et al. (2009), and was developed and applied as part of a Cluster of Excellence funded by the German Research Foundation, which deals with data-driven collaboration between domains from the development, production, and usage phases of products. This specific version of the HSE methodology was developed in a research group that considers the market and customer perspective in research on agile product development by integrating the needs and requirements of customers into the development of socio-technical system solutions. The use case for this research is designing and developing a vehicle assistance system that warns drivers of cyclists in a blind spot. The procedure and the results from the individual phases of the HSE process are summarized below, demonstrating the applicability of the HSE methodology in the context of agile product development.

Example: Ideation Phase

During the ideation cycle of the HSE methodology, as applied by the research group in the cluster of excellence, the focus is on generating innovative ideas and concepts for a vehicle assistance system to alert drivers to cyclists in their blind spots. This phase is crucial in blending the market and customer perspectives with agile product development, emphasizing the integration of customer needs and requirements into the design of socio-technical solutions. This first phase focuses on an initial understanding of the use, design, and value space, acknowledging the safety risks associated with vehicle blind spots and the increasing need for enhanced cyclist detection to prevent accidents in urban traffic.

For this given use case, a brainstorming session was started on possible design approaches for the assistance system. The brainstorming sessions were conducted to explore a wide range of design elements, encouraging creativity and leveraging the interdisciplinary expertise within the project team. Mind mapping and sketching are employed to visualize and discuss possible solutions and how they might interact within the socio-technical traffic system. Due to the COVID-19 pandemic, these sessions were conducted online, using Miro as a collaboration platform. The outcome of the ideation cycle was an initial overview of the design space for the assistance system, offering different design elements enhancing driver awareness of cyclists in blind spots.

Example: Structuring Phase

In the structuring cycle, the focus was on setting up the exploration infrastructure for the explorative prototyping of the driver assistance system. This is crucial for preparing and aligning the concepts and design elements gained from the idea phase for practical implementation, whereby the focus here was primarily on the direct integration of customers in the agile development of sociotechnical solutions.

To this end, the most promising elements from the initial design space that promise a significant reduction in accidents were identified. Consequently, 28 potential design elements were selected for implementation. The design space included visual, acoustic, and haptic elements. Warning lights were integrated as visual design elements. The acoustic design included various warning signals that could be coupled with a voice warning to draw attention to the immediate danger a cyclist poses in the blind spot.

The door's resistance force or torque was identified regarding the haptic design. Tangible XR, a VR environment enhanced with haptic components, was used to implement the exploration infrastructure. Tangible XR connects virtual and real worlds through haptic elements that enable touch and physical interaction (Flemisch et al., 2020). This technology uses visual, acoustic, and haptic channels to create an immersive user experience. This is particularly important for human-system integration, enabling better interaction between humans and machines. Using Tangible XR and haptic feedback, complex systems can be made more "tangible" to users and other stakeholders in the design process, allowing requirements and needs to be better identified and addressed (Boy, 2016). A system model and a real picture of the implemented exploration infrastructure are shown in Figure 10.5.

Meyer et al. (2021) describe a setup with a physical door model made of aluminum profiles at the center that enables physical interaction with the car door. The door model can be rotated to simulate the door opening from the perspective of the vehicle interior. An actuator attached to the door shape exerts resistive forces when the door is rotated, enabling physical interaction with this frame. The user perceives these forces as a haptic impulse. Parallel to the physical interaction, the user experiences a virtual reality in which an augmented version of the car door is visualized. The virtual representation is generated by a Unity-based system that runs on a Windows computer and communicates with a Linux/Ubuntu-based system via an ROS interface. This system controls the actuator and provides data on the force and angle of the door movement.

An exploration panel is also displayed in the virtual environment, allowing the user to interact directly with elements of the design space. The exploration panel lets users activate or deactivate various visual and acoustic warning signals. They can also manipulate the actuator's force and resistance, allowing the user's haptic experience in the virtual environment to be adapted directly. This highly interactive and adaptable infrastructure provides a unique platform for exploring and evaluating different haptic properties of the car doors' opening mechanism. By providing fast feedback and the ability to customize in real time, the infrastructure enables efficient tuning of the haptic qualities to the user's needs and preferences without needing physical changes to the prototype. This approach simplifies the development process and opens new possibilities for designing the user experience in the automotive sector.

Example: Prototyping Phase

The exploratory prototyping phase focused on the participatory and exploratory design of concepts for the assistance system. The exploration infrastructure in the structuring cycle was used for this purpose. Participants were first introduced to the context of use and the situation in which the system would be used (Figure 10.6, left). They experienced the critical situation of the approaching cyclist and were instructed to verbalize their needs for the system's functionalities using the Thinking Aloud method. Following this activity, a semi-structured interview can be conducted to gain deeper insights into the users' mental models. The needs and functionalities mentioned were revisited and scrutinized to capture as many insights and ideas as

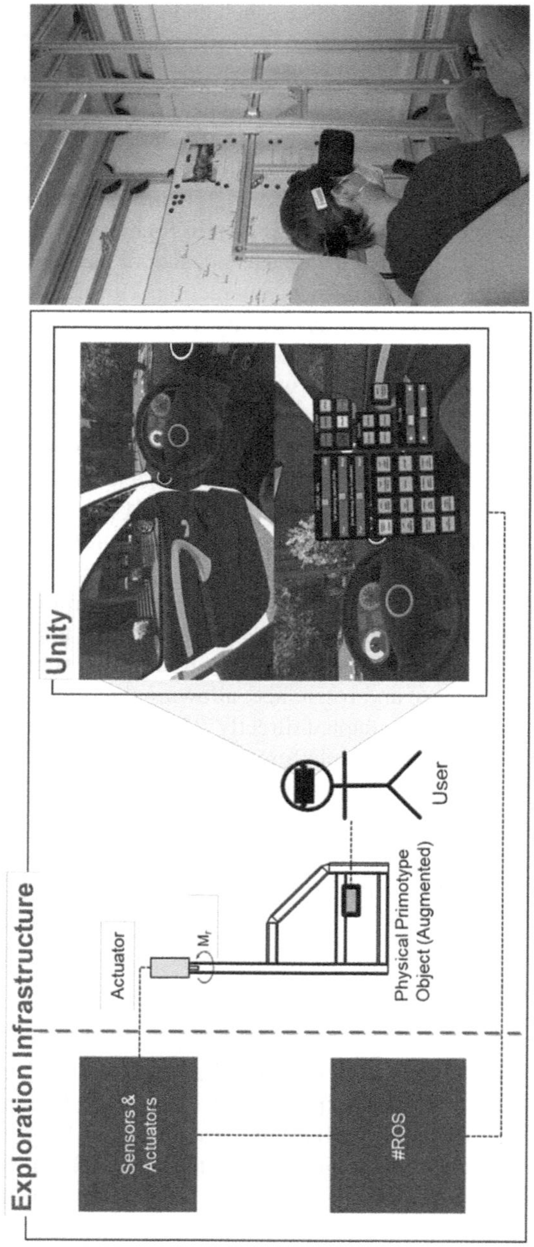

FIGURE 10.5 Exploration infrastructure. (Based on Meyer et al. 2021.)

FIGURE 10.6 Exploratory stakeholder-driven design using the exploration panel.

possible. The participants were then introduced to the prepared design elements from the structuring cycle and the possibility of trying them out and configuring them via the exploration panel (Figure 10.6, right). The playful nature of experiencing, designing, and adapting the system is emphasized so that the participants are encouraged to actively take on the role of designers and create a system concept. After this step, a further interview was conducted to explain the design decisions made by the participants. This process ensures that not only are design concepts captured, but also the underlying considerations. This helps to identify which design elements have potential and which may need to be refined or omitted in subsequent prototyping cycles.

Evaluating the users' decisions provided insights into the preferred and less popular design features. A survey was also conducted to understand the underlying thought models behind the design decisions. New ideas that emerged during the design process were captured and incorporated into the design revisions. Participants' feedback on the design process was positive regarding the opportunity to develop a concept within the exploration sandbox and test it directly in the virtual environment. The direct involvement with the prototype gave the participants an impression of what the final implementation of the assistance system could look like. The feedback also contained valuable suggestions for improving the exploration sandbox's usability and expanding the design elements. The feedback indicates that the method used is practical for developing new products and system concepts with users or other interest groups (Preutenborbeck et al., 2024). This iterative approach ensured that the design and development of the assistance system are agile and responsive to user and market needs, leading to a solution that is both innovative and practical.

OUTLOOK: HOW HUMAN–SYSTEMS DIALECTICS, IDEATION, AND EXPLORATION COULD BE REFINED TO SOLVE THE CHALLENGES OF THE 21ST CENTURY

Human–systems dialectics and HSE are methodological frameworks that aim to address the challenges of the 21st century by focusing on the interaction and cooperation between humans and systems. It involves using system models that incorporate the perspectives of various stakeholders, such as users, designers, and engineers. These models evolve from initial ideas to more concrete concepts, ultimately leading to the implementation or adaptation of socio-technical systems. The thinking and acting in tension fields, i.e., balancing opposing goals and values, and integrating design, use, and value space, opens possibilities for alternative system variants. By balancing constructive and critical approaches, the Collingridge dilemma can be resolved, allowing for exploration of a wide range of possible outcomes. The human–system dialectics, combined with exploration, is crucial for addressing the challenges of the 21st century and achieving greater agility and sustainability in system design. Combining creative design methods with systems engineering and ergonomics, HSE promotes a participatory design process that considers the complexity of socio-technical systems. The HSE process involves cycles of ideation, structuring, and exploratory prototyping, allowing for the systematic development of detailed, real-world implementations.

The fundamental methodological framework is described in this chapter, but it is important to develop human system dialectics and HSE further as part of HSI. How can we ensure that all relevant stakeholders are effectively involved? Which methods and tools could best support the HSE process? By collaborating with the scientific community and the general public, the HSE paradigm can be further developed to develop interactive human–machine systems that meet the increasing challenges of the 21st century, such as the climate crisis or geopolitical conflicts.

REFERENCES

Boehm, B. (1979): Guidelines for Verifying and Validating Software Requirements and Design Specifications. In: *Euro IFIP*. https://robertohernandez18.wordpress.com/wp-content/uploads/2014/08/usccse79-501.pdf

Boehm, B. W. (1988): A Spiral Model of Software Development and Enhancement. *Computer* 21 (5), pp. 61–72. https://doi.org/10.1109/2.59

Booher, H.R. (2003). Introduction: Human Systems Integration. In Handbook of Human Systems Integration, Booher, H.R. (Ed.). John Wiley & Sons, Inc. ISBN:9780471020530, https://doi.org/10.1002/0471721174

Boy, G. A. (2016): *Tangible Interactive Systems: Grasping the Real World with Computers*. Human–Computer Interaction Series. Springer International Publishing, Switzerland. https://ebookcentral.proquest.com/lib/kxp/detail.action?docID=4510834

Boy, G. A. (2020): *Human–Systems Integration*. CRC Press: Taylor & Francis Group, Boca Raton, FL, USA. https://doi.org/10.1201/9780429351686

British Design Council (2004): *History of the Double Diamond – Design Council*. British Design Council. Available online at www.designcouncil.org.uk/our-resources/the-double-diamond/history-of-the-double-diamond/

Collingridge, D. (1982): *The Social Control of Technology*. Repr. London: Pinter.

Cooley, M. (1989): Human-Centred Systems. In K. S. Gill, H. Rosenbrock (Eds.): *Designing Human-Centred Technology*. The Springer Series on Artificial Intelligence and Society. London: Springer London, pp. 133–143.

Elliot, T. S. (1934): The Rock. Available online at https://bentley.umich.edu/elecrec/d/duderstadt/Speeches/JJDS6/jjd1341.pdf, checked on April 12, 2024.

Flemisch, F.; Baltzer, M.; Abbink, D.; Siebert, L. C.; van Diggelen, J.; Herzberger, N. et al. (in press): Towards a Dynamic Balance of Humans and AI-based Systems within Our Global Society and Environment – Holistic Bowtie Model of Meaningful Human Control over Effective Systems. In Giulio Mecacci, Daniele Amoroso, Luciano Cavalcante Siebert, David Abbink, Jeroen van den Hoven, and Filippo Santoni de Sio (Eds.) *Multidisciplinary Research Handbook on Meaningful Human Control Over*. Edward Elgar Publishing. https://doi.org/10.4337/9781802204131

Flemisch, F. (2003): Was kommt nach User Centered Design? (Skizze auf der Suche nach Leitmotiven für das Entwerfen und Gestalten). In: *5. Berliner Werkstatt Mensch-Maschine Systeme "Entwerfen und Gestalten*," vol. 5.

Flemisch, F. O.; Adams, C. A.; Conway, S. R.; Goodrich, K. H.; Palmer, M. T.; Schutte, P. C. (2003): The H-Metaphor as a Guideline for Vehicle Automation and Interaction. NASA. Available online at https://ntrs.nasa.gov/citations/20040031835, checked on May 13, 2024.

Flemisch, F.; Schindler, J.; Kelsch, J.; Schieben, A.; Damböck, D. (2008): Some Bridging Methods towards a Balanced Design of Human–Machine Systems, Applied to Highly Automated Vehicles. In: *Applied Ergonomics International Conference*. Las Vegas. https://elib.dlr.de/57623/1/2008AEIFlemischEtAlBalancedDesignAndAnalysis.pdf

Flemisch, F. O.; Preutenborbeck, M.; Baltzer, M.; Wasser, J.; Kehl, C.; Grünwald, R. et al. (2022): Human Systems Exploration for Ideation and Innovation in Potentially Disruptive Defense and Security Systems. In G. Adlakha-Hutcheon, A. Masys (Eds.): *Disruption, Ideation and Innovation for Defence and Security*, vol. 39. Cham: Springer International Publishing (Advanced Sciences and Technologies for Security Applications), pp. 79–117.

Gerritsen, J. (2019): Problem and Solution Space (and What So Many People Do Wrong). Available online at https://jochemgerritsen.com/2019/10/problem-space-and-solution-space/

Haberfellner, R.; de Weck, O.; Fricke, E.; Vössner, S. (2019): *Systems Engineering*. Cham: Springer International Publishing.

Hegel, G. W. F. (1845): *Werke*. www.suhrkamp.de/werkausgabe/georg-wilhelm-friedrich-hegel-werke-in-20-baenden-mit-registerband-w-40

Hollnagel, E.; Woods, D. D. (2005): *Joint Cognitive Systems*. CRC Press. Taylor & Francis Group, Boca Raton, FL, USA.

Horkheimer, M.; Adorno, T. W. (2004): *Briefe und Briefwechsel*. 1. Aufl. Edited by Christoph Gödde. Frankfurt/M.: Suhrkamp.

Kahneman, D. (2013): *Thinking, Fast and Slow*. First paperback edition. New York: Farrar Straus and Giroux (Psychology/Economics).

Kant, I. (1781): Kritik der reinen Vernunft. www.gutenberg.org/ebooks/6342

Lewin, K. (1951): *Field Theory in Social Science: Selected Theoretical Papers* (Edited by Dorwin Cartwright). Oxford, England: Harpers.

Lindemann, U. (2009): *Methodische Entwicklung technischer Produkte*. Berlin, Heidelberg: Springer.

Luhmann, N. (1984): *Soziale Systeme. Grundriß einer allgemeinen Theorie. 1. Auflage*. Frankfurt am Main: Suhrkamp (Suhrkamp-Taschenbuch Wissenschaft).

Meyer, R.; Preutenborbeck, M.; Usai, M.; Briele, K.; Witzel, A.; Flemisch, F. O. (2021): Increasing Agility: A Tangible XR for Rapid Prototype Development, Design Exploration and Usability Assessment. In S. Trzcielinski, B. Mrugalska, W. Karwowski, E. Rossi, M. Di Nicolantonio (Eds.): *Advances in Manufacturing, Production Management and Process Control*, vol. 274. Cham: Springer International Publishing (Lecture Notes in Networks and Systems), pp. 85–91.

Molter, H. (2017): Wenn Wahrheit die Erfindung eines Lügners ist, dann ist Heinz von Förster ein Lügner. Online. Available online at https://systemagazin.com/wp-content/uploads/2018/07/Wahrheit-12.pdf, checked on April 12, 2024.

NASA (2013): Why Go to Space. Available online at www.nasa.gov/humans-in-space/why-go-to-space, updated on 9/22/2023, checked on May 13, 2024.

Norman, D. A. (Ed.) (1986): *User-Centered System Design. New Perspectives on Human–Computer Interaction*. Hillsdale, NJ: Erlbaum.

Popper, K. (1968): *Logic of Scientific Discovery*. London: Hutchinson.

Preutenborbeck, M.; Wolf-Monheim, F.; Carey, E.; Canu, P.; Herzberger, N.; Usai, M.; Flemisch, F. (2024). Stakeholder-driven design and exploration of sociotechnical systems using virtual reality. *Procedia CIRP* 128, pp. 274–279. https://doi.org/10.1016/j.procir.2024.04.006

Reckwitz, A. (2021): *Das Ende der Illusionen. Politik, Ökonomie und Kultur in der Spätmoderne*. 8. Auflage 2021. Berlin: Suhrkamp (Edition Suhrkamp, 2735).

Reichardt, C. S.; Rallis, S. F. (1994): Editors' Notes. *New Directions for Program Evaluation* 1994 (61), pp. 1–3. https://doi.org/10.1002/ev.1662

Rosa, H. (2015): *Social Acceleration. A New Theory of Modernity*. Paperback edition. New York: Columbia University Press (New directions for critical theory).

Royce, W. (1970): Managing the Development of Large Software Systems. In: *Proceedings, IEEE WESCON*. https://github.com/tpn/pdfs/blob/master/Managing%20the%20Development%20of%20Large%20Software%20Systems%20-%201970%20(waterfall).pdf

Schieben, A.; Heesen, M.; Schindler, J.; Kelsch, J.; Flemisch, F. (2009): The Theater-System Technique. In A. Schmidt, A. Dey, T. Seder, O Juhlin (Eds.): *Proceedings of the 1st International Conference on Automotive User Interfaces and Interactive Vehicular Applications. AutomotiveUI'09: International Conference on Automotive User Interfaces and Interactive Vehicular Applications, Essen, Germany*. New York, NY: ACM, pp. 43–46.

Versteegen, G. (2002): Vorgehensmodelle. In G. Versteegen (Ed.): *Software Management*. Berlin, Heidelberg: Springer (Xpert.Press), pp. 29–61.

Vom Brocke, J.; Hevner, A.; Maedche, A. (2020): Introduction to Design Science Research. In J. Vom Brocke, A. Hevner, A. Maedche (Eds.): *Design Science Research Cases*. Cham: Springer International Publishing (Progress in IS), pp. 1–13.

Weick, K. E.; Sutcliffe, K. M.; Obstfeld, D. (2005): Organizing and the Process of Sensemaking. *Organization Science* 16 (4), pp. 409–421. https://doi.org/10.1287/orsc.1050.0133

Züst, R. (2004): *Einstieg ins Systems Engineering. Optimale, nachhaltige Lösungen entwickeln und umsetzen*. 3. Aufl., vollst. Neu bearb. Zürich: Verl. Industrielle Organisation.

11 Design Space Exploration Approach for the Design of Complex Systems

Audrey Abi Akle

INTRODUCTION

Regardless of the emergence of digital tools, calculations, and simulations, design will remain primarily a human activity for problem-solving. Design concerns people, processes, and artifacts: all human-made goods, services, and sociotechnical systems. It is worth recalling that Herbert Simon had defined design as the "science of the artificial." Design encompasses the diversity of needs, design robustness, resistance to variations, expected performance, optimization, decision-making, solutions, customization, usage, user behaviors, preferences, and satisfaction. Fortunately, numerous approaches exist to meet the expectations and challenges that design must address. Indeed, as Boujut expressed it, design is a "marvelous phenomenon by which human beings can send some of their fellow men to walk on the Moon or to travel at the speed of sound across the oceans, but also to solve simple everyday problems," in Papalambros (2015).

Today, we can rely on more than thirty design methods, such as axiomatic design, concurrent engineering, safety by design, user experience design, or waterfall design. Although very different, all these approaches are often presented on the same level, as if they would not meet the exact expectations of the designer's teams. The work of Vance et al. (2017) has allowed the classification of the main approaches into three levels: environment, process, and method, on the one hand, and considering the complexity of the artifact to be designed on the other hand. Designing complex systems often brings challenges regarding the system's structure (subsystems, system of systems, and numerous interactions), project organization and duration, and the large number of specialized stakeholders involved in the design process. Traditional methods, whether linear or iterative, have limitations as they are based on a so-called point-based design paradigm: a single design taking shape throughout the development process, inevitably resulting in a lack of flexibility regarding design options.

To counter this limitation, engineers and designers apply so-called set-based design approaches. Initially developed by Toyota to support the multidisciplinary design of complex systems in a concurrent approach, set-based design now extends to a broader spectrum of design methods, such as Design Space Exploration (DSE).

 DOI: 10.1201/9781003666431-13

DSE is a design method that involves the systematic and iterative analysis of many solutions derived from numerical simulations. The distinctive feature of this approach lies in the engineer's interactive dimension with numerical simulation and optimization models. Here, the engineer is directly reintegrated into the simulation loop, thus breaking with the often-opaque traditional design processes. Indeed, the method is based on three major iterative phases: (1) the knowledge discovery and insights gain to the design problem, (2) narrowing down or the "optimization stage" (a loop with the first phase allowing the reduction of epistemic uncertainty), and (3) the evaluation of alternatives and choices based on robustness indicators.

In this chapter, we will present the entire method and pertaining concepts, as well as various application cases of DSE in different industrial domains: an automotive case with the configuration of a new vehicle (Ferguson et al., 2005; Donndelinger et al., 2006); an aeronautics case with wing design (Simpson et al., 2008); an aerospace case with satellite selection (Stump et al., 2004; Jordan et al., 2008); an additive manufacturing case with trajectory optimization for "filling" (Xiong et al., 2019; Cazaubon, 2021); and an electrical energy case with social innovation design (Samir, 2023).

The conclusion of this chapter will involve a critical analysis of DSE for the development of sociotechnical systems, demonstrating how this method aligns with the Human Systems Integration (HSI) approach.

DESIGN SPACES OPTIMIZATION APPROACHES

Set-Based Design and Design Space Principles

The initial decisions in product development have the most significant impact on overall product quality and project cost. Many approaches aim to reduce development cycle time. In terms of design strategy, this often translates into proposing the right solution as quickly as possible (Toche et al., 2020). In complex product development, companies encourage teams to promptly deliver a feasible concept refined through multiple iterations (Sobek & Ward, 1996). This is known as "point-based" design: a single solution is proposed and progressively improved until all stakeholders are satisfied.

In contrast, set-based design (SBD) emphasizes flexibility in requirements and design choices. Rather than prematurely selecting one solution, SBD involves identifying and exploring several options in parallel, gradually discarding less promising ones. This deferral of commitments allows assumptions to be validated first, improving outcomes. SBD is a sophisticated approach to developing robust systems through extensive exploration and feasibility assessment. Designers contribute from diverse perspectives, combining their sets to optimize the overall design. Set-Based Concurrent Engineering (SBCE), observed at Toyota, was formalized by Sobek and Ward into 11 principles grouped into three broad categories (Sobek et al., 1999).

In this context, engineers discuss a range of acceptable parameter values rather than seeking a single optimum. SBD allows sets of possibilities to converge toward the best outcome gradually. It is a convergence process, where multiple feasible options are juxtaposed rather than modifying a single solution successively. This prevents premature narrowing of the design space, fostering innovation. However, supporting

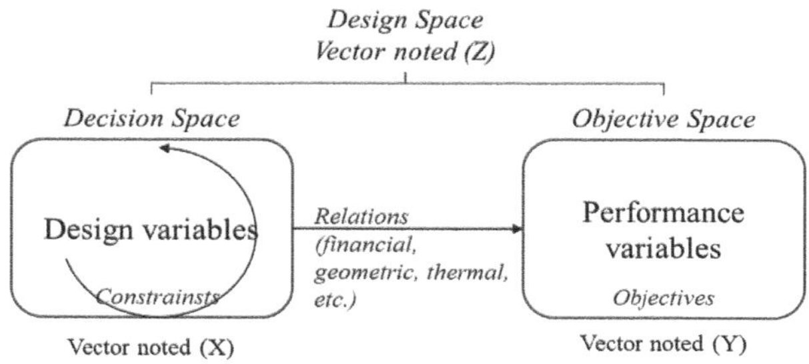

FIGURE 11.1 Illustration of the composition of the design space.

multiple alternatives throughout development can increase costs. By enabling broad exploration and awareness of the design space, set-based design extends the classical approach and encourages more profound discovery.

In design engineering, we refer to the design space as Z [X Y], the combined set of two spaces: the decision space and the objective space (see Figure 11.1). Here, X represents the design parameters, forming the design parameter space or decision space. Y represents the performance variables, creating the performance space or objective space. Therefore, a design point is a potential solution (i.e., feasible and compliant with constraints) within the design space. In other words, the design space often denotes a parametric space containing the set of solutions to a multi-objective optimization problem, where the criteria are represented by vectors X (input) and Y (output).

Multi-Objective Optimization Problem

Indeed, a standard multi-objective optimization problem attempts to find the design variables X that optimize a vector objective function f(X) over the feasible design space S. The determination of a set of non-dominated solutions, the Pareto optimum solutions, or non-inferior solutions X^* can achieve a compromise among several objective functions. The problem formulation is defined as follows:

Minimize:	f(X)	(1)
Subject to:	gi (X) ≤ 0, i = 1, 2, … m	(2)
	$X_{inf} \leq X \leq X_{sup}$	(3)

where $f(X) = [f1, f2, f3, \ldots, fk]^T$: $R^n \rightarrow R^k$, with f_i (X): $R^n \rightarrow R$ as a vector with the values of objective functions to be minimized. X is the vector that contains the design variables, also called decision variables, defined in the space R^n. X_{inf} and X_{sup} are the design variables' lower and upper bounds. gi(X): $R^n \rightarrow R$ represents the ith inequality constraint function. Equations 11.2 and 11.3 define the region of feasible solutions, S, in the decision variable space. The constraints gi(X) are "less than or equal" functions because "greater than or equal" functions may be converted to the first type if they are multiplied by –1. Similarly, the problem deals with the "minimization" of functions f(X), given that function "maximization" can be transformed into the former by multiplying it by –1.

Various approaches and methods exist to move from a space composed of infinite potential solutions to one or a set of optimal solutions. We suggest describing the major families of design optimization and positioning ourselves within this field, particularly by considering the designer's role in the process.

Design optimization approaches fall into two main categories: single-objective and multi-objective optimization (Yannou et al., 2013). As mentioned, exploring solution spaces typically involves optimization or multi-criteria decision problems, where solutions are computed and a suitable compromise is found automatically or interactively.

In automatic approaches, a finite set of "best" solutions is presented directly to the designer. The designer intervenes at the start, defining the problem, and at the end, by selecting one solution. This traditional approach includes well-known methods like weighted sum of preferences, indifference points, or weight compensation (Scott & Antonsson, 2005). The process is considered "semi-automatic" when importance weights are specified directly. Although this adds some designer input, the optimization remains opaque primarily to them.

In contrast, interactive optimization integrates the designer into the process. Two paradigms stand out: uncertainty reduction and Design by Shopping. In uncertainty reduction, the designer progressively narrows variable ranges, gaining knowledge and reducing uncertainty toward an optimal solution (Paredis et al., 2007; Malak et al., 2009). In early design stages, constraint programming proves helpful for this purpose (Yannou & Hamdi, 2004; Yannou et al., 2009). Here, the designer contributes throughout the optimization.

Design by Shopping involves active exploration of solutions and variable interactions. The design space is refined through discovery, insight, and awareness (Zhang et al., 2012). This paradigm places the designer at the core of the simulation loop, directly engaging them in the process.

Beyond process and intervention, the two paradigms differ in strategy (Fathianathan & Panchal, 2009). Uncertainty reduction follows a bottom-up or extended bottom-up approach (Canbaz, 2013), while Design by Shopping combines bottom-up and top-down strategies. In this approach, the designer navigates both the objective and decision spaces, engaging in analysis and synthesis.

As Eder (2009) explains, analysis involves understanding system behavior through decomposition and causality, often using mathematical models. Synthesis, by contrast, aims to find structures that fulfill performance requirements—balancing contradictions and exploring alternatives with creativity to select the most fitting solution.

Design by Shopping Paradigm

Balling (1999) showed that the traditional design process—formulating the problem, modeling it, and executing an optimization algorithm—often leaves designers unsatisfied. This dissatisfaction stems mainly from poor problem formulation and the lack of active designer involvement. In this model, the designer is reduced to an input-output role, and creativity is limited. As Zhang et al. (2012) noted, designers and people often do not know what they want until they see specific design patterns. The objectives and constraints used in optimization frequently fail to reflect real stakeholder desires: "People do not know what they want until they see some designs."

The traditional process is opaque; designers struggle to understand models or "see" feasible solutions. However, before deciding, designers must learn from alternatives and build intuition about the links between parameters and performance. Like consumers, they want to "shop" to understand trade-offs and explore viable options. Design by Shopping addresses this by allowing designers to explore the design space before optimizing and selecting a solution. Through exploration, designers develop realistic expectations and build understanding. This phase helps them connect parameters to performance variables by directly interacting with possible solutions. Optimization becomes an iterative activity guided by exploration. Designers first eliminate infeasible or poor options, using algorithms to steer toward better ones. This paradigm places designers back at the heart of the simulation loop. Exploration fosters "informed" decision-making, based on insights gathered during the process (Abi Akle et al., 2019). Crucially, it shifts focus from finding the correct answer to excluding the wrong ones. Design by Shopping encourages preference construction after visualizing the design space: designers gain control by seeing possibilities first, then selecting optimal solutions (Rangavajhala, 2007). According to Feather et al. (2008), this approach avoids assuming that selection criteria are fixed from the outset. Designers can explore and learn even when the problem is not fully defined. This paradigm is embodied in DSE (O'Hara et al., 2007), which centers the designer in the process. It offers greater flexibility and increases satisfaction by aligning the process with designers' thinking and work.

DESIGN SPACE EXPLORATION

DSE discovers and evaluates design alternatives during system development. As Kang et al. (2011) noted, it serves multiple purposes: rapid prototyping, optimization, and system integration. DSE generates a set of prototypes, whose simulation enhances understanding of design decisions by accounting for complex system dynamics. With relevant metrics, DSE can also optimize designs by eliminating inferior ones and identifying optimal candidates. Finally, it helps find valid assemblies and configurations that meet global design constraints.

DSE is recognized as a visual and intuitive method for formulating, refining, and solving multi-criteria design problems. However, it must be executed with care, given the vast number of possible alternatives, ranging from millions to potentially infinite options in complex systems.

According to Kang et al. (2011), a practical DSE framework relies on three core components:

- *Exploration Methodology:* Once inferior designs are filtered out, designers are often left with many similar candidates. They may only want to explore distinctive or personally relevant options.
- *Analysis:* The framework should use machine-assisted tools to identify viable candidates and verify them against constraints. It must also manage the computational load of solving numerous complex constraints.
- *Representation:* A robust representation of the design space is essential. Valid designs must satisfy numerous constraints involving arithmetic, Boolean logic, or data type conditions, sometimes over infinite domains. The representation must capture this complexity effectively.

Exploration Methodology

This section presents the primary exploration methods available in the scientific literature. Following a review of the various perspectives by researchers, we propose our representation of the DSE methodology.

Simpson et al. (2008) describe the typical approach to DSE in three main phases: building the model, conducting experiments, and exploring. Here, the "explore" activity includes optimization.

Yan et al. (2012) propose a "work-centered" approach that emphasizes the integration of data-driven computational algorithms with a user-centered design process. In their vision, visualization is the interface between data, models, and the user. Their approach aligns with visual analytics, in which they incorporate algorithm manipulation through direct interactions on the graph for DSE.

Kirov et al. (2015) describe the exploration methodology as a workflow. Their model merges exploration and optimization into a single flow, allowing designers to progress through the main flow by performing the following steps: run more simulations, update requirements, change the model (new parameters), and change technology. The model seems primarily focused on the process "from problem definition to optimal solution," while exploration is an activity integrated in a retroactive loop to the main flow.

In their work, Fuchkina et al. (2018) consider exploration an integral part of the design process. According to them, the exploration stage occurs after modelling and analysis but before the decision-making stage. This use of the term "exploration" differs from other scientific references, particularly due to the separation between the exploration and decision-making activities. The term "exploration" in design is defined by evaluating alternatives and solving multi-objective problems. From the model proposed by Fuchkina et al. (2018), we note the evolution of input and output data spaces at different stages of a data-driven design process. The proposed modelling represents the evolution of the design space after each stage of the process. It is important to note that this representation may imply that the spaces are presented. They are different from each other, and they are presented as evolving throughout the process.

Finally, we will note the works of Neubauer et al. (2020) and Balasubadra et al. (2017), which describe the DSE method as an optimization process in which a space

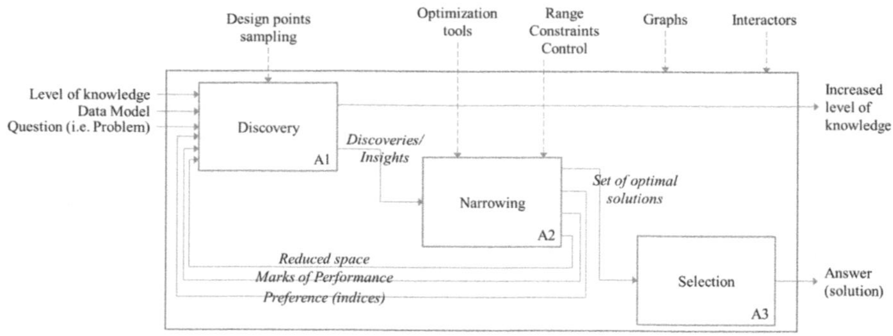

FIGURE 11.2 Design space exploration process (Abi Akle et al., 2019).

is defined, and apply a search for Pareto-optimal solutions. Neubauer et al. (2020) refer to this approach as a filtering process.

To present the DSE, we propose describing the process and the significant phases of exploration. DSE is a designer-centric process in the Design by Shopping paradigm. Therefore, we suggest a modelling of the exploration method focused on the designer's activity. Based on the work of Miller et al. (2013), exploring the design space comprises three main phases: (1) discovery: acquiring knowledge and understanding of the problem; (2) narrowing: active pursuit of a design by eliminating sets, exploring limits, highlighting preferences, etc.; and (3) selection: checking satisfaction (see Figure 11.2). Based on the example provided by Stump et al. (2009), we identify that the discovery and narrowing phases are inseparable, while the selection phase can be performed independently of the two others. Indeed, when designers reach this stage, they know and understand the design problem, and only Pareto optimal solutions remain in the design space.

Thus, narrowing down into the design space is no longer necessary. It is appropriate to compare performance solutions to identify those that are a good trade-off, and select a solution that adequately satisfies preferences. In addition, this phase can be performed by an additional person ("supra-decision-maker"), i.e., disconnected from the first two phases, when space is defined as optimal with a limited number of solutions that require inter- and intra-criteria comparisons.

Analysis and Algorithms

The design space analysis relies mainly on multi-objective optimization using genetic algorithms (GAs) and the approximation of the Pareto front.

Pareto Optimality Principles

As presented in section "Multi-Objective Optimization Problem" a standard multi-objective optimization problem attempts to find the X design variables that optimize a vector objective function f(X) over the feasible design space S.

Indeed, the objectives on the performance variables are often in conflict, making it difficult or even impossible to obtain an ideal solution. It is often necessary to make a compromise. The determination of a set of non-dominated solutions, the Pareto optimum solutions or non-inferior solutions X^* can achieve a compromise among several objective functions.

The principle of Pareto optimum is based on an economic theory: an economic state in which it is no longer possible to improve one individual's welfare without degrading that of at least another.

We use the principle of dominance to capture these optimal solutions in the Pareto sense.

The dominance is defined as: *y* dominates *z* if and only if $\forall i \in [1 \ldots n]$, $f_i(y) \leq f_i(z)$ and $\exists k \in [1 \ldots n] | f_k(y) < f_k(z)$

Dominance in the Pareto sense is defined as: One solution $x^p . \in X$ is Pareto-optimal if and only if $\nexists$ x $\in$ X such that x dominates x^p, and where: we consider a multi-objective problem: $\min(F(x) = (f_1(x), f_2(x), \ldots, f_n(x)), n \geq 2$

- $x \in X$
- X is the decision space
- Y is the objective space (or performance space)
- And Y = F (X)

Engineers use algorithms, particularly GAs, to capture and approximate the Pareto front.

GAs for DSE

The primary role of the algorithms used in DSE is based on two needs:

- Populate the design space
- Capture the Pareto front

GAs, defined by Goldberg (1989), are search algorithms that mimic the behavior of natural selection. GAs aim to discover the optimal solution to a problem (e.g., minimizing a function) by creating a collection ("population") of potential solutions ("individuals") to the problem. Through mutation and recombination (crossover) operations, better solutions are hoped to emerge from the current set of potential solutions.

The crossover and mutation are applied to explore the search space by constructing new individuals from individuals of the previous generation, while selection favors individuals with high fitness. GAs are iterative algorithms that follow the classic pattern of evolutionary algorithms.

In the context of DSE, we focus on a multi-objective evolutionary algorithm (MOEA). Many GAs are based on the same principle. The most well-known and used principles in DSE are:

- Niched Pareto Genetic Algorithm (NPGA)
- Strength Pareto Evolutionary Algorithm (SPEA) and SPEA2
- Multi-Objective Genetic Algorithm (MOGA) and MOGA II
- Non-Dominated Sorting Genetic Algorithm (NSGA) and NSGA II

Tools and Visual Exploration

This chapter is interested in graphs representing the design space and points. There are existing tools for exploring design points and the design space using different graphs, namely: the ARL Trade Space Visualizer (Stump et al., 2004), the VIDEO tool (Kollat & Reed, 2007), the LIVE tool (Yan et al., 2012), and the Rave tool (Daskilewicz & German, 2012). The ARL Trade Space Visualizer (Stump et al., 2004) is used to visualize design parameters and performance variables with Scatter Plot Matrix, Parallel Coordinates Plot, and Bar Charts. The VIDEO tool (Kollat & Reed, 2007) is used to visualize on two graphs separately the objective space (i.e., performance variables) and decision space (i.e., design parameters) with the three-dimensional (3D) Scatter Plot using color and size glyphs. The LIVE tool (Yan et al., 2012) is, on the one hand, used to visualize the input variables (design parameters) on one side and the output variables (performance variables) on the other in two graphs.

On the other hand, a tree structure is displayed that represents the classification of input variable combinations. Both spaces (input/output) are projected in a two-dimensional (2D) Scatter Plot where color and shape glyphs (cluster creation) can be added. The Rave tool (Daskilewicz & German, 2012) is a computational framework designed specifically as a research platform for design decision-support methods. It provides a "blank canvas" to arrange graphs, tables, images, text, and user interface controls (automatically linked together). This tool proposes several graphs to visualize design points, such as line plot, scatter plot, 3D scatter plot, scatter plot matrix, parallel coordinates plot, bar charts, and density scatter plot. Rave also provides continuous visualization.

These graphs help represent multidimensional sets of data with an unlimited number of alternatives (design points). These graphs are adapted to this method, mainly when a designer generates a sample of design points.

APPLICATION CASES OF EXPLORATION

As stated, DSE methodology is used for design optimization with complex projects. We identify various application cases of DSE in different industrial domains: an automotive case with the configuration of a new vehicle (Ferguson et al., 2005; Donndelinger et al., 2006); an aeronautics case with wing design (Simpson et al., 2008); an aerospace case with satellite selection (Stump et al., 2004; Jordan et al., 2008); an additive manufacturing case with trajectory optimization for "filling" (Xiong et al., 2019; Cazaubon, 2021); and an electrical energy case with social innovation design (Samir, 2023).

Preliminary Vehicle Design Case

In their works, Ferguson et al. (2005) and Donndelinger et al. (2006) present the use of DSE methodology to define balanced and compatible sets of vehicle specifications in the early stages of vehicle development. In these works, the authors use MOGA to find Pareto-optimal design solutions. Ferguson et al. (2005) present the mapping between the performance and design space to determine new automobile design configurations for a given set of performance specifications. The advantage presented is that the automobile's design information is obtained without repeating the system analysis. Donndelinger et al. (2006) propose the application of Technical Feasibility Model to assess technical feasibility and optimality of specified combinations of vehicle performance targets. Their study explores the relationships between vehicle mass, vehicle performance measures (such as acceleration, fuel efficiency, and interior roominess), and high-level vehicle design parameters (such as overall exterior dimensions, occupant positions, and powertrain selection). Finally, they showed that DSE methodology allows identifying opportunities for refining, relaxing, improving, or prioritizing performance specifications.

The use case is described as follows: "Suppose that a vehicle manufacturer is planning its next-generation entry into a hypothetical midsize car market segment." After careful analysis of marketplace conditions, the vehicle program manager has developed an initial set of targets for measures of the vehicle's interior accommodation, acceleration, and fuel economy. After applying the proposed model, they concluded that the vehicle program manager was given a detailed overview of feasible and efficient combinations of performance objectives within the boundaries set by the initial performance targets. A refined selection of vehicle performance targets was made from the set defined, considering criteria beyond the technical feasibility domain. The program team confidently proceeded with the vehicle development process, ensuring that the designed and produced vehicle would stay within its specified test weight class and achieve the intended performance.

Wing Design Problem Case

In the work of Simpson et al. (2008), DSE is applied to the design of an aircraft wing. This study focuses on the impact of constraint handling on the trade space exploration process. The authors consider three different constraint handling methods: (1) no constraint handling, (2) manual constraint handling, and (3) automatic constraint handling. The wing design problem involves sizing the wing of an aircraft to minimize its cost, subject to constraints on range, buffet altitude, and take-off field length. Six design variables (inputs) were used to size the wing. The optimization problem is described entirely in the paper by Simpson et al. (2008).

Satellite Selection Case

In their study, Stump et al. (2004) demonstrate how DSE allows a decision-maker to form a posteriori preference and use it to select a preferred satellite.

The data was used to illustrate the ATSV and was generated using a conceptual model of a Mars space probe, with rules based on satellite design textbooks and the Mars Odyssey mission characteristics (Stump et al., 2004). The 2001 Mars Odyssey satellite's primary functions are to measure the elements and minerals on the surface of Mars, to search for the presence of water, and to measure radiation levels that human-crewed missions to Mars would experience. The primary considerations for this analysis revolve around evaluating (1) the impact of propellant quantity on mission ΔV, total spacecraft mass, and cost, and (2) the influence of the reaction wheel choice on the probe's capability to pivot its cameras between different targets. Furthermore, these considerations are interlinked through the overall mass properties of the probe. The data set for this use case includes 5000 designs, each with 12 design variables, forming a 12-dimensional satellite trade space populated by 5000 unique design configurations. Designers could explore the design candidates and visualize different structures by highlighting different regions within the design space. Different relative weights between dry mass and slew time highlight different Pareto optimal designs, aiding the designers in selecting the most preferred design. The entire study is presented in the work of Stump et al. (2004).

Additive Manufacturing Case

More recently, we find the use of DSE methodology for optimizing trajectories during the metal printing of mechanical parts. The new capabilities offered by additive manufacturing make it possible to manufacture increasingly complex parts, often requiring the manipulation of specific input parameters to obtain a printed part close to the desired part (designed one). On one side, Xiong et al. (2019) present the use of DSE to optimize the 3D printing process of a part. In Xiong et al. (2019), the authors use a Bayesian network classifier for a reasoning framework to explore the design space in the embodiment design stage, while the Gaussian process regression model is used as the evaluation function for an optimization method to exploit the design space in detailed design. This cost-effective data-driven approach is demonstrated in designing a customized ankle brace with a tunable mechanical performance using a highly stretchable design concept with tailored stiffnesses. Conversely, Cazaubon (2021) applies the DSE methodology to define which trajectories of the printing nozzle are the most favorable to obtain a printed part closest to the "designed" part.

Social Innovation in the Energy Sector Case

The last application case concerns the design of social innovations in the energy sector (SIE) proposed by Samir et al. (2023). These innovations deal with encouraging citizens' involvement. To support decision-making in SIE design, the authors propose a multi-objective decision-support model based on the definition and exploration of the SIE design space. The model has the particularity to integrate objective functions related to economic, environmental, and social perspectives, to ensure that selected solutions are tailored to the needs of citizens. In their work, Samir et al. (2023) show how to include the social dimension in the optimization problem and define it in the design space. They illustrated the use of their method by solving a photovoltaic self-consumption problem.

CONCLUSION

In this chapter, we presented the entire approach of DSE with related concepts, as well as various application cases of DSE in different industrial domains.

To encompass the approach, we have addressed the concepts of SBD, multi-objective optimization problems in design, and design space. These concepts are crucial for understanding and conceptually positioning DSE. The methodology is described to allow any reader to grasp it. We particularly delve into the design of the shopping paradigm and the framework of the DSE approach. Finally, we have outlined the approach in three parts: methodology, analysis, algorithms, and visual tools. To close the chapter, we illustrated using the DSE through five industrial application cases.

We have shown that the DSE is a robust alternative to optimization-based methods for defining, refining, and resolving multi-objective design challenges. This approach has the advantages of making the design process of complex systems more flexible by reintegrating the designers/engineers directly into the simulation loop.

This approach is efficient for the design of complex systems, as evidenced by the presented applications. However, we acknowledge that this approach is unsuitable for designing simpler systems with few stakeholders and describing systems with fewer subsystems and subcomponents. Indeed, implementing such approaches requires extensive and specialized knowledge in algorithm development and mathematics. This task demands a significant upfront development time before addressing the problem resolution.

Today, some industrial companies sell software environment solutions that integrate the necessary components for deploying this method, such as Dassault Systèmes with its 3DEXPERIENCE suite or ESTECO Engineering with its modeFRONTIER software. However, the field of application is vast: system design in automobiles and aeronautics, the design of mechanical parts using additive manufacturing processes (Industry 4.0), or even the design of social innovation in the energy sector, integrating the user and the citizen as decision variables for the designer.

Traditionally used to identify optimal solutions within a technically defined design space, DSE can be enriched by the contributions of HSI, which introduces criteria related to human performance, safety, ergonomics, and cognitive load. Integrating scenarios from SBD provides a narrative framework to anticipate real-world interactions between users and systems in diverse contexts. These scenarios help identify specific human-centered requirements, which are then translated into measurable indicators that structure or constrain the explored design space. This combined approach fosters a co-evolution of technical solutions and user practices by incorporating iterative evaluations based on scenario feedback. It also considers multidimensional trade-offs by crossing technical and human performance through multi-criteria analysis tools. Usually, DSE applications are found in industrial contexts and are often associated with numerical or quantitative optimization problems. More recently, the approach has been combined with scenario-based design methods. DSE has been used to explore multiple renovation scenarios for a hospital building using a digital twin to reduce energy consumption (Bikai et al., 2022).

Additionally, we are currently developing a design method for social housing buildings that integrates renewable energy sources (Gartzia et al., 2024). This method uses usage scenarios within an SBD approach (Yannou et al., 2013). Many contributions and developments are still needed to popularize the methodology with increasingly efficient and turnkey tools.

REFERENCES

Abi Akle, A., Yannou, B., & Minel, S. (2019). Information visualisation for efficient knowledge discovery and informed decision-making in design by shopping. *Journal of Engineering Design*, 30(6), 227–253.

Balasubadra, K., Shanthi, A. P., & Srinivasan, V. P. (2017). Hybrid Design Space Exploration methodology for application-specific system design. *IJNCAA*, 73, 102–111.

Balling, R. (1999, May). Design by shopping: A new paradigm? In *Proceedings of the Third World Congress of Structural and Multidisciplinary Optimization (WCSMO-3)* (Vol. 1 , pp. 295–297) Buffalo, NY, USA. Buffalo, NY: International Society for Structural and Multidisciplinary Optimization (ISSMO).

Bikai, M., Merlo, C., & Abi Akle, A. (2022, March). User-Centred Design for digital twin to support decision-making for renovation of hospital building towards nZEB. In *2e Ecole d'Ingénierie Système de l'AFIS, Alès, France.*

Canbaz, B. (2013). Preventing and resolving design conflicts for a collaborative convergence in distributed set-based design. Doctoral dissertation, École Centrale Paris.

Cazaubon, V. (2021). Procédé de fabrication additive par dépôt de fil fondu: modèle, méthodes et stratégies pour la correction de défaut de pièce mécanique. Doctoral dissertation, Université de Bordeaux.

Daskilewicz, M. J., & German, B. J. (2012). Rave: A computational framework to facilitate research in design decision support. *Journal of Computing and Information Science in Engineering*, 12(2), 021005.

Donndelinger, J., Ferguson, S., & Lewis, K. (2006). Exploring mass trade-offs in preliminary vehicle design using Pareto sets. In *Proceedings of the 11th AIAA/ISSMO Symposium on Multidisciplinary Analysis and Optimization, Portsmouth, VA*. Reston, VA: American Institute of Aeronautics and Astronautics (AIAA). AIAA Paper No. AIAA-2006-7056.

Eder, W. E. (2009). Analysis, synthesis, and problem solving in design engineering. In *DS 58-2: Proceedings of ICED 09, the 17th International Conference on Engineering Design, Vol. 2, Design Theory and Research Methodology, Palo Alto, CA, USA, 24–27 August 2009* (pp. 13–24). Glasgow, UK: The Design Society.

Fathianathan, M., & Panchal, J. H. (2009). Modelling an ongoing design process utilizing top-down and bottom-up design strategies. *Proceedings of the Institution of Mechanical Engineers, Part B: Journal of Engineering Manufacture*, 223(5), 547–560.

Feather, M., Cornford, S., Hicks, K., Kiper, J., & Menzies, T. (2008). Application of a broad-spectrum quantitative requirements model to early-lifecycle decision making. *IEEE Software*, 1673–1680.

Ferguson, S., Gurnani, A., Donndelinger, J., & Lewis, K. (2005). A study of convergence and mapping in preliminary vehicle design. *International Journal of Vehicle Systems Modelling and Testing*, 1(1), 192–215.

Fuchkina, E., Schneider, S., Bertel, S., & Osintseva, I. (2018). Design Space Exploration framework. In *eCAADe* (Vol. 36, pp. 367–376).

Gartzia, S., Abi Akle, A., Hamwi, M., & Lizarralde, I. (2024). A state of the art of renewable energy integration in collective social housing in France: A matrix for identification of new development scenarios during the planning phase. In *CONFERE 2024, July 2024, Porto, Portugal.*

Goldberg, D. E. (1989). *Evolutionary Algorithm in Search, Optimization, and Machine Learning* (Vol. 8). Reading, Addison-Wesley.

Jordan, D. D., Spencer, D. B., Simpson, T. W., Yukish, M. A., & Stump, G. M. (2008). Optimal spacecraft trajectories via visual trade space exploration. In *AAS/AIAA Space Flight Mechanics Meeting* (pp. 27–31).

Kang, E., Jackson, E., & Schulte, W. (2011). An approach for compelling Design Space Exploration. In *Foundations of Computer Software. Modeling, Development, and Verification of Adaptive Systems: 16th Monterey Workshop 2010, Redmond, WA, USA,* March 31–April 2, 2010, Revised Selected Papers 16 (pp. 33–54). Springer Berlin Heidelberg.

Kirov, D. A., Passerone, R., & Ozhiganov, A. A. (2015). A methodology for Design Space Exploration of real-time location systems. *Научно-технический вестник информационных технологий, механики и оптики*, 15(4), 551–567.

Kollat, J. B., & Reed, P. (2007). A framework for visually interactive decision-making and design using evolutionary multi-objective optimization (VIDEO). *Environmental Modelling & Software*, 22(12), 1691–1704.

Malak Jr, R. J., Aughenbaugh, J. M., & Paredis, C. J. (2009). Multi-attribute utility analysis in set-based conceptual design. *Computer-Aided Design*, 41(3), 214–227.

Miller, S. W., Simpson, T. W., Yukish, M. A., Bennett, L. A., Lego, S. E., & Stump, G. M. (2013). Preference construction, sequential decision making, and trade space exploration. In *ASME 2013 International Design Engineering Technical Conferences and Computers and Information in Engineering Conference* (pp. V03AT03A014-V03AT03A014). American Society of Mechanical Engineers.

Neubauer, K., Beichler, B., & Haubelt, C. (2020). Exact design space exploration based on consistent approximations. *Electronics*, 9(7), 1057.

O'Hara, J. J., Stump, G. M., Yukish, M. A., Harris, E. N., Hanowski, G. J., & Carty, A. (2007, April). Advanced visualization techniques for trade space exploration. In *Collection of Technical Papers—AIAA/ASME/ASCE/AHS/ASC Structures, Structural Dynamics and Materials Conference* (Vol. 2, pp. 2137–2142).

Papalambros, P. (2015). Design science: Why, what, and how. *Design Science*, 1, E1. doi:10.1017/dsj.2015.1

Paredis, C., Malak Jr, R. J., & Aughenbaugh, J. M. (2007). Multi-attribute utility analysis in set-based conceptual design.

Rangavajhala, S. (2007). *Decision Making in Design under Uncertainty with Multiobjective Robust Design Optimization* (p. 205). Troy, NY, Rensselaer Polytechnic Institute.

Samir, B., Abi Akle, A., & Lizarralde, I. (2023). A multi-objective decision-making approach to support the design of social innovations in the energy sector. *Proceedings of the Design Society*, 3, 313–322.

Scott, M. J., & Antonsson, E. K. (2005). Compensation and weights for trade-offs in engineering design: Beyond the weighted sum. *Journal of Mechanical Design*, 127(6), 1045–1055.

Simpson, T. W., Carlsen, D. E., Congdon, C. D., Stump, G., & Yukish, M. A. (2008). Trade space exploration of a wing design problem using visual steering and multi-dimensional data visualization. In *The 4th AIAA Multidisciplinary Design Optimization Specialist Conference*.

Sobek, D. K., & Ward, A. C. (1996). Principles from Toyota's set-based concurrent engineering process. In *Proceedings of the 1996 ASME Design Engineering Technical Conferences and Computers in Engineering Conference, DETC 1996*. American Society of Mechanical Engineers.

Sobek II, D. K., Ward, A. C., & Liker, J. K. (1999). Toyota's principles of set-based concurrent engineering. *MIT Sloan Management Review*.

Stump, G., Simpson, T. W., Donndelinger, J. A., Lego, S., & Yukish, M. (2009). Visual steering commands for trade space exploration: User-guided sampling with example. *Journal of Computing and Information Science in Engineering*, 9(4), 044501–1.

Stump, G. M., Yukish, M., Simpson, T. W., & O'Hara, J. J. (2004, March). Trade space exploration of satellite datasets using a design by shopping paradigm. In *Aerospace Conference, 2004. Proceedings. 2004 IEEE* (Vol. 6, pp. 3885–3895). IEEE.

Toche, B., Pellerin, R., & Fortin, C. (2020). Set-based design: A review and new directions. *Design Science*, 6, e18.

Vance, J. K., Giambalvo, J., & Hoffenson, S. (2017). Navigating the common approaches to product development. In *DS 87-9 Proceedings of the 21st International Conference on Engineering Design (ICED 17) Vol. 9: Design Education, Vancouver, Canada, 21–25 August 2017.*

Xiong, Y., Duong, P. L. T., Wang, D., Park, S. I., Ge, Q., Raghavan, N., & Rosen, D. W. (2019). Data-driven Design Space Exploration and exploitation for design for additive manufacturing. *Journal of Mechanical Design*, 141(10), 101101.

Yan, X., Qiao, M., Li, J., Simpson, T. W., Stump, G. M., & Zhang, X. (2012, January). A Work-Centered Visual Analytics Model to support engineering design with interactive visualization and data-mining. In *System Science (HICSS), 2012 45th Hawaii International Conference on* (pp. 1845–1854). IEEE.

Yannou, B., & Hamdi, A. (2004). Truss dimensioning with an uncertainty reduction paradigm. In *DS 32: Proceedings of DESIGN 2004, the 8th International Design Conference, Dubrovnik, Croatia.*

Yannou, B., Troussier, N., Chateauneuf, A., & Boudaoud, N. (2009). Dimensioning a product in preliminary design through different exploration techniques. *International Journal of Product Development*, 9(1), 140–163.

Yannou, B., Yvars, P. A., Hoyle, C., & Chen, W. (2013). Set-based design by simulation of usage scenario coverage. *Journal of Engineering Design*, 24(8), 575–603.

Zhang, X., Simpson, T., Frecker, M., & Lesieutre, G. (2012). Supporting knowledge exploration and discovery in multi-dimensional data with interactive multiscale visualisation. *Journal of Engineering Design*, 23(1), 23–47.

Section III

Modeling and Simulation Approaches

12 A Virtual Human-Centered Design Method to Support New Developments in Vehicle Automation

Thierry Bellet and Jean-Charles Bornard

INTRODUCTION: NEW DESIGN CHALLENGES FOR VEHICLE AUTOMATION

With the recent progress in vehicle automation, a new generation of Advanced Driving Aid Systems (ADAS) is progressively emerging. Compared to the previous generation primarily focused on information or warning delivery, new ADAS can now take control of the car, which is a key change in the human–machine interaction (HMI) issue (Hoc et al., 2009). From a driving task under the human driver's full responsibility, we are moving toward a *co-managed* activity under the authority of a *joint cognitive* Human–Machine System (Hollnagel & Woods, 2005). Faced with this disruptive evolution, the decision and the way of introducing vehicle automation to assist, support, or replace the human driver must not only consider the technological capabilities themselves, but also the drivers' needs and expectations, their driving abilities and individual characteristics, and ultimately, their acceptance of driving aids based on partial or complete automation (Kyriakidis et al., 2019;Madigan et al., 2017; Nordorff et al., 2018; Bellet and Banet, 2023).

In this general context, a Human-Centered Design approach is critical due to the inherent risks associated with the driving task, the variability of traffic situations, the potential dangerousness for misuses of ADAS or other "ironies" of automation (Bainbridge, 1983) and – last but not least – the question of responsibility in case of accident (Bellet et al., 2019). However, integrating end-users' needs into the design process is not always straightforward, especially when designing entirely innovative systems. This is because ergonomics evaluations involving real humans necessitate developing expensive mock-ups or prototypes, which require significant financial resources, effort, and time.

Recent progress in virtual prototyping tools for the development of future ADAS could support this effort (Son et al., 2018; Singh et al., 2021; Donà & Ciuffo, 2022). However, the key challenge is not only to simulate the technical components of ADAS (from car sensors to algorithms of vehicle automation), but also the Human–Machine System as a whole, to assess how future fundamental drivers will effectively use

DOI: 10.1201/9781003666431-15

ADAS. Alongside the simulation of technical systems, which primarily focuses on ADAS *effectiveness* (i.e., accuracy and completeness of specified goals to be achieved through this system, according to ISO 9241-11:2018), it is in addition necessary to develop a virtual model of future end-users capable of dynamic interactions with these driving aids. Such a virtual driver is crucial for evaluating ADAS *efficiency* (i.e., resources and efforts required from the user to achieve adequate results through this ADAS). In this aim, the challenge is to develop a virtual simulation tool able to support human-based simulations (i.e., using a digital twin of the human driver) for (1) assessing critical scenarios and accident risks for unassisted drivers, and then (2) estimating potential benefits of future ADAS when these driving aids will assist the virtual driver. These integrated human–system simulations are expected to reduce the cost and increase the efficiency of the design process of future driving assistances, which are increasingly complex and thus expensive to develop, test, validate, and certify.

This chapter aims to present an example of an integrated simulation platform, developed at the Gustave Eiffel University, based on a cognitive simulation model of the car driver. The general principle of this simulation tool is to use a virtual model of future end-users to consider the needs and characteristics of human drivers from the earliest stages of the design process (i.e., even when ADAS are still at the stage of a technological concept). Considering the user's needs from the beginning of the technology design is essential to develop truly "cooperative" driving aids (Bellet et al., 2011). Traditionally, this consideration by the ergonomist relies on the involvement of fundamental human drivers using mock-ups or simulated environments that allow them to "apprehend the future" by interacting with ADAS in a simulated but sufficiently *tangible* (cf. Boy, 2020) and realistic manner. However, this method is very time-consuming and resource-intensive. Indeed, aside from the need for a simulation environment, experiments must also be designed to immerse the user in future usage situations. Moreover, each potential technological solution, along with all its gradual developments, must be subjected to specific tests, which implies conducting a large number of experiments, often sequentially (the results of a first experiment conditioning technological evolutions to be then evaluate in a second one, and so on), thus increasing the time required.

Furthermore, humans quickly become fatigued and learn from previous experiences. Therefore, many participants must be involved, and the panel must be frequently renewed to avoid/control these biases. This is burdensome and often very costly, so the number of experiments is generally more limited than would be needed. Using a driver model can be an efficient solution to address these difficulties. The aim is not to systematically replace human participants by a model, but to use a virtual driver for part of the tests, thus limiting the involvement of real humans to key stages of the design process, where their opinions are mandatory (e.g. to assess their *Satisfaction* related to the adequacy of the ADAS to their real needs, that is the third of the three pillars of *systems usability* in the ISO norm 9241-11:2018). However, for some phases of ergonomic specifications of ADAS and their evaluation, using a driver model can streamline the work, so real humans are only involved with technological solutions that have been pre-validated by the virtual driver.

In the next section of this chapter, we will present a computational model of the driver (so-called COSMODRIVE), which is interfaced with a simulation tool dedicated to the virtual design of ADAS. After having introduced the theoretical foundations in cognitive sciences supporting this model, some examples of driving errors simulated through COSMODRIVE will be presented and discussed to show the potential benefits of such model-based simulations (as a virtual "*Human-in-the-Loop*" approach) for proposing ergonomic requirements of future ADAS and their HMI. Then, we will present a Virtual Human-Centered Design platform (so-called V-HCD), combining COSMODRIVE and simulated ADAS, aimed at supporting the virtual prototyping and evaluation of future driving aids through driver model-based simulations at the earliest stages, followed by immersive experiments involving real humans at the final stages. Finally, future perspectives about using the V-HCD simulation for modeling traffic situations in a systemic way (i.e., *system of systems* simulation; cf. G.A. Boy's introduction) will be discussed in the conclusion, as a future enhancement of this HSI approach.

COGNITIVE MODELING AND SIMULATION OF THE CAR DRIVER

To support a virtual human-centered design method, it is necessary to have a computational model capable of simulating driving behaviors and the cognitive processes underlying them. This is the case of the COSMODRIVE model (*COgnitive Simulation MOdel of the DRIVEr*) developed at Gustave Eiffel University to simulate the mental activities carried out when driving a car, from perception to action. Synthetically, this model can be defined as a virtual driver able to pilot a virtual car within a virtual road environment, through a dynamic and iterative "Perception–Cognition–Action" regulation loop (Figure 12.1).

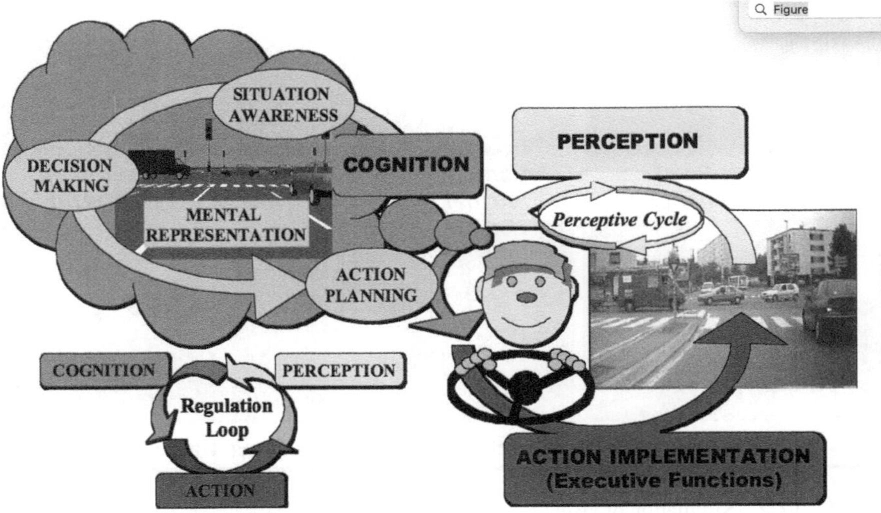

FIGURE 12.1 Driving activity as a dynamic "Perception–Cognition–Action" regulation loop, from the COSMODRIVE theoretical model.

Computational Implementation of COSMODRIVE

At the perceptive level, two complementary processes are implemented in the Perception Module of COSMODRIVE (Bornard et al., 2012). The first one, *Perceptive Integration*, is a "data-driven" process (i.e., bottom-up processing) in charge of simulating the cognitive integration of environmental pieces of information into the Cognition Module, according to their saliencies for the human driver's eyes. The second process is the *Perceptive Exploration* (referring to Neisser's theory of *perceptive cycle* [1976]), which is a "knowledge-driven" process (i.e., top-down) responsible for dynamically exploring the road scene according to the drivers' situation awareness, risk assessment, intentions, and decisions to be made. From these two processes, COSMODRIVE can dynamically explore the road environment with a "*Virtual Eye*," thereby simulating fundamental drivers' visual scanning performed when driving a car. These visual strategies are a set of *fixation points,* which are "outputs" of the COSMODRIVE model, similar to empirical data collected from eye-tracking systems among real humans (e.g., bottom central view of Figure 12.2).

Perceptive data collected by the virtual eye and processed by the Perception Module are then integrated into the Cognition Module of COSMODRIVE. The key elements of this Cognition module are *Mental Representations* (cf. Bellet et al., 2009), corresponding to the driver's *Situation Awareness* (SA) as defined by Endsley (1995; "*the perception of the elements in the environment within a volume of time and space, the comprehension of their meaning, and the projection of their status shortly*"). Mental representations of the traffic situation are dynamically formulated in working memory through a matching process between, from the one

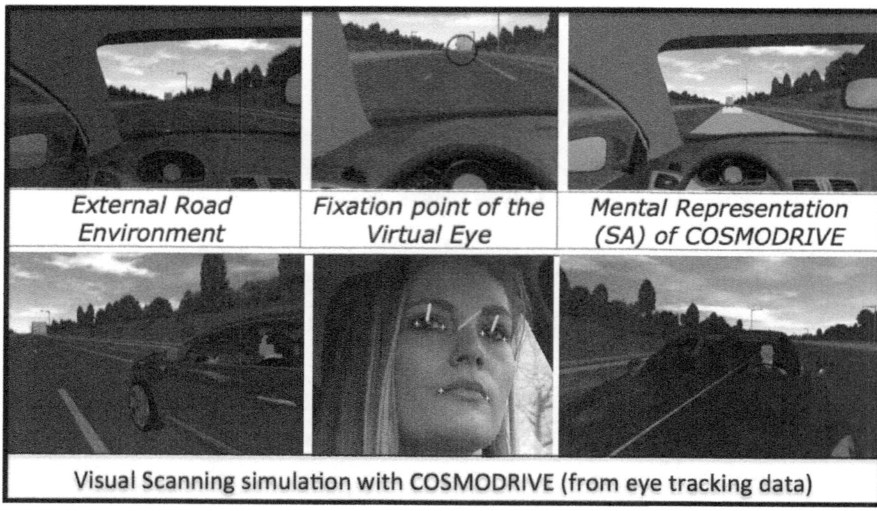

FIGURE 12.2 Simulation of visual scanning with COSMODRIVE and mental representation built in the cognition module (i.e., situation awareness) from the data processed by the perception module.

hand, information perceived/extracted in the external environment and, from the other hand, *operative* (Ochanine, 1977) driving knowledge, that are modeling in COSMODRIVE as "*Driving Schemas*" and "*Envelop Zones*" (Bellet et al., 2009, 2011), respectively inspired by Minsky's *Frames* (1975) and Hall's *Proxemics Theory* (1966).

These mental representations provide ego-centered and goal-oriented mental models of the driving situations. They take the form of three-dimensional visual-spatial models of the road environment, liable to be mentally "*deployed*" by the human cognitive system, to support *anticipation* through a cognitive process of *mental simulation*, and thus provide *expectations* about future situational states. This cognitive process of anticipation, based on cognitive deployments, is illustrated in Figure 12.3. In this traffic situation, the driver/model is following a truck that suddenly brakes. Facing this unexpected event, the virtual driver must decide between braking and implementing a Lane Change (LC) manoeuver to overtake the truck. To support this decision-making process, the virtual driver (like real humans) mentally explores alternative behaviors to assess their respective feasibility and their related levels of risk (for instance, by immediately implementing an LC maneuver with the current speed, versus after a braking maneuver aiming to reduce the ego-car velocity). From the results of these cognitive deployments, COSMODRIVE will decide by selecting the most appropriate maneuver and/or the least critical behavior. Then, this planned behavior is provided to the action module to be effectively performed by the virtual driver through specific actions implemented on the vehicle pedals and/or steering wheel.

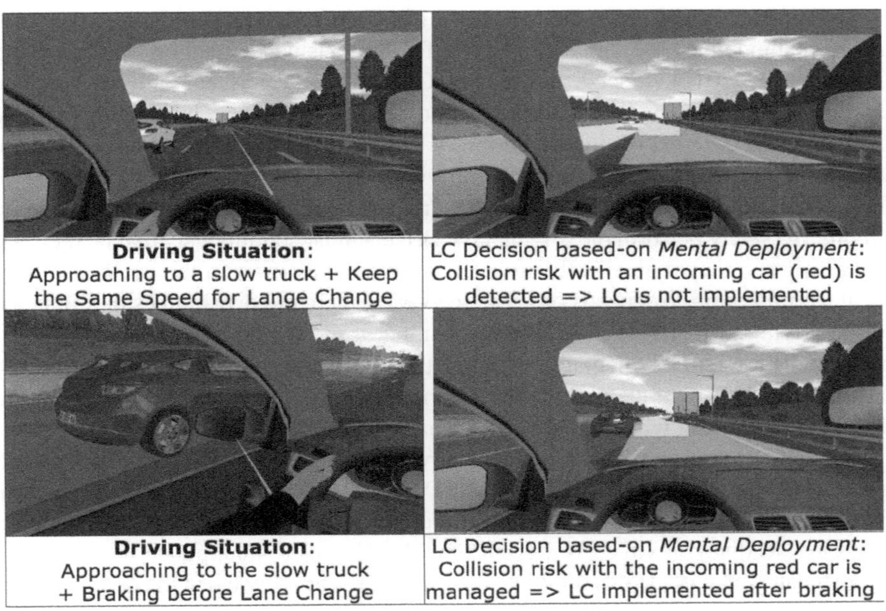

FIGURE 12.3 Decision-making simulation with the COSMODRIVE model.

This cognitive process of anticipation, based on both "implicit" and "explicit" mental simulations (Bellet et al., 2009) is a core function of the human cognitive system in dynamic environments. It plays a key role in drivers' risk assessment to evaluate situational danger and support drivers' decision-making and driving behavior planning to manage situational risk adequately.

COSMODRIVE Model Validity and Calibration for Simulation

To use a driver model in a simulation platform as a "digital twin" of real humans, it is essential that the model is validated with empirical data and that its validity limits are well understood. Indeed, a model remains a model: it simplifies reality. Moreover, when what it models is as complex as a human being, it goes without saying that the model can only account for a small part of that complexity. However, it is possible to evaluate its validity and limitations by comparing the driver model's decisions, behaviors, and driving performances with the way fundamental human drivers behave.

In the simulations presented in this chapter, the version of COSMODRIVE used is firstly based on both analyses of real driving activity on highways, conducted with various driver profiles invited to pilot an instrumented vehicle (Georgeon, 2008; Georgeon et al., 2012; Mathern, 2012; Paris, 2014; Paris et al., 2015), and a series of experimental protocols carried out on a driving simulator. These last set of experiments provided empirical validation for cognitive processes such as situation awareness (Bellet et al., 2009), decision-making based on mental deployment associated with the envelope zones theory (Bornard, 2012; Bornard et al., 2016), and the perceptive, cognitive, and behavioral analysis of lane-changing maneuvers on highways (Deniel, 2019; Deniel et al., 2019).

Furthermore, these simulations are also based on a driving simulator experiment aimed at modeling and simulating the effects of visual distraction while driving (Bellet et al., 2012). Indeed, the aim of COSMODRIVE is not only to simulate drivers' perception, cognition, and optimal driving performance, but also to simulate human errors in terms of misperception of events, erroneous situational awareness, dangerous decision-making, or critical driving behaviors, that are liable to occur in driving conditions. It is typically when the driver is involved in a secondary task while driving. To safely perform the driving task, drivers must continuously update their SA and also when they dynamically progress on the road. However, the content of their mental representation depends on the perceptive and cognitive resources they allocate to the driving task and on their visual scanning of the road environment. In case of a visual distraction, the mental model updating will be negatively impacted, particularly for long off-road glances and/or if an unexpected event suddenly occurs in the driving environment while the driver is visually distracted.

Figure 12.4 presents an example of erroneous SA due to this type of off-road glances, as simulated with the COSMODRIVE virtual driver. In this driving scenario, like the traffic situation presented in Figure 12.3, the driver follows a truck with a safe time headway of 1.8 s. At a given time, he decides to manipulate the car radio. Unfortunately, the truck ahead initiates an emergency braking when the driver is visually distracted (the model's virtual eye focuses on the car radio, as shown in the

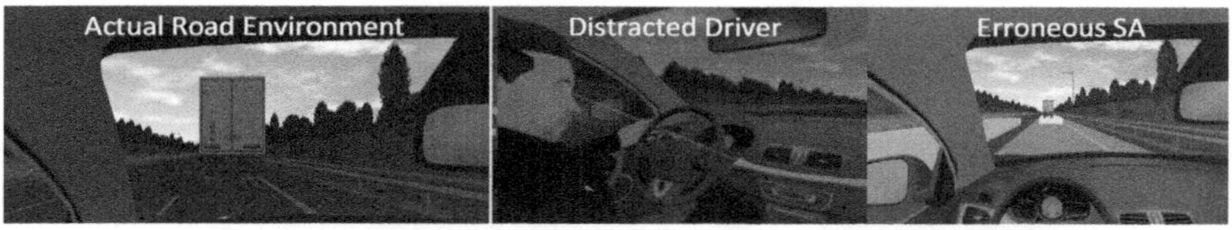

FIGURE 12.4 Simulation of erroneous situation awareness due to a visual distraction.

central view of Figure 12.4). As a result of this unexpected event, after 1 s of visual distraction (corresponding to the correct view of Figure 12.4), the driver's mental representation is not correctly updated. In the mental model, the truck ahead still seems far away (reflecting the driver's *expectation* based on the truck's previous stable velocity), compared to the objective reality (shown in the left view of Figure 12.4). At this precise moment, we have a typical example of a "near-death" situation where the driver is unaware of the high criticality of the current driving situation.

Finally, to support the simulation presented in the next section, COSMODRIVE was also specifically calibrated by using Reaction Time (RT) measures collected among a sample of 21 experienced drivers (12 men and nine women), middle-aged (mean age was 34.9 years old; standard deviation [S.D.] 4.7). These data were collected on a driving simulator, for the same truck following scenario. While driving, participants also had to perform a secondary visual distraction task. When the truck started to brake, a warning was delivered, and they had to stop immediately. Their secondary task was to intervene on the vehicle commands to avoid the crash (by braking or changing lanes). Results showed that the mean RT of drivers of this profile was 0.9 s (S.D. of 0.14) when warned to both (1) stop their off-road glance, (2) make their decision, and (3) start to implement their actions on vehicle pedals and/or the steering wheel. This mean value was used to calibrate COSMODRIVE accordingly to perform the simulation presented in the next Section FROM COGNITIVE MODELING TO DRIVING BEHAVIORS AND ERRORS SIMULATION. Of course, another calibration of the model would be required to simulate other profiles of drivers, like older adults or novice drivers who generally have longer RTs.

From Cognitive Modeling to Driving Behaviors and Errors Simulation

COSMODRIVE-based simulations may be used to study accident risks for an unassisted driver, which could help specify future ADAS accordingly. Figure 12.5 illustrates these potentialities by exploring the respective effects of different durations of a visual distraction in this driving scenario.

For instance (Figure 12.5 A), in the case of a distracted driver, until 0.5 s after the truck begins braking, it should be possible (for an experienced middle-aged driver) to avoid the frontal collision by performing an emergency braking. By contrast, in the case of a longer visual distraction of 0.75 s or more (refer to Figure 12.5 B), model-based simulation indicates that this emergency braking is not sufficient to avoid the crash.

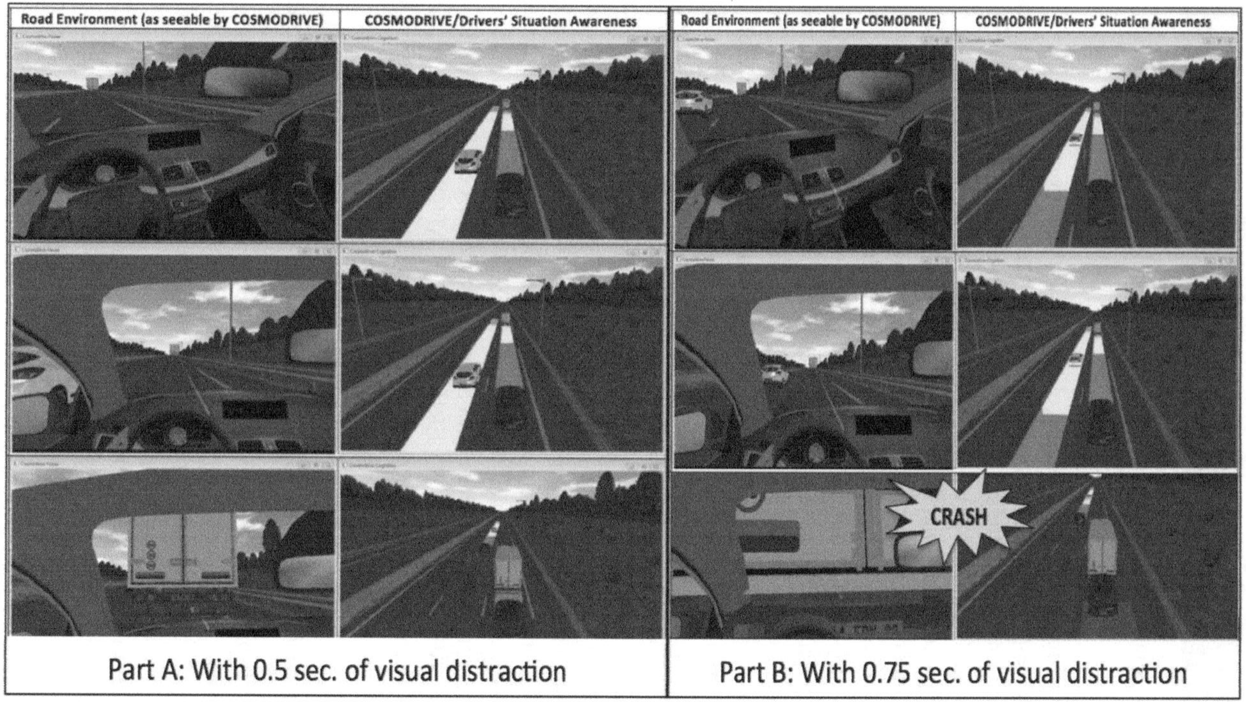

FIGURE 12.5 COSMODRIVE-based simulations assessing the frontal collision risk for a visually distracted driver 0.5 s (part A) vs. 0.75 s (part B) after the truck brakes.

Similarly, COSMODRIVE-based simulations may explore alternative reactions that a driver might implement to avoid a frontal collision, such as changing lanes to overtake the truck (combined or not with braking). In the simulated scenario, however, the LC maneuver was challenging due to two lateral vehicles approaching in the left lane (a white car, followed by a red car), which were progressively overtaking the ego-car piloted by our virtual driver. The following two figures present examples of COSMODRIVE simulations performing such an LC. Like the emergency braking, these simulations assessed how appropriate an LC maneuver could be, depending on the duration of the driver's visual distraction. In case of a distracted driver during 0.5 s after the truck starts braking, the COSMODRIVE-based simulation presented in Figure 12.6 indicates that an LC could be a possible behavior. However, risky and challenging to perform (i.e., the LC must be executed immediately behind the red car, or combined with a strong braking to avoid a frontal collision with the truck).

By contrast, in the case of a visual distraction during 0.75 s (refer to Figure 12.7) the virtual driver cannot perform a safe LC (including when combined with braking).

FIGURE 12.6 COSMODRIVE-based simulation assessing the lateral collision risk for a visually distracted driver during the first 0.5 s after the truck's braking.

FIGURE 12.7 COSMODRIVE-based simulation assessing the lateral collision risk for a visually distracted driver during the first 0.75 s after the truck's braking.

This is because the red car and the truck are too close to the ego-car, preventing the driver from performing this maneuver safely. Simulations indicate that a crash will occur systematically in all variations of this scenario, resulting in either a lateral collision with the red car, a frontal collision with the truck, or both. Moreover, collisions will occur for longer durations of off-road glances, confirming that LC is only appropriate for the shortest visual distractions.

FROM HUMAN MODELING TO "HUMAN-IN-THE-LOOP"

Use of Model-Based Simulation Results to Assess Risks and Identify Human Drivers' Needs

In this section, we will use a driver model like COSMODRIVE as a *digital twin* of future end users to ergonomically specify a future ADAS that can prevent accidents caused by visual distraction while driving. If we consider the truck-following scenario presented in the previous section, simulations based on the driver model provide two key results:

- Using these simulations allows us to identify situations where the driver's visual distraction may cause an accident, depending on when it occurs and, more importantly, its duration after the truck starts braking. For example, suppose the driver has not returned their eyes to the road 0.75 s after the truck starts braking. In that case, they will no longer be able to avoid the accident in the situation simulated by this scenario, whether by braking (refer to Figure 12.6) or by changing lanes to overtake the truck from the left (refer to Figure 12.7).
- According to the available time budget, generating a warning could be appropriate from 0 to 0.5 s after the truck starts braking, but implementing an Automatic Emergency Braking (AEB) by the ADAS will be required in case of a more persistent off-road glance (definitely, if persistent at 0.75 s).

Moreover, model-based simulations can provide additional insights into the risks associated with alternative driving behaviors (likely to be adopted by drivers) and the potential severity of accidents, depending on the duration of drivers' visual distraction when warned by an ADAS in this traffic scenario. Regarding the braking maneuver (refer to Table 12.1), it could be a viable solution for experienced drivers if they return their gaze to the road within 0.75 s. In other words, future ADAS should ensure drivers refocus on the road before this critical deadline. If the distraction lasts longer, human-initiated emergency braking will not avoid a frontal collision with the truck. Furthermore, the longer the visual distraction persists, the greater the accident's severity.

Regarding the LC maneuver, COSMODRIVE-based simulations (summarized in Table 12.2) assessed it as a possible but risky behavior, feasible only if the driver refocuses on the road within 0.5 s. If the visual distraction persists for 0.75 s, a collision with either the lateral car and/or the front truck will inevitably occur, though with varying degrees of severity (*Severe* with the truck, *Minor* with the red car).

TABLE 12.1
Collision Risks Simulation, Accident Severity, and Requirements towards ADAS, According to the Driver's Distraction Duration (Before Braking)

Persistence of Off-Road Glance after the Warning	Accident Risk and Severity (Ego-Car Velocity at the Collision Time with the Truck)	Requirements toward ADAS and HMI
Eyes on-road at 0.50 s (or less)	Possible and safe maneuver	Warning is appropriate
Eyes on-road at 0.75 s	Minor crash (4.4 m/s)	Automatic braking required
Eyes on-road at 1.15 s	Sever crash (10.8 m/s)	Automatic braking required
Eyes on-road at 1.75 s	Fatal crash (16.8 m/s)	Automatic braking required

TABLE 12.2
Collision Risks Simulation, Accident Severity, and Requirements towards ADAS, According to the Distraction Duration (before Changing Lanes)

Persistence of off-road glance after the warning	Accident Risk and Severity (Ego-Car Velocity at the Collision Time with Truck)	Requirements towards ADAS and HMI
Eyes on-road at 0.25 s (or less)	Possible, but risky maneuver	Warning could be appropriate
Eyes on-road at 0.50 s	Highly critical maneuver	Automatic braking required
Eyes on-road at 0.75 s	Crash (severe /minor)	Automatic braking required
Eyes on-road at 1.15 s	Crash (severe/minor)	Automatic braking required

As illustrated through the preceding two tables, model-based simulations can provide a set of functional specifications, decision rules, and HMI requirements of future ADAS to assist human drivers safely and adaptively (i.e., according to their abilities in the traffic conditions). These requirements concern the respective time budgets for delivering warnings *versus activating* an AEB system for collision avoidance or mitigation. The most important one is the necessity to warn the driver about the frontal collision risk in a very short time (less than 0.5 s), due to their increased reaction time when visually distracted, and because the simulation results show that a driver has less than 1.5 s as a whole to react appropriately. For longer visual distractions, it will not be possible for a human driver to manually avoid the accident by braking, changing lanes, or combining these maneuvers. Of course, the traffic conditions studied in our virtual simulations are the most critical variations of the initial generic scenario. However, if they occur, the preceding requirements must be respected. In addition, regarding the time constraint, if the ADAS cannot evaluate the collision risk, make its decision, and warn the driver in less than 0.75 s, the activation of the AEB system must be systematic to avoid or mitigate the collision.

Virtual Human Centered Design (V-HCD) of future ADAS

Simulations of unassisted human drivers are insufficient to support the virtual human-centered design and evaluation of future ADAS. The next step involves the virtual prototyping ADAS systems, then evaluating their benefits through human model-based simulations (i.e., when the ADAS assists the virtual driver). To achieve this objective, COSMODRIVE was integrated with an ADAS simulation platform supporting embedded systems' virtual prototyping and testing (Gruyer et al., 2006; Bellet et al., 2012). The functional architecture of this integrative V-HCD is shown in Figure 12.8.

The COSMODRIVE virtual driver integrated in the V-HCD consists of three modules: a *Perception Module* (simulating human drivers' perceptive functions via a *Virtual Eye* adapted from a virtual camera), a *Cognition Module* (aiming to simulate drivers' situation awareness, risk assessment and decision-making processes), and

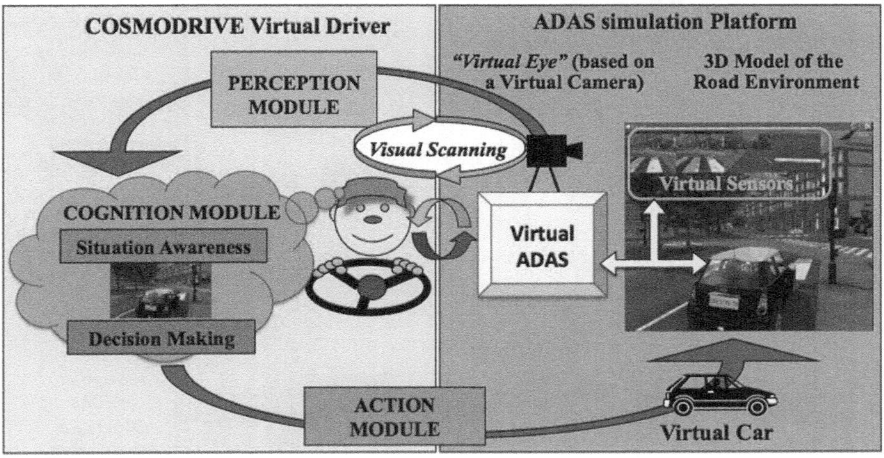

FIGURE 12.8 The V-HCD platform interfaces the COSMODRIVE virtual driver and an ADAS simulation platform.

an *Action Module* (which simulates drivers' actions on virtual car controls, such as pedals, steering wheel or indicators). This virtual driver allows for simulating drivers' activities and behaviors in a virtual road environment. When considered within the framework of the "*V life cycle process*" commonly applied in system engineering (cf. INCOSE, 2023), using a driver model is a key component of a *Human–System Integrative* approach, as presented in Figure 12.9.

This approach enables dynamic simulations of future drivers' interactions with virtual vehicles and their simulated ADAS, allowing for exploring human factor issues at various stages of the design process. During the earliest stages, as discussed in COGNITIVE MODELING AND SIMULATION OF THE CAR DRIVER, these virtual simulations may help designers identify critical scenarios for unassisted driving (refer to Circle no. 1 in Figure 12.9). These critical scenarios refer to driving situations where the human drivers' reliability (assessed through the digital twin's performance and driving error simulations) seems insufficient to avoid accidents or adequately manage risks. These simulations determine ergonomic specifications for human drivers' needs regarding future ADAS (as discussed in FROM HUMAN MODELING TO "HUMAN-IN-THE-LOOP").

During the ADAS prototyping phase (Circle 2), which occurs later in the design process, virtual technical solutions can be iteratively evaluated using COSMODRIVE (cf. Circle 3). These evaluations focus on the critical scenarios initially identified through model-based simulations to assess ADAS effectiveness and efficiency in those specific traffic conditions. Finally, once an ADAS prototype has been virtually validated with COSMODRIVE, evaluations with fundamental human drivers can be implemented in a driving simulator. A specific feature of the V-HCD at this level (cf. Circle 4) is the option to reuse the initial critical scenarios one last time, on a driving simulator interfaced with the V-HCD software, to assess how effectively the ADAS

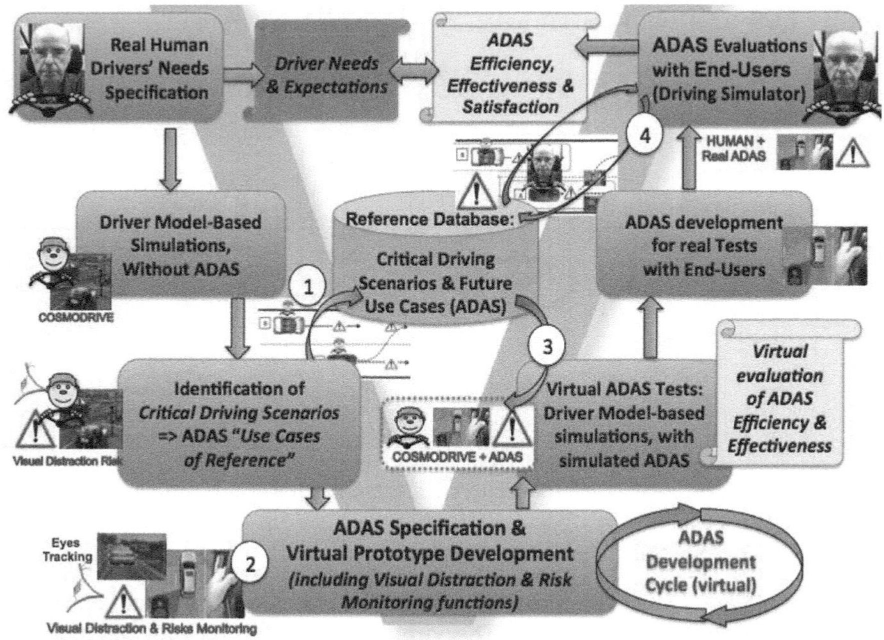

FIGURE 12.9 The V design process of ADAS, as supported by COSMODRIVE in the V-HCD.

can manage the visual distraction risk for fundamental human drivers experiencing these traffic situations.

Unlike the *System Engineering* methods commonly used in the automotive industry for ADAS design, the *Human–System Integrative* approach supported by the V-HCD platform allows designers to better account for human factor issues throughout all stages of the development cycle. The key innovation introduced by this platform is the ability to use a virtual model of future end-users from the beginning of the design process. During the earliest phases, COSMODRIVE-based simulations may assist designers in assessing humans' performance, errors, and accident risks for unassisted driving scenarios. From the critical driving scenarios identified at this level (as "*Use Cases of reference*" for future driving aids), functional and technical requirements can be established to guide the development of future ADAS solutions.

Then, the V-HCD platform can also support engineering teams in the virtual prototyping of ADAS. At this stage, the initial *Use Cases of reference* can be used as *test scenarios* to iteratively evaluate the *effectiveness* (regarding situational risks liable to be managed) and *efficiency* (in terms of task and effort required from the human driver) of different simulated ADAS. These evaluations allow them to assess the potential benefits of each alternative system in terms of road safety. As these evaluations are based on a virtual driver, they can be easily replicated at every phase

FIGURE 12.10 Comparative evaluation of ADAS with COSMODRIVE vs. real humans, from the V-HCD software interfaced with the real car of a driving simulator.

of the V-design cycle, including the earliest stages focused on the conceptual definition of future ADAS.

However, the goal of this approach is not to replace evaluations with real users. Instead, it allows evaluations when involving real humans is not feasible. Ultimately, once virtual prototyping and validation phases are completed, final assessments with real humans should be conducted to evaluate the ADAS's suitability and its acceptance by future end-users. To support this last stage of the evaluation process (e.g., to measure end-users' level of *Satisfaction* towards the ADAS, according to their own needs), the V-HCD platform was interfaced with a driving simulator (Figure 12.10) for implementing immersive experiments with real humans inside an actual car (i.e., a Peugeot 308).

Additionally, this final step allows for comparing virtual simulation results from COSMODRIVE with fundamental drivers' performances (both observed and measured under identical driving scenarios simulated through the V-HCD software). This comparison aids in further developing and validating the COSMODRIVE model, progressively improving its accuracy and enhancing its ability to support the future development of new human–machine systems.

CONCLUSION AND PERSPECTIVES: USE OF THE V-HCD TO PROGRESS FROM THE "AUTONOMOUS" TO THE "HETERONOMOUS" VEHICLE

Facing the recent disruptive advances in vehicle automation, allowing ADAS to take control of the cars, a new set of challenges has emerged concerning their technical design, development, evaluation, and certification. To address these challenges, the automotive industry has progressively developed virtual prototyping methods and tools to assist system engineers. However, technical validation is only part of the equation. The other crucial aspect is assessing how these systems will be used (or misused) by fundamental drivers, and how they will interact with other vehicles and humans inside or outside the car. To better anticipate these human factor issues, a new generation of virtual testing methods, supported by integrative simulation platforms that include both simulated ADAS and human models of future users, must be employed before real systems are introduced on public roads. These tools are essential for predicting potential risks to end users, enhancing the effectiveness, efficiency, and benefits of future ADAS for road safety, and preventing new types of accidents that may emerge from vehicle automation, as occurred when automated systems were introduced in aviation (Young et al., 2007). The V-HCD integrative platform described in this chapter has been designed to meet these objectives. By combining virtual simulations of (1) end-users, (2) ADAS, and (3) driving scenarios that can be shared by both human factors experts and system engineers, this platform fosters collaboration around a shared *Human–System Integration* vision. Although additional efforts are required to implement a human-in-the-loop approach throughout the V-design process entirely, the V-HCD platform was already used for the virtual design of future adaptive HMIs dedicated to automated vehicles (e.g., Bellet et al., 2021), by focusing experiments with end users on the most efficient solutions that were virtually pre-validated with COSMODRIVE.

From a perspective, the V-HCD platform could also be used for complex traffic simulations involving several drivers in interaction. In this context, the objective would not simply be to use a driver model like COSMODRIVE interacting with a virtual ADAS to pilot a virtual vehicle, but to multiply this human–machine system "entity" to simulate traffic scenarios where ADAS and several drivers are interacting with each other. This brings us directly into the multi-agent simulation of a *system* (traffic) *of systems* (humans–machines) discussed in G.A. Boy's introduction, which would allow us to go beyond the ergonomics design of ADAS and enter the realm of traffic engineering. The aim would be then to analyze how, through the virtual simulation of this "*system of systems*," the gradual introduction of ADAS into a varying number of vehicles could impact the safety and efficiency of the road "organizational" system.

A second perspective would focus on developing a multi-agent *Metaverse* to support the human-centered design of artificial intelligence (HCAI) algorithms responsible for driving automated vehicles (Bellet et al., 2021; Bingley et al., 2023). In this context, using a driver model like COSMODRIVE would enable the simulation of varied human behaviors and realistic driving errors (such as dangerous driving maneuvers or unexpected decisions made by other drivers). Allowing the automated driving system to experience such critical situations in a simulated but realistic environment would enable the artificial intelligence (AI) to learn how to manage factual human errors before facing them on the road. To achieve this aim, the challenge will be to enhance the V-HCD as a metaverse composed of multiple COSMODRIVE instances to expose the AI to a diversity of driving behaviors and strengthen its ability to react effectively to dangerous driving behaviors of others, just as human drivers must do. Finally, at the intersection of the two previous perspectives, we are currently developing cognitive simulation models inspired by COSMODRIVE, allowing the simulation of other road users in the V-HCD, such as pedestrians, cyclists, or motorcyclists. The goal is not simply to simulate their "standard behaviors" but also human errors, poor decisions, and dangerous risk-taking, whether intentionally or not, which could lead to critical interaction situations. Again, the aim would be to train the AI to respond effectively when the automated system faces dangerous behaviors from other users.

These perspectives cannot be considered without improving the COSMODRIVE model, which, as we pointed out in COSMODRIVE MODEL VALIDITY AND CALIBRATION FOR SIMULATION, has only been validated for a limited number of driving scenarios. It is also necessary to calibrate this model to simulate different drivers' profiles. This work is currently carried out in the European project BERTHA (https://berthaproject.eu), for which we are developing a decision-making module according to the ages, genders, and driving styles (*aggressive* vs. *cautious*) of human drivers. In parallel, we are also working on interactions between pedestrians and automated vehicles (AVs). This work has allowed us to collect experimental data from young and elderly participants to study their decision-making when crossing a road when an AV is approaching (Bellet et al., 2022). However, a new step is underway in the French project PRéVISION, using a virtual reality headset, to study the actual crossing behavior in front of an AV. From this last project, it is expected to develop a new version of COSMODRIVE capable of simulating pedestrians (i.e., *COSMO-Pedestrian*).

Finally, beyond this roadmap specific to the COSMODRIVE model, the ambition is also to use the V-HCD platform to design future automated vehicles and solve specific ethical issues faced by engineers working on vehicle automation. The challenge would be to move toward developing vehicles that are not "*Autonomous*,"but "*Heteronomous*." In the sense of Levinas (1961), a heteronomous vehicle would be a "*deeply responsible*" vehicle that would act not only in its own interest but also with concern for the needs and vulnerabilities of other road users and those of its occupants. Unlike an autonomous vehicle, which focuses primarily on optimizing its performance, a heteronomous vehicle would need to incorporate an ethical responsibility toward others, leading it to prioritize the safety and well-being of all humans, whether inside or outside the vehicle. To design and train such a heteronomous vehicle, the *V-HCD metaverse* – as a multi-agent simulation – could be used to both simulate passengers and other road users, identify their respective vulnerabilities, simulate ethical dilemmas, and thus train the AI to make responsible decisions that minimize risks for all humans in interaction, even if it means making compromises for the passenger. Therefore, in the case of conflicts between the needs of passengers and those of external road users, the heteronomous vehicle should be able to assess the vulnerability of each, anticipate behaviors, and establish a hierarchy of responsibilities. Trained in this way, a heteronomous vehicle could then adopt an "*empathetic*" driving (toward both its passengers and other road users) and act according to an "*ethic of shared responsibility*," respecting the specificities, needs, and vulnerabilities of all humans interacting in the current traffic situation.

ACKNOWLEDGMENTS

The developments of the COSMODRIVE model and the V-HCD platform presented in this chapter were successively supported in the past by the four European projects funded by the European Union: *ISI-PADAS (Integrated Human Modelling and Simulation to support Human Error Risk Analysis of Partially Autonomous Driver Assistance Systems,* 2008–2011, Grant Agreement GA218552, https://cordis.europa.eu/project/id/218552), *HoliDes (Holistic Human Factors and System Design of Adaptive Cooperative Human–Machine Systems, 2013–2016,* www.holides.eu, GA332933), *HFAuto (Human Factor for Automated Driving*, 2013-2017, GA605817, https://cordis.europa.eu/project/id/605817/reporting/fr) and ***SUaaVE*** *(SUpporting acceptance of automated VEhicle,* 2019–2022, GA814999, https://cordis.europa.eu/project/id/814999).

Currently, these research activities are continuing (regarding pedestrian modeling in the V-HCD) under the French National project ***PRéVISION*** (*Plateforme de Réalité Virtuelle pour l'étude de la Sécurité des futures Interactions véhicule autonome-piétons*) funded by the Délégation à la Sécurité Routière of the French Ministère de l'Intérieur. Moreover, regarding driver modeling, *under the* ***BERTHA project*** *(GA101076360) funded by the European Union (Views and opinions expressed are, however, those of the authors only and do not necessarily reflect those of the European Union or the European Climate, Infrastructure and Environment Executive Agency [CINEA]. Neither the European Union nor the granting authority can be held responsible).*

REFERENCES

Bainbridge, L. (1983). Ironies of automation. *Automatica*, 19(6), 775–779.

Bellet, T. (2011). Analysis, modeling, and simulation of human operators' mental activities. In G. A. Boy (Ed.), *The Handbook of human–machine interaction: A human-centered design approach*. Ashgate, pp. 23–52.

Bellet, T., Bailly-Asuni, B., Mayenobe, P., & Banet, A. (2009). A theoretical and methodological framework for studying and modelling drivers' mental representations. *Safety Science*, 47, 1205–1221.

Bellet, T., & Banet, A. (2023). UTAUT4-AV: An extension of the UTAUT model to study intention to use automated shuttles and the societal acceptance of different types of automated vehicles. *Transportation Research Part F: Traffic Psychology and Behaviour*, 99, 239–261.

Bellet, T., Banet, A., Petiot, M., Richard, B., & Quick, J. (2021). Human-centered AI will support the adaptive management of human–machine transitions with vehicle automation. *Information*, 12(1), 13.

Bellet, T., Cunneen, M., Mullins, M., Murphy, F., Pütz, F., Spickermann, F., Braendle, C., & Baumann, M. F. (2019). From semi to fully autonomous vehicles: New emerging risks and ethico-legal challenges for human–machine interactions. *Transportation Research Part F: Traffic Psychology and Behaviour*, 63, 153–164.

Bellet, T., Hoc, J. M., Boverie, S., & Boy, G. A. (2011). From human–machine interaction to cooperation: Towards the integrated Copilot. In C. Kolski (Ed.), *Human–computer interaction in transport*. Ashgate, pp. 129–156.

Bellet, T., Laurent, S., Bornard, J. C., Hoang, I., & Richard, B. (2022). Interaction between pedestrians and automated vehicles: Perceived safety of yielding behaviors and benefits of an external human–machine interface for elderly people. *Frontiers in Psychology*, 13, 1021656.

Bellet, T., Mayenobe, P., Bornard, J. C., Gruyer, D., & Claverie, B. (2012). A computational model of the car driver interfaced with a simulation platform for future Virtual Human-Centred Design applications: COSMO-SIVIC, *Engineering Applications of Artificial Intelligence*, 25, 1488–1504.

Bingley, W. J., Curtis, C., Lockey, S., Bialkowski, A., Gillespie, N., Haslam, S. A., Steffens, N., Ko, R. K. L., Wiles, J., & Worthy, P. (2023). Where is the human in human-centered AI? Insights from developer priorities and user experiences. *Computers in Human Behavior*, 141, 107617.

Bornard, J. C. (2012). Développement d'un modèle du conducteur automobile: de la modélisation cognitive à la simulation numérique. Doctoral dissertation, Université de Bordeaux.

Bornard, J. C., Bellet, T., Mayenobe, P., Gruyer, D., & Claverie, B. (2012). Modelling visual distraction effects on driver's perception and cognition. *International Journal of Human Factors Modelling and Simulation*, 3(1), 48–65.

Bornard, J. C., Sassman, M., & Bellet, T. (2016). Use of a computational simulation model of drivers' cognition to predict decision making and behaviour while driving. *Biologically Inspired Cognitive Architectures*, 15, 41–50.

Boy, G. A. (2020). *Human-systems integration: From virtual to tangible.* Boca Raton: CRC Press, Taylor & Francis Group.

Deniel, J. (2019). Conscience du risque et automatisation de la conduite: de l'analyse des processus cognitifs en situation de changement de voie à la conception centrée sur l'humain d'un système d'automatisation human-like. Doctoral dissertation, Université de Bordeaux.

Deniel, J., Bornard, J. C., Claverie, B., & Bellet, T. (2019). Innovative methodological framework for virtual human centered design: An application to automated driving effects on driver's decisions-making. *Proceeding of the HSI2019 Conference, Biarritz, 11–13 September 2019.*

Donà, R., & Ciuffo, B. (2022). Virtual testing of automated driving systems. A survey on validation methods. *IEEE Access*, 10, 24349–24367.

Endsley, M. R. (1995). Toward a theory of situation awareness in dynamic systems. *Human Factors*, 37(1), 32–64.

Georgeon O. (2008). *Analyse de traces d'activité pour la modélisation cognitive*. Doctoral dissertation. Université Lumière Lyon 2, France.

Georgeon, O., Mille, A., Bellet, T., Mathern, B., & Ritter, F. E. (2012). Supporting activity modelling from activity traces. *Expert Systems*, 29(3), 261–275.

Gruyer, D., Royere, C., du Lac, N., Michel, G., & Blosseville, J. M. (2006). SiVIC and RT-MAPS Interconnected platforms for the conception and evaluation of driving assistance systems. *Proceedings of the ITS World Congress, London, UK, October.*

Hall, E. T. (1966). *The hidden dimension.* New York: Doubleday.

Hoc, J. M., Young, M. S., & Blosseville, J. M. (2009). Cooperation between drivers and automation: Safety implications. *Theoretical Issues in Ergonomics Science*, 10, 135–160.

Hollnagel, E., & Woods, D. D. (2005). *Joint cognitive systems: Foundations of cognitive systems engineering*. London: CRC Press.

INCOSE (2023). *Systems engineering handbook: A guide for system life cycle processes and activities*, version 5.0. Hoboken, NJ: Wiley, ISBN: 978-1-119-81429-0. https://sebokwiki.org/wiki/Vee_Life_Cycle_Model.

ISO 9241-11 (2018). *Ergonomics of human–system interaction – Part 11: Usability: Definitions and concepts*. Geneva: International Organization for Standardization.

Kyriakidis, M., de Winter, J. C., Stanton, N., Bellet, T., van Arem, B., Brookhuis, K., Martens, M. H., Bengler, K., Andersson, J., Merat, N., Reed, N., Flament, M., Hagenzieker, M., &

Happee, R. (2019). A human factors perspective on automated driving. *Theoretical Issues in Ergonomics Science*, 20, 1–27.

Levinas, E. (1961). *Totalité et infini: essai sur l'extériorité*. La Haye: Martinus Nijhoff.

Madigan, R., Louw, T., Wilbrink, M., Schieben, A., & Merat, N. (2017). What influences the decision to use automated public transport? Using UTAUT to understand public acceptance of automated road transport systems. *Transportation Research Part F: Traffic Psychology and Behavior*, 50, 55–64.

Mathern, B. (2012). *Découverte interactive de connaissances à partir de traces d'activité: Synthèse d'automates pour l'analyse et la modélisation de l'activité de conduite automobile*. Doctoral dissertation, Université Claude Bernard-Lyon I.

Minsky, M. (1975). A framework for representing knowledge. In P. H. Winston, (Ed.), *The psychology of computer vision*. New York: McGraw-Hill, pp. 211–277.

Neisser, U. (1976). *Cognition and reality: Principles and implications of cognitive psychology*. San Francisco: W.H. Freeman.

Nordhoff, S., de Winter, J., Kyriakidis, M., Van Arem, B., & Happee, R. (2018). Acceptance of driverless vehicles: Results from a Large Cross-National Questionnaire Study. *Journal of Advanced Transportation*, 2018, 1–22.

Ochanine, V. A. (1977). Concept of operative image in engineering and general psychology. In B. F. Lomov, V. F. Rubakhin, & V. F. Venda (Eds.), *Engineering psychology*. Moscow: Science Publisher.

Paris, J. C. (2014). *Ingénierie cognitive pour l'aide à la conduite automobile de la personne âgée: analyse et modélisation de l'activité de conduite en situation naturelle pour la conception de fonctions de monitorage*. Doctoral dissertation, Université de Bordeaux.

Paris, J. C., Bellet, T., Marin-Lamellet, C., Cour, M., Boverie, S., & Claverie, B. (2015). Assistance aux conducteurs âgés: Analyse de l'activité pour la conception de futures fonctions de monitorage de la conduite automobile. *RTS. Recherche, transports, sécurité*, 2014(04), 227–243.

Singh, H., Midlam-Mohler, S., & Tulpule, P. (2021). Simulation-based virtual testing for safety of ADAS Algorithms – Case studies. *SAE Technical Paper 2021-01–0114*. https://doi.org/10.4271/2021-01-0114.

Son, T. D., Hubrechts, J., Awatsu, L., Bhave, A., & Van der Auweraer, H. (2018). A simulation-based testing and validation framework for ADAS development. Proceedings of 7th Transport Research Arena (TRA 2018), *April 16–19, Vienna, Austria*.

Young, M. S., Stanton, N. A., & Harris, D. (2007). Driving automation: Learning from aviation about design philosophies. *International Journal of Vehicle Design*, 45, 323–338.

13 Scenario-Based Systems Integration

Dimitri Masson and Guy André Boy

INTRODUCTION

Scenario-Based Design (SBD) and Human System Integration (HSI) can be associated with creating a comprehensive framework for addressing various kinds of complexity of a sociotechnical system (STS). This integration will be called scenario-based systems integration (SBSI). It facilitates a human-centered approach, ensuring systems are technically robust and adapted to operational needs and realities. This chapter explores the connection between SBD and HSI practices, focusing on the Concept of Operations (CONOPS), use cases, function allocation, and the early integration of verification and validation through humans in the loop simulations. In the specific context of STS, we are particularly interested in the following aspects: multi-agent collaboration, multi-perspective integration, and multi-level detail.

Developing an integrative approach to SBSI for human-centered STS design requires considering human and machine agents or systems. We typically represent people and technology as human and machine systems/agents.[1] In both cases, an agent has a structure and a function. Developing a scenario requires describing the various human and machine agents interacting with each other as a narrative. Each agent is assigned appropriate functions, enabling it to perform the prescribed tasks.

SBD supports a multi-agent perspective by providing a structured framework for capturing and analyzing interactions among agents within an STS. SBD lets design teams explore different scenarios, helping them understand how agents interact and behave in specific situations. Scenarios force us to represent the various agents in declarative terms, describing their objectives, functions, and capabilities, as well as in procedural terms (Boy, 2020, 2023). For example, scenarios can be developed using Business Process Model and Notation (BPMN) representation (White & Bock, 2011).

SBD involves a variety of stakeholders, including manufacturers, technical experts, and safety managers, who contribute their views on how the system will operate under different conditions and from various angles. A large variety of viewpoints is generated, and synthetic scenarios are produced to enhance safety, efficiency, and comfort, for example. Designing scenarios enables eliciting operational issues of robustness, flexibility, costs, and safety, to cite a few. It also allows addressing training needs and ethical concerns. SBD explicitly addresses the holistic aspects of the STS being designed.

DOI: 10.1201/9781003666431-16

The first scenario rehearsal usually focuses on the leading agents' dialogue. Subsequent rehearsals gradually incorporate increasingly complex features. SBSI is an incremental process that enables design teams to involve operational players and progressively familiarize themselves with the solutions under development. More specifically, agents, relationships between agents, and generated events get incrementally more transparent and precise.

For example, "Alice, a human operator, initiates a verification process, and an alarm occurs. She verifies the sensor data to determine whether this is a false positive. She muted that specific alarm and followed the recommended procedure, following the P-005 procedure to verify the carbon dioxide T6493 on Deck 4. At 08:00, another alarm rings in the control room, and so on."

Levels of detail are progressively adapted to satisfy the meaning of the scenario being developed. For example, "Alice initiates the software-based verification process connected to the main production process, etc." Each task or event can be broken down into a sub-scenario, like the process used in hierarchical task modeling. Each task requires a function to be performed to produce an activity. This function can be described through a scenario that aims to fulfill it.

This incremental composition/decomposition process allows the system to be developed more concretely in an operational sense, considering normal, abnormal, and emergency situations. At this point, the design team discovers the system-of-systems nature of the STS and necessarily creates teams of teams to describe scenarios appropriately. Indeed, various aspects emerge, such as technological, organizational, and human.

SCENARIO-BASED DESIGN FOUNDATIONS

The scenario concept covers various artifacts depending on the field that produces or uses them. Etymologically, the term "scenario" comes from the word scena, "scene," and describes a set of events that happen to specific agents (Baek et al., 2022). More commonly, a scenario can be seen as the story of a play or a movie. Futurists consider that "a scenario is a product that describes some possible future state and/or tells the story about how such a state might emerge. The former are referred to as end state or even day in the life scenarios; the latter are chain (of events) scenarios or future histories" (Bishop et al., 2007).

A scenario is not created out of the blue. It belongs to a project and a set of other scenarios. It evolves during the project. In the realm of systems engineering (SEs), scenarios play a pivotal role in the definition of various concepts. Scenarios enable the definition of a CONOPS that explains how an organization will function to achieve its missions, goals, and objectives, serving as a basis for guiding the characteristics of future activities and associated systems. Scenarios emphasize the impact of the system being developed. It forces the elicitation of a mandatory background and stakeholders' requirements. In addition, as the CONOPS describes how a system will be used during operations, it supports detailing its human-centered capabilities, functionalities, and risks. Scenario development forces clear definitions of use cases and contexts of operations.

A scenario needs a goal. Otherwise, it is just a story (Baek et al., 2022). This objective depends on the current project (e.g. the specific functionality of the system under development that the scenario is designed to clarify) and on a method that depends on the point in the life cycle at which the scenario is developed (e.g., at the beginning of a project, the goal of a scenario might be to help define the scope of the functionality, while in the end, the goal might be to validate the functionality). For example, a scenario could be developed for requirements discovery, validation, system specification, system design, coding, integration, testing, operations, and maintenance (Alexander & Maiden, 2005).

The Simulation Interoperability Standards Organization delineates three distinct scenario types within the SE process (Guide for Scenario Development (GSD) Product Development Group, 2018). Operational scenarios, formulated by subject matter experts (SME) using a domain-specific language, serve as authoritative blueprints, describing the crucial real-world elements for simulation representation. These scenarios establish mandatory prerequisites and provide an overview of the dynamic characteristics essential to achieving staff operational objectives in a simulation environment. Conceptual scenarios, offering a comprehensive depiction of the simulated world, are integral to subsequent stages of simulation environment engineering. Designed primarily by experts in modeling and simulation (M&S), collaborative efforts with operations people and SMEs guarantee precise scenarios, aligned with operational requirements, and provide the necessary details for development. Executable scenarios, serving as detailed specifications for initializing and executing a simulation environment, are orchestrated by operations personnel with potential support from M&S experts or SMEs. This transition from conceptual to executable scenarios strives for direct accessibility by member applications, presenting the required level of detail in a specified format for effective implementation.

In SBSI, knowledge representation models play a crucial role in supporting the capture and communication of the essential aspects of the envisioned system. The Cooperative Requirements Engineering with Scenarios (CREWS) framework argues that the form of a scenario has two facets: description and presentation (Rolland et al., 1998). The "description" facet encompasses the medium (e.g., text, graphics, and images) and the notation, representing a certain degree of formality. The presentation facet deals with the interactivity and the animation of the scenario.

One of the most popular forms of scenario representation is the narrative form, which can take the shape of pure text or operational personnel stories. There are various degrees of formality ranging from free form (Carroll, 1999; Rosson & Carroll, 2002) to more structured representation such as user stories (Cohn, 2004), which provide a structured template ("As a [type of user], I want [some goal] so that [some reason]") that helps maintain a user-centric perspective throughout the project lifecycle, ensuring that design and development efforts are focused on operational needs and experiences or use cases are more structured version of the pure textual version (Cockburn & Cockburn, 2008).

Graphic storytelling is another effective way of representing scenarios, with tools such as storyboards providing a visual representation of the sequence of events or interactions within a scenario. Storyboards can help stakeholders visualize the operational personnel's journey and understand the flow of activities more intuitively, making identifying potential design improvements and requirements easier.

Diagrams are also commonly used to represent scenarios, but their notation is more formal. The first version of this type of diagram dates to the 1920s, with flow charts (Gilbreth & Gilbreth, 1921). A flow diagram represents the scenario in the form of an algorithm, allowing not only linear stories to be described, but also branched and looped stories. It shows the stages: tasks or activities to be carried out as rectangular boxes, and decision points as diamonds. The flow of information is represented by arrows moving from one box to the next.

Unified Modeling Language/Systems Modeling Language (UML/SysML) has several diagrams that enable the representation of the dynamicity of systems and thus can be used to model scenarios (Fowler, 2018). UML/SysML-style sequence diagrams use-case diagrams (Cockburn & Cockburn, 2008), activity diagrams (Jarraya et al., 2009), and sequence diagrams are popular choices. These diagrammatic representations help illustrate the interactions between different system components, the flow of information, and the sequence of actions within a scenario. Process modeling techniques such as BPMN and customer journey mapping offer additional ways to represent scenarios by capturing the high-level processes and touchpoints involved, enabling a holistic view of the system behavior and interactions during operations. Even more formal models can be used, such as formal modeling languages like Petri nets (PN) and Monterey Phoenix (MP) (Auguston et al., 2015), which offer a rigorous, mathematical method for representing scenarios. PN, for instance, can capture concurrency, synchronization, and resource allocation aspects of system behavior. At the same time, MP provides a specialized language and tool for modeling and reasoning about behavior, particularly in complex systems like those in the defense domain.

Hierarchical task modeling is another valuable representation model for scenarios where the focus is less on the linear flow of events but on providing a structured approach to decomposing complex tasks into simpler subtasks. Hierarchical task models help in understanding the cognitive aspects of operational personnel interactions. They can guide the design of user interfaces and system functionalities to support efficient task completion, with cognitive task analysis techniques like ConcurTaskTrees (CTT), a diagrammatic notation for specifying task models (Paterno et al., 1997).

SBSI IN STS DEVELOPMENT

STS development supported by SBSI involves progressively improved scenarios, which evolve along two axes: objective and formalism. The purpose of scenarios starts by describing the existing situation and then progresses to present hypothetical solutions until precisely detailing the chosen solution. The form of the scenarios will also change; the formalism used must evolve during the development of the sociotechnical system to allow flexibility at each stage. Less formal models are used for exploratory purposes, then formalized to define, anchor, and validate conceptual solutions. The evolution is not linear on either axis; we can move from a very formal "as-is" model to an informal description of potential solutions that will be refined and formalized iteratively. Similarly, conceptual scenarios can evolve in parallel with the ongoing development of operational scenarios, influencing each other. Alternatively, simulation scenarios may potentially reframe the problem.

An HSI team is crucial throughout the SBSI process (Rippy, 2021). SMEs are pivotal in designing and analyzing sociotechnical systems, providing essential insights into operational and design aspects. Operations-dedicated subject matter experts (O-SMEs) are key in eliciting procedural knowledge, identifying current process inefficiencies, and pinpointing improvement areas (Hoffman et al., 1995; Lewicki et al., 1987). This early stage of As-Is analysis is critical for understanding existing operational workflows and user interactions.

Meanwhile, specialist design-focused subject matter experts (D-SMEs) provide an in-depth analysis of design principles, architecture, and technical system constraints, guiding the strategic direction of potential innovations and ensuring technical feasibility. Their declarative knowledge is essential for informing analysis of the current state (As-Is), shaping the desired scenario (To-Be) development, and integrating information into system requirements and design specifications.

However, this knowledge can be difficult due to its often-tacit nature. Effective strategies include structured interviews, participatory design sessions, and cognitive task analysis (Clark et al., 2007; Boy, 1998). Integrating O-SME and D-SME knowledge is essential to achieve a holistic understanding of the system, enabling the development of STSs that are efficient and adapted to operational personnel's needs. Finally, the To-Be scenarios are transformed into simulation scenarios for simulation with human intervention.

Starting with As-Is Scenarios

Initiating scenario development for STSs design requires a foundational understanding of the STS's purpose, the operational personnel it serves, and the overall operational context. This initial conceptual phase is critical for ensuring that subsequent design efforts align with operational personnel's needs and STS goals. A methodical approach to qualitative and quantitative data collection and analysis forms the basis of this understanding, facilitating the identification of key tasks, inherent challenges, and potential opportunities for system improvement.

During the conceptual phase, it is essential to have human factors specialists to acquire the operational personnel's needs and to identify O-SMEs as early as possible. Scenarios are the synthesis of this data acquisition. In addition, it is essential to identify the agents involved in the STS and characterize their goals, capabilities, and interactions in the scenarios. Many data collection techniques, such as semi-structured interviews with stakeholders, operational personnel, system operators, and decision-makers, can be used. These fundamental techniques uncover insights about system use and operational personnel expectations. Deploying surveys allows data to be collected from a broad operational personnel base, providing quantitative insights into operational personnel satisfaction, system usage patterns, and feature requests. Direct observation and ethnographic studies offer rich, contextual insights into operational personnel interactions with STSs, thus revealing nuances that surveys and interviews might overlook.

Data analysis highlights trends, operational personnel's needs, and obstacles encountered when interacting with the system. The business acumen of O-SMEs, combined with the conceptual vision of D-SMEs, is essential for decoding the influence of various factors (cultural, social, and environmental) on operational personnel interactions within the STS. The following steps are advised:

- *Determining Essential Operational Tasks:* It is pivotal to identify the primary tasks operational personnel undertake within the STS and the processes they follow. This step is instrumental in directing the focus of scenario development toward enhancing the operational personnel's experience concerning these vital tasks.
- *Identifying Operational Challenges:* This involves recognizing the issues operational personnel encounter, which could stem from the system's design, external influences, or inadequate training. Pinpointing these challenges is essential for crafting scenarios to reduce or remove such obstacles.
- *Recognizing Opportunities for Enhancement:* Opportunities for innovation or enhancement can emerge from operational personnel feedback, observed makeshift solutions by operational personnel, or emerging technology trends. These insights are valuable for guiding the development of new features or improvements in the system.

Anticipating To-Be Scenarios

Selected scenarios now describe a To-Be projection. The team uses designers, engineers, and D-SMEs to produce these scenarios. Regarding SE, we move on to the development phase, which primarily uses virtual prototypes. The generation of these new scenarios is an iterative process involving the analysis of existing scenarios (As-Is for the first iteration) and then the evolution of the proposed scenarios.

Merging insights from observational and ethnographic methodologies with the specialized knowledge of O-SMEs and D-SMEs significantly enriches scenario development. This collaboration begins with an exhaustive data collection phase, combining direct observation and ethnographic research to thoroughly document operational personnel interactions and the contexts in which they occur. The operational knowledge from O-SMEs offers invaluable context on existing workflows and operational personnel interactions, while D-SMEs bring a deep understanding of the system's design and technical infrastructure. Examining existing documentation, such as system logs, operational personnel feedback, and helpdesk tickets, furnishes a historical view of the system's performance and prevalent operational personnel issues. This review helps pinpoint usage patterns or persistent challenges the scenario development must address, ensuring the resulting scenarios are grounded and aligned with the system's long-term objectives.

The *Processus Decisionnel* (French for Decision-Making Process, PRODEC) methodology (Boy et al., 2024) proposes an analysis of cognitive functions, known as Cognitive Functions Analysis (CFA; Boy, 1998), which makes it possible to elicit functions from task and activity analyses, to break down these functions, and to classify them systematically. The CFA approach is a declarative modelling method based on operational experience. By grouping and classifying elicited functions, we can assimilate a smaller number of more generic functions and define their validity contexts and resource requirements. CFA can be done directly from interviews with experts and observations of human and machine agents/systems. It becomes very complex when there are many agents/systems interacting with each other. This process results in emergent functions associated with various agents/systems (structures) that link prescribed tasks to performed activities. Each elicited cognitive function is declared with a role, a context of validity, and a set of resources that are functions themselves.

Experiencing Simulation Scenarios

Simulation scenarios are a pivotal phase in the SBSI of STSs. They are indispensable in the early and iterative testing stages of the System of Interest (SoI). They facilitate the identification of emergent functions, novel properties, and potential operational interactions critical to the system's development and refinement.

Early testing within the design process is fundamental, enabling stakeholders to explore and validate the SoI in a controlled yet dynamic environment. Simulation scenarios mimic real-world interactions and processes, offering a practical and efficient means to evaluate system performance, usability, and effectiveness. This proactive engagement with the SoI ensures that potential issues are identified and addressed well before the full-scale deployment, saving time, resources, and effort in later stages. "An executable scenario is a specification of a specific scenario containing the level of detail necessary for initialization and execution of a simulation environment" (Guide for Scenario Development (GSD) Product Development Group, 2018).

This definition underlines the importance of precision and detail in scenario development, so that each scenario provides a solid framework for testing and analysis. Executable scenarios form the backbone of simulation-based testing, enabling stakeholders to systematically assess SoI performance from different angles.

A core component of human-in-the-loop simulation (HITLS) scenarios is incorporating the human element in virtual testing. This approach emphasizes the active involvement of operations personnel in the HITLS process, allowing for real-time feedback and interaction within the STS at stake. HITLS-supported testing is instrumental in uncovering operations-centric insights, emergent behaviors, and potential system limitations that may not be apparent through automated testing alone. It fosters a deeper HSI understanding and contributes to developing an intuitive, efficient STS aligned with operational personnel expectations. Such testing must start as early as possible; the essence of SBSI lies in implementing what-if scenarios that can be effectively supported, for example, by the Wizard of Oz (WoZ) methodologies (Cooke et al., 2020).

The WoZ technique (Kelley, 1983) is a research and design-based method where participants interact with a system that they believe to be autonomous, but which is being operated or partially operated by an unseen human, the "wizard." This approach allows researchers and designers to simulate the functionality of a system that has not yet been fully developed, facilitating the early exploration of operational personnel interactions, responses, and overall experience. The technique derives its name from the character in L. Frank Baum's novel, which uses illusion to appear more powerful than it is, is analogous to the simulated functionalities of the system being tested. The WoZ method is beneficial in human–computer interaction (HCI) and usability testing, where understanding how operational personnel interact with new technologies is crucial.

In the STS context, the WoZ technique can be exceptionally valuable, given the complexity and interconnectedness of human and technical components within these systems. STS design challenges often revolve around optimizing social and technical elements to achieve system goals effectively. By employing WoZ, designers and researchers can simulate parts of the STS that involve complex human–technology interactions before these system components are fully operational. This approach enables emerging behaviors, system requirements, and potential design flaws to be identified and explored early in the development process, ensuring that the final system is better adapted to operational personnel's needs and system objectives.

The following steps are advised:

- *Incremental Testing and Complexity Enhancement:* Begin with early-stage testing to establish a baseline understanding of system interactions and operational personnel experiences. Gradually increase the tangibility and complexity of simulations, employing methods such as the WoZ to simulate advanced system functionalities before they are fully developed. This approach allows for iterative refinement based on operational personnel feedback and system performance insights.
- *Integrating Human-in-the-Loop Testing:* Implement HITLS to closely examine and evaluate human–system interactions in real time. This direct involvement of operational personnel in the testing process ensures that the system's design and functionalities are aligned with human capabilities, expectations, and needs, enhancing the overall operational personnel experience and system usability.
- *Aligning Simulations with Future Aspirations:* Design and conduct simulation scenarios that mirror the envisioned To-Be state of the system. These forward-looking simulations should reflect the anticipated improvements and functionalities of the system, allowing stakeholders to assess the feasibility, effectiveness, and operational personnel's acceptance of proposed changes and innovations in a controlled environment.

DISCUSSION

This chapter highlights the indispensable role played by SBD and HSI in the development of complex STSs, underlining the profound value of multidisciplinary collaboration toward SBSI.

The significance of SBSI stems from its roots in HCI techniques and its expansion into SEs. Grounding the design process in human experiences allows for a more comprehensive understanding of the role of humans within these systems. It extracts procedural knowledge from SMEs and stakeholders, enabling a better understanding of complex system problems and facilitating modeling dynamic and complex STSs.

The PRODEC methodology addresses both procedural scenario development and declarative configuration design using an SBSI approach, where As-Is scenarios are initially developed and discussed to elicit emergent phenomena and properties of the STS at stake, leading to a creative session where new system prototypes are proposed for an implementation of To-Be scenarios. Prescribed tasks are "injected" into the STS and HITLS, uncovering emerging functions and possibly forgotten structures that will be considered in the subsequent design phase.

Because of the convergence of diverse expertise (e.g., engineering, design, operations personnel, and ethicists), SBSI excels in promoting the creation of systems that are not only at the cutting edge of technological progress but also deeply rooted in human values and considerations. This collaborative approach is essential, as it ensures that all aspects of HSI are treated with the utmost care, facilitating the development of operations-centric STSs that are adaptable and ready to navigate the complexities of an ever-changing technological landscape.

In addition, the transition from SBD to SBSI requires a radical shift from flexible narratives to formal models, which improves the accuracy of descriptions, facilitates integration with existing formal representations used in SEs, and extends the application of SBD to the entire system life cycle. The introduction of key concepts such as CONOPS, which is expected to become Concept of Use (CONUSE), marks a significant step forward in SE, establishing an essential link between SBD and various SE aspects, including use cases, function allocation, and the early integration of verification and validation through human-intervention simulations. Given the inherent complexity and emerging properties of STSs, early and thorough human testing is imperative to ensure that the system meets its intended purpose and operational personnel requirements and can adapt to the unpredictable dynamics of sociotechnical environments.

SBSI highlights the importance of adopting holistic perspectives in developing STSs by fostering an environment where multidisciplinary insights and expertise converge. This approach guarantees that these systems are as versatile and dynamic as the societies they are designed to serve, bridging the gap between technological innovation and human-centric design, and ensuring that STSs remain resilient and relevant in the face of technological evolution and societal change.

NOTE

1 The term "system" used in systems engineering expresses nearly the same concept as the term "agent" used in artificial intelligence. In this chapter, we will use alternatively one or the other depending on the context. In this section, for example, it is more appropriate to use the term "agent."

REFERENCES

Alexander, I.F. & Maiden, N. (2005). *Scenarios, Stories, Use Cases: Through the Systems Development Life Cycle*. Wiley, Hoboken, USA.

Auguston, M., Giammarco, K., Baldwin, W.C., Crump, J. & Farah-Stapleton, M. (2015). Modeling and Verifying Business Processes with Monterey Phoenix. *Procedia Computer Science, 2015 Conference on Systems Engineering Research*, 44 (January): pp. 345–53. doi: 10.1016/J.PROCS.2015.03.055.

Baek, Y.-M., Cho, E., Shin, D. & Bae, D.-H. (2022). Literature Review to Collect Conceptual Variables of Scenario Methods for Establishing a Conceptual Scenario Framework. arXiv abs/2205.08290. doi: 10.48550/ARXIV.2205.08290.

Bishop, P., Hines, A. & Collins, T. (2007). The Current State of Scenario Development: An Overview of Techniques. *Foresight*, 9 (1): pp. 5–25. doi: 10.1108/14636680710727516.

Boy, G.A. (1998). Cognitive Function Analysis (Contemporary Studies in Cognitive Science and Technology), Praeger, CT, USA.

Boy, G.A. (2020). *Human Systems Integration: From Virtual to Tangible*. CRC Press, Taylor & Francis Group, Boca Raton, FL, USA. www.taylorfrancis.com/books/9780429351686

Boy, G.A. (2023). An Epistemological Approach to Human Systems Integration. *Technology in Society Journal (Open Access)*, p. 102298. https://doi.org/10.1016/j.techsoc.2023.102298

Boy, G.A., Masson, D., Durnerin, E. & Morel, C. (2024). PRODEC for Human Systems Integration of Increasingly Autonomous Systems. *Systems Engineering Journal,* 27(5), 805–826. Hoboken. https://doi.org/10.1002/sys.21751

Carroll, J.M. (1999). Five Reasons for Scenario-Based Design. *Proceedings of the Hawaii International Conference on System Sciences*, 123. doi: 10.1109/HICSS.1999.772890.

Clark, R. E., Feldon, D., vanMerrienboer, J., Yates, K, and Early, S. (In Press for August 2007). Cognitive Task Analysis. In Spector, J. M., Merrill, M. D., van Merriënboer, J. J. G., & Driscoll, M. P. (Eds.) *Handbook of research on educational communications and technology* (3rd ed.). Mahwah, NJ: Lawrence Erlbaum Associates.

Cockburn, A. & Cockburn, L. (2008). *Writing Effective Use Cases*. Pearson Education India.

Cohn, M. (2004). Effective user stories. Extreme Programming and Agile Methods – XP/Agile Universe 2004, Proceedings, Springer, p. 208, https://doi.org/10.1007/978-3-540-27777-4_35

Cooke, N., Demir, M. & Huang, L. (2020). A Framework for Human-Autonomy Team Research. In *Engineering Psychology and Cognitive Ergonomics. Cognition and Design*, edited by Don Harris and Wen-Chin Li, pp. 134–1346. Springer International Publishing, Cham. https://doi.org/10.1007/978-3-030-49183-3_11

Fowler, M. (2018). *UML Distilled: A Brief Guide to the Standard Object Modeling Language*. Addison-Wesley Longman Publishing Co., Inc., Boston, MA, USA, ISBN: 978-0-321-19368-1.

Gilbreth, F.B. & Gilbreth, L.M. (1921). Process Charts: First Steps in Finding the One Best Way to Do Work. *Journal of Fluids Engineering*, 43: pp. 1029–1043.

Guide for Scenario Development (GSD) Product Development Group (2018). *Guideline on Scenario Development for Simulation Environments*. Simulation Interoperability Standards Organization (SISO), Orlando.

Hoffman, R.R., Shadbolt, N.R., Burton, A.M. & Klein, G. (1995). Eliciting Knowledge from Experts: A Methodological Analysis. *Organizational Behavior and Human Decision Processes*, 62 (2): pp. 129–158. https://doi.org/10.1006/obhd.1995.1039

Jarraya, Y., Debbabi, M. & Bentahar, J. (2009). On the Meaning of SysML Activity Diagrams. In *16th Annual IEEE International Conference and Workshop on the Engineering of Computer-Based Systems*, pp. 95–105.

Kelley, J.F. (1983). An Empirical Methodology for Writing User-Friendly Natural Language Computer Applications. In *Proceedings of the SIGCHI Conference on Human Factors in Computing Systems – CHI'83*, pp. 193–96. ACM Press, Boston, MA. https://doi.org/10.1145/800045.801609

Lewicki, P., Czyzewska, M. & Hoffman, H. (1987). Unconscious Acquisition of Complex Procedural Knowledge. *Journal of Experimental Psychology: Learning, Memory, and Cognition*, 13 (4): pp. 523–30. doi: 10.1037/0278-7393.13.4.523.

Paterno, F., Mancini, C. & Meniconi, S. (1997). ConcurTaskTrees: A Diagrammatic Notation for Specifying Task Models. In *Human–Computer Interaction INTERACT*'97: IFIP TC13 International Conference on Human–Computer Interaction, *14–18 July 1997, Sydney, Australia*, pp. 362–369. Springer.

Rippy, L.O. (2021). *NASA Human Systems Integration Handbook*. Document # 20210010952, Langley Research Center, VA, USA.

Rolland, J.-P., Parker, W., & Stumpf, H. (1998). A psychometric examination of the French translation of the NEO-PI-R and NEO-FFI. *Journal of Personality Assessment*, 71, 269–291.

Rosson, M.B. & Carroll, J.M. (2002). Scenario-Based Design. In *The Human–Computer Interaction Handbook: Fundamentals, Evolving Technologies and Emerging Applications*, pp. 1032–1050. L. Erlbaum Associates Inc., USA.

White, S.A. & Bock, C. (2011). *BPMN 2.0 Handbook Second Edition: Methods, Concepts, Case Studies, and Standards in Business Process Management Notation*. Future Strategies Inc. ISBN 978-0-9849764-0-9.

14 MBHSI4Rail
Model-Based Human System Integration for Railway

Marc Sango and Guy André Boy

INTRODUCTION

In traditional systems engineering approaches, humans are considered as actors who interact with the technical system (machine) through human–machine interfaces. Today, with the emergence of autonomous systems and artificial intelligence, needs have evolved toward human and machine collaboration and cooperation to achieve a common goal. Therefore, the concept of Human–Machine Teaming (HMT) combines perspectives from systems engineering, artificial intelligence, and human-centered design to achieve Human–Systems Integration (HSI) through the development of an integrative system that encapsulates human and machine attributes and properties (Boy & Morel, 2022).

A concrete example of HMT is the teamwork between the train driver and the semi-automated train operation system. The introduction of the European Rail Traffic Management System/European Train Control System (ERTMS/ETCS) and Automatic Train Operation (ATO) will enable the transition from non-automated operation (Grade of Automation 1 – GoA1) to unattended operation (GoA4). The definition of GoA arises from apportioning responsibility for the given functions of railway operations between operational staff and involved technical railway systems. Table 14.1 defines the operation principles for each GoA level.

The ERTMS/ATO system and its subsystem specifications are underway in European railway sector (ERA ETCS, 2023) and (ERA ATO, 2023). While the systems and user-interface specifications are still document-centric, some railway projects are moving towards applying Model-Based Systems Engineering (MBSE) to manage system complexity and ensure consistency between all system interfaces (Sango et al., 2023). Recent research has tried to extend the MBSE modeling languages to support modeling human tasks (Amokrane-Ferka et al., 2021) and human–machine teams (Miller et al., 2020). However, these languages are theoretical or underspecified for specific domains, such as the railway domain (Sango et al., 2024). This raised our first research question.

 DOI: 10.1201/9781003666431-17

TABLE 14.1
Grades of Automation High-Level Description

GoA	GoA Name	Train Operator	Description
GoA1	Non-automated train operation	Train driver in the cab	The train is driven manually; but protected by the automatic train protection (ATP). This GoA can also include providing advisory information to assist manual driving.
GoA2	Semi-automated train operation	Train driver in the cab	The train is driven automatically, stopping is automated but a driver in the cab is required to start automatic driving of the train, the driver can operate the doors (although this can also be done automatically), the driver is still in the cab to check the track ahead is clear and carry out other manual functions. The driver can take over in emergency or degraded situations.
GoA3	Driverless train opera tion	Train attendant on- board the train	The train is operated automatically including automatic departure, a train attendant has some operational tasks, e.g., operating the train doors (although this can also be done automatically) and can assume control in case of emergency or degraded situations.
GoA4	Unattended train operation	No staff on- board competent to operate the train	Unattended train operation: all functions of train operation are automatic with no staff on-board to assume control in case of emergencies or degraded situations.

Research Question 1

What is a suitable railway domain-specific modeling language of MBSE that considers human factors at the early design phase? It should be the first step toward bridging the language gap between technical system architects and human factors specialists by relying on the same modeling language for system functions and human tasks. This modeling language should also help other engineering specialists, such as safety engineers, perform their work for safety assurance.

With the rapid advancement of MBSE, Model-Based Safety Analysis (MBSA) (Heimdahl et al., 2005) has been gaining more attention for its ability to verify whether a design solution adequately addresses the safety scenarios. Many MBSA approaches have been developed over the past two decades (Gudemann & Ortmeier, 2010; Yakymets et al., 2018), and a systematic review can be found in this paper (Suna et al., 2022). In general, the MBSA approaches start with a given design model. Faults are modeled to achieve a safety model, and finally, verification is conducted to see whether the given properties (functional or safety, deterministic or probabilistic) are satisfied. However, there is still an open challenge. Suppose specific safety scenarios are not considered in the given design model, or the properties do not correctly reflect the hazard. In that case, the results of MBSA cannot be fully trusted for safety assurance. In addition, a human activity (what is effectively performed) is often divergent from a human task (what is prescribed to be performed). This raises the second research question.

Research Question 2

What is the appropriate model-based design methodology that can support the analysis of safety scenarios of semi-automated driving systems? This methodology should help identify the scenarios caused by technical component failures and human factors.

The remainder of the chapter is structured to address these two questions and discuss the method's application in a case study.

A META-MODEL FOR HUMAN–SYSTEMS INTEGRATION IN RAILWAY

This section addresses research question 1 by extending a domain-specific language for railway system architecture (Sango et al., 2023) to support modeling human-centered and safety scenarios. Today, in the railway sector, terms are not consistently used across systems engineering, human factors engineering, safety engineering, and their related model-based practices (MBSE: Model-Based Systems Engineering, MBHSI: Model-Based Human–System Integration, and MBSA).

Figure 14.1 presents the graphical and high-level view of the proposed metamodel mapped in these three model-based practices. At the top of the metamodel, a

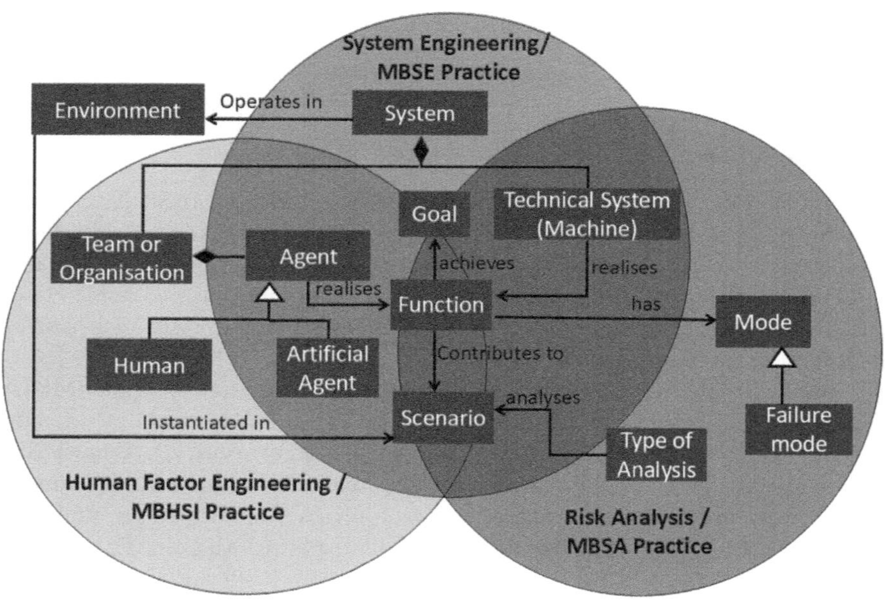

FIGURE 14.1 Human–systems integration metamodel using a block definition diagram.

"System" is defined according to the railway standard EN 50126-1 as a set of interrelated elements considered in a defined context as a whole and separated from their environment (EN 50126-1, 2017). Consistent with this definition, the system is comprised of a "Technical System (Machine)" such as a train or infrastructure, and an "Organization or Team", which controls the technical system. The team is potentially comprised of other teams and multiple agents. An "Agent" is an entity capable of perceiving its environment, processing the perceived environment to determine a desired action, and performing the desired action in response to the environment. An agent can be "Human" or "Artificial" in function to interact with the machine.

The agents cooperate in a structured way by adjusting their behaviors to achieve a common "Goal" within the goal hierarchy. Each goal is measured using different metrics (Damacharla et al., 2018). To reach one or more business goals, a "Function" represents either a technical function that can be performed by the technical system or a task performed by an agent or multiple agents (humans or artificial agents). The functions contribute to a given "Scenario" representing envisioned systems applications within a specified "Environment". Our general scenario-based approach is based on the following steps:

- *Step 1:* Functional analysis and function allocation to technical systems and human procedures.
- *Step 2:* Construction and selection of prescribed scenarios for specific perspectives, such as reliability, availability, safety, and security.
- *Step 3:* Preliminary hazard analysis and identification of consequences.
- *Step 4:* Human-in-the-loop simulations of selected scenarios on suitable simulators.
- *Step 5:* Observation and analysis of activities effectively performed during simulation compared to functions prescribed to enable the elicitation of emergent properties of the corresponding HSI early in the design phase.

The following section presents the model-based methodology from a safety perspective.

HUMAN-IN-THE-LOOP SIMULATION OF SAFETY SCENARIOS

This section addresses the second research question by extending the state of practices of MBSE, MBSA, and MBHSI to support the safety performance assessment of design scenarios. Figure 14.2 presents the methodology for the performance assessment of safety scenarios.

First, the system theory provides a generic formalism $(u(t), x(t), p) \rightarrow (\dot{x}, y(t))$ to formalize the controlled systems, where the control is x, the system state, p is the set of parameters, and y is the output. This generic construct must be properly constrained for system components within the design state (controlled systems, controllers) and those outside the design scope (i.e., environment or human behavior that the designer has no control over).

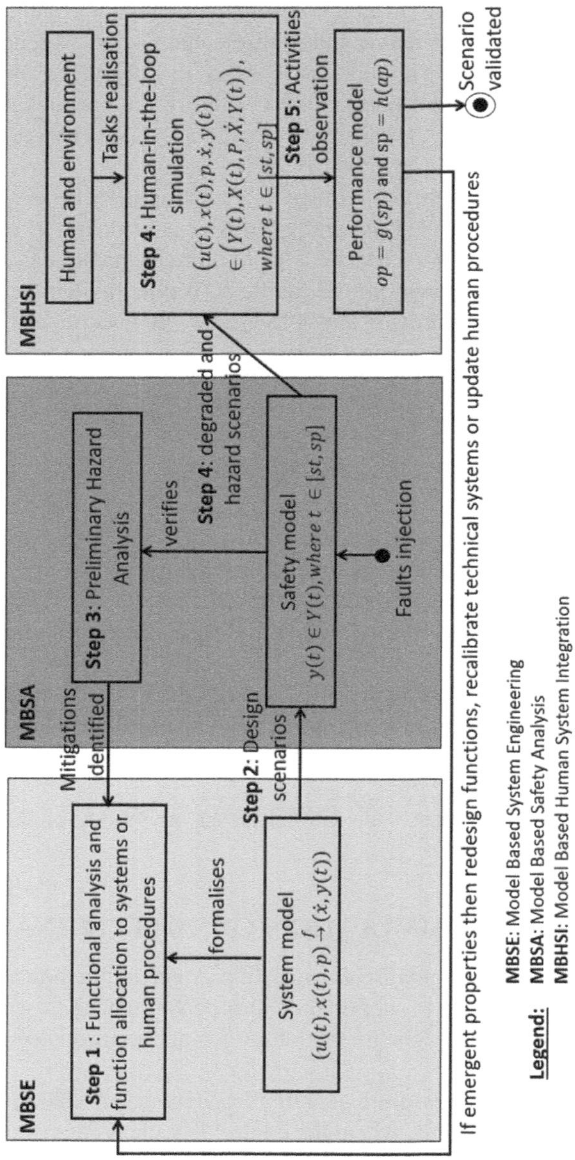

FIGURE 14.2 Methodology for safety performance assessment of design scenarios.

Second, for specific scenarios within the design space, the prescriptive constraints are defined on the output behavior, i.e., $y(t) \in Y(t)$, where $t \in [st, sp]$. If a design model is not properly constrained, it can lead to hazardous scenarios. Existing MBSA methods allow constraining the output of functions considered in the given design solution (e.g., ETCS signalization display or ATO functions) with failure modes, such as loss, erroneous, or untimely (erroneous start time st, stop time sp, or evolution in time windows [st, sp]). However, for functions that designers have no complete control over (i.e., detecting obstacles on the rail, bad weather, and practical human activities such as engaging or disengaging ATO), we need one more step to validate the assumptions on the environment and human tasks.

Finally, each model constraint may have assumptions on the environment and the human to be validated during simulation or operational testing. Indeed, a model is an abstraction of a real-world system. It increases our understanding by evaluating various appropriate metrics and proposing standard metrics to benchmark human–machine teams. In our approach, we used three levels of metrics: operational performance (op), system performance (sp), and agent performance (ap). The high-level performance can be derived from low-level performance: $op = g(sp)$ and $sp = h(ap)$.

Operational performance op is a high-level metric that enables system performance analysis in a specific context and measures the distance between an expected performance (i.e., prescribed task) and effective performance (i.e., activity). It is based on the analysis of sp data categorized according to three indicators: task completion (i.e., did the agent successfully complete the assigned mission?); resource availability (i.e., did the agent meet resource management guidelines, e.g., energy management, time management?); risk management (i.e., did the agent maintain an appropriate level of safety relative to the level of risk of the mission?).

ATO GOA2 CASE STUDY AND SAFETY SCENARIOS IDENTIFICATION

Figure 14.3 shows the reference architecture of ATO GoA2. At GoA2 (semi-automated train operation), ATO equipment onboard the train (ATO-OB) shall drive the train automatically by controlling traction and braking while the driver engages it. The ATO-OB drives the train to respect the journey profile (including departure and arrival time, ATO inhibition zone, etc.) provided by ATO track-side equipment (ATO-TS) without infringing the safe limits imposed by European Rail Traffic Management System/European Train Control System (ERTMS/ETCS).

The ATO-TS interfaces with the Traffic Management System (TMS, located at the infrastructure manager's control center), which supports dispatcher and signaler operations such as traffic planning, traffic regulation, and/or route setting. The ATO-OB can synthesize an ATO operational speed profile or time information in the Driver Advisory System (DAS). This information can be optionally displayed as advice to the driver in GoA1.

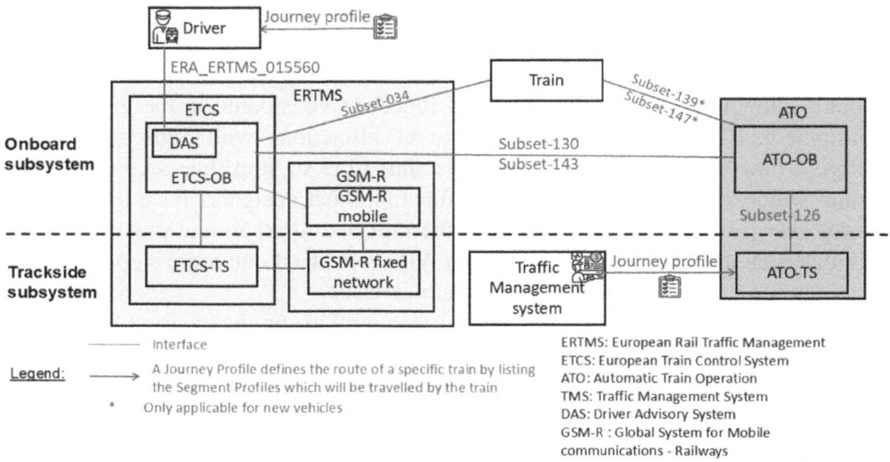

FIGURE 14.3 ATO GoA2 over ETCS Reference Architecture. (Adapted from ERA: ATO, 2023.)

Step 1: Functional Analysis

According to the ATO system requirement specification (ERA ATO, 2023), ATO provides a set of non-safety functions related to speed control, accurate stopping, door opening and closing, and other functions traditionally assigned to a driver, while the safety of operation is still ensured by ETCS (ERA ETCS, 2023), with regard to the speed and distance limits and also by other safe systems.

The ATO-TS shall inform the ATO-OB using the journey and speed profiles if it is impossible to apply traction, e.g., due to passing through a powerless section. The ATO-OB is not responsible for checking the vigilance of the driver's activity control function. This is the responsibility of an external system. The driver is responsible for verifying that it is safe to begin automatic driving before starting automatic driving (e.g., passenger exchange completed, no obstacles).

Step 2: Nominal Operational Scenario

In GoA1, the driver is responsible for train movement following the indications from ETCS, observing the physical environment (e.g., trackside signals), using the timetable information and route information provided by a Traffic Management System (TMS) located in the infrastructure manager's control center. The driver executes the mission provided by a TMS located in the railway undertaking's control center and is responsible for the rolling stock. In GoA1, a DAS is an optional assistance system that synthesizes timetable information into advice, such as the drivers' optimal speed or time information.

In GoA2, the responsibility of train movement following the indications from ETCS is shared between the train driver and ATO. During the mission, the decision

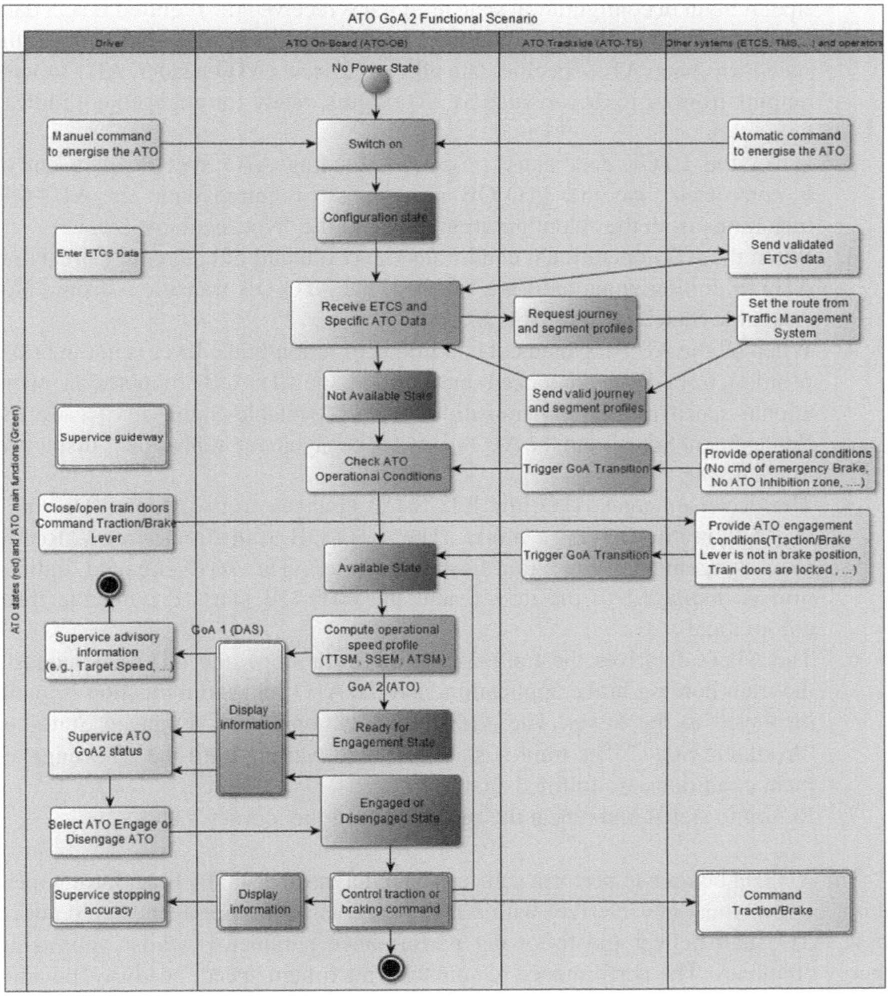

FIGURE 14.4 ATO GoA2 nominal operational scenario using BPMN[1] graph.

to engage the ATO or not is still in the train driver's hands. Once engaged, ATO computes the operational speed profile and generates the train's output traction/braking command accordingly. When ATO is involved, the train driver's mission changes to supervision from train operation. Figure 14.4 shows the nominal operational scenario of ATO. The states of ATO are presented in red, ATO GoA2 functions are in green, and the functional interfaces with other functions are grey. The scenario is based on the following steps.

1. The ATO on-board (ATO-OB) is powered on. After the switch-on, the ATO-OB transitions from "No Power State" to "Configuration State." The ATO-OB

shall remain in configuration state until it has received the required ETCS data (Driver ID, Train Running Number, train category, length, maximum train speed, etc.) and ATO-specific data entry and view (ATO header, ATO inputs/outputs from or to driver, such as ATO status, ready for engagement indication, etc.).

2. When the ETCS data entry process (including ATO specific data entry) is completed, and the ATO-OB receives the required data, the ATO-OB transitions from the "Configuration" State to the "Not Available State."
3. When the ATO operational conditions (no command of emergency break, no ATO inhibition zone, etc.) are fulfilled, the ATO-OB transitions from "Not Available State" to "Available State."
4. When all the ATO engagement conditions (traction/brake lever is not in brake position, train doors are locked) are fulfilled, the ATO-OB computes the operational speed profile and transitions from "Available State" to "Ready for Engagement State," and the ATO ready for engagement indication is displayed to the driver.
5. The driver engages ATO; the ETCS-OB changes from "Manual Driving" to "Automatic Driving" mode. The ATO-OB transitions from "Ready for Engagement State" to "Engaged State." The ATO-engaged indication is displayed to the driver, and the ATO-OB starts driving the train automatically.
6. The ATO-OB drives the train. When the train stops, the ATO-OB requests the train holding brake application, and the ATO selected indication is again displayed to the driver. The ATO-OB transitions from "Engaged State" to "Available State." The train is stationary and waiting until the ATO engagement conditions are fulfilled again.
7. Return to step 4 and repeat the sequence until the journey's end.

The ATO on board can perform self-tests and diagnostics at any time during operation, but this must not interfere with ATO operation. From a performance point of view, ATO shall deliver the following performance parameters whilst optimizing energy efficiency. The performance parameters (maximum speed, headway, punctuality, stopping accuracy, required reaction times, and riding comfort) shall be specified for each application by the railway operators.

One of the operational performances we want to measure is the safety of operational performance, i.e., the task completion in required reaction times and the appropriate level of safety.

Step 3: Preliminary Hazard Analysis with MBSA

Hazard identification and classification are fundamental activities in safety risk management in the railway domain. A hazard is a condition that could lead to an accident. An accident means an unwanted or unintended sudden event or a specific chain of such events with harmful consequences. Classification of railway accidents, both in terms of causes and effects, is a valuable aid in studying rail accidents to help prevent similar ones from occurring in future systems.

In terms of accident effects (consequences), railway accidents can be classified into three categories: collisions (collisions with trains, collisions with people on track, collisions with obstructions on the track), derailments (in plain track, curves, and junctions), and others (fires, explosions and release of hazardous chemicals, including sabotage/terrorism, people falling from trains).

In terms of accident causes, railway hazards can be classified into three categories: human hazard (drivers' errors such as passing signals at danger, excessive speed, failure to stop at required positions such as level crossings; signalmen's errors such as allowing two trains into same occupied block section, incorrect operation of signals, points or token equipment); technical system failures (mechanical, electronic or software failures of rolling stock or infrastructure subsystems); and natural causes (track obstruction or damage by landslides, avalanches, floods, trees, fog or snow that obscure signals or the current position of the train, wet leaves (or their remains) making the tracks slippery).

Based on the above general hazards and interviews with a group of drivers, it appears that one of the critical hazards during the journey of ATO GoA2 is the "failure to stop at required positions." Indeed, when the ATO operational conditions are lost, the ATO disengages, the driver is informed, and he/she shall take control of the train. The failure mode analysis of ATO and driver functions involved in the nominal operational scenario of Figure 14.4 shows that the failure of the following functions of ATO and drivers could lead to failure to stop at required positions.

- Failure of "Supervise ATO on-board status" or "Supervise guideway" due to the driver's hypovigilance or interruption by third parties.
- Failure of "Disengaged ATO" due to the reaction time of the driver or due to the ATO technical failure.
- Failure of "Control traction or braking command" due to undetected (or detected too late) reduced braking performance due to adherence problems.

The preliminary hazards analysis results are recommendations regarding safety barriers or human procedures for each critical function identified. The analysis of the most frequent hazards shows that the harmful consequences generally come from specific chains of hazards or hazardous scenarios (Sun et al., 2023).

Step 4: Hazardous Operational Scenario

We constructed the hazardous operational scenarios from the hazards identified in the previous phase. Figure 14.5 shows a hazardous scenario constructed for "failure to stop at required positions." When ATO-OB is in an engaged state, and any of ATO's operational conditions is lost, ATO-OB transitions to a disengaged state. The "ATO disengaging" indication is displayed to the driver; then, the driver shall react within a period to take complete control of the train.

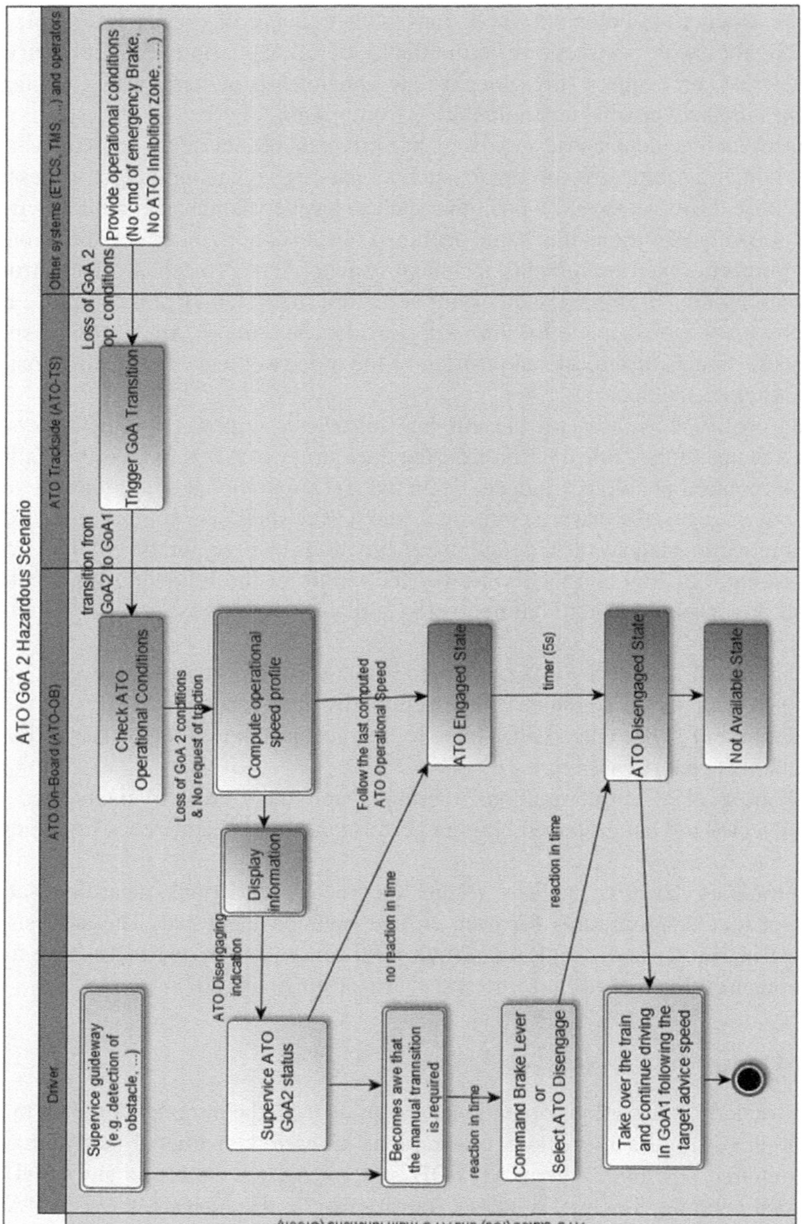

FIGURE 14.5 ATO GoA2 hazardous operational scenario.

For the performance parameters (availability and safety), the current specification requires the following conditions:

- *Safety:* If the driver applies the brake using the "*transmission balise-locomotive*" (TBL, i.e., beacon-to-locomotive transmission) or selects "ATO Disengage" before 5 seconds, the ATO-OB transitions from DE to NA State, and the ETCS-OB leaves the automatic driving mode.
- *Availability:* If ATO-OB recovers ATO operational conditions before 5 seconds, the ATO-OB transitions from DE to EG state and continues to drive the train automatically.
- *Safety:* If the driver does not react and the ATO operational conditions are not recovered after 5 seconds, the ATO-OB remains disengaged and applies the full-service brake.

From a safety perspective, did the driver maintain these required reaction times relative to the hardened operational context (e.g., multiplication of threats)? To answer this question, we performed a driver-in-the-loop simulation better to explore the potential emergent functions and function allocation.

Step 5: Human-in-the-Loop Simulation and Activities Analysis

From the hazardous scenarios identified, we conducted a series of Human-in-the-Loop Simulation (HITLS) based on dense and realistic scenarios contextualizing (with and without the virtual system of perception, audible warnings for ATO disengagement, and/or other systems), as illustrated in Figure 14.6. The simulation shows that the driver may not react before this threshold of 5 seconds regarding the driving context or external environment events (information simultaneously).

Observing activities and interviews with drivers shows that adapted training, an intuitive driver–machine interface, and an integrated safety assistant system are essential to ensure the safe operation of these semi-automated trains. The general recommendations derived are:

- *Training with Driving Simulators:* Using them to train drivers helps prepare them for real-world situations without risks. Training planning should consider complex driving context environment events, such as weather conditions, which can lead to train adherence hazards.
- *Design Driver–Machine Interface with Drivers' Feedback:* the most used driver commands, such as touchscreens, buttons, or levers for acceleration, braking, and ATO engaged/disengaged, shall be placed directly in front of the driver within easy reach of the driver to minimize movement and reduce fatigue.
- *Design Alert Systems (Audio or Visual) by Avoiding Workload and Stress:* the frequency and the type of information (no safety and safety alerts) shall be clear with easily understandable graphics and sounds.

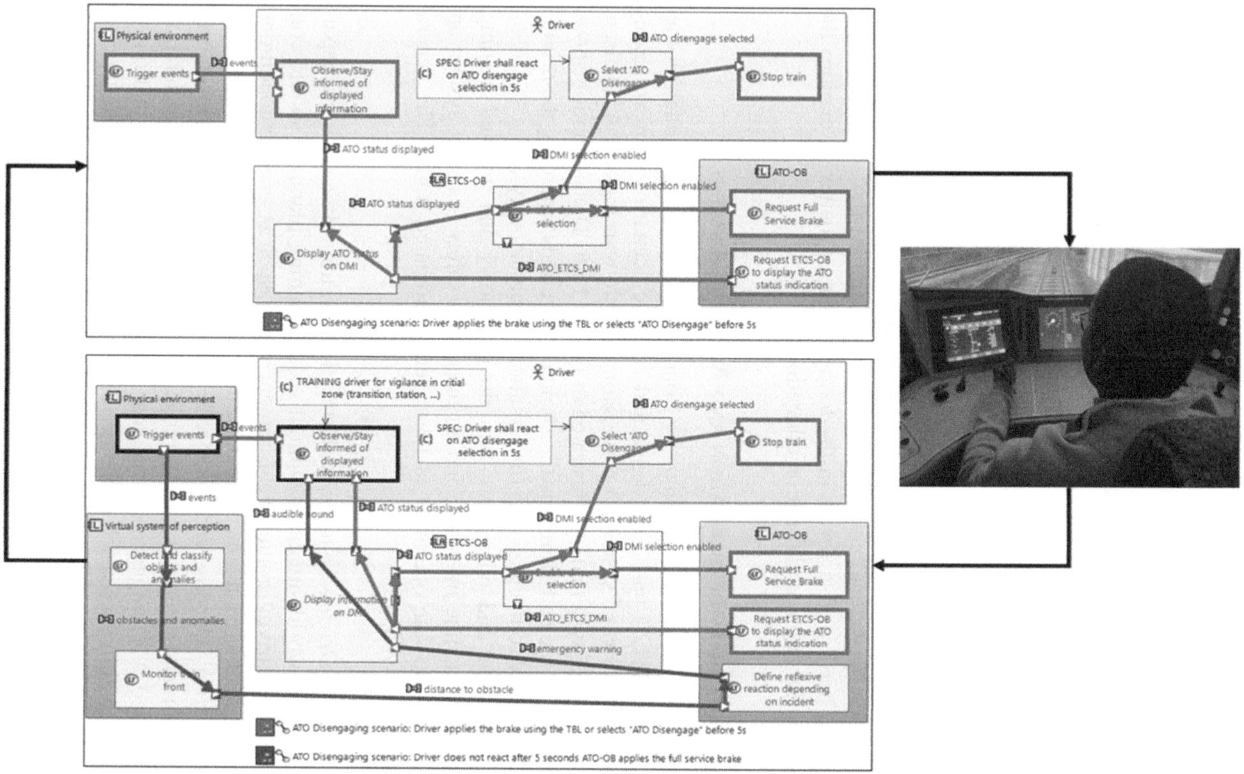

FIGURE 14.6 Example of task analysis (left side with disengaging scenario in blue chain), activity analysis (right side), and emergent scenario identification (in red).

RELATED WORKS AND DISCUSSION

HITLS has been discussed in several studies to evaluate HMT performance and reliability. Trust is one of the key elements that support teamwork. Boyce et al. (2015) conducted an HILTS to test how displaying transparency information affects operator trust. The study showed that including transparency information in the display for a robotic agent can increase trust. Enhancing the understanding of the automation process can also increase the trust in the automation (Lee et al., 2004). Recent research proposes a relational framework that emphasizes that the goal is not to maximize or calibrate trust but to support a process of trusting through automation responsibility (Chiou & Lee, 2023).

HITLS assesses single pilot incapacitation (Castro et al., 2023). HITLS can still elicit interactions across organizations and domains with low-fidelity simulations. In the MOHICAN project, the PRODEC method is used and validated to support the integration of pilots and virtual assistance (Boy & Morel, 2022). The HITLS supported the observation of activities in HMT. Applying the PRODEC method for semi-automated trains initially focused on people (drivers) task analysis (Sun et al., 2023).

In the TOP model, Technology, Organizations, and People should be studied rationally and consistently (Boy, 2020). The TOP model shown in Figure 14.7 supports the transition from the Railway AS-IS system to the TO-BE System, integrating new technology for automation (ATO, FRMCS, etc.) while addressing the innovation challenges (e.g., how to accelerate the interoperability of rail signaling systems in the EU at lower cost?) and scalability issues (e.g., is ERTMS suitable for small railway lines?) (Sango et al., 2023). In digital transformation, success requires observing and analyzing various activities using virtual modeling and simulation to

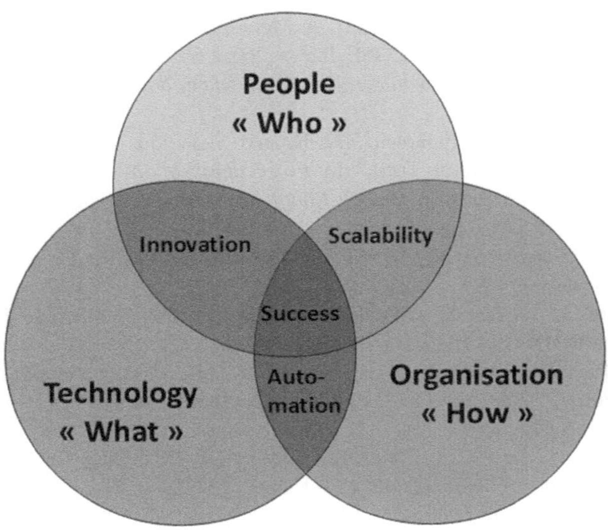

FIGURE 14.7 TOP model with automation, innovation, and scalability for success.

identify emerging properties of technologies under development, and not waiting to discover them at operational time.

CONCLUSION

This chapter addresses two main questions that can help progress in implementing human systems integration in the railway domain. The first contribution is the human–systems integration metamodel proposal, which bridges the language gap between technical systems engineering practice (MBSE) and human factor engineering practice (MBHSI). It relies on the same modeling language for system functions and human task analysis, and then helps construct the design scenarios for expected performances. The second one is the methodology for safety performance assessment of design scenarios. It is based on two main steps: identification of safety scenarios caused by technical system failures using an MBSA practice and the analysis of safety-critical scenarios by using human-in-the-loop simulation. The first experimentation in a semi-automatic train operation system shows that the MBHSI has become possible at the earliest design stages by using scenario-based design and virtual simulation to understand human activities. Consequently, advanced visualization techniques, data acquisition, and cognitive low-level measurement techniques, such as eye tracking and electrocardiograms, are needed to increase tangibility and elicitation of human teaming metrics, such as trust and collaboration.

NOTE

1 Business Process Model and Notation.

REFERENCES

Amokrame-Ferka K., Hein A. M. (2022). A Meta-Model for Integrating Human Factors into System Engineering. INCOSE International Symposium Volume 32 Issue S1 – Supplement: *Human System Integration Conference, San Diego, USA*. https://doi.org/10.1002/iis2.12876

Boyce, M.W., Chen, J.Y.C., Selkowitz, A. R. and Shan G.L (2015). "Effects of Agent Transparency on Operator Trust." In *Proceedings of the Tenth Annual ACM/IEEE International Conference on Human-Robot Interaction Extended Abstracts*, 179–180. New York, NY, USA

Boy, G.A. (2020). *Human Systems Integration: From Virtual to Tangible*. CRC Press–Taylor and Francis Group, USA.

Boy, G.A., Morel, C. (2022). The Machine as a Partner: Human–Machine Teaming Design Using the PRODEC Method. *Work* 73(s1): S15–S30.

Castro, A.L., Reis R.J.N., Mallozzi G., Vizioli, A.D.B. (2023). Assessing Single Pilot Incapacitation Using Low-Fidelity Simulation. Journal of Physics: Conference Series 2526 (1): 012081.

Chiou, E.K., Lee, J. D. (2023). Trusting Automation: Designing for Responsivity and Resilience. *Human Factors* 65 (1): 137–165.

Damacharla, P., Javaid, A.Y., Gallimore, J.J., Devabhaktuni, V.K. (2018). Common Metrics to Benchmark Human–Machine Teams (HMT): A Review. *IEEE Access* 6: 38637–38655.

EN-50126-1. (2017). Railway Applications – The Specification and Demonstration of Reliability, Availability, Maintainability and Safety (RAMS) – Part 1: Generic RAMS Process.

ERA ATO. (2023). *System Requirements Specification*, Subset-125.

ERA ETCS. (2023). *System Requirements Specification* – ERTMS/ETCS Reference Architecture, Subset-26-2.

Gudemann, M., Ortmeier, F. (2010). A Framework for Qualitative and Quantitative Formal Model-Based Safety Analysis. In *Proceedings of the 12th IEEE High Assurance Systems Engineering Symposium, San Jose, CA, USA* , pp. 132–141. https://ieeexplore.ieee.org/document/5634319

Joshi, A., Heimdahl, M.P.E, Miller, S.P. Whalen, M.E (2006). *Model-Based Safety Analysis: Final Report.* NASA Technical Report. NASA/CR-2006-213953. https://ntrs.nasa.gov/citations/20060006673

Lee, J.D., Hoffman J. D., Hayes E.. (2004). Collision Warning Design to Mitigate Driver Distraction. In Proceedings of the SIGCHI Conference on Human Factors in Computing Systems, 65–72. Vienna Austria.

Miller, M.E., McGuirl, J.M., Schneider, M.F., Ford, T.C. (2020). Systems Modeling Language Extension to Support the Modeling of Human-agent Teams. *Systems Engineering* 23: 519–533.

Sango, M., Gautier, P.E., Cheron, C. (2023). Towards the Modelling of European Railway System Architecture. *Transportation Research Procedia* 72: 860–867.

Sango, M., M., Sun, Y., Cappi, C., Ammad, N. (2024). *SNCF System Engineering Handbook v1.0.* SYNAPSES, Network of SNCF Scientific and Technical Experts on System Engineering.

Sun, Y., Sango, M., Barros, A., Boy, G.A. (2023). Preliminary Safety-Critical Scenario Analysis of Semi-Automated Train Operation. Report.

Suna, M., Gauthamb, S., Elksb, C., Fleming, C. (2022). Characterizing the Identity of Model-based Safety Assessment: A Systematic Analysis. *arXiv preprint arXiv:2212.05401.*

Yakymets, N., Sango, M., Dhouib, S., Gelin, R. (2018). Model-Based Engineering, Safety Analysis, and Risk Assessment for Personal Care Robots. In *IEEE/RSJ International Conference on Intelligent Robots and Systems (IROS), Madrid, Spain.*

15 Modeling and Human-in-the-Loop Simulation

Jawahar Bhalla

INTRODUCTION

Systems thinking, modelling, and simulation (STM&S) are intrinsic to human cognition. Since the dawn of consciousness, we have built mental models through experiential learning and used internal simulations to interpret our world, anticipate outcomes, and guide decisions (Bhalla, 2018). From weather prediction to Monte Carlo simulations and interplanetary mission planning (NASA, 2023), M&S empowers us to navigate complexity and extend human capability beyond natural limits.

STM&S forms the backbone of systems engineering (SE), enabling the transformation of complex concepts into safe, sustainable, ethical, and practical solutions – now increasingly realised through model-based systems engineering (MBSE) (INCOSE, 2022; Pickard et al., 2018; Rogers & Mitchell, 2021; Wang et al., 2020). It is also foundational to developing AI/ML-enabled systems, where simulation accelerates learning and validation (Bhalla et al., 2023a, 2023c).

This chapter focuses on the role of human-in-the-loop (HITL) simulation solutions, which are critical societal enablers that support:

- *Exploration* of unsafe, high-cost, or time-constrained scenarios.
- *Focus* – by abstracting complexity to relevant elements.
- *Visualization* of dynamic behaviors and emergent patterns.
- *Motivation* – through engaging, immersive learning.
- *Control* over timing, scale, and variation.
- *Practice* – via safe, repeatable experiential learning.

The chapter begins by contextualizing HITL simulation in modern systems, introduces a categorization framework, and explores core concepts including mental models, SA, model fidelity, and the risks of unintended emergent behaviors. It concludes with a synthesis of key insights and references for further exploration.

DOI: 10.1201/9781003666431-18

BACKGROUND – THE CRITICALITY OF HITL SIMULATION SOLUTIONS

A NEED FOR "FAMILIARITY"

The use of HITL simulations can be traced back to antiquity, but their modern utility is most notably entrenched in high-risk domains such as defence and aviation. The value of immersive, experiential preparation was powerfully articulated by Carl von Clausewitz (1874), the 18th-century Prussian military strategist, who wrote in *On War*:

> *... it is of immense importance that the soldier should not have to encounter in War those things which, when seen for the first time, set him in astonishment and perplexity; ... if he has only met with them one single time before, even by that he is half acquainted with them ...*

Clausewitz's insight underscores a timeless principle: familiarity with uncertainty is a force multiplier. In today's terms, military forces call this mission rehearsal (MR) – the deliberate use of HITL simulations to enhance SA by immersing personnel in realistic, high-threat operational environments. These simulations are not merely about practice – they are designed to build cognitive readiness, reduce surprise, and increase confidence under stress. Whether in defense, healthcare, or transportation, the rationale remains the same: when lives are on the line, familiarity through simulation is not a luxury – it is a necessity.

INNOVATION IN AVIATION

Flight simulation stands today as one of the most advanced and enduring applications of HITL simulation, seamlessly integrating cutting-edge technology with socio-technical principles through human–systems integration (HSI). However, its roots lie in the unforgiving early years of aviation, when high accident rates exposed the limits of trial-and-error learning. Visionaries like Otto Lilienthal, the "Glider King" and often regarded as the Father of Flight, tragically died in a gliding accident in 1896. The Wright Brothers, just seven years later in 1903, made history with their first powered flight, built on hundreds of iterative wing designs and personal test flights, often at significant personal risk (Dodson & Miklosovic, 2005).

In those pioneering days, there were no flight simulators. Training occurred in the cockpit, midair – essentially on-the-job learning in life-or-death conditions. Unsurprisingly, injury and fatality rates were alarmingly high, often triggered by incorrect control responses from novice pilots under stress. For instance, early aircraft like the Antoinette (pictured in the upper left of Figure 15.1) featured dual control wheels – one for pitch, one for yaw/roll – requiring the pilot to make quick, precise, and nonintuitive movements with each hand during turbulence or emergency descent. Without margin for error, hesitation or confusion could prove fatal. This unforgiving environment sparked the innovation of flight simulation – a means to build familiarity, muscle memory, and cognitive resilience before taking to the skies. And so was

FIGURE 15.1 Antoinette aircraft (left) and the "French Barrel with Wings," or the first flight simulator (right).

born the first flight simulator. Fondly labelled as the "*French half-barrel with wings*," the Antoinette Trainer (Figure 15.1) was developed for the Antoinette aircraft.

The Antoninette trainer was successfully used to train novice aviators to develop muscle memory through practice in a safe environment, to control the aircraft in a high-stress, instinctive, and correct manner in an emergency (Hayes & Langlois, 2005), and heralded the birth of HITL (flight) simulation in a safety-critical context. What began as a safety imperative became a cornerstone of modern aviation training – and a model for HITL simulation across high-risk sectors.

Contemporary HITL Simulation Solutions

Flight simulation remains the most mature, regulated, and widely adopted application of HITL simulation. Flight simulation training devices (FSTDs) support a full training continuum – from theoretical knowledge transfer, systems familiarisation, operational competence, and integrated MR.

FSTDs achieve this by replicating aircraft systems and environmental conditions, such as weather and time of day, with levels of fidelity tailored to specific training outcomes. A full flight simulator (FFS), certified to Level-D (zero flight time [ZFT] standard), offers the highest fidelity. A FFS features a full-scale replica cockpit, dynamic motion platforms, visual systems, and software capable of emulating real-world flight scenarios with such precision that no additional in-aircraft training is required for pilot certification (EASA, 2020).

FSTD accreditation standards now form the backbone of civil aviation training globally, with regulatory bodies such as EASA (2020), the FAA (Authority, 2021), and CASA (2001) adopting structured qualification frameworks. These standards trace back to Page (2021), who pioneered flight simulation training at Qantas during the jet age (Qantas' acquisition of B-707 aircraft in the 1960s) and helped develop the first international FSTD accreditation frameworks. His legacy continues through the Ray Page Lifetime Achievement Award and the formation of Simulation Australasia.

Today, the impact of HITL simulation extends well beyond aviation, with growing adoption in other safety-critical sectors – such as healthcare, transport, and defence – for training, exploration, experimentation, and certification (Erdemir et al., 2020; Motola et al., 2013). Moreover, while simulation technology has evolved dramatically, its value lies in its ability to create immersive, purpose-driven SA – the cornerstone of all HITL simulations.[1]

CONTEXTUALIZING HITL SIMULATIONS

Categorizing Simulation Solutions

Contemporary HITL simulations are often associated with virtual reality (VR), augmented reality (AR), or gaming environments. These technologies span a continuum – from fully digital simulations (VR) to the overlay of contextual digital information on the real world (AR), enhancing user understanding and interaction. This continuum is illustrated in Figure 15.2, with VR on the left (fully virtual), AR on the right (digitally augmented real world), and a central zone commonly referred to as mixed reality (MR) or extended reality (XR) – where physical and virtual environments coexist and interact, such as through AR glasses (e.g., HoloLens) or VR headsets.

While many HITL applications focus on individual users, defense and other complex domains have driven the need for multi-participant, interoperable simulations, supporting team, and mission-level training. This need led to developing a 2 × 2 simulation categorization framework, classifying participants and environments as real or simulated (Australian Government-Department of Defense, 2018; United States Department of Defense, 2011).

This framework defines four primary simulation types – live, virtual, constructive, and replicated (Figure 15.3). The label "Replicated" is personal, as an established/accepted label has not been used for this (emerging) category, with contemporary defense focus primarily on L–V–C categories.

Note: HITL simulation solutions map to the L, V, and R categories, while one could consider a "C" (constructive) simulation to map to a human-*on*-the-loop (HOTL) solution.

FIGURE 15.2 A VR/AR HITL simulation continuum.

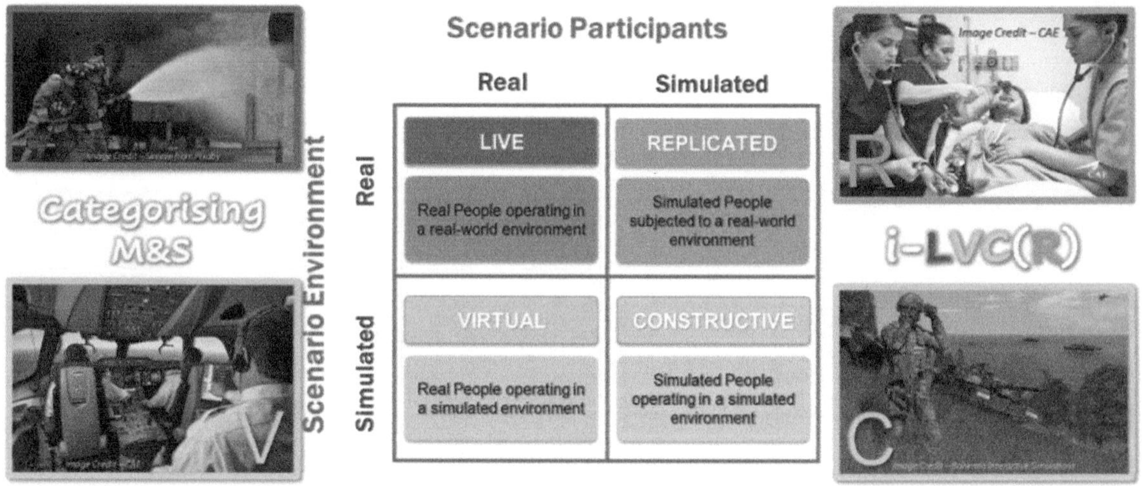

FIGURE 15.3 Four-part categorization of simulation solutions [LVC(R)].

Live Simulation Solutions

Live simulations involve real people operating in real-world environments, often blending learners and role-players to achieve outcomes such as training, rehearsal, validation, or review. Examples include:

- Firefighting and emergency response drills.
- The defense team and tactics training are on live ranges with authentic or replica equipment.
- Mental health scenarios, where role-play helps practitioners navigate complex, sensitive interactions.
- Reliability, availability, and maintainability demonstrations (RAMD) – testing actual systems in operational settings to validate supportability and performance.

Key advantages include being experiential by nature, bridging theory and practice through hands-on, immersive learning in safety-critical domains with flexibility to adjust the scenario through live props, participants, and aspects of the environment. Key challenges include varying objectivity through scenarios and a risk of consistent repeatability across sessions. The roles participants play are influenced by their backgrounds and individual mental models at the time, which will vary across participants and multiple scenarios (i.e., temporarily). Additionally, those aspects of the environment that are not controllable (props) will also change (naturally), depending on the setup and the extant ambient conditions.

Virtual Simulation Solutions

Virtual simulations immerse human participants in digitally created environments for training, experimentation, design validation, or entertainment. These solutions replicate complex systems and scenarios, enabling safe, repeatable, cost-effective learning and analysis. A FFS is a leading example – delivering accredited, multi-sensory immersion (sight, sound, motion) at a fidelity high enough for ZFT pilot certification. At a more focused level, part task trainers (PTTs) provide targeted simulation of specific tasks or procedures, much like the historical Antoinette Trainer (refer to Figure 15.1). Beyond training, virtual simulations are invaluable in SE, allowing designers, users, and stakeholders to interact with and refine digital prototypes of a System of Interest (SoI) before physical development. These virtual prototyping environments accelerate decision-making, reduce risk, and support iterative design through HITL feedback.

Key advantages include measurable objective fidelity (accredited against reference data), consistency, and repeatability of performance outcomes. Key challenges include a lack of flexibility, risk of negative consequences (if not accredited or used outside of their accredited fidelity scope), a dependency on the enabling technology (with regard to performance and obsolescence), and ongoing operational utility (performance degradation).

Constructive Simulation Solutions

Constructive simulations model environments and participants entirely in a digitally constructed space, with minimal or no real-time human interaction. Often described as HOTL, these simulations are ideal for decision support, enabling users to analyze, experiment, and optimize scenarios faster than real-time. In defense, this includes the following.

Computer-generated forces (CGF), where simulated entities replicate tactical behaviors in environments such as Virtual Battlespace Simulation (VBS) (from Bohemia Interactive) or VR-Forces (by Mak Technologies). In broader systems contexts, constructive simulation underpins operational research (OR) and operational analysis (OA) tools, such as FlexSim and Arena, used for modelling and optimizing complex systems like manufacturing lines, healthcare operations, logistics networks, and supply chains.

Key advantages include running large-scale, repeatable "what-if" scenarios, faster-than-real-time execution, supporting time-sensitive decision-making, and parametric control to explore trade-offs and optimize performance. Key challenges are related primarily to the modelling approach adopted (stochastic vs. deterministic), as that choice provides for a mix of outcomes and should be carefully chosen depending on the context and access to quality parametric data.

Replicated Simulation Solutions

Replicated simulations place simulated participants – such as mannequins or crash-test dummies – into real environments with human operators to support training, prototyping, and experimentation without harming humans. Examples of replicated simulation solutions include advanced medical mannequins used in clinical training or automotive crash-test dummies for vehicle safety testing. Though increasingly recognized, this simulation category is still emerging and lacks a formal, universally accepted label.

Key advantages include enabling safe, repeatable practice on simulated humans in controlled settings, reducing ethical and physical risks when testing high-stakes scenarios, and being valuable for emergency response, healthcare, and human-factors engineering. Key challenges include limitations due to the fidelity of simulated physiological responses, potential for negative training if realism is insufficient or misleading, and a lack of formal accreditation standards, which creates ambiguity in validation.

Integrated Live-Virtual-Constructive-(Replicated) Simulation Solutions

Integrated live-virtual-constructive-(replicated) [iLVC(R)] simulations integrate real and simulated environments and participants across all four simulation categories, enabling cohesive, joint immersive experiences.

The core benefit of iLVC(R) is its ability to deliver complex, multi-domain team training and MR that would be impossible or unsafe to conduct in the real world. This makes it invaluable for high-threat, safety-critical scenarios and evaluating system design modifications (e.g., via digital twins) in an SE context. Key challenges include model fidelity mismatches across federated components, which can cause unintended behaviours (e.g., differing vehicle dynamics across systems), integration difficulties due to proprietary architectures, inconsistent data formats, limited standardisation, and intellectual property (IP) protection concerns, which can hinder open interoperability.

Progress has been made through standards such as Distributed Interactive Simulation (DIS) (IEEE, 2015) and the High-Level Architecture (HLA) (IEEE, 2010), which enable meaningful, semantically consistent state exchanges while safeguarding IP. The Common Database (CDB) standard (Saeedi et al., 2017) also facilitates shared synthetic environments across federated systems.

Blended/Mixed Simulation Solutions

Blended/mixed simulation solutions are an alternative approach to iLVC(R), where information is overlaid in one solution (as opposed to integrating multiple stand-alone simulation types) across two or more of the four simulation categories to "augment" the immersive experience. These solutions are widely used for training, familiarisation, experimentation, and remote support, and are increasingly seen in both professional and consumer domains. Examples include AR headsets like Microsoft HoloLens and apps like Pokémon Go, where digital content is contextualized within the user's real-world view.

Key advantages include enabling high-fidelity, in-situ immersion without the sensory side effects often associated with VR, utility in real-time support scenarios (e.g., surgery or equipment maintenance), and remote learning/ training, blending live context with virtual guidance. Key challenges include a lack of standards for accreditation, fidelity assessment, and model interoperability, subjective fidelity evaluations, limited cross-platform integration, and regulatory lag, with technology outpacing governance.

UNDERSTANDING HITL SIMULATION SOLUTIONS

Systems Thinking, Modelling and Simulations

Systems thinking (ST) is foundational to SE and underpinned by M&S, such as MBSE. The INCOSE SE Handbook (Walden et al., 2023) recognizes ST as "*a perspective that sharpens our awareness of wholes and how the parts within those wholes interrelate*." Senge (1997) defines ST as a "*conceptual framework, a body of knowledge and tools*" that help make "*full patterns clearer*" and to "*change them effectively*." A personal view is that ST is a conceptual framework built through experiential learning to make sense of the world we live in, deal efficiently and effectively with the challenges we perceive, and create the reality we desire. From an experiential standpoint, ST helps us make sense of complexity by answering three core questions: **What is happening**? **How is it happening**? Moreover, most importantly – **Why**?, while M&S brings these insights to life, making ST actionable, particularly in domains like MBSE and HITL simulations.

A powerful metaphor for ST is the Iceberg Model, which illustrates different depths of understanding:

- **What** we see: observable events or symptoms.
- **How**: patterns of behaviour over time, shaped by system structure.
- **Why**: underlying mental models that drive design and decision-making.

To respond effectively, especially in safety-critical and complex domains, we must dive below the surface – moving from reaction to prediction, structure to transformation. These concepts are further explored in the authors' INCOSE IS 2024 paper (Bhalla et al., 2024), focusing on the role of mental models and M&S in enhancing SA in HITL contexts.

Definitions – What Are Models and Simulations

While there are many definitions of models and simulation (United States Department of Defense, 2011; INCOSE, 2022), safety-critical systems must be defined broadly, in terms of purpose and suitability. A practical definition (Bhalla, 2018) is:

- A model is a physical, mathematical, or logical abstraction of reality for a specific purpose or need.
- A simulation executes that model over time, enabling dynamic understanding.

A model captures static structure, representing a system, process, or phenomenon with defined scope and level of detail (abstraction). Conversely, a simulation brings that model to life, transforming static information into experiential knowledge through temporal interaction. The tool that runs a simulation is called a simulator.

This is exemplified using a GPS for driving direction in a car: the underlying map and algorithms form the models, and the real-time guidance based on vehicle motion is the simulation, delivering actionable SA.

Contextualizing Mental Models in HITL Simulations

Mental models are the deeply ingrained assumptions, beliefs, and internal frameworks that shape how we perceive the world and make decisions (Senge, 1997). Endsley (1995) describes them as the cognitive mechanisms that allow humans to understand system functions, interpret current states, and predict future behaviour. Maturana's concept of ***autopoiesis*** emphasizes that we do not merely perceive reality – we construct it through interaction and interpretation: "*we do not perceive the world we see; we see the world we perceive*" (Maturana et al., 1973). Mental models are critical to navigating life – they define our SA and shape our judgments, responses, and learning. One of the most important goals of HITL simulation is to build and refine these mental models, helping individuals develop a more profound understanding and more effective behaviours in complex systems.

We form mental models through experiential learning – sensing, interpreting, and assigning meaning. For instance, learning a new language begins with meaningless sounds (data), which through repeated exposure become syllables (information), then words and grammar (knowledge), and finally nuanced communication (understanding and wisdom). This DIKUW progression – data, information, knowledge, understanding, wisdom – is fundamental to human cognition and systems thinking (The Deming Cooperative, 2004).

HITL simulations are uniquely effective at accelerating this transformation. They provide immersive, safe environments that enable repeated, context-rich experiences – bridging knowledge to wisdom through lived understanding. Designing HITL simulations, therefore, should focus on enabling rich SA and experiential learning that foster meaningful, lasting mental model development.

The Centrality of Situational Awareness

Situational awareness (SA) is a foundational capability in high-stakes environments, recognized since World War I as critical to aviation safety (Endsley, 1995; Pitts et al., 1990). SA is especially vital in complex, dynamic operations, where timely, informed decisions determine outcomes. Endsley (1995) defines SA as:

> *The perception of elements in the environment within a volume of time and space, the comprehension of their meaning, and the projection of their status shortly.*

This model is commonly represented in three levels (Figure 15.4) (Bhalla et al., 2023c):

1. *Perception:* Identifying key elements in the environment (Level 1).
2. *Comprehension:* Interpreting their meaning using internal mental models (Level 2).
3. *Projection:* Anticipating future states to inform decision-making (Level 3).

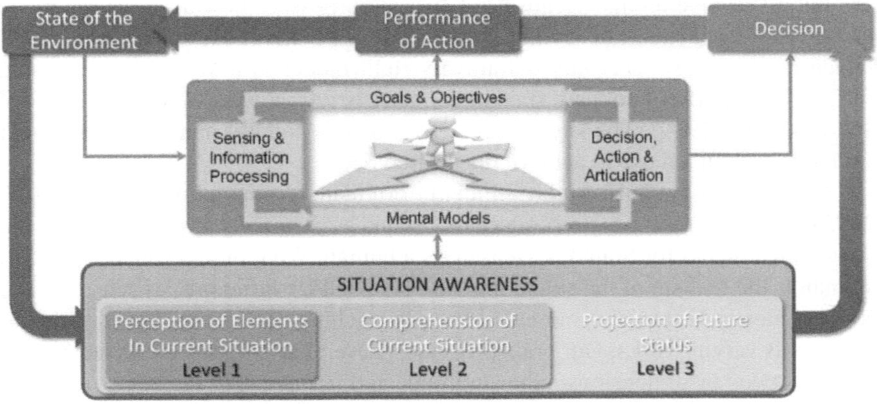

FIGURE 15.4 A model of human SA. (Based on Endsley, 1995.)

These levels form a continuous SA cycle – from sensing, interpretation, decision-making, and action, which modifies the environment, restarting the loop.

Importantly, SA is a state of knowledge, not merely the process to achieve it. Prior experiences and mental models shape it. A person may have complete information but still make poor decisions due to inadequate comprehension or weak mental models. Conversely, even perfect SA may not yield optimal action if decision-making knowledge is lacking.

In HITL simulations, the goal is to build and reinforce mental models that improve SA. Simulations must go beyond presenting raw data – they must enable comprehension and foster intuitive, experience-based anticipation. This is especially crucial in sociotechnical systems (STS) design and evaluation, where human–system interaction depends heavily on SA. Effective HITL simulation design must consider:

- *Sensory Fidelity:* How accurately stimuli reflect real-world cues.
- *Cognitive Alignment:* How well the simulation supports mental model formation.
- *Action-Feedback Loops:* How responsive and meaningful the simulation's feedback is to user input.

By stimulating SA through realistic, immersive, and feedback-rich experiences, HITL simulations foster familiarity and confidence, enabling safer, more effective human performance.

So, when we consider the use or design of an HITL simulation solution, we must consider the "fidelity" of the models and resulting simulations from the perspective of the human with regards to sensory perceptions, mental models, clarity in the availability of options, and their ability to effect chosen actions with appropriate impact and feedback from the environment.

Defining Model Fidelity

Model fidelity refers to how closely a model replicates the real world. Definitions vary, but all revolve around "realism relative to purpose" – how accurately a model mirrors real-world states, behaviors, or perceptions (Clark et al., 2000; Gross, 1999; Walden et al., 2023). The only definitively right model of a system is the system itself, as George Box famously stated, "*all models are wrong, but some are useful*" (Box & Draper, 1919). The key is sufficient realism – fidelity matched to the HITL simulation's intended use to achieve the desired SA.

A Useful Analogy: HiFi Audio – Fidelity originates from audio systems – *High-Fidelity (HiFi)* sound aims to reproduce real sound as accurately as possible. Consider abstracting a model for Sound, as sensed by a human. What makes sound "HiFi" for a human is the realism of the sound pressure level (SPL) variations, as sensed by our auditory sensory organs, i.e., our ears.[2] Sounds reaching our ears are "sensed" by our eardrums as varying SPLs. Our brain processes these SPL variations detected by each ear across two key dimensions – the amplitude and frequency of the SPL waves. The latter is translated by our brains into pitch, while the former registers as the loudness level of the sound heard. Further, as we have two ears, our brain senses SPL variations between our ears and establishes a sense of direction through triangulation. So, were we to consider a "reference model architecture" for sound as sensed by a human (consider here a live, virtual, or replicated HITL simulation), we would need to consider abstractions across these three key dimensions – frequency (pitch), amplitude (loudness), and direction (spatial location).

So, we could abstract a "representative" sound model, similar to the real world, at the right frequency (pitch) but at a lower amplitude (volume) and with no directional information (i.e., monophonic). Alternatively, we could increase the fidelity of our sound model by matching amplitude (loudness), frequency (pitch), and direction, to provide for truly HiFi immersion. In so doing, we would be increasing the abstracted sound model's "level of detail" across all three dimensions to match the variations in the SPL of the real sensed sound. This "level of detail" or granularity of the sound model, across each applicable dimension, is what we typically term its "resolution." This is precisely what digital "surround sound" processing techniques, such as "Dolby Digital," attempt to deliver in movie theatres, or for that matter, in home theatre sound systems. This (amplitude, frequency, direction) sound is a reference sound model architecture that manufacturers of audio systems instantiate through innovation in digital signal processing and speaker construction technologies to (re)create spatial audio to represent real-world sounds (such as music recordings) or to create immersive audio environments (such as surround-sound systems for movie theatres and home-theatre systems). Figure 15.5 (revised from Bhalla et al., 2023b) depicts this concept of fidelity for sound models in terms of resolution (increasing to the right) across each of the three key dimensions of amplitude, frequency, and direction for three example use-cases – (1) a monophonic "Lift Floor Tone" (yellow), (2) a HiFi "Home Theatre" (green), and (3) a "Full Flight Simulator" (blue).

Model fidelity can therefore be considered a function of two key aspects – breadth of scope (in the example of sound – amplitude, frequency, and direction) and depth of resolution (granularity or detail needed to model each applicable scope-element). The alignment of a model's abstraction with its intended purpose is a measure/indicator of its fidelity (suitability to the need), which itself is therefore a relative term ("hence Gorge Box's quote of "*all models are wrong, but some are useful*"!(Box & Draper, 1919)).

A HITL Simulation Fidelity Framework

In HITL simulations, fidelity is fundamentally tied to the human sensory system. While many human senses exist, the five core senses – sight, hearing, touch, smell, and taste – provide a practical foundation for designing simulation experiences. In a HITL simulation, Fidelity hinges on how accurately and effectively the simulation stimulates the relevant human senses for a given use case. HiFi is not about overwhelming realism but targeted realism – stimulating the right senses to the right degree to enable immersive, effective interaction, thereby correcting SA for a particular purpose.

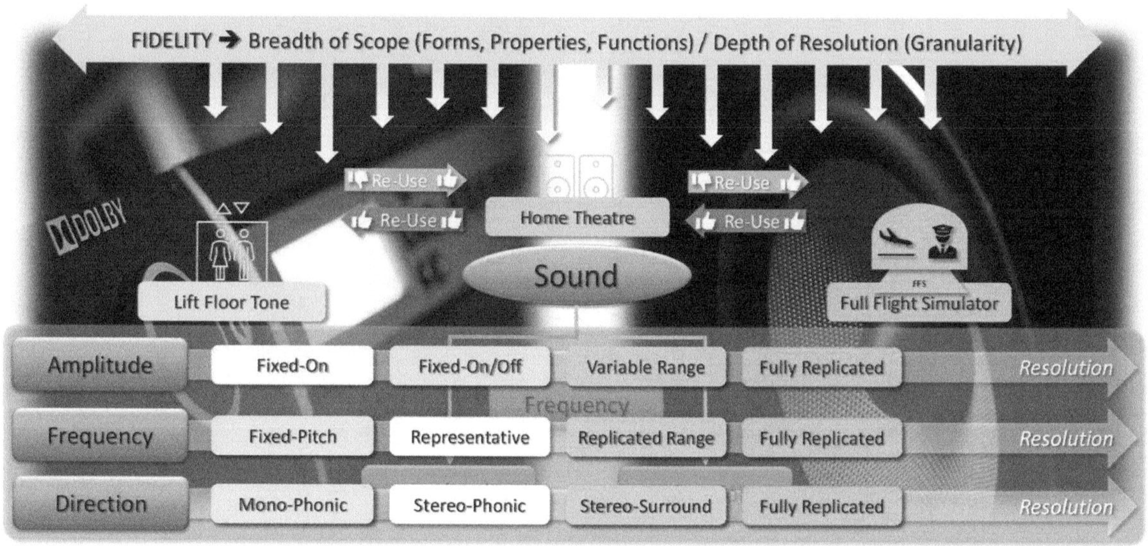

FIGURE 15.5 Sound models of varying fidelity suited to specific purposes (Bhalla et al., 2023b).

To determine the appropriate level of fidelity, we must:

- Analyze each use case from a sensory perspective.
- Identify which senses are involved and how.
- Define scope (which senses to stimulate) and resolution (detail for each stimulation).

This framework enables purposeful design, tailoring sensory fidelity to achieve the desired level of SA, engagement, or training effectiveness for the human in the loop.

"Emergence" in a HITL Simulation Solution Context

A cautionary note (concerning fidelity) related to the systems concept of "emergence" and potential for adverse outcomes in HITL simulations. When models of known fidelity (defined and compatible abstractions) interact, the resulting emergence is typically positive and expected. However, negative emergence can occur when incompatible models interact across mismatched abstractions, leading to unintended and potentially harmful outcomes.

As systems engineers, our goal is to minimize unintended emergence by ensuring:

- Each model's scope and fidelity are aligned with the intended use case.
- The underlying data used for model development is accurate and consistent.
- The simulation's execution architecture synchronizes models correctly and predictably.

A typical example of negative emergence is motion sickness in flight or VR/AR simulations, caused by temporal mismatches in sensory cues. Humans perceive motion slightly before related visual and auditory cues in the real world. If a simulation presents visual cues before motion cues – or omits motion entirely – this sensory misalignment can trigger "Visually Induced Motion Sickness" (VIMS). VIMS can cause some users to experience severe nausea and disorientation. To prevent such outcomes, HiFi HITL simulations, such as a Level-D FFS, must undergo strict latency testing to ensure the correct timing and order of sensory cues, especially that motion precedes sight (EASA, 2020).

CONCLUSION

HITL simulation solutions are no longer optional in safety-critical domains – they are essential. As this chapter has explored, HITL simulations provide an unparalleled means to safely explore, train, rehearse, experiment, understand, and operate complex scenarios that are otherwise too costly, dangerous, or impractical to experience firsthand. Through purposeful M&S – guided by ST principles – HITL enables the experiential formation of accurate mental models and enhanced SA, vital for effective decision-making and human–system interaction.

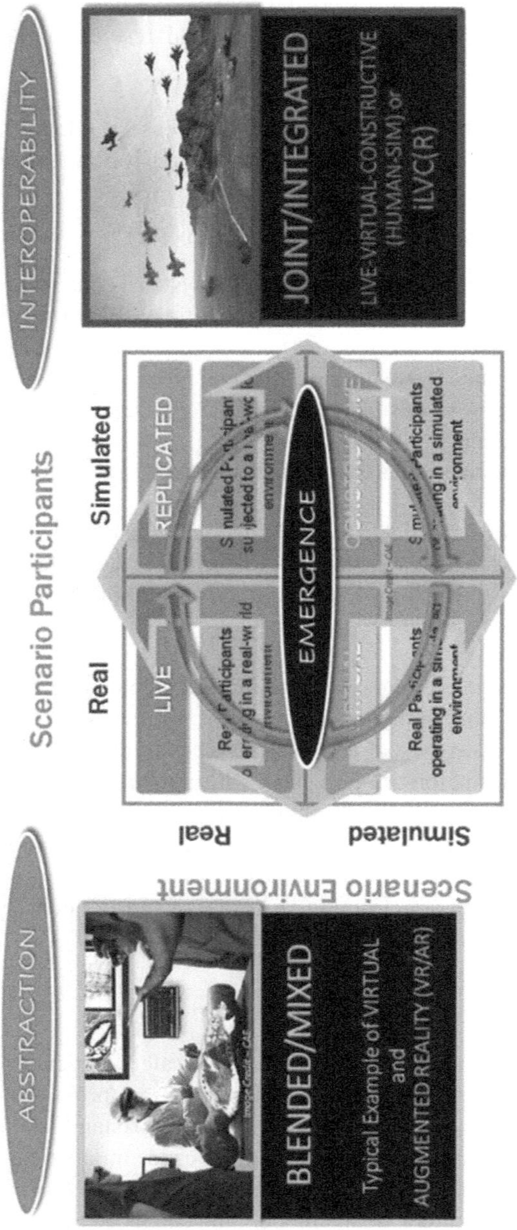

FIGURE 15.6 Simulation types across LVC(R) categories.

A categorization framework across live, virtual, constructive, and replicated (LVC(R)) domains provides context and clarity in how simulation solutions can be structured. Further, hybrid approaches such as integrated LVC(R) and blended/mixed simulations extend these capabilities to complex, multi-domain scenarios, where interoperability, fidelity alignment, and responsiveness become crucial.

Crucially, the fidelity of an HITL simulation must be aligned not with absolute realism, but with purpose-driven sufficiency. As shown through examples such as HiFi audio modelling, model fidelity is a function of scope and resolution, anchored in human sensory perception. The proposed HITL Simulation Fidelity Framework emphasizes tailoring fidelity to human senses and experiential learning needs, ensuring HITL simulations generate immersive, effective, and safe cognitive environments. However, HiFi HITL simulation integration carries the risk of unintended emergent outcomes, especially when combining models of mismatched abstraction. Attention to model compatibility, system execution timing, and user-centered design is essential to avoid adverse effects such as VIMS.

Ultimately, the design and application of HITL simulation solutions must remain grounded in the SE discipline, adapted to support human-centric, experiential learning across diverse sectors – from defense and aviation to healthcare, transport, and beyond. As technology evolves, the future of simulation lies not just in digital sophistication but in its ability to meaningfully connect people, models, and context to create understanding and improve outcomes.

NOTES

1 "The material presented in this chapter builds on the authors' personal learning and experiences over a career spanning over 30 years in Aerospace Systems Engineering, with over 24 of these spanning the acquisition and sustainment of the first high-fidelity Level-D defence full flight and mission rehearsal systems for most ADF aircraft, across Services between 1996 and 2019 (Bhalla, 2018)). The author was the recipient of the Ray Page Lifetime Achievement award in 2021. "

2 *Note:* We do "feel" sound as well, especially at higher power and lower frequencies; however, for simplicity here we will limit our sensing of sound to our ears.

REFERENCES

Australian Government-Department of Defense (2018). Live, Virtual, and Constructive Simulation and Synthetic Environments. Defence Science and Technology Group. Retrieved 1 June 2023 from www.dst.defence.gov.au/publication/live-virtual-and-constructive-simulation-and-synthetic-environments

Bhalla, J. (2018, 30 April 2018). Tailoring Systems Engineering for Modelling & Simulation [Book Chapter Conference Paper]. *Systems Engineering Test and Evaluation Conference 2018, Sydney, Australia.* Engineers Australia. https://search.informit.org/doi/abs/10.3316/INFORMIT.626638214419987

Bhalla, J., Cook, S. C., & Harvey, D. J. (2023a). A Conceptual Framework for the SE of AI-Intensive Systems (SE4AI). *Considering Data through the Lifecycle INCOSE International Symposium 2023, Honolulu, HI, USA*. Wiley.

Bhalla, J., Cook, S. C., & Harvey, D. J. (2023). Towards a systems framework for the assurance of maritime autonomous systems. Australian Journal of Multi-Disciplinary Engineering, 19(1), 89–108. https://doi.org/10.1080/14488388.2023.2235735

Bhalla, J., Cook, S. C., & Harvey, D. J. (2024). *One Model to Rule Them All and through Emergence Bind Them. INCOSE International Symposium 2024, Dublin, Ireland.*

Box, G. E., & Draper, N. R. (1919). Essentially, All Models Are Wrong, But Some Are Useful. *Statistician*, *3*(28), 2013.

CASA (2001). *Operational Standards and Requirements – Approved Flight Simulators (FSD-1).* CASA Certification Specifications for Aeroplane Flight Simulation Training Devices (2020). Retrieved from www.easa.europa.eu/document-library/easy-access-rules/easy-access-rules-aeroplane-flight-simulation-training-devices

Clark, P., Ryan, P., & Zalcman, L. (2000). *Advanced Distributed Simulation for the Australian Defense Force* (DSTO-GD-0255). Melbourne, VIC, Australia: DSTO Aeronautical and Maritime Research Laboratory. Retrieved from https://apps.dtic.mil/sti/tr/pdf/ADA385312.pdf

Clausewitz, G. C. V. (1874). On War (Vol. II). Project Gutenberg Ebook. Retrieved from www.gutenberg.org/files/1946/1946-h/1946-h.htm#chap08

Dodson, M. G., & Miklosovic, D. S. (2005). A Historical and Applied Aerodynamic Study of the Wright Brothers' Wind Tunnel Test Program and Application to Successful Manned Flight. *Fluids Engineering Division Summer Meeting*, 1, 269–278. ASME https://asmedigitalcollection.asme.org/FEDSM/proceedings-abstract/FEDSM2005/269/312755

Endsley, M. R. (1995). Towards a Theory of Situation Awareness in Dynamic Systems. *The Journal of the Human Factors and Ergonomics Society*, *37*, 32–64. https://doi.org/10.1518/001872095779049543

Erdemir, A., Mulugeta, L., Ku, J. P., Drach, A., Horner, M., Morrison, T. M., Peng, G. C. Y., Vadigepalli, R., Lytton, W. W., & Myers Jr, J. G. (2020). Credible Practice of Modeling and Simulation in Healthcare: Ten Rules from a Multidisciplinary Perspective. *Journal of Translational Medicine*, *18*(1), 369. https://doi.org/10.1186/s12967-020-02540-4

Flight Simulation Training Device Qualification Guidance – Advisory Circulars (2021). www.faa.gov/about/initiatives/nsp/ac/

Gross, D. C. (1999). Report from the Fidelity Implementation Study Group. *Fall Simulation Interoperability Workshop Papers.*

Hayes, M., & Langlois, R. G. (2005). Atlas: A Novel Kinematic Architecture for Six DOF Motion Platforms. *Transactions of the Canadian Society for Mechanical Engineering*, *29*, 3–4. https://cdnsciencepub.com/doi/abs/10.1139/tcsme-2005-0047

IEEE. (2010). IEEE Standard for Modeling and Simulation (M&S) High Level Architecture (HLA) – Object Model Template (OMT) Specification. In IEEE Std 1516.2-2010 (Revision of IEEE Std 1516.2-2000) (pp. 1–110). IEEE.

IEEE. (2015). IEEE Standard for Distributed Interactive Simulation (DIS) – Communication Services and Profiles. In IEEE Std 1278.2-1995 (pp. 1–42): IEEE.

INCOSE. (2022). *Systems Engineering Vision 2035* (INCOSE, Ed.). www.incose.org/publications/se-vision-2035

Maturana, Humberto R., and Francisco J. Varela. (1980). Autopoiesis and Cognition: The Realization of the Living. Boston Studies in the Philosophy and History of Science 42. Dordrecht: D. Reidel Publishing Company.

Motola, I., Devine, L. A., Chung, H. S., Sullivan, J. E., & Issenberg, S. B. (2013). Simulation in Healthcare Education: A Best Evidence Practical Guide. AMEE Guide No. 82. *Medical Teacher*, *35*(10), e1511–e1530. https://doi.org/10.3109/0142159X.2013.818632

NASA. (2023). *NASA Science – Mars Exploration.* The Mars Exploration Program and the Jet Propulsion Laboratory. Retrieved 25 May 2023 from https://mars.nasa.gov/

Page, R. (2021). *Zero Flight Time – The Story of Ray Page & the Revolution in Simulation* (S. Australasia, Ed.). Simulation Australasia & Shooting Star Press.

Pickard, A. C., Beasley, R., & Nolan, A. J. (2018). The Systems Engineering Incompetency Framework. *INCOSE International Symposium*, *28*(1), 503–517. https://doi.org/10.1002/j.2334-5837.2018.00496.x

Pitts, J., Kayten, P., & Zalenchak, J. (1990). The National Plan for Aviation Human Factors. In (pp. 529–540). Springer Berlin Heidelberg. https://doi.org/10.1007/978-3-662-02933-6_38

Rogers, E. B., & Mitchell, S. W. (2021). MBSE Delivers Significant Return on Investment in the Evolutionary Development of Complex SoS. *Systems Engineering*, *24*(6), 385–408. https://doi.org/10.1002/sys.21592

Saeedi, S., Liang, S., Graham, D., Lokuta, M. F., & Mostafavi, M. A. (2017). Overview of the OGC CDB Standard for 3D Synthetic Environment Modeling and Simulation. *ISPRS International Journal of Geo-Information*, *6*(10), 306. https://doi.org/10.3390/ijgi6100306

Senge, P. M. (1997). The Fifth Discipline. *Measuring Business Excellence*, *1*(3), 1.

Britannica, The Editors of Encyclopaedia. (2024). Otto Lilienthal. In Encyclopaedia Britannica. www.britannica.com/biography/Otto-Lilienthal

The Deming Cooperative (2004). 2004 A Day with Russell Ackoff Part 4 of 5 [Speech].

United States Department of Defense (2011). *Modeling and Simulation (M&S) Glossary*. Alexandria, VA: US DOD.

Walden, D., Shortell, T., Roedler, G., Delicado, B., Odile, M., Yew-Seng, Y., & Endler, D. (2023). *Systems Engineering Handbook – A Guide for System Life Cycle Processes and Activities* (5th ed.). Wiley.

Wang, H., Li, H., Tang, C., Zhang, X., & Wen, X. (2020). Unified Design Approach for Systems Engineering by Integrating Model-Based Systems Design with Axiomatic Design. *Systems Engineering*, *23*(1), 49–64. https://doi.org/10.1002/sys.21505

16 Advancing Human-Computing in Model-Based Systems Architecting

Romain Pinquié

HUMAN-CENTERED MODEL-BASED SYSTEMS ARCHITECTING

SYSTEMS ARCHITECTING

Systems architecting is distinct from systems engineering. Engineering design is primarily a scientific deductive process focused on quantifiability, utilizing analytic tools for technical optimization. Conversely, architecting is more of an artistic inductive process centered on qualitative value, employing nonquantitative tools to ensure stakeholder satisfaction (Maier & Rechtin, 2009). The architect, who works for a client and collaborates with a builder, engages in a joint exploration of requirements that have only been vaguely stated. He designs satisfactory and feasible solutions to an ill-structured problem where neither goals nor means are well understood.

In contrast, the engineer, who collaborates with the architect and works for a builder, seeks an optimal design solution to a clearly defined set of requirements. System architecting creates a system architecture to ensure the resulting system is "fit for purpose." This involves striving for fit, balance, and compromise among the tensions of client needs, resources, technology, and various stakeholder interests. Architecture transforms stakeholder needs into a strategic blueprint that establishes the framework for a unified vision, while engineering design converts that blueprint into tangible, buildable realities. Although there is no distinct boundary between engineering design and architectural design, architectural design sometimes consciously influences engineering design, while at other times, it unconsciously arises from engineering design. Good architecture anticipates dependencies and organizes them to enable design to proceed confidently, thus avoiding rework cycles. Design then builds upon that foundation to bring the concept to life. In essence, systems architecture serves as the front end of systems engineering and can generally be summed up as the art or science of constructing systems for human use (Rechtin, 1991).

DOI: 10.1201/9781003666431-19

Model-Based Systems Architecting

Before the 1950s, information was tied to physical media – books, records, and photos. Claude Shannon's work in the 1950s introduced systematic digitization, replacing atoms with bits. This shift sparked the digital revolution, forming the foundation of Industry 4.0.

In line with this overall trend, companies engaged in developing technological systems focus much of their efforts on establishing a model-based systems engineering (MBSE) approach. Model-based systems architecting (MBSA) is frequently used interchangeably with MBSE. Is MBSA a subset of MBSE, a superset of MBSE, or are they entirely distinct? As systems architecting serves as the front end of systems engineering, MBSA represents the front end of MBSE. This chapter defines MBSA as the formalized application of modeling to support the top-down recursive definition of expected functions, imposed design constraints, requirements, and architecture, along with the continuous early verification, validation, and traceability of architecture descriptions.

Despite the lack of empirical evidence (Campo et al., 2023; Henderson et al., 2022; Henderson & Salado, 2021), practitioners perceive that the paradigm shift from a document-centric to a model-based approach enhances various aspects, including a more comprehensive capture of architectural knowledge, improved communication, better information sharing, enhanced consistency, traceability, systems understanding, complexity management, and an increased capacity for data reuse. However, studies (McDermott et al., 2020; Chami & Bruel, 2018; Cloutier, 2015; Herzog et al., 2014; Maurandy et al., 2012; Andersson et al., 2010) indicate that the complexity of MBSE methods and tools makes adopting MBSE difficult and time-consuming.

Human-Centred Computing for MBSA

When architecting and engineering technical systems, architects and engineers often view human considerations as an afterthought, whereas "*Human Systems Integration (HSI) is a transdisciplinary sociotechnical and management approach of systems engineering used to ensure that system's technical, organisational, and human elements are appropriately addressed across the whole system lifecycle, service, or enterprise system*" (Boy & Kennedy, 2023). Adopting an HSI approach for MBSA consists of representing and integrating human aspects in and within the overall model-based system architecture.

Human-in-the-loop simulation, virtual human-centered design, and, in a broader sense, human-centered computing (HCC) are critical aspects of HSI as they holistically integrate people and systems. HCC is not only human–computer interaction (HCI). HCC operates in human contexts rather than creating contexts that humans must learn and understand. It is a sort of HSI bridging the gaps between human sciences (e.g., sociology, human factors, cognitive science, ergonomics, psychology) and computer sciences (e.g., HCI, information science, computer graphics, ubiquitous computing, artificial intelligence), where the convergence of multiple disciplines involved in interdisciplinary research enables crossing the perceptual line between interaction with virtual and real worlds.

HCC for MBSA is a visionary computing application for system architects. It enhances figurative tangibility, allowing actors to grasp abstract concepts mentally. This facilitates communication and cooperation across disciplines and introduces flexibility in human–system integration (Boy, 2020).

HCI FOR MBSA

Needs of System Architects

Two main categories of stakeholders engage with modeling interfaces for MBSA: systems architects with a software engineering background and architects from other engineering disciplines, such as project management, business analysis, and related fields. As discussed in the next section, commercial MBSA modeling interfaces have not been designed with the needs of systems architects who lack a software engineering background in mind. Before examining the limitations of modeling interfaces for systems architecting, let us first explore the needs of a systems architect.

With the widespread use of SysML-like modeling languages and software, systems architecture is often reduced to systems modeling. We must not forget that a system architect is not a systems modeler but a conductor. He elicits and validates the stakeholders' needs, including system functions and design constraints. He develops and validates a complete and correct set of system requirements before designing candidate architectures that correspond to the decomposition of the system into a set of related conceptual subsystems and encompass the definition of the behavioral and structural properties. Finally, the architect selects the preferred architectural design based on key value drivers and top program objectives, verifies that it meets the system requirements, and conducts a final validation against the stakeholders' needs. The exact process is recursively performed for each subsystem until a make, buy, or reuse decision is reached. This iterative and recursive systems architecting process requires a multidisciplinary core team that collaborates face-to-face and remotely during synchronous co-design sessions or asynchronously through integrated data repositories. Furthermore, the system architect continuously conducts trade-offs, traceability, [...], and coverage analysis between each core activity while maintaining the requirements and architecture design through transdisciplinary technical and business design reviews.

Reference materials such as lecture notes, handbooks, tutorials, and case studies often illustrate key systems architecting activities in an idealized way. The process starts with clearly defining stakeholders' needs, omitting the methods and techniques essential for eliciting these needs. Conversely, specific industries, like MedTech combination products, compel drug delivery system manufacturers to create devices based on assumed requirements. Pharmaceutical clients typically do not express their needs, especially regarding drug properties, until they reach Phase II of clinical research for a preclinical prototype (Menshenin et al., 2023). The architecture design activity involves breaking the system into subsystems linked by internal interfaces, shown as abstract blocks and arrows. While this corresponds with systems architects' cognitive framework, they often dismiss commercial MBSA software in favor of conventional text processors, spreadsheets, and tools like Visio or PowerPoint. Why is that?

A review of systems architecting literature shows it involves more than decomposing systems into abstract subsystems and modeling architectures. The definition of system topology must also include component placement (Sanchez et al., 2021). The selection of configurable components (Broodney et al., 2012), the allocation of space and the routing of interfaces between subsystems (Formentini et al., 2022), the preliminary sizing of subsystems (Sanchez & Liscouët-Hanke, 2020), and safety analysis (Bamrah et al., 2025) are also core tasks a system architect achieves. These tasks are rarely discussed, if ever, in the primary documentation resources dealing with MBSA. The comparison of these system architecture tasks with the capabilities of conventional MBSA conceptual modeling software and language reveals a significant gap that partly explains the limited adoption of MBSA tools by systems architects engaged in developing complex technological systems. In many instances, MBSA toolsets are mandated by top management, who have been persuaded by MBSA marketing.

According to the Systems Modeling Language (SysML) v2 Request For Proposal (RFP) Working Group repository, the preliminary driving requirements for the System Modeling Environment require "*providing flexible and rich visualization*" through "*advanced visualization techniques*" to " *enable efficient and intuitive use by a broad range of users with diverse skills.*" The Object Management Group (OMG) Working Group acknowledges several limitations for SysML V1 model visualization, including that "*the learning curve for the SysML language and tools is quite steep.*" In response to this adoption barrier, the usability requirements of SysML V2 specify that "*usability must be a primary consideration for the next-generation modeling language and tools.*" The Model Visualization Working Group[1] emphasizes usability, stating that it "*is a critical driver for the ultimate acceptance and use of SysML V2 and is fundamental to enabling a more cost-effective application of MBSE. In particular, the usability goals aim to reduce the learning curve for new users, improve productivity for constructing models for both new and experienced users, and enhance the visualization capability and interaction with the model for a diverse set of stakeholders who consume the model data.*" Despite these good intentions, SysML V2 remains a diagrammatic and heavily codified language that systems architects will likely struggle to adopt.

Limits of Modeling Interfaces for MBSA

Since the 2000s, companies have commercialized MBSA software like Catia Magic, Rhapsody, Capella, and OPC Cloud. All these offer similar modeling capabilities and utilize model-based diagrammatic representations. Model-based indicates that a modeler employs a modeling language to create models that a processor can verify. Despite growing model-based solutions, many architects still use Microsoft Office or Google software to create drawings instead of formal models. Additionally, there is a lack of empirical evidence supporting an MBSE and MBSA approach (Henderson & Salado, 2021).

Systems Engineering Is Software Engineering.

Many SysML experts come from software engineering or embedded systems. Despite attempts to standardize graphical modeling for systems engineers, software engineering's influence is still problematic. Why does SysML define seven property types (value, part, reference, constraint, flow, participant, classifier behavior) and four port types (port, proxy port, full port, flow port) or three connector types (connector, binding connector, item flow)? These concepts are not in the systems architecting knowledge base, except for flow and port, and lack shared understanding within the systems architecting community. SysML – a Unified Modeling Language (UML) profile – reflects a software-centric view of systems. Why should all systems engineers learn SysML? How many hours have been wasted as non-engineers are forced to adopt it? Modeling tools must focus on systems engineering fundamentals, enabling architects to utilize their expertise. However, vendors aim for incomplete, "flexible" solutions.

Lack of Convergence

Another issue is the lack of convergence. Companies acquire flexible commercial modeling languages and software, tailoring them to their needs. Consequently, software editors, consultants, and manufacturers develop their architectural frameworks, using SysML to create new domain-specific languages for engineering systems. If the underlying principles of systems engineering exist (Walden & International Council on Systems Engineering, 2023), why is there no standard approach for MBSE, or at least more convergence? Furthermore, why do vendors have the power to dictate the situation? Is it because the common element is marketing MBSE, and the individual value propositions, which can be very valuable, are local creations?

Poor Diagrammatic Visual Notations

When defining graphical conceptual modeling notations, most efforts focus on designing semantics, often overlooking or underestimating the role of visual syntax. The lack of theoretical rigor in designing visual dialects results in undesirable properties. For instance, in SysML, the fundamental concept of function in systems engineering can be represented by a block, an activity, or an operation. Consequently, there is a deficiency in semiotic clarity due to the prevalence of symbol redundancy. Moreover, MBSE diagrams are overly homogeneous. We call them "camouflage models," where many elements are identical, making extracting anything significant complicated. The poor discriminability of conceptual modeling notations, primarily consisting of shape variants – mostly rectangles – further complicates this issue (Petre, 1995), and connecting visually similar lines can lead to confusion (Patig, 2004). Moreover, the poor quality of graphical conceptual modeling notations undermines acceptance among nonspecialist audiences (McLeod, 2018; Moody, 2009), whereas MBSE is a multidisciplinary approach that integrates various disciplines.

Perceptual and Conceptual Integration Processes

MBSE visual notations typically do not support the perceptual and conceptual cognitive processes involved in diagrammatic reasoning. In diagrammatic representations, each diagram corresponds to a single, consistent perspective. Information about the system is spread across multiple diagrams, requiring the architect and stakeholders to navigate these different diagrams to find relevant information dispersed among various displays and integrate it into a cohesive representation. This complicates generating and refining hypotheses about the system of interest (Kim et al., 2000). The limited physical size of viewing surfaces directly impacts engineers' ability to work effectively. Indeed, not all the required information can be displayed simultaneously. Engineers must selectively view glimpses serially, over time, through the restricted "keyhole" (Woods, 1984), the small display surface provides. Grasping the entire system through multiple diagrams is challenging for complex systems, as the number of diagrams surpasses a system architect's span of control. The limited capacity of human working memory makes it nearly impossible for a system architect to effectively retain all information from every diagram to reason about the system in question. This issue calls for systems modeling interfaces that increase the visual momentum, *"the extent to which an interface supports a practitioner in transitioning between various information-seeking activities that are required for understanding and exploring work domains"* (Bennett & Flach, 2012). It exists at different levels. It can measure the user's ability to extract and integrate information across multiple screens, within individual screens, and display it on a screen. At a low level of structure, it involves transitions within a screen (i.e., between collections of displays or windows — the "view" level) or within a display (i.e., between the visual elements of a display itself – the "form" level).

Esoteric Graphical Representations Acting as a Barrier to Communication

As the benchmarking study by the Systems Engineering Research Center noted, "We usually must cut and paste diagrams into PowerPoint slides; the tools and models are too complicated to use directly in a meeting. Often, the diagrams are too busy to support meetings. Few engineers know how to use a diagram to tell a simple story" (McDermott et al., 2020). Architecture description involves capturing the architecture to share and communicate with people involved in the system. As we move to digital modeling tools, we should not expect all the systems architecture core team members to understand the details of these tools. The tools should conceal their implementation and offer a domain-appropriate palette for engineers to use. Experts often create highly codified models that nonexperts must comprehend effectively to achieve their goals. However, business stakeholders typically relate to documents, presentations, and spreadsheets, finding it difficult to quickly extract meaning from an unfamiliar graphical presentation. Thus, "practitioners spend inordinate effort transposing models from rigorous tools to non-structured formats to overcome the acceptance problem. The connection to the source repository is lost in the process, destroying the derived output's integrity, reusability, maintainability, and currency when the source changes" (McLeod, 2018). Why are successful MBSE initiatives confined to the R&D department? Because "Non-MBSE personnel get confused during the review by certain symbology needed by the modeler" (McDermott et al., 2020). Indeed, the survey report highlights several usability issues, such as "MBSE models are too technical, typically meaning that the audience does not easily comprehend the collection of symbols and connections and their meaning."

MBSE languages are difficult to adopt because the meaning of symbolic MBSE visual notations is purely conventional, relying on an arbitrary mapping between concepts and their representations. It must be learned, even though it asserts that it is domain-independent and suitable for multidisciplinary activities such as systems architecting. Studies (Moeser et al., 2015; Grundel & Abulawi, 2016; Schumacher & Inkermann, 2022) indicate a lack of acceptance due to the level of abstraction being too distant from the mental schemas of the diverse group of people involved in the system development. Decoding conventional representations requires substantial learning effort; poorly designed or overloaded notations can worsen this. The high level of abstraction in the symbols that form MBSE diagrams allows for encoding all concepts. However, the lack of adoption by target users raises concerns about collaboration and integration goals within a project team. Visual metaphors, which are homomorphic representations of the information they convey, are crucial when diagrams serve as cognitive artefacts that support collaborative work (Blackwell, 2001).

The World of Systems versus the World of Geometry

In manufacturing complex engineered systems, we observe two concurrent worlds: the realm of systems engineering and the domain of the digital mock-up (Hick et al., 2023; Schumacher & Inkermann, 2022). Both decompose a system into subsystems and define internal interfaces between them, but they are not integrated across all levels (humans, processes, methods, modeling languages, and software). Various individuals, using different tools and techniques, developed both architectural descriptions. Conversely, a common underlying engineering logic should be supported by an MBSA toolkit, enabling architects to swiftly navigate among different views (e.g., function, behavior, structure) and switch between preferred representations from two-dimensional (2D) to three-dimensional (3D), symbolic to iconic, and monocular to stereocular.

Global System, Product System, and Industrial System Architectures

MBSA modeling interfaces focus on designing product architecture. Once the product architecture (e.g., aircraft architecture) matures, another team develops the industrial system (e.g., a final assembly line). This can lead to local product optimization but a suboptimal industrial system. Achieving a globally optimal solution requires trade-offs while co-designing product and industrial system architectures to maximize performance, efficiency, and overall business outcomes. To support this high-level co-design for identifying product-process couplings and making trade-off decisions, MBSA modeling interfaces should allow a global architect to integrate digital artefacts representing both product and industrial system architectures in a single virtual environment using systems of systems principles.

This review of MBSA software highlights the lack of empathy for systems architects. Designing MBSA interfaces requires a human-centered approach to meet the needs of product, process, and global systems architects.

IMMERSIVE HUMAN-CENTRED MBSA

The Future of MBSA Is Immersive

According to the futuristic visions (2020 and 2035) of MBSE specified by NASA (Weiland, 2021) and INCOSE (International Council on Systems Engineering, 2022), virtual reality (VR) will be a crucial technology for implementing a human-centered MBSE. As MBSA is a subset of MBSE, it will also be immersive.

On the Importance of Iconic and 3D Visuals

Empirical studies show that iconic graphics are preferred over symbolic signs to reduce cognitive load (Masri et al., 2008), facilitate recognition and memorization (Irani & Ware, 2003), and make models more accessible to those new to notations (Petre, 1995; Britton & Jones, 1999; Masri et al., 2008). Generally, humans favor concrete shapes over abstract objects (Bar & Neta, 2006; Petre, 1995). With the current MBSA notations, pragmatic solutions have been envisaged to inject iconic representations into symbolic descriptions. For example, Figure 16.1 shows that modelers can enrich diagram blocks with images (Moody, 1996; Masri et al., 2008) or technical sketches (Moeser et al., 2015; Grundel & Abulawi, 2016) and combine abstract and concrete representations using more flexible environments such as PowerPoint (Bayer, 2018).

In mechanical engineering, codified 2D drawings have been replaced by 3D digital models that aid in interdisciplinary activities like technical design reviews. The semantic transparency of 3D representations enables easy inference of meaning. Furthermore, creating interactive 3D visuals for MBSA enhances collaboration between disciplines in systems architecture (Deshmukh et al., 2015). According to NASA and INCOSE, MBSE's future will be immersive. However, few have attempted to replace 2D diagrams with immersive 3D graphics. VR and 3D graphics offer new ways to encode domain knowledge in visuals that require less decoding for those unfamiliar with MBSE notations. This is already evident in product geometry. However, these outcomes and other known VR use cases (e.g., marketing, accessibility evaluation, maintenance training, assembly/dismantling) pertain to concrete objects, not abstract concepts (e.g., operational scenarios, functions, requirements, conceptual interfaces, and blocks) relevant to systems engineers. We can identify two alternatives for visually encoding these abstract concepts in VR.

On one hand, iconic representations can replace symbolic ones. For example, earlier works have employed the metaphors of the city, solar system (Oberhauser & Lecon, 2017), or archipelago of islands (Schreiber et al., 2019) to encode software architecture concepts modeled with UML. A pioneering application of this approach in system architecture is the graphical modeling of operational scenarios for the defense industry (Korfiatis et al., 2015), but without objective proof of the added value of these new modes of interaction. More recently, it was argued that human-centric immersive 3D visuals should sometimes replace 2D diagrams to enhance communication and participation in multidisciplinary systems architecture co-design activities (Pinquié et al., 2023).

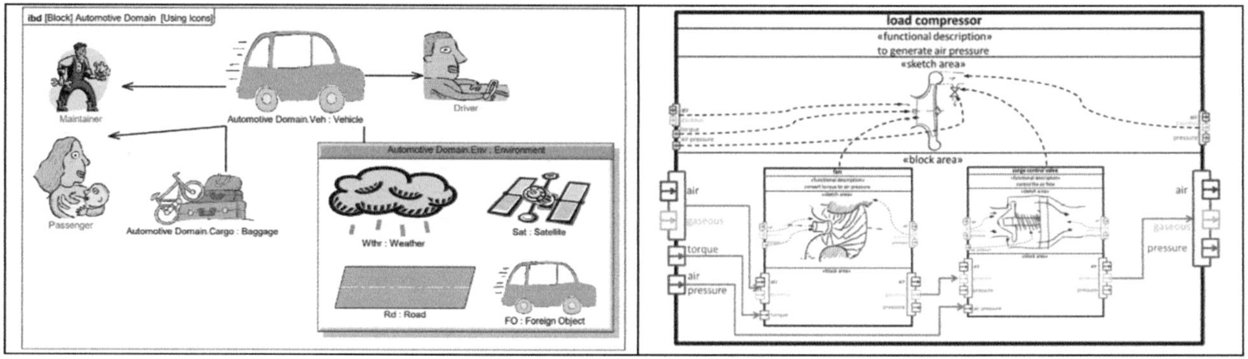

FIGURE 16.1 SysML enriched by pictures (left) and sketches (right) (Grundel & Abulawi, 2016).

On the other hand, one can encode diagrams with symbolic visual 3D metaphors (Romero et al., 2021) or create diagrammatic representations in a stereoscopic 3D virtual environment while preserving the metaphors (Oberhauser, 2022; Yigitbas et al., 2022).

Despite pioneering efforts in using VR for conceptual modeling, knowledge about its impact on designers' cognitive and business activities remains limited due to a lack of experimentation. We still lack ergonomic criteria to frame the design and validation of immersive environments, aiming to enhance the performance and satisfaction of engineered systems architects, like ergonomic guidelines for 2D interfaces.

TOWARD AN INDUSTRIAL MBSE METAVERSE

Industrial Metaverse

The nature of systems engineering work is becoming more distributed with the rise of telework. New digital forms of organizations, such as decentralized autonomous organizations, are emerging and providing a different way to work together. According to Lockheed Martin Corp, MBSE will transition to the Metaverse (Cesena et al., 2022).

The Metaverse is an interoperable network of virtual worlds, featuring persistent, 3D, real-time, immersive environments that blur the line between real and virtual. It is designed for socializing, working, learning, transactions, playing, and creating. The European Commission projects that the global virtual worlds market will grow from EUR 27 billion in 2022 to over EUR 800 billion by 2030 (European Commission. Directorate General for Communications Networks, Content and Technology, 2022). In specific sectors like the automotive industry, forecasts predict a rise from EUR 1.9 billion in 2022 to EUR 16.5 billion by 203012. eXtended Reality (XR) technologies, including VR and augmented reality (AR), are key components of virtual worlds. Their development will significantly benefit the job market, with an estimated 860,000 new XR-related jobs in Europe by 2025. Research and innovation in immersive human-centered MBSA contribute to an exciting vision of industrial Metaverses being co-imagined by software editors (e.g., Nvidia, Microsoft, IBM) and manufacturers (e.g., BMW, General Electric). Still, their substance and form have yet to be invented.

Industrial Metaverse for MBSA

In the MBSA Metaverse, stakeholders involved in systems architecting will naturally co-design rich 3D architecture descriptions, such as the system life cycle (Figure 16.2), system contexts including external entities and interfaces, and architecture designs.

Utilizing a 3D stereoscopic virtual world provides wider visual variables. The extensive use of 3D representations enhances the semantic transparency of modeled elements and facilitates perceptual and conceptual reasoning processes as defined in Kim et al. (2000). An unlimited virtual 3D space links relevant visual items across various perspectives. This improves perceptual integration by allowing exploration of model elements in one environment without disrupting cognitive processes during

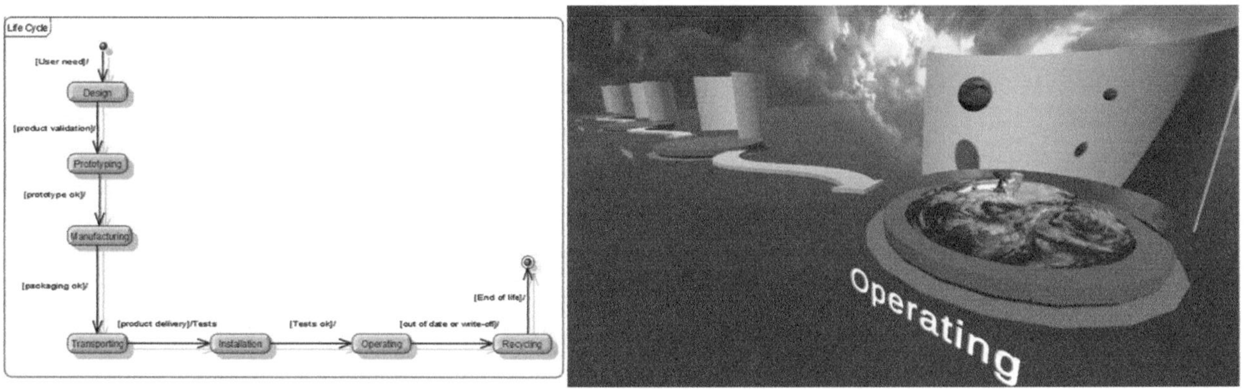

FIGURE 16.2 System life cycle model in SysML (left) and VR (right).

view switches. It also aids conceptual integration, supporting hypothesis generation and refinement about the target system using contextual information from a "*long shot*," which presents the complete system from diverse viewpoints. MBSE software solutions provide 2D planar graphs and desktop-based visualization capabilities with limited interactions, such as filtering and expandable nodes. This restricts designers from exploring large amounts of data simultaneously, especially in a conventional office setup with one or two small desktop screens. In such a configuration, "*even simple operations often required practitioners to navigate multiple display windows and integrate information (sometimes presented in different formats)*" (Bennett & Flach, 2012). Figure 16.3 illustrates how VR can enhance visual momentum, allowing for exploring various interconnected digital artefacts.

Most 3D architecture descriptions will be dynamic to emulate the behavioral properties of the system and its environment. The continuous early verification and validation of architecture descriptions will rely on immersive human-in-the-loop co-simulations of specification and design models in as-experienced environmental conditions (Romero et al., 2022). These realistic, human-centered virtual simulations enable testing of nominal and critical scenarios, allowing for the injection of randomness that reflects potential unintended operational situations without creating physical prototypes beforehand. The early integration of VR during the architecture definition paves the way for a seamless implementation of digital twins, effectively addressing the bidirectional integration challenges between the physical and digital realms throughout the system life cycle.

The MBSA Metaverse will also serve as a comprehensive platform to explore digital threads from a bird's-eye view. Digital threads are seamless flows of linked digital artefacts, with their consistency actively managed throughout the system life cycle. These threads can be examined in an immersive virtual environment, allowing systems engineers to navigate an integrated systems engineering digital data hub (Figure 16.3). The visual metaphors that represent MBSA data are flexible enough to shift from one representation to another based on the concerns and cognitive abilities of end-users, particularly the system architect.

Designing complex engineered systems in a fully immersive virtual world could be feasible in 10–15 years if the hardware matures enough to eliminate usability issues. This desirability remains debatable, requiring a deeper understanding of the cognitive, psychological, and perceptual mechanisms involved in complex systems engineering. HCI for MBSA should be visually rich, highly interactive, and adaptable to enable asymmetric collaboration among businesses, architects, and various disciplines over the next decade. Users must be able to switch on demand from 3D immersive visuals to 3D desktop-based visuals according to the task. Both formats will support cross-platform collaboration (desktop, touch tables, wall displays, headsets, CAVE systems, etc.) with appropriate interaction modalities, including controllers, mouse, keyboard, multitouch, gestures, and speech. Architects will transition between immersive 3D environments and traditional 2D software as needed, without facing interoperability issues or learning multiple modeling languages and software.

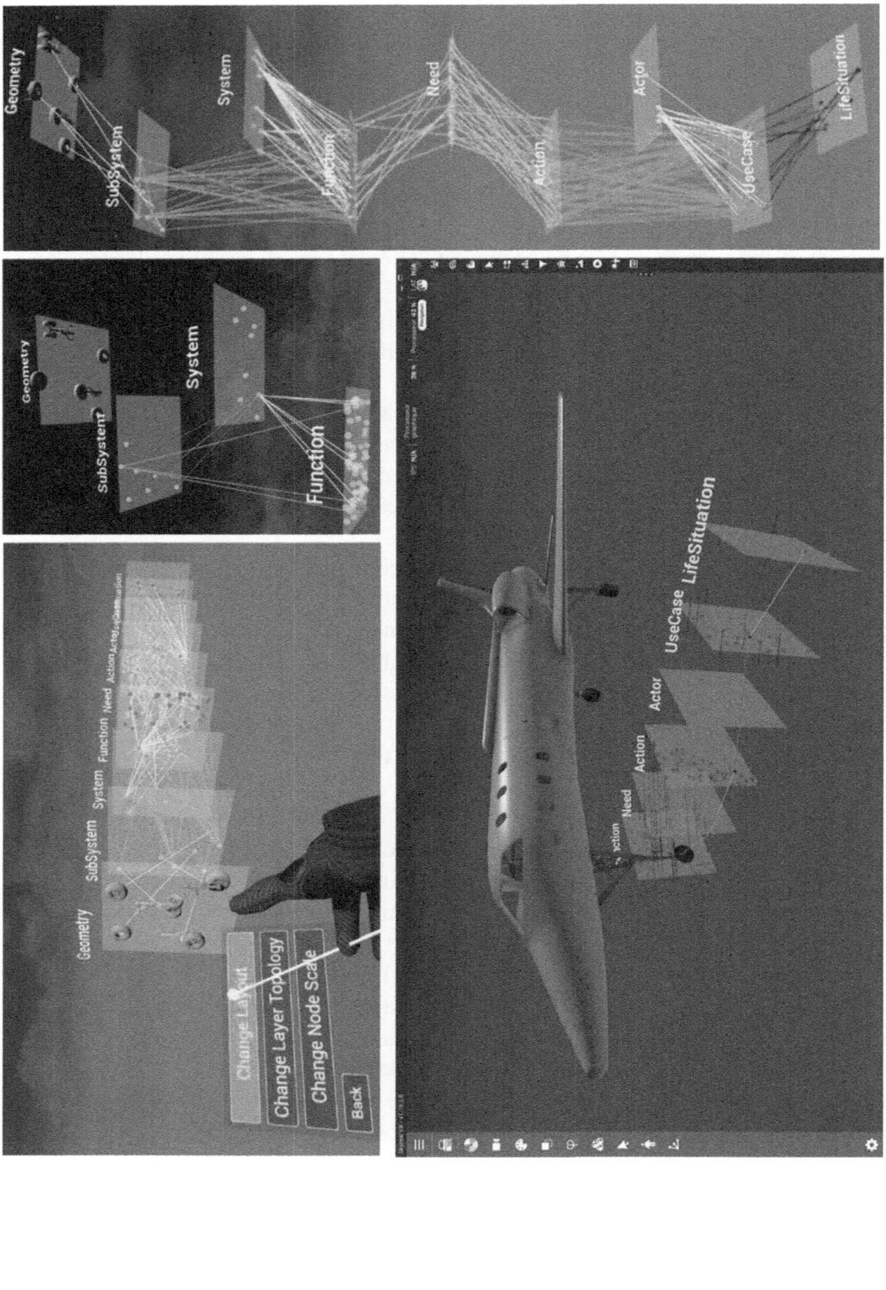

FIGURE 16.3 Exploration of MBSA digital threads in VR.

CONCLUSION

This chapter discusses the limits of conventional diagrammatic modeling interfaces for system architecture tasks. Flexible, human-centered interaction with rich 3D visuals is needed to enhance users' understanding of abstract concepts. Developing a MBSA Metaverse represents a significant step toward utilizing VR in system engineering. However, the success of this virtual environment depends on its practical utility for system architects. Thus, while envisioning this virtual world, attention must focus on ergonomic design, considering cognition, psychology, perception, user experience, and more. We can only harness VR's transformative power in advancing HCC in engineered systems architecture through such meticulous consideration.

NOTE

1 Model Visualization Focus Area www.omgwiki.org/OMGSysML/doku.php?id=sysml-roadmap:model_visualization_working_group

REFERENCES

Andersson, H., Herzog, E., Johansson, G. & Johansson, O. (2010). Experience from introducing unified modeling language/systems modeling language at Saab Aerosystems. *Systems Engineering*, *13*(4), 369–380. https://doi.org/10.1002/sys.20156

Bamrah, P., Liscouët-Hanke, S., Tfaily, A. & Tamayo, A. (2025). Zonal safety and particular risk analysis for early aircraft design. *Journal of Aircraft*, *62*(2), 254–269. https://doi.org/10.2514/1.C037774

Bar, M. & Neta, M. (2006). Humans prefer curved visual objects. *Psychological Science*, *17*(8), 645–648. https://doi.org/10.1111/j.1467-9280.2006.01759.x

Bayer, T. (2018). Is MBSE helping? Measuring value on Europa Clipper. In *2018 IEEE Aerospace Conference Proceedings*, Big Sky, MT, USA pp. 1–13. https://doi.org/10.1109/AERO.2018.8396379

Bennett, K. B. & Flach, J. M. (2012). Visual momentum redux. *International Journal of Human–Computer Studies*, *70*(6), 399–414. https://doi.org/10.1016/j.ijhcs.2012.01.003

Blackwell, A. F. (2001). Pictorial representation and metaphor in visual language design. *Journal of Visual Languages & Computing*, *12*(3), 223–252. https://doi.org/10.1006/jvlc.2001.0207

Boy, G. A. (2020). *Human–Systems Integration: From Virtual to Tangible*. CRC Press, Taylor & Francis Group.

Boy, G. A. & Kennedy, G. (2023). *Human Systems Integration: A Primer*. INCOSE.

Britton, C. & Jones, S. (1999). The untrained eye: How languages for software specification support understanding in untrained users. *Human–Computer Interaction*, *14*(1–2), 191–244. https://doi.org/10.1080/07370024.1999.9667269

Broodney, H., Dotan, D., Greenberg, L. & Masin, M. (2012). Generic approach for systems design optimization in MBSE. *INCOSE International Symposium*, *22*(1), 184–200. https://doi.org/10.1002/j.2334-5837.2012.tb01330.x

Campo, K. X., Teper, T., Eaton, C. E., Shipman, A. M., Bhatia, G. & Mesmer, B. (2023). Model-Based Systems Engineering: Evaluating perceived value, metrics, and evidence through literature. *Systems Engineering*, *26*(1), 104–129. https://doi.org/10.1002/sys.21644

Cesena, J. R., Schloemer, J. D., McInnis, K., Montes, J. D., Viveros, R. R., Kalkbrenner, M. D. & Flint, B. (2022, October 24). Migrating MBSE to the metaverse. In ASCEND 2022, *Las Vegas, Nevada*. Online. https://doi.org/10.2514/6.2022-4386

Chami, M. & Bruel, J.-M. (2018). A survey on MBSE adoption challenges. In *INCOSE Systems Engineering Conference of the Europe, Middle-East and Africa Sector EMEASEC, 2018* (November). Wiley. https://incose.onlinelibrary.wiley.com/doi/abs/10.1002/iis2.70036

Cloutier, R. (2015). Current modeling trends in systems engineering. *INSIGHT*, *18*(2), 10–13. https://doi.org/10.1002/inst.12013

Deshmukh, M., Wolff, R., Fischer M.P., Flatken, M. & Gerndt, A. (2015). *Interactive 3D Visualization to Support Concurrent Engineering in the Early Space Mission Design Phase*. In: Challenges in European Aerospace, CEAS Air & Space Conference. Challenges in European Aerospace, 7–11. Delft, Niederlande.

European Commission. Directorate General for Communications Networks, Content and Technology. (2022). *VR/AR Industrial Coalition: Strategic Paper*. Publications Office. https://data.europa.eu/doi/10.2759/197536

Formentini, G., Bouissiere, F., Cuiller, C., Eric, P. & Claudio, D. (2022). CDFA method: A way to assess assembly and installation performance of aircraft system architectures at the conceptual design. *Research in Engineering Design*, *33*(1), 31–52. https://doi.org/10.1007/s00163-021-00378-5

Grundel, M. & Abulawi, J. (2016). SkiPo – a sketch and flow based model to develop mechanical systems. *INCOSE International Symposium*, *26*(1), 399–414. https://doi.org/10.1002/j.2334-5837.2016.00168.x

Henderson, K., McDermott, T., Van Aken, E. & Salado, A. (2022). Towards developing metrics to evaluate digital engineering. *Systems Engineering*, 26(1) 3–31. https://doi.org/10.1002/sys.21640

Henderson, K. & Salado, A. (2021). Value and benefits of Model-Based Systems Engineering (MBSE): Evidence from the literature. *Systems Engineering*, *24*(1), 51–66. https://doi.org/10.1002/sys.21566

Herzog, E., Hallonquist, J. & Naeser, J. (2014). 4.5.1 Systems modeling with SysML – an experience report. *INCOSE International Symposium*, *22*(1), 600–611. https://doi.org/10.1002/j.2334-5837.2012.tb01359.x

Hick, H., Sanladerer, S., Trautner, J., Ryan, K., Piguet, J., Wilking, F., Horber, D., Faustmann, C., Kranabitl, P., Kollegger, S., Bajzek, M., Schleich, B. & Wartzack, S. (2023). Combining system models and CAD for change scenario management. *INCOSE International Symposium*, *33*(1), 117–132. https://doi.org/10.1002/iis2.13012

INCOSE (Ed.). (2023). *INCOSE systems engineering handbook*. Wiley.

International Council on Systems Engineering. (2022). *Systems Engineering Vision 2035: Engineering Solutions for a Better World*.

Irani, P. & Ware, C. (2003). Diagramming information structures using 3D perceptual primitives. *ACM Transactions on Computer–Human Interaction*, *10*(1), 1–19. https://doi.org/10.1145/606658.606659

Kim, J., Hahn, J. & Hahn, H. (2000). How do we understand a system with (so) many diagrams? Cognitive integration processes in diagrammatic reasoning. *Information Systems Research*, *11*(3), 284–303. https://doi.org/10.1287/isre.11.3.284.12206

Maier, M. W. & Rechtin, E. (2009). *The Art of Systems Architecting* (3rd ed). CRC Press.

Masri, K., Parker, D. & Gemino, A. (2008). Using iconic graphics in entity-relationship diagrams. *Journal of Database Management*, *19*(3), 22–41. https://doi.org/10.4018/jdm.2008070102

Maurandy, J., Gill, E., Helm, A. & Stalford, R. (2012). Cost–benefit analysis of SysML modelling for the atomic clock ensemble in space (ACES) simulator. In *22nd Annual*

International Symposium of the International Council on Systems Engineering, INCOSE 2012 and the 8th Biennial European Systems Engineering Conference 2012, EuSEC 2012, vol. 3, 1726–1745.

McDermott, T. A., Hutchison, N., Clifford, M., Aken, E. V., Salado, A. & Henderson, K. (2020). *Benchmarking the Benefits and Current Maturity of Model-Based Systems Engineering across the Enterprise – Results of the MBSE Maturity Survey.* System Engneering Research Center (SERC), Stevens Institute of Technology, March, 124.

McLeod, G. (2018). Facilitating design and use of effective visual languages in enterprise modelling and information systems. In *CEUR Workshop Proceedings*, vol. 2234, pp. 15–30.

Menshenin, Y., Pinquié, R. & Chevrier, P. (2023). Holistic perspective to the drug–device combination product development challenges. In *IFIP International Conference on Product Lifecycle Management, Montreal, Canada.*

Moeser, G., Albers, A. & Kumpel, S. (2015). Usage of free sketches in MBSE raising the applicability of Model-Based Systems Engineering for mechanical engineers. In *1st IEEE International Symposium on Systems Engineering, ISSE 2015 – Proceedings*, pp. 50–55. https://doi.org/10.1109/SysEng.2015.7302511

Moody, D. (1996). Graphical entity relationship models: Towards a more user understandable representation of data. In *Lecture Notes in Computer Science (including subseries Lecture Notes in Artificial Intelligence and Lecture Notes in Bioinformatics), 1157 LNCS*, pp. 227–244. https://doi.org/10.1007/bfb0019926

Moody, D. (2009). The physics of notations: Toward a scientific basis for constructing visual notations in software engineering. *IEEE Transactions on Software Engineering*, *35*(6), 756–779. https://doi.org/10.1109/TSE.2009.67

Oberhauser, R. & Lecon, C. (2017). Immersed in software structures: A virtual reality approach. In *Tenth International Conference on Advances in Computer–Human Interactions (ACHI 2017)*, pp. 181–186.

Oberhauser, R. (2022). VR-SysML: SysML model visualization and immersion in virtual reality. In *International Conference of Modern Systems Engineering Solutions (MODERN SYSTEMS 2022).*

Patig, S. (2004). Measuring expressiveness in conceptual modeling. In *Lecture Notes in Computer Science (including subseries Lecture Notes in Artificial Intelligence and Lecture Notes in Bioinformatics), 3084*, pp. 127–141. Springer, Berlin, Heidelberg. https://doi.org/10.1007/978-3-540-25975-6_11

Peter Korfiatis, Robert Cloutier, and Teresa Zigh. (2015). Model-Based Concept of Operations Development Using Gaming Simulation. Simul. Gaming 46, 5, 471–488. https://doi.org/10.1177/1046878115571290

Petre, M. (1995). Why looking isn't always seeing. *Communications of the ACM*, *38*(6), 33–44. https://doi.org/10.1145/203241.203251

Pinquié, R., Wang, H. & Noel, F. (2023). Human-centric co-design of model-based system architecture. *Procedia CIRP*, *119*, 146–151. https://doi.org/10.1016/j.procir.2023.02.131

Rechtin, E. (1991). *Systems Architecting: Creating and Building Complex Systems.* Prentice Hall.

Romero, V., Pinquié, R. & Noël, F. (2021). An immersive virtual environment for reviewing model-centric designs. *Proceedings of the Design Society*, *1*(August), 447–456. https://doi.org/10.1017/pds.2021.45

Romero, V., Pinquié, R. & Noël, F. (2022). A user-centric computer-aided verification process in a virtuality-reality continuum. *Computers in Industry*, *140*, 103678. https://doi.org/10.1016/j.compind.2022.103678

Sanchez, F. & Liscouët-Hanke, S. (2020). Thermal risk prediction methodology for conceptual design of aircraft equipment bays. *Aerospace Science and Technology, 104*, 105946. https://doi.org/10.1016/j.ast.2020.105946

Sanchez, F., Liscouët-Hanke, S. & Tfaily, A. (2021). Improving aircraft conceptual design through parametric CAD modellers – a case study for thermal analysis of aircraft systems. *Computers in Industry, 130*, 103467. https://doi.org/10.1016/j.compind.2021.103467

Schreiber, A., Nafeie, L., Baranowski, A., Seipel, P. & Misiak, M. (2019). Visualization of software architectures in virtual reality and augmented reality. In *2019 IEEE Aerospace Conference*, pp. 1–12. https://doi.org/10.1109/AERO.2019.8742198

Schumacher, T. & Inkermann, D. (2022). Heterogeneous models to Support Interdisciplinary Engineering-Mapping Model Elements of SysML and CAD. *Procedia CIRP, 109*, 653–658. https://doi.org/10.1016/j.procir.2022.05.309

Weiland, K. J. (2021). *Future Model-Based Systems Engineering Vision and Strategy Bridge for NASA*. October. www.sti.nasa.gov

Wilkinson, M. K. & Rabbets, T. J. (2020). *Don't Panic!: The Absolute Beginner's Guide to Architecture and Architecturing* (Qinetiq special edition). INCOSE.

Woods, D. D. (1984). Visual momentum: A concept to improve the cognitive coupling of person and computer. *International Journal of Man–Machine Studies, 21*(3), 229–244. https://doi.org/10.1016/S0020-7373(84)80043-7

Yigitbas, E., Gorissen, S., Weidmann, N. & Engels, G. (2022). Design and evaluation of a collaborative UML modeling environment in virtual reality. *Software and Systems Modeling,* 22, 1397–1425. https://doi.org/10.1007/s10270-022-01065-2

Section IV

Human–Automation Interaction and Performance

17 Pilot-Virtual Assistant Integration

Issues of Operational Performance, Trust, and Collaboration

Chloé Morel and Guy André Boy

INTRODUCTION

The MOHICAN project is taking place within the "Man–Machine Teaming" scientific program launched in March 2018 by the *Direction Générale de l'Armement* (DGA, French MOD) and led by Dassault Aviation and Thales. The objective of Man–Machine Teaming (MMT) projects was to create an industrial ecosystem that would enable innovations to be detected, evaluated, and, ultimately, matured and integrated into the development of future combat aircraft, particularly for the renovation of the Rafale by 2030 and then into the Future Combat Air System (FCAS). A team of researchers and research engineers from the FlexTech Chair at ESTIA Institute of Technology and fighter pilots from SYNAPSE Defense carried out the MOHICAN project.

The MOHICAN approach is based on the latest advances in participatory human-centered design (HCD), leading to better human–systems integration (HSI). HSI is the transdisciplinary technical and management process that ensures that the human elements of a system are adequately addressed and integrated. It differs from human–computer interaction (surface and late) in its founding approach to the recursive and HCD of a sociotechnical system throughout its life cycle.

Studying the performance of teaming between the fighter pilot and their virtual assistant (VA) was a seldom explored topic. Literature, particularly in the aeronautics and automotive sectors, shows that human–machine trust and collaboration are complex concepts, loosely articulated around a cohort of difficult parameters to qualify and quantify. Within MOHICAN, we explored the appropriateness and usability of many physical, informational, human, environmental, and organizational parameters already considered by previous studies on automation and autonomy. We had to create an innovative methodological approach, which is also one of the study's results. This approach involves prototyping a VA sufficiently elaborate to adequately test the basic metrics, criteria, and measures of teaming performance.

DOI: 10.1201/9781003666431-21

STATE-OF-THE-ART – TRUST AND COLLABORATION

MOHICAN focuses on the integration, interaction, and performance of cognitive systems/agents (onboard systems and pilots) in advanced military fighter aircraft, and more specifically on evaluating trust and collaboration. Therefore, we started using a state-of-the-art mix of several sources, including cognitive engineering, automation, artificial intelligence, advanced cockpit research and development, cognitive psychology, computer-assisted collaborative work, and sociotechnical systems research. Much work is already invested in human–machine teaming (HMT) across multiple fields worldwide. This state-of-the-art applies not only to air combat but also to studies and research on trust and collaboration in other sectors, such as the automotive industry. In all cases, trust and collaboration are complex factors that require the study of human factors such as situation awareness, decision-making, and action (i.e., the human model SADMAT: Situation Awareness, Decision Making & Action Taking). SADMAT is a single-agent model that can be used at various levels of granularity of a multi-agent system. Situation awareness has already been studied, along with trust in automated systems (Wickens, 1995), and can be correlated. Wickens studied how to design systems for SA and trust in automation and proposed a synthesis from a combined set of experiments in the laboratory, pilot surveys, and accident/incident analysis. He deduced issues of over-trust (complacency), mistrust, workload, situation awareness, and perceived control. His recommendations were to simplify design, improve display design, improve training, and generate appropriate corporate policy.

Trust is a vibrant topic that has been explored for a long time in many fields (e.g., psychology, sociology, human factors, philosophy). According to French et al. (2018), interest in trust most often appears in situations of uncertainty and vulnerability. In addition, trust is necessary when an element of risk arises from the possibility that the trustee will fail to complete the task (Hardin, 2006). Atkinson (2012) is one of the leading scientists working on trust in human–machine systems, where machines are increasingly autonomous. He insisted on the fact that autonomous agents are not tools, but partners having the following properties: (1) appropriate reliance and interdependency, (2) delegation of authority, (3) initiative (taking and relinquishing), (4) social interaction and personalization, (5) agent self-motivated behavior, and (6) ethical behavior by and with agents.

Atkinson, former Research Director at JPL-Caltech, claimed, “Our trust in automation today is based on confidence built over many years in developing highly reliable, very complex systems.” Examples of complex systems are spacecraft, airplanes, planet-wide information systems, and so on. We have many tools and techniques to verify, validate, test, and evaluate complex systems. These tools are not perfect, but usually good enough. Consequently, we increasingly rely on complex, automated systems, but only in proportion to our trust in them (optimal utility requires “appropriate” reliance). This all means that trust in complex systems relies on familiarity and maturity of practice. We can talk about intersubjectivity. Boy (2013, 2020) developed three Systemic Interaction Models (SIM) that include: (SIM 1) supervision; (SIM 2) mediation; and (SIM 3) cooperation by mutual understanding. SIM 3 is about intersubjectivity, where any agent can build a model of its environment to interact with it safely, efficiently, and comfortably. In MOHICAN, what should be the mental model of a fighter pilot of their VA? What should be the cognitive model of the VA of a fighter pilot interacting with it? What should be their levels of granularity?

Atkinson (2012) stated the following question: "Can we reverse-engineer human interpersonal trust and enable [machine] agents to use this social interface appropriately?" He divided the concept of "trust" into three concepts: Trustworthiness, Trustability, and Trusting. Trustworthiness requires having the necessary and sufficient qualities needed for a person to give trust (e.g., competence). Trustability requires manifestation of trustworthiness qualities so they can be observed or inferred, directly or indirectly (e.g., behaviors, signals, communications, reference, reputation). Trusting is about becoming reliant (i.e., dependent on another agent for something of value).

Summarizing, trust is about three points: (1) competence, predictability, and transparency; (2) reputation (stable signals and behaviors); and (3) reliability (anticipate, guide, enable teamwork). Trust involves stable interaction among agents, minimal conflicts among agents, and the predictability of interactions. Based on Rasmussen's (1986) model of human behavior, we can say that trust is a matter of getting appropriate machine feedback:

- skill-based feedback on information entered (e.g., when the pilot enters the value of a parameter into the machine via the cockpit user interface, the machine should immediately inform the pilot that the entered information is well-received and considered for treatment);
- rule-based feedback on information processed (e.g., showing a watch to mention that the treatment is ongoing, and how long); and
- knowledge-based feedback on repercussions on the environment (e.g., effect(s) of action taken) – at this level, the human operator is assumed to be knowledgeable and well-trained on the system they oversee.

This is a matter of control. People trust a system when they know who (i.e., pilot) or what (i.e., VA) is in charge. Trust is also a matter of the reliability of the system. People tend to distrust a system when they experience too many failures. This is a general rule (i.e., systems can be scored concerning successes and failures – in the same way, several Web systems are scored by communities of people). This is the same for human errors. When people commit too many (same) errors, they tend to distrust the system. This could be caused by their level of proficiency with the system or the quality of the system itself. This relation between trust and reliability fosters more research efforts in tangibility issues: physical tangibility (i.e., trust at the human hand level – ability to grasp physical objects); figurative tangibility (i.e., trust at the human mind level – ability to grasp abstractions and concepts).

Trust is also intimately related to cooperation and collaboration. The distinction between cooperation and collaboration is that in cooperation, the goal of every participant is the same, but their interests are individual. Instead, in collaboration, people come together for a common interest and the same objective.

Driskell and Salas (1992) experimentally showed that collectively oriented team members are more likely to attend to other team members' task inputs and improve their performance during team interaction than egocentric team members. Therefore, training for collaboration is a key issue. The question now applies to collaboration between pilots and VAs. What kind of training should we develop to adapt these new co-workers? Is the virtual assistant egocentric? What type of machine learning should be implemented?

Salas et al. (2008) put to the forefront the importance of shared cognition, the measurement of shared cognition, advances in team training, the use of synthetic task environments (STEs) for research, factors influencing team effectiveness, models of team effectiveness, a multidisciplinary perspective, and training and technological interventions designed to improve team effectiveness. Challenges faced in the coming decades include an increased emphasis on team cognition, reconfigurable, adaptive teams; multicultural influences; and the need for naturalistic study and better measurement. They elicited the following teamwork issues:

- Shared cognition matters in team performance.
- Shared cognition can be measured.
- Team training promotes teamwork and enhances team performance.
- STEs provide context for research.
- Team performance can be modeled.
- Factors that influence team performance have been identified.
- Well-designed technology can improve team performance.
- The field of teams, teamwork, and team performance belongs to many disciplines.

Salas et al. (2008) studies suggest that team training interventions are a viable approach that organizations can take to enhance team outcomes. They help improve cognitive outcomes, affective outcomes, teamwork processes, and performance outcomes. Moreover, results suggest that training content, team membership stability, and team size moderate the effectiveness of team training interventions. In the MOHICAN context, a strong suggestion is to recommend the definition of appropriate training on human–machine teaming (HMT), where the machine is mainly a VA.

TRUST AND COLLABORATION MODEL

Materials to be used for the accomplishment of MOHICAN are to be found at the intersection of several disciplines contributing to the genesis of HSI, including human factors and ergonomics, human-centered design, human-systems integration, cognitive science, cognitive engineering, organization sciences, multi-agent systems, artificial intelligence, and others. These disciplines offer various approaches, methods, tools, and models that can be used independently or integrated to form a new approach to human–machine performance assessment.

Our approach to the performance monitoring problem is multi-agent. Consequently, it deals with generic multi-agent models that investigate trust and collaboration. More specifically, types and levels of human interaction with automation (Parasuraman et al., 2000), distributed cognition (Hutchins, 2020), multi-agent systems (Ferber, 1999), human–robot interaction (Fong et al., 2006), SIM (Boy, 2019), organization automation (Boy & Grote, 2009), computer-supported cooperative work (Grudin, 1988), trust in automation (French, Duenser & Heathcote, 2018), and other relevant models. This state-of-the-art aimed to provide a sufficient background to define air-combat performance assessment metrics from two perspectives: trust in and collaboration with virtual assistant systems (VAS).

Our multi-agent model aimed to represent the information and processing involved in measuring the performance of trust and collaboration between a pilot and a VA. These models allow for the definition of metrics to support the elicitation of qualitative and/or quantitative data. "Metrics are the enabling tools to devise a benchmark in any system and serve as an evaluation platform for assessing the performance, along with the verification and validation of a system" (Damacharla et al., 2018). In the MOHICAN project, our primary concerns were trust and collaboration, which we tried translating into useful qualitative metrics for HMT.

The aim was to highlight, for a set of scenarios (the model's domain of use), the cognitive and physical functions {Fc} = {Role, Validity Context, Necessary Multi-Agent Resources} within the cockpit. Then, with the support of low-level metrics {mi} obtained through the collection of objective and subjective data, we evaluated criteria {Cj} (e.g., usability, explainability on the VA side, as well as workload and stress on the human side) that characterize the multi-agent system of the human–machine team with high-level metrics {Tk} (e.g., operational performance, trust, and collaboration). Operational performance is a generic high-level metric that measures the distance between an expected performance (i.e., prescribed task) and effective performance (i.e., activity). For example, a flight trajectory can be the prescribed task that defines the expected performance, and the actual trajectory resulting from the flying activity is the effective performance. The flight is perfect when the distance between the two trajectories is lower than an appropriate threshold. However, a discrepancy should be analyzed and understood when this distance exceeds this threshold.

Several iterations are necessary to ensure that the model used is appropriate (for the job). Sometimes, we must change the initial model to a more sophisticated or simpler one. Without subject matter experts, the model method does not work. In fact, not only should the results be assessed by experts, but the metrics model should also be assessed. To do this, we used the PRODEC method developed as part of the FlexTech Chair.

THE PRODEC METHOD

Based on the analysis of trust and collaboration, we have used the PRODEC method to create this human–machine team performance model. This was the first time this method had been used in a project. This method recommends the development of scenarios (i.e., a procedural analysis of tasks) and progressively discovers configurations (i.e., a declarative analysis of the structures and functions implemented

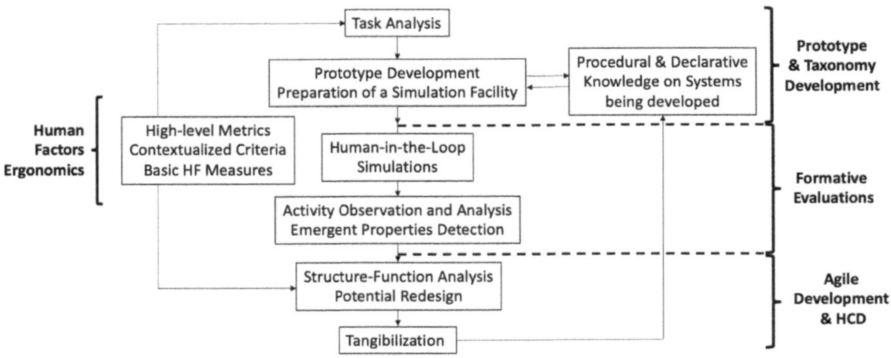

FIGURE 17.1 The PRODEC process (Boy et al., 2024).

and/or to be implemented). In this logic, the Business Process Model & Notation (BPMN; White & Bock, 2011) tool covered the procedural part, and the Cognitive Function Analysis (CFA) method covered the declarative part. As articulating the procedural and declarative aspects requires activity analyses, it is necessary to implement human-in-the-loop simulations (HITLS), as a flight test team would do in a virtual setting (i.e., in simulation). Several iterations are necessary to ensure that the model used is appropriate for the work and can be improved. Analyzing these incrementally developed task graphs allows the discovery of functions that were not initially anticipated. These are called "emergent features." It may also turn out that "emergent structures," forgotten in the architecture of the sociotechnical system under development, appear as integral parts. These "discoveries" of emerging functions and/or structures discovered during HITLS contribute incrementally to the improvement of the declarative scenarios (the DEC part, for "declarative") expressing the configurations of the sociotechnical system under development.

In summary, the PRODEC process (Figure 17.1) starts with the elicitation of procedural knowledge in the form of an AS-IS analysis based on operations-dedicated subject matter experts' experience, followed by the elicitation of underlying declarative knowledge in the form of an AS-IS analysis of the system, leading to the creation and development of a TO-BE virtual prototype, based on design-dedicated subject matter experts' experience, and subsequent HITLS with the collection of human factors data (Boy et al., 2024). Consequently, tasks and activities can be compared using both BPMN diagrams, which can lead to a modification of the previous task analysis and generate recommendations for the next design iteration. Deviations between tasks and activities enable incremental discovery of emergent properties and features that will be reintroduced in the subsequent design iteration (both tasks and the VA were modified).

This method allows for the step-by-step capture of the expertise of human operators during the design and development of the assistance system functions. The articulation of scenario design and meaningful criteria for evaluating the designed system contributes to a progressive improvement of the system itself. It allows for defining and testing the quality and efficiency of the human–machine integration very early in the design process of a new device (e.g., a VA) or system of systems.

There are two main tools to carry out the PRODEC process (Figure 17.1): (1) elicitation from operations experts and (2) direct observation. Knowledge elicitation (e.g., through interviews) from experts leads to task analysis during the early stages of engineering design. Verbalizing is unnatural, so interviews help elicit knowledge (Vermersch, 2019). It allows a design team to explore a system's context of use, users' needs, how a system can be evaluated, information considered by a human operator during the execution of a task, or mental representations of the person being interviewed, for example. Direct observation, followed by activity analysis, is possible if you have easy access to existing systems (for large, complex systems, operations can be challenging to monitor). Observation is a field method that involves collecting observable data from human operators. The data collected are behaviors, verbalizations, interactions between colleagues, and with manual and technological tools. Observation enables understanding the operators' working environment without interfering and is the best way to get away from individual interpretations and access non-verbal behaviors that people are unaware of. Direct observation is typically performed in the HITLS. The activity must then be confronted, condensed, compiled, and consolidated to redesign tasks from which we deduce the function.

OPERATIONAL CONTEXT

The operational context of the MOHICAN project has already been presented (Morel & Boy, 2023). To explore trust and collaboration between the human and the machine on most of the operational tasks that could be entrusted to the system of systems (e.g., latest generation fighter aircraft, crew), the operational scenario selected mixed planned air-to-ground attack sequences, dynamic targeting phases, and finally air-to-air interception phases. The crews were immersed in three complex missions in a semi-permissive environment, where they could not avoid considering the enemy threat nor coordinating with friendly forces.

All simulations were carried out in SYNAPSE Defense Laboratories. To set up the simulations, we selected the DCS World flight simulator, combined with the third party program Tacview, a universal flight analysis tool that automatically exports trajectory elements and flight data. DCS World had the advantage of an F14 module offering a game mode (single seater + AV called Jester) giving the pilot the choice of distributing tasks within the cockpit (management of navigation systems, heading, altitude, speed, delegation of tactical functions, etc.) via a context menu controlled by a head tracker or by voice recognition. This VA was then improved with each iteration. Indeed, the first setup provided by DCS World flight simulator required the development of complementary functions to enhance the maturity of the VA. Three versions of the VA have been developed (JESTER+/JESTER BASIC/ JESTER ADVANCED). Figure 17.2 shows a picture of the experimental setup with the JESTER ADVANCED version of the VA. Some VA functions using a voice recognition system were simulated using a Wizard of Oz technique (e.g., Air Defense Package Leader, C2 personnel).

FIGURE 17.2 Experimental setup of the MOHICAN project (Morel & Boy, 2023).

To highlight the performance of trust and collaboration between the human and its VA, we increased the difficulty level of missions during human-in-the-loop iterations to increase the pilots' workload. Three pilots from SYNAPSE Defense were involved in our simulations. Five test iterations were necessary to vary the difficulty level of the scenario, evolve the performance model of the human–machine team, and enable progressive improvement of the VA. To increase the level of difficulty of the experiments, the operational context has been hardened to make it iteratively more complex (e.g., targets more difficult to identify, multiplication of threats and their technicality, reduction of margins (e.g., time, fuel)). Due to COVID, only one pilot could complete all five iterations, while the other two completed the first three. Each iteration lasted around 30–40 minutes. A pre-test interview was carried out during the interview to define the subjects' profile (age, qualification, combat experience, single-seater/two-seater experience, number of hours of simulator training, etc.). Each participant was previously trained to fly the aircraft selected for the tests and the different versions of the VA. Four human factor tools were used to analyze the experiments: the Tobii Pro Lab eye-tracker was used to analyze pilot activity via eye-tracking. Bonita studio for business process modeling in BPMN format. A Garmin heart rate belt to measure the instantaneous heart rate in beats per minute (BPM). Finally, the Observer XT is used for video sensor synchronization, encoding, and analysis of pilot behavior.

RESULTS

Initially, the mission was divided into specific flight phases based on the accounts of three expert pilots. These flight phases corresponded to generic permanent or temporary processes activated according to the context and tasks. Three generic permanent processes have been identified (e.g., piloting) and eight temporary processes (e.g., ground-to-air threat avoidance). A task analysis was carried out for each multi-agent process and described in a BPMN model. These first BPMNs were based on experience with two-seaters and correspond to the AS-IS BPMNs.

After this task analysis, we carried out the physical and cognitive structure–function analysis. In total, 21 functions were identified and classified according to

Endsley's SA model and described in a situation awareness, Decision Making & Action Taking (SADMAT format). Each function was defined by its role, context of validity, and valuable human and/or machine resources. For example, the "Acquiring Information" function enables the pilot to retrieve data via the weapons, geographic information systems, and other systems. This function can be evaluated from various perspectives, including accuracy, time, workload, utility, etc. Among the cognitive resources associated with this function, we can name perception. Among the physical resources related to the context to which it appeals, we can name the information from the Geographic Information System (GIS), the weapon system, the targeting pod, and so on. Of course, this depends on the context and resources available.

From this representation of human knowledge, a distribution of tasks between humans and virtual assistants was proposed that corresponds to the BPMNs TO-BE (Figure 17.3). Then the resulting activity was analyzed via HITLSs while considering the complexity of the operational context. Five HITLS were carried out where the difficulty of the scenarios increased from one simulation to the next, and functions of the VA were upgraded three times (JESTER+/JESTER BASIC/JESTER ADVANCED) in an agile manner. The pilot's activity was recorded for further activity analysis, and an HMT performance analysis was performed based on a performance model and a new structure-function analysis, leading to progressive tangibilization of the domain ontology. Tangibility is taken in the sense defined in Boy (2016). In many cases, the physical simulation facility had to be modified incrementally using rapid prototyping to improve the VA design. During the simulation phase, we collected human factors data (e.g., eye tracking, heart rate, video) and meaningful operational parameters regarding fighter pilots' activity (e.g., risk management, mission effectiveness, and operational margin).

After each simulation and for each pilot, a new BPMN was created. The comparison of tasks and activities re-instantiated in a BPMN-CFA process form enabled us to explore and eventually discover emergent functions and structures that were reintroduced in the subsequent design iteration (both tasks and the VA were modified). Figure 17.4 presents the analysis of the first iteration of simulations with the JESTER+ version of the VA. Through BPMN-CFA analysis, we identified the emergent function of reminding the pilot of the safe altitude and heading. This function has been added to the next VA version, JESTER BASIC. It has been analyzed that in the case of an operational context without significant difficulties, the pilot's level of trust and collaboration toward the VA for this function was high. On the other hand, in the case of an increase in the difficulty of the scenario and the mobilization of significant cognitive resources on the part of the pilot, the level of trust decreased sharply. The pilot's trust in his onboard VA is a complex, dynamic, and multidimensional concept that is highly context- and time-dependent (i.e., the VA can quickly gain or lose the pilot's trust during a mission).

At each iteration, our trust and collaboration performance model, consisting of the core measures, intermediate criteria, and high-level metrics, was challenged and updated, considering their relevance, validity (meaning and content), and ease of implementation. After five iterations, we improved the method by evaluating intermediate criteria and combining them to assess degrees of trust and collaboration. All

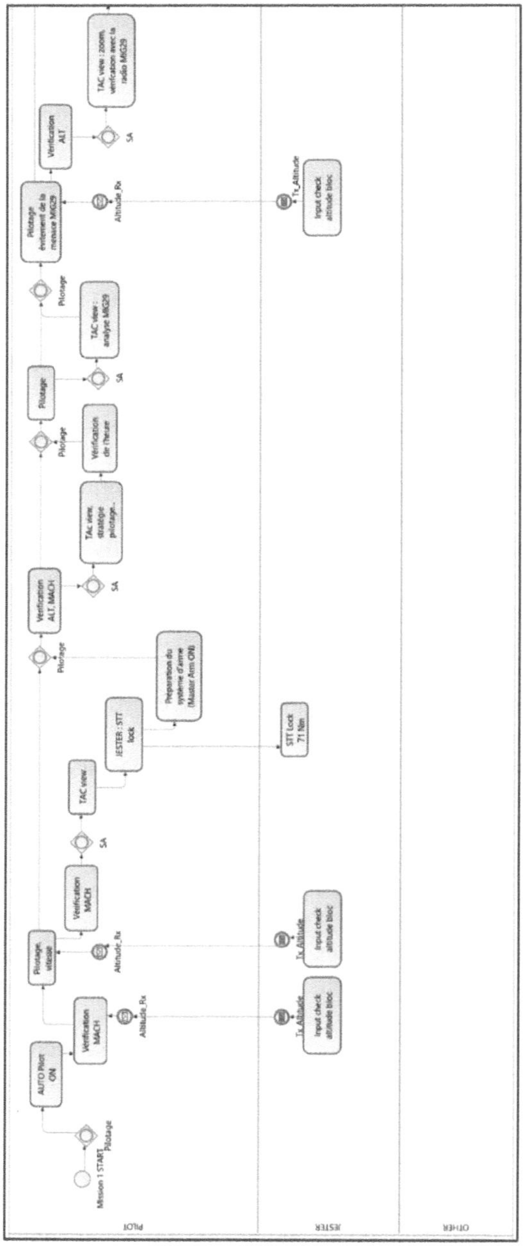

FIGURE 17.3 Example of TO-BE task analyses using BPMN graphs.

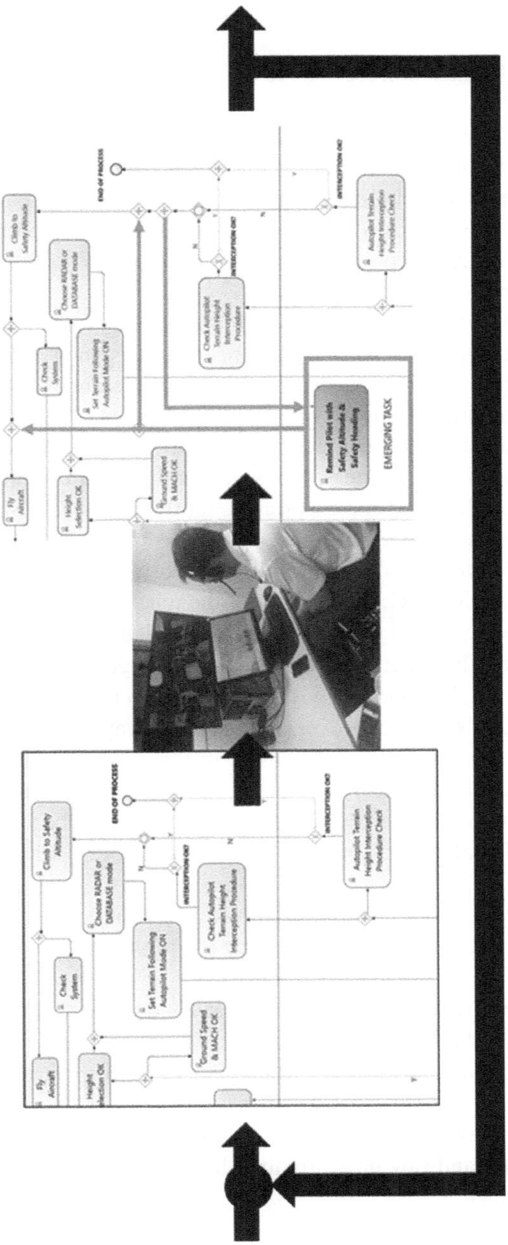

FIGURE 17.4 Example of task (left side) and activity (right side) analyses using BPMN graphs, and identification of emergent properties (in blue on activity side) (Morel & Boy, 2023).

requirements and their measures were grouped into an analysis grid to facilitate analyzing interactions between the pilot and his VA.

Below is the list of the trust and collaboration criteria and subjective measures analyzed that were already described in Boy and Morel (2022):

- *Relevance:* Relevance corresponds to the ability of the machine to respond in a practical and relevant way by providing added value to the pilot.
- *Transparency:* The pilot's trust is based on the pilot's perception/understanding of the operation of the VA. Knowledge about the VA is crystallized in the format of a mental model. According to Zuboff (1988), they can establish trust only when the operator has successfully constructed a mental image of the machine. This criterion comprises two sub-criteria: perception of information and comprehension of information.
- *Flexibility/Adaptability:* The adaptive characteristic of cooperative systems is a preponderant element of trust. This concerns the capacity of the VA to adapt to the context (environment, task) and the pilot (tolerance to the pilot's errors, to his actions/manipulations).
- *Usability:* Usability is the degree to which pilots can use the VA to achieve defined goals with effectiveness, efficiency, and satisfaction in a specified use context. The standardized scale System Usability Scala (SUS; Brooke, 1996) was used to evaluate this criterion.
- *Quality of Feedback:* A VA that explicitly shares what it is doing can only facilitate cooperation (i.e., this quality of the machine being an element of its knowledge- cooperation). If the VA explains what it is doing, the pilot will easily detect possible interferences with their sub-goals.
- *Perceived Task Reduction:* The VA is supposed to improve performance and facilitate the pilot's task. However, it does not always promote the task (e.g., by inducing a new supervision subtask). In that case, the VA must not complicate the overall task, at the risk that the pilot prefers manual control to automatic control and loses the benefits of automation. The standardized scale NASA-TLX (Task Load Index; Hart & Staveland, 1988) was used to evaluate this criterion.
- *Absence of Discomfort:* The discomfort is introduced by using the VA compared to the realization of the task by the pilot alone.

The methods used to evaluate these criteria were self-confrontation interviews, post-test interviews, and standardized questionnaires. The questionnaires developed for the MOHICAN project were inspired by Rajaonah's (2006) post-test questionnaire on trust.

Other trust and collaboration criteria were also measured objectively using eye-tracking data, heart rate, and raw data provided by the flight simulator that evaluate the operational performance:

- *Effectiveness:* Effectiveness is the ability to achieve objectives, regardless of the resources allocated to the operation. The interaction with the VA must be effective to be efficient. Several measures have been implemented for the analysis of effectiveness:

- Operational performance measurement by expert pilots.
- Number and type of interaction between the VA and the pilot for each subtask. We determined if the information transmitted by the VA was effectively processed by the pilot using the eye-tracker (i.e., analysis of the pilot's actions in the simulator, double check, verbalization, etc.).

- *Efficiency:* The interaction with the VA must be time-efficient. The measure was the total interaction time between the pilot and the VA (reaction time, interaction processing in seconds). Raw data provided by the flight simulator was used for this purpose.
- *Reliability/Robustness:* Trust is about the reliability of the system. People tend to distrust a system when it has too many failures. This criterion was analyzed by identifying the VA's potential bugs and/or functional defects.
- *Situation Awareness/Mental Workload:* Measuring trust and collaboration performance only makes sense in a specific context. This notion of context is difficult to formalize. The way we attempted to formalize context iteratively was by clustering the various operational sequences (BPMN), expert air-combat knowledge elicitation, as well as pilots' activity and simulation context data acquisition and interpretation, which enabled the definition of an index evaluating pilots' theoretical workload in real time. This indicator is an algorithm based on the system, in this case, the aircraft, the context, its environment, and finally, the pilot's tasks, integrating the mobilization of these cognitive resources. He was implemented in the latest version of the VA (JESTER ADVANCED). Interestingly, heart rate (captured during the simulation activity) measurements overlapped significantly with the data resulting from this index. We also correlated this data with the subjective scale called *Situation Awareness Rating Technique* (SART) (Endsley, 1988). We cannot declare that this index and heart rate are correlated, because the data generated comes from experiments involving a few fighter pilots. However, the persistence of the similarity in the temporal evolution of the index and heart rate during our last two cycles of experimentation represents an exciting line of enquiry not only for the analysis of trust and collaboration between a pilot and an AV, but also for other work. Due to confidentiality reasons, the algorithm of this index cannot be presented in detail.

CONCLUSIONS AND PERSPECTIVES

The MOHICAN project was to provide a solution for evaluating the performance of HMT within a combat aircraft cockpit. Focusing on measuring the performance of trust and collaboration, this model was to highlight, for a range of scenarios and contexts (the model's domain of use), the cognitive and physical functions (Role, Validity Context, Multi-Agent Resources required) within the cockpit, and the associated measurement parameters. We achieved this objective using the PRODEC method.

Initially based on the analysis of a two-seater combat cockpit (the reference), the model was revised, then experimented with and iterated via HITLSs, notably to incorporate the emerging tasks and cognitive functions of a single-seater cockpit with virtual assistance.

As planned, we modeled trust and collaboration performance using the PRODEC method by analyzing operational performance criteria (risk management, mission effectiveness, and operational margin) and trust and collaboration performance criteria (e.g., effectiveness, efficiency, flexibility, reliability, relevance, transparency, workload, situational awareness). These criteria are assessed based on low-level measures obtained through collecting objective data (e.g., eye-tracking, military operational performance measurement, heart rate, video encoding) and subjective data (e.g., elicitation interviews, observations, standardized scales).

The PRODEC method, therefore, enabled us to develop metrics in an agile way, in conjunction with the development of a VA, in a relevant and meaningful way. This method could not be used without the involvement of fighter pilots within an HCD team.

During the project, we also defined an index capable of contextualizing the pilot's theoretical workload, thanks to real-time algorithms prioritizing the load of tasks in progress and the level of exposure to threat. These indices, built by business experts, are differentiating factors for choosing teaming trust and collaboration evaluation logics adapted to the circumstances. This study compared them using a physiological measure, heart rate. However, to validate the concept, further experiments should be carried out using other physiological measures, such as heart rate variability, pupillary diameter, ultrasound measurement of blood flow, electroencephalography, infrared spectroscopy (NIRS) to measure the rate of blood oxygenation in the brain, etc.

ACKNOWLEDGMENTS

The authors warmly thank the partner Synapse Défense for its expertise on the MOHICAN project as well as the Man–Machine Teaming (MMT) program launched by the Direction Générale de l'Armement (DGA, French DOD) and led by Dassault Aviation and Thales.

REFERENCES

Atkinson, D. (2012). *Human–Machine Trust.* Invited speech at IHMC (Florida Institute for Human and Machine Cognition).

Boy, G.A. (2013). Dealing with the Unexpected in Our Complex Sociotechnical World. In *Proceedings of the 12th IFAC/IFIP/IFORS/IEA Symposium on Analysis, Design, and Evaluation of Human–Machine Systems. Las Vegas, Nevada, USA*; 2013. Also, chapter in Millot, P. & Boy, G.A. (eds.) *Risk Management in Life-Critical Systems.* Wiley.

Boy, G.A. (2019, November). *Cross-Fertilization of Human–Systems Integration and Artificial Intelligence: Looking for Systemic Flexibility.* INCOSE AI4SE.

Boy, G.A. (2020). *Human Systems Integration: From Virtual to Tangible.* CRC-Taylor & Francis Press, Miami, FL.

Boy, G.A. & Grote, G. (2009). Authority in Increasingly Complex Human and Machine Collaborative Systems: Application to the Future Air Traffic Management Construction. In the Proceedings of the 2009 International Ergonomics Association World Congress, Beijing, China.

Boy, G.A. & Morel, C. (2022). The Machine as a Partner: Human–machine Teaming Design Using the PRODEC Method. *Work*, 73(Preprint), pp. S15–S30.

Boy, G.A., Masson, D., Durnerin, E. & Morel, C. (2024). PRODEC for Human Systems Integration of Increasingly Autonomous Systems.

Brooke, J. (1996). Sus: A "Quick and Dirty" Usability. *Usability Evaluation in Industry*, 189(3), 189–194.

Damacharla, P., Javaid, A.Y., Gallimore, J.J. & Devabhaktuni, V.K. (2018). Common Metrics to Benchmark Human–Machine Teams (HMT): A Review. *IEEE Access*, 6, pp. 38637–38655. doi: 10.1109/ACCESS.2018.2853560.

Driskell, J.E. & Salas, E. (1992). Collective Behavior and Team Performance. *Human Factors*, 34(3), pp. 277–288.

Endsley, M., Selcon, S.J., Hardiman, T. & Croft, D.G. (1998). Comparative Analysis of SAGAT and SART for Evaluations of Situation Awareness. *Human Factors and Ergonomics Society Annual Meeting Proceedings*, 42(1), 82–86. doi: 10.1177/154193129804200119.

Ferber, J. (1999). *Multi-Agent Systems: An Introduction to Distributed Artificial Intelligence.* Addison-Wesley Longman Publishing Co., Inc., Boston, MA.

Fong, T., Steinfeld, A., Kaber, G., Lewis, M., Scholtz, J., Schultz, A. & Goodrich, M. (2006). Common Metrics for Human–Robot Interaction. In *HRI'06, March 2–4, Salt Lake City, Utah, USA.*

French, B., Duenser, A. & Heathcote, A. (2018). *Trust in Automation – A Literature Review.* CSIRO Report EP184082. CSIRO, Australia.

Grudin, J. (1988). Why CSCW Applications Fail: Problems in Designing and Evaluating Organizational Interfaces. In *CSCW '88: Proceedings of the 1988 ACM Conference on Computer-Supported Cooperative Work*, pp. 85–93. ACM Digital Library. ACM O-89791-282-9/88/0085. https://doi.org/10.1145/62266.62273

Hardin, R. (2006). *Trust.* Polity, Cambridge, UK.

Hart, S.G. & Staveland, L.E. (1988). Development of NASA-TLX (Task Load Index): Empirical and Theoretical Research Results. In *Advances in Psychology* (Vol. 52, pp. 139–183). Advances in Psychology, Elsevier, North Holland.

Hutchins, E. (2020). The Distributed Cognition Perspective on Human Interaction. In *Roots of Human Sociality* (pp. 375–398). Routledge.

Morel, C. & Boy, G.A. (2023). Evaluation de la performance, de la collaboration et de la confiance au sein d'une équipe pilote-assistant virtuel dans un cockpit d'un avion de combat simulé. In Symposium ÉPIQUE. Arpege Science Publishing *Paris.*

Parasuraman, R., Sheridan, T.B. & Wickens, C.D. (2000). A Model for Types and Levels of Human Interaction with Automation. *IEEE Transactions on Systems, Man, and Cybernetics – Part A: Systems and Humans*, 30(3), pp. 286–297.

Rajaonah, B. (2006). Rôle de la confiance de l'opérateur dans son interaction avec une machine autonome sur la coopération humain-machine (Doctoral dissertation, Paris 8).

Rasmussen, J. (1986). *Information Processing and Human-Machine Interaction: An Approach to Cognitive Engineering.* North-Holland, Amsterdam, ACM Digital Library.

Salas, E., Cooke, N.J. & Rosen, M.A. (2008). On Teams, Teamwork, and Team Performance: Discoveries and Developments. *Human Factors*, 50(3), pp. 540–547. doi: 10.1518/001872008X288457.

Salas, E., DiazGranados, D., Klein, C., Burke, C.S., Stagl, K.C., Gerald, F., Goodwin, G.F. & Halpin, S.M. (2008). Does Team Training Improve Team Performance? A Meta-Analysis. *Human Factors: The Journal of the Human Factors and Ergonomics Society*, 50, p. 903. doi: 10.1518/001872008X375009.

Vermersch, P. (2019). L'Entretien d'Explicitation [Elicitation Interview]. ESF Sciences Humaines, Paris, France.

White, S.A. & Bock, C. (2011). *BPMN 2.0 Handbook Second Edition: Methods, Concepts, Case Studies, and Standards in Business Process Management Notation*. Workflow Management Coalition (WfMC), Lighthouse Point, FL, USA, Future Strategies Inc. ISBN 978-0-9849764-0-9.

Wickens, C.D. (1995). Designing for Situation Awareness and Trust in Automation. In *IFAC Proceedings Volumes*, Vol. 28, Issue 23, pp. 365–370. Elsevier, North Holland.

Zuboff, S. (1988). *In the Age of the Intelligent Machine: The Future of Work and Power*. New York, USA, Basic Books, Inc.

18 Quantitative Measures of Performance of Standard Operating Procedures (SOPs)

Lance Sherry

ROLE AND IMPORTANCE OF STANDARD OPERATING PROCEDURES

Standard Operating Procedures (SOPs) are the "glue" that holds the operator, automation, and complex plant (e.g., vehicle, power generator, manufacturing process) together. SOPs define the detailed step-by-step instructions for mission operations to meet enterprise safety and efficiency targets (Figure 18.1). Crucially, SOPs define the sequence of steps during abnormal and emergency scenarios.

By defining SOPs, the enterprise and regulators can ensure best practices are adopted across operations to maximize safety and efficiency. The SOPs also tailor operations to meet an enterprise's operational policies and cost structure. Standardizing SOPs also enables fungible crew assignment, and a single, shared operator's mental model amongst crew members (Barshi et al., 2016; Deghani & Wiener, 1997).

In addition, SOPs are the basis for crew training and crew evaluation. SOPs are also used to establish legal liability in the event of an accident investigation.

Ironically, in many industries, the plant (e.g., aircraft) and automation are certified by the regulatory authorities (e.g., airworthiness certification), and the operators are licensed and qualified by the regulatory authorities. Still, the SOPs are loosely regulated (Barshi et al., 2016; Deghani & Wiener, 1997; Kortenkamp et al., 2008).

SOPs are generally approved following "inspection" and "vetting" by subject matter experts (SMEs) (Barshi et al., 2016). Sometimes, the SOPs are evaluated in a human-in-the-loop simulator (Degani & Wiener, 1997). However, these human-in-the-loop tests are time and cost-prohibitive and cannot cover *all* the plausible scenarios, especially the boundary conditions in non-normal scenarios.

For all these reasons, the operation of SOPs must be monitored using available flight data to ensure that the procedures can be executed promptly in *all plausible operational scenarios*. Further, to enhance the SOP approval process, the SOP performance can be simulated before the publication and release of the SOPs.

This chapter describes a method to be used in the Human–Systems Integration process (Boy, 2020) for quantifying the performance of SOPs in field operations.

DOI: 10.1201/9781003666431-22

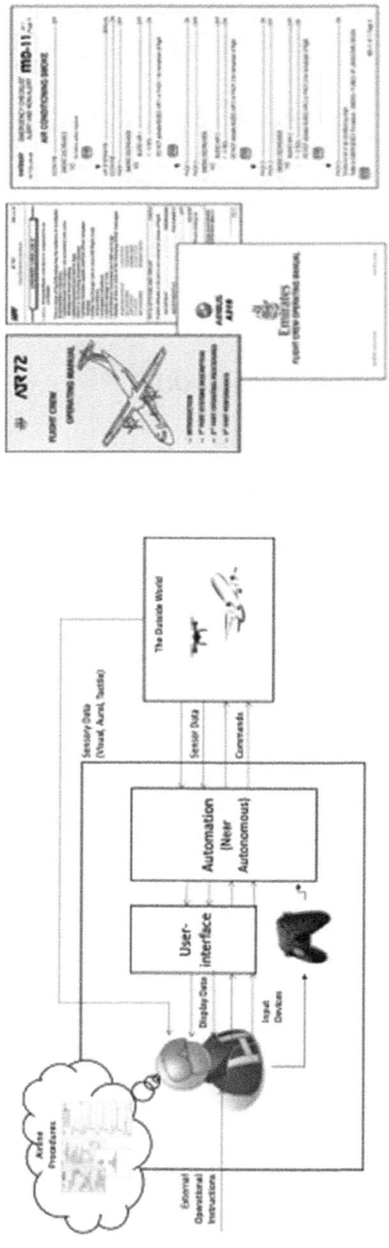

FIGURE 18.1 SOPs are the "glue" that holds the operator, automation, and complex plant (e.g., vehicle, power generator, manufacturing process) together.

The method also describes how SOPs can be modeled for integration and simulation during system design and development.

The chapter is organized as follows: second section defines terms. Third section describes the quantitative metrics for measuring SOP performance. Fourth section describes the method for analyzing operations in the field. Fifth section describes the process for simulating SOPs in the development phase. Sixth section provides conclusions.

DEFINITION OF SOP AND SOP STEPS

SOPs comprise more than one SOP Step (Figure 18.2). SOP Steps define the specific physical motor actions the operator performs (e.g., move lever, push button, enter data). SOP Steps can also define additional information such as the triggering conditions, decision-making required to determine the appropriate action, and verification that automation and/or plant accepted the Action and responded appropriately (Kortenkamp et al., 2008; Bushatah & Sherry, 2021).

A canonical model of SOP Steps includes the following elements (Figure 18.3):

1. Triggering conditions
2. Decision-making is required to determine the appropriate action
3. Actions performed by the operator (e.g., move lever, push button, enter data).
4. Verification that automation accepted the action and responded appropriately
5. Verification that the plant responded appropriately to the action

1. *Triggering conditions* define the situation (or emerging situation) that initiates the SOP Step. Triggering conditions cognition is cued by visual (e.g., cockpit display, out-the-window), aural (e.g., air traffic control instruction, crew

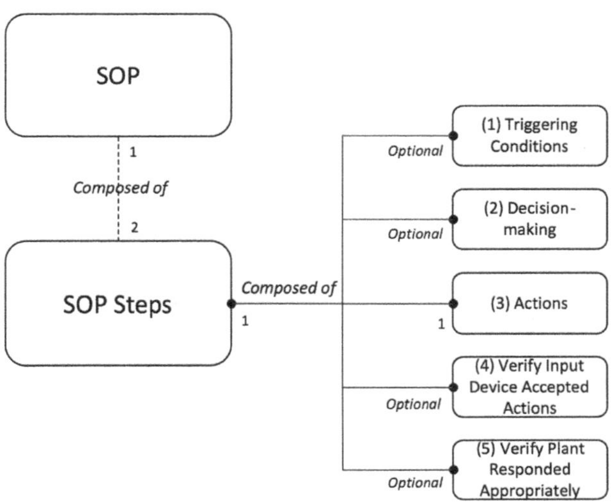

FIGURE 18.2 Canonical model of SOPs. SOP is composed of SOP Steps. SOP Stems have an Action as well as other optional elements.

call-out, engine noise), and haptic cues (e.g., stick-shaker, aircraft vibration). Triggering conditions can also be cued by a memorization item.

2. *Decision-making* defines the cognition required to decide on the appropriate Action. Decision-making cognition is cued by visual (e.g., cockpit display, out-the-window), aural (e.g., air traffic control instruction, crew call-out, engine noise), and haptic cues (e.g., stick-shaker, aircraft vibration). A memorization item can also cue decision-making. Decision-making is limited to the cognition to determine the appropriate Action and is not part of the cognition associated with the Triggering Conditions element. In this way, the decision-making element is a separate cognition. Most of the SOP Steps lack a decision-making element (i.e., only one appropriate action).
3. *Actions* are the operator's physical motor skills (e.g., move the lever, push the button, enter data). Actions may include call-outs, moving objects, or looking up information. Actions are cued by visual (e.g., label on input device), aural (e.g., call-out), and haptic cues (e.g., stick-shaker). A memorization item can also cue action.
4. *Verification that automation accepted the Action* defines the cognition to confirm that the automation accepted the operator input. This cognition is cued by visual (e.g., backlit buttons, cockpit display, out-the-window), aural (e.g., button push "click," air traffic control instruction, crew call-out, engine noise), and haptic cues. Many times, this element is instantaneous. In some cases, there may be a time lag between the time of the action and the confirmation from the input device, and the operator will be required to wait. This element is often self-evident and not explicitly stated in the written SOP.
5. *Verification that the plant responded appropriately* defines the cognition to confirm that the plant (i.e., the aircraft) has reacted to operator input (e.g., maneuver completed, airspeed returns to safe operating regime). This cognition is cued by visual (e.g., aircraft states on cockpit display, terrain out-the-window), aural (e.g., air traffic control instruction, crew call-out, engine noise), and haptic cues (e.g., stick-shaker stops).

Written SOP Steps vs. Complete SOP Steps

By definition, every SOP Step will be composed of all five elements. In some cases, however, some aspects of the SOP Step may be self-obvious and are not always specified in the written SOP Step. For example, a checklist may not need to show the Trigger Conditions for SOP Steps executed in a strict sequence. In addition, verification may occur instantaneously as part of the action (e.g., flip a switch). Specifying every element of each SOP step would result in very long, cumbersome, and difficult-to-read SOP documents. As a result, variants of the canonical SOP Steps are commonly used.

Trigger-Decision-Action-Wait/Verify SOP Steps

SOP Steps executed during operations are situation-dependent and therefore require specification of a Triggering Condition *and* Decision-making (Table 18.1). These SOP Steps may include the Verification element.

TABLE 18.1
Example Trigger-Decision-Action SOP Steps

After initiating takeoff roll, apply half forward stick if the crosswind is at or below 20 knots.

Trigger	Decision	Action
After initiating takeoff roll	If the crosswind is at or below 20 knots	Apply half forward stick

Push the LOC switch when on an intercept heading to the localizer course with the PFD localizer pointer displayed. Verify the armed roll flight mode annunciation is LOC.

Trigger	Decision	Action/Verify
When on an intercept heading to the localizer course	With PFD localizer pointer displayed	push LOC switch. Verify the armed roll flight mode annunciation is LOC

TABLE 18.2
Example Trigger-Action SOP Steps

When FUEL PMP STAB L and R messages are displayed … **push both stabilizer tank L and R switches off.**

Trigger	Action
When FUEL PMP STAB L and R messages are displayed …	push both Stabilizer tank L and R switches off

Pilot Flying: Above 10,000 feet … position Inboard Landing Light Switches OFF.

Trigger	Action
Pilot Flying: Above 10,000 feet …	position Inboard Landing Light Switches OFF

At VR … call "ROTATE"

Trigger	Action
At VR …	call "ROTATE"

Trigger-Action SOP Steps

SOP Steps executed during operations are situation-dependent and therefore require specification of a triggering condition (Table 18.2). These SOP Steps may include the Verification elements.

TABLE 18.3
Example Action-Only and Action-Only with Wait-Verify SOP Steps

STANDBY POWER selector .... BAT. Verify EICAS advisory messages BAT DISCH MAIN and BAT DISCH APU display. Messages may take up to 3 minutes to display
Action/Wait-Verify
STANDBY POWER selector .... BAT. Verify EICAS advisory messages BAT DISCH MAIN and BAT DISCH APU display Messages may take up to 3 minutes to display

Action-Only SOP Steps

Every SOP Step is, by definition, defined by an Action (Table 18.3).

Many SOPs are used to configure the automation/plant before or after operations, or to check the status before moving onto the next mission phase. These "checklist" SOPs are composed of a long list of Actions. The triggering condition is generally omitted as it is implicit that the SOP Steps shall be conducted in the sequence once the prior SOP Step is completed.

QUANTITATIVE MEASURES OF SOP PERFORMANCE

SOPs and their SOP Steps are conducted in a time-sensitive operational environment.

The SOP/SOP step is completed when the objective of the SOP/SOP Step is achieved *within the required time window* (Bushatah & Sherry, 2021; Kourdali & Sherry, 2017). Safety and efficiency impacts may occur when a procedure is not completed within the required window.

Some SOP/SOP Steps are directly related to the plant's immediate safety, such as normal operations at the edge of the operational envelope and abnormal or emergency procedures. These procedures must be completed before the onset of the hazardous state for the plant (e.g., aircraft aerodynamic stall flight regime).

The tempo of the operations also constrains other procedures. For example, a "descent checklist" must be completed before the aircraft approaches the descent phase of the mission (i.e., top-of-descent).

The challenge for the design of procedures is that *all* procedures must be designed to be completed in a timely manner, as defined by the operational requirement for *all* plausible operational scenarios.

By definition, each procedure has a defined Available Operational Time Window (AOTW) in which the procedure must be completed (Figure 18.4).

The AOTW is not a fixed time window; it is inherently variable. The AOTW varies due to specific environmental and operational circumstances. For the case of an aircraft, aircraft performance, aircraft configuration, aircraft weight (e.g., fuel, passengers, and cargo), atmospheric conditions (e.g., wind, temperature), air traffic and air traffic control, and cockpit workload all contribute to different AOTW for the same SOP/SOP Step on other flights.

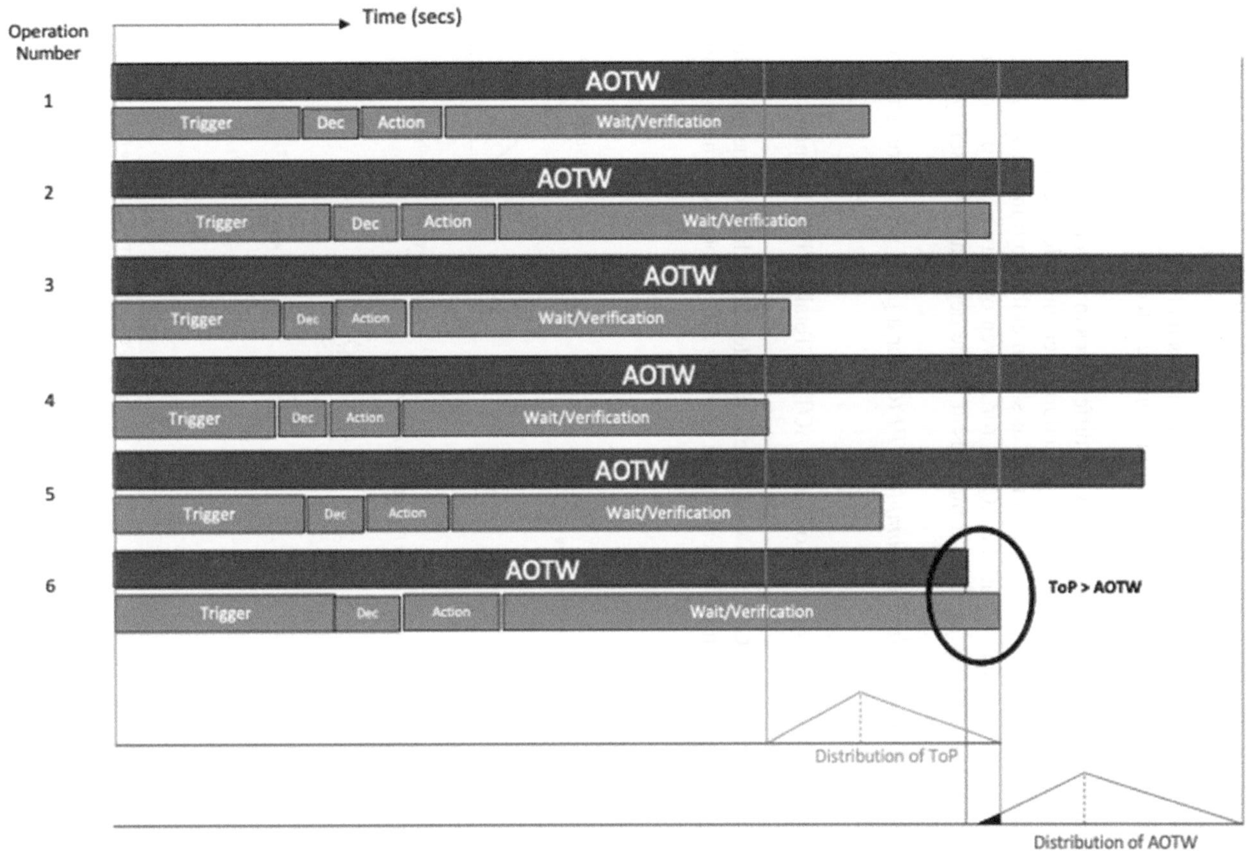

FIGURE 18.3 Repetition of the same procedure on different days yields variability in AOTW (light gray) and variability in ToP (dark gray). In operation #6, the procedure is not completed in a timely manner (i.e., ToP > AOTW).

The variability in AOTW is illustrated by the differences in the length of the time bars in the chart presented in Figure 18.3. The AOTW can be presented by a triangular distribution at the bottom of Figure 18.4 for a given procedure over multiple operations.

Likewise, the time to complete a procedure, *Time on Procedure (ToP)*, is also likely to exhibit variability. In addition to some factors that affect AOTW, ToP is also impacted by cockpit workload, crew coordination, crew technique, and other human factors such as fatigue.

The variability in ToP is illustrated by the differences in the length of the time bars on the chart in Figure 18.3. For a given procedure over multiple operations, the ToP can be presented by a triangular distribution, as shown at the bottom of the figure.

When the ToP is greater than the AOTW for a given operation, the procedure is not completed in a timely manner. It may result in the degradation of safety and/or efficiency (Figure 18.4).

The ***Probability of Failure to Complete (PFtC)*** for a given procedure is the percentage of ToP > AOTW:

PFtC (procedure i) = Count of ToP > AOTW / Total AOTW/ToP pairs

Although achieving a PFtC of zero would be desirable, depending on the procedure and the circumstances, it may be reasonable to allow a PFtC below some threshold (e.g., 0.1% or 1%).

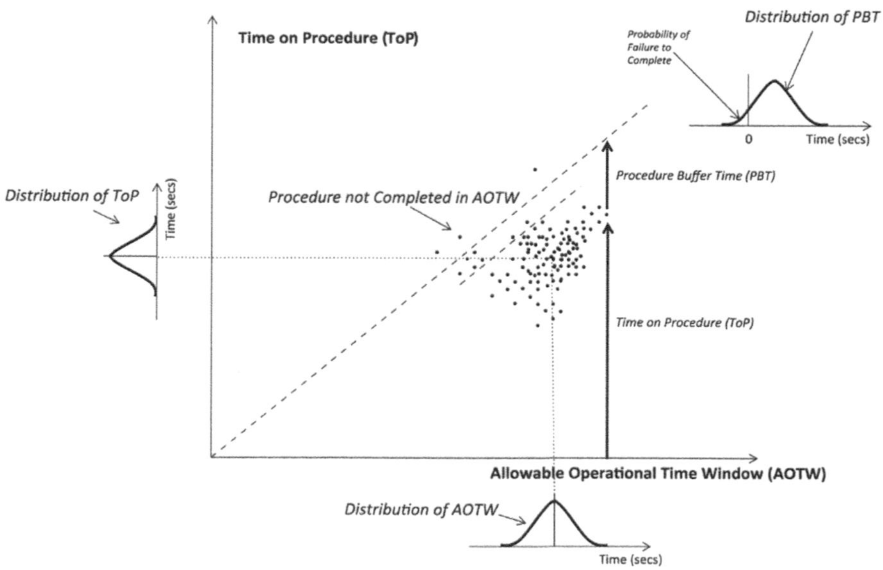

FIGURE 18.4 AOTW/ToP Scatter Plot for a specific procedure. AOTW/ToP pairs above the diagonal line represent the scenarios when the procedure was not completed promptly (i.e., ToP > AOTW).

An alternative visualization of AOTW and ToP over many operations is shown in the scatter plot (Figure 18.5). Each AOTW and ToP pair is shown as a point on the scatter plot. The diagonal line represents ToP equal to AOTW. When the AOTW/ToP pair is below the diagonal line, the ToP is less than the AOTW, and the procedure was completed with time to spare. When the AOTW/ToP pair is above the diagonal line, the ToP is greater than the AOTW, and the procedure was not completed in a timely manner.

The PFtC (procedure i) is the count of AOTW/ToP data points above the diagonal line (i.e., AOTW=ToP) divided by the total number of AOTW/ToP data points.

The Procedure Buffer Time (PBT) differs between AOTW and ToP. When PBT is positive, the procedure is completed on time. When PBT is negative, the procedure is completed too late.

It should be noted that this methodology is *agnostic* to the actual actions that the operator takes when executing the SOP Steps. Completion of the SOP is measured only when the correct endpoint of the procedure is achieved, independent of the Actions taken by the operator to get there. Slips, mistaken actions, and corrections add time to the ToP and are not explicitly tracked. A never-completed procedure (e.g., prematurely stopped) is assigned a ToP of infinity (999 seconds). Understanding what the operator did or did not do requires an alternate methodology.

ANALYZING SOP OPERATIONS IN THE FIELD

The use of AOTW and ToP has proven to be a valuable tool to monitor operations [6, 7]. An example of SOPs for takeoff of an airliner is illustrated in Figure 18.5.

During takeoff operations of a modern airliner, the flaps on the wings are extended for takeoff to provide extra lift force as the vehicle climbs out from the runway. Once the aircraft is safely airborne, the flaps are retracted. During takeoff, the Flap Retraction Procedure Step retracts the flaps while the aircraft climbs and *accelerates*. The Flaps must be fully retracted before the aircraft accelerates through a Do Not Exceed Speed Limit.

The Procedure for Flap Retraction is shown below. ToP and AOTW start when the speed achieves 126 knots (#1). ToP ends when the Pilot Monitoring confirms "CHECKED" (#9). AOTW ends when Speed reaches 142 knots (#10).

1. *SPEED ACHIEVED* (e.g., 126 knots)
2. *Pilot Flying:* observes (e.g., 126 kts) on PFD
3. *Pilot Flying:* call-out "FLAPS 18"
4. *Pilot Monitoring:* silently checks (e.g., 126 kts) on the PFD
5. *Pilot Monitoring:* selects the flaps lever to 18^0
6. *Pilot Monitoring:* monitors flaps transition on the flaps position indicator
7. *Pilot Monitoring:* confirms "FLAPS 18"
8. *Pilot Flying:* checks flaps position 18° on the flaps position indicator
9. *Pilot Monitoring:* confirms "CHECKED"
10. *SPEED REACHES* Maximum Flap Extension for 24° (e.g. 142 knots)

Analysis of a hypothetical airline's operations data using this method is shown in Figure 18.6. The introduction of airline operations at a new airport results in ToP >

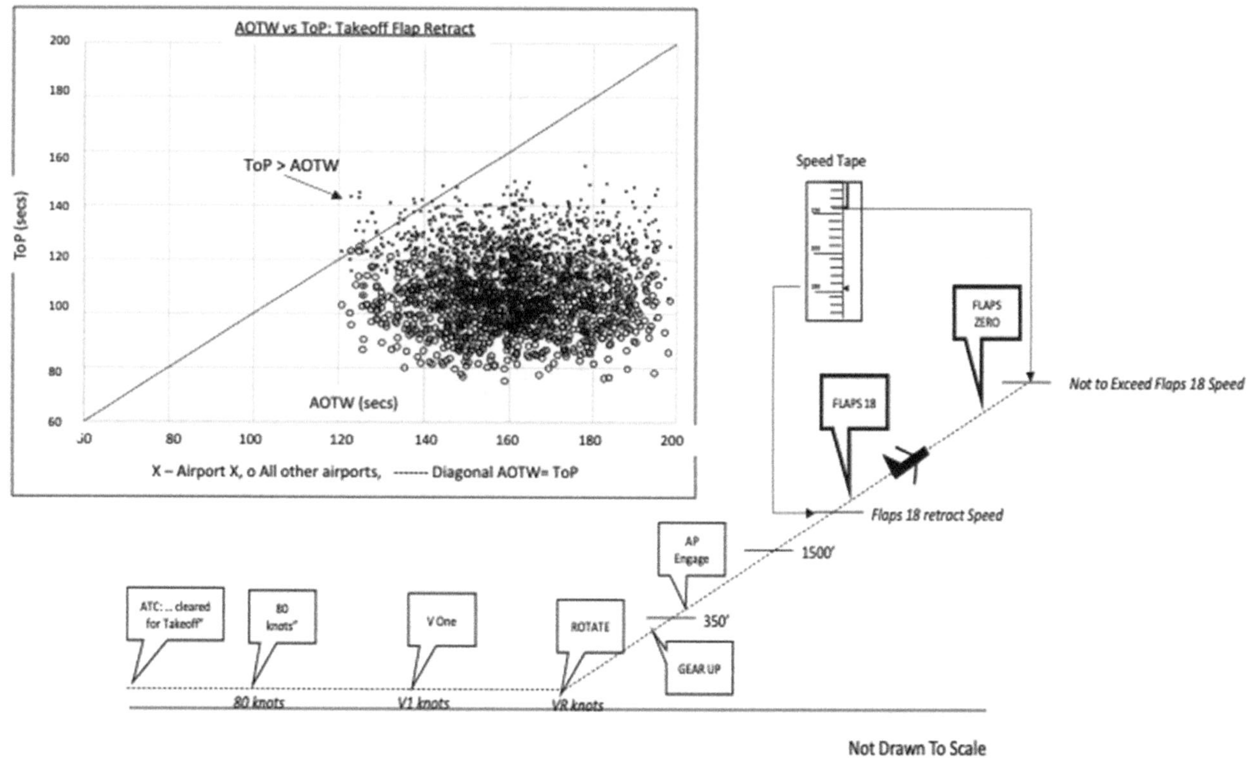

FIGURE 18.5 SOP for a jet airliner. Flaps must be retracted before reaching Flaps 18, not to exceed speed. AOTW vs. ToP Scatter Plot. Operations show ToPs exceeding AOTW, resulting in an unacceptable Probability of Failure to Complete (PFtC) = 5.6%.

AOTW (shown by the "X"s in the AOTW vs. ToP chart). Takeoffs from Airport X exhibited the same AOTW times (122, 180, 203 seconds) but longer ToP (95, 136, 154), resulting in a PFtC of 5.6%.

Based on this evidence, it was discovered that when unusual wind conditions prevailed, air traffic control used a runway with a departure procedure that required a turn shortly after takeoff to avoid a mountain. With the increased workload in the cockpit, the ToP increased such that it sometimes exceeded AOTW 5.6% of the departures.

ANALYZING SOPS BY SIMULATION FOR APPROVAL

Before approval and publishing, SOPs must be vetted. In addition to being reviewed by SMEs and/or human-in-the-loop testing in a cockpit simulator, it may also be prudent to simulate the performance, considering the *variability* in the AOTW and the ToP. Using the method described in this chapter (Bushatah & Sherry, 2021; Kourdali & Sherry, 2016; Schmidt et al., 2021)

TABLE 18.4
SOP Steps for Engine Failure in Cruise above Engine-out Max Altitude with AOTW and ToP Time Distributions

	SOP Steps	Trigger Condition	Action
AOTW		From Engine Fail to Decelerate to stall speed TRIANG(220, 300, 450)	
ToP	PF: Initiate SOP	Aircraft Engine Fails and Aircraft starts to decelerate to stall speed	PF: Call "Engine Fire, Severe Damage, or Separation Recall Items"
		TRIANG(0.03, 5.2, 24.3)	TRIANG(0.068,0.07,0.073)
	PM: Identify Failure by Calling "No N2 Rotations on Engine #1"	PF Call out	PM: Identify Failure by Calling "No N2 Rotations on Engine #1"
		TRIANG(0.03, 5.2, 24.3)	TRIANG(0.068,0.07,0.073)
	PF: Disconnect Autothrottle	PM confirms Engine-out	PF: Disconnect Autothrottle by pressing A/T Disconnect Button
		TRIANG(0.06, 121, 250)	TRIANG(0.67,0.75, 0.83)
	PM: Send "PANPAN/ MAYDAY" to ATC	Autothrottle Disconnected	PM: Type Entry Send "PANPAN/MAYDAY" to ATC
		TRIANG(0.06, 121, 250)	TRIANG(0.068,0.07,0.073)

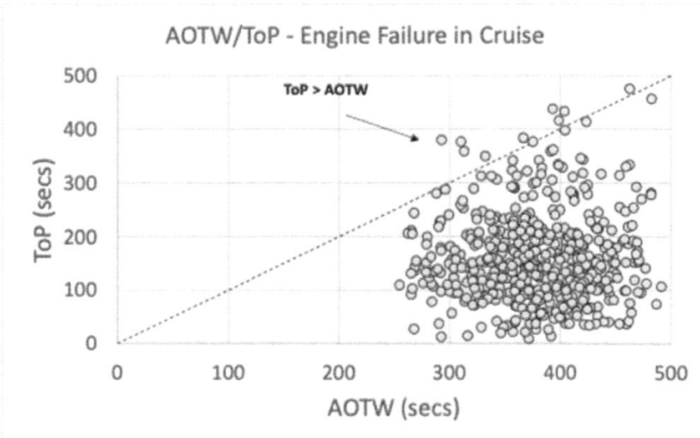

FIGURE 18.6 AOTW/ToP Scatter Plot. AOTW/ToP pairs above the dashed line occur when ToP > AOTW.

Example Analysis of Simulated SOP Performance for Design

An example analysis is provided for an "Engine Failure in Cruise SOP." When an airliner engine failure occurs in the cruise phase, the first order of business is to disconnect the Auto Throttle and to alert air traffic control with a PAN–PAN message (Table 8.4). These actions must be taken before the aircraft decelerates into an aerodynamic stall, resulting in uncontrolled descent and roll. The time to aerodynamic stall represents the AOTW. The SOP Steps and the time distribution for each step used to calculate the ToP are shown in Table 18.4, which represents the ToP.

The AOTW/ToP Scatter Plot based on 1000 simulated scenarios results in a PFtC of 6.4% (Figure 18.6). If this were unsatisfactory, the SOP Steps would need to be adjusted to reduce the ToP. Note: the AOTW shown is only for deceleration from Max Endurance (i.e., slowest speed) and varies depending on aircraft weight and wind conditions.

CONCLUSIONS

The AOTW/ToP method provides a quick technique to analyze the performance of the SOPs in the field. The analysis can be conducted using flight data from most airlines and aircraft operators. When ToP exceeds AOTW above a threshold, further investigation must be undertaken to evaluate why the AOTW is shortened or the ToP is extended.

The AOTW/ToP method is also used to assist in vetting SOPs during the equipment design phase or developing new SOPs. The AOTW and ToP are generated from a simulation model. The simulation uses appropriate time distributions collected from similar operations. Adjustments to the SOP Steps and/or the AOTW can be made to achieve the desired PFtC (i.e., Probability of Failure to Complete).

REFERENCES

Barshi, Immanuel, Robert Mauro, Asaf Degani, Loukia Loukopoulou, 2016, *Designing Flightdeck Procedures* (NASA/TM-2016-219421). Washington, DC: NASA.

Bashatah, Jomana A., Lance Sherry, 2021, A Model-Based Approach for the Qualification of Standard Operating Procedures, in *2021 Integrated Communications Navigation and Surveillance Conference (ICNS), Dulles, VA, USA*, pp. 1–10. doi: 10.1109/ICNS52807.2021.9441587.

Boy, G. A., 2020, *Human–Systems Integration: From Virtual to Tangible* (1st ed.). CRC Press, Boca Raton, USA. https://doi.org/10.1201/9780429351686

Degani, Asaf, Earl Wiener, 1997, *Procedures in Complex Systems: The Airline Cockpit*, Vol. 27, Alberta, Canada: IEEE, pp. 302–312.

Kortenkamp, D., R. P. Bonasso, D. Schreckenghost, K. M. Dalal, V. Verma, L. Wang, 2008, A Procedure Representation Language for Human Spaceflight Operations, pp. 8.

Kortenkamp D., Bonasso R. P., Schreckenghost D., Dalal K., Verma V., Wang L. (2008). A procedure representation language for human spaceflight operations. In The 9th International Symposium on Artificial Intelligence, Robotics and Automation in Space (i-SAIRAS-08). Pasadena, CA, Feb 26–29, 2008.

Kourdali, Houda, Lance Sherry, 2016, A Systems Engineering Method and Simulation of Standard Operating Procedures, in *Paper Presented at 2016 HCIAero, Paris, France*.

Kourdali, Houda, Lance Sherry, 2017, *Available Operational Time Window: A Method for Evaluating and Monitoring Airline Procedures*, Vol. 11, SAGE Publications, pp. 371–381.

Schmidt, Tim André, Houda Kerkoub Kourdali, Jim Nixon, 2021, Evaluating Process-Based and Crew-Centred Approaches to Procedure Design in Aviation: Workload and Performance Changes in Go-around Manoeuvres. *Applied Ergonomics*, 90. https://doi.org/10.1016/j.apergo.2020.103244

19 Onboard Context-Sensitive Information System[1]

Wei Tan

INTRODUCTION

Commercial aviation is an ultra-safe system (Amalberti, 2001; Ilbert et al., 2007; Amalberti, 2012). A complex system becomes ultra-safe when fatal accidents per hour are minimal (e.g., 10^{-6}). This is due to the development of appropriate technology, organizational setups, and training. Complexity is now managed instead of being controlled. Regulations lead to operational documents that aviation personnel must comply with and use; onboard technical operational documents include procedures, checklists, charts, and performance data. Flight crews follow Standard Operating Procedure (SOP) to complete tasks and ensure safety. In abnormal situations, such as system malfunctions or extreme weather conditions, abnormal procedures are typically used for troubleshooting and recovery. Operational documentation is regularly improved using experience feedback for normal, abnormal, and emergency situations (Ramu et al., 2004). The onboard documents can be categorized into four kinds of documents: flying documents, which are related to all flight operations; systems documents, which include systems' theory, principles, and controls; navigation documents, which are the charts that pilots use on the flight deck; and performance documents, which provide operational data for all flight phases such as takeoff, landing, and go-around (Tan, 2014).

Pilots are familiar with paper-based manuals, which are easy to use, tag, mark, and retain, even if they are heavy and challenging to carry. Nobody can permanently remember all procedures and technical knowledge under time pressure. Therefore, several instruments, panels, and displays have been developed and improved to assist the flight crew in executing tasks and actions. Airbus's Electronic Centralized Aircraft Monitoring (ECAM) and Boeing's Engine Indication and Crew Alerting System (EICAS) were among the first onboard information systems directly connected to flight parameters. They were developed during the mid-80s, considering basic parameters for supporting failure detection and recovery. They have been proven extremely useful for providing comprehensive information on the state of the aircraft in an integrated way, unlike previous flight decks, where pilots had to constantly check many instruments.

DOI: 10.1201/9781003666431-23

However, these systems did not cover all possible cases. This is why onboard paper-based documentation is still required. The aeronautical community is currently working on transferring related information from paper to computer support (e.g., Airbus's onboard information system and Boeing's electronic flight bags). These new tools cover aircraft technical information, operating manuals, performance calculations, and mission management information. They are context-free databases. However, computer support enables interconnectivity among relevant information (i.e., hypertext links) and between cockpit information and flight parameters (i.e., context-sensitivity).

This chapter presents the Onboard Context-Sensitive Information System (OCSIS) on a tablet wirelessly connected to the relevant cockpit parameters. OCSIS enables and requires new information formatting (e.g., the page concept is no longer applicable). In addition, OCSIS's internal information is structured according to context. We completed the first set of formative evaluations, including cognitive walkthrough and workload and situation awareness (SA) assessments.

METHODOLOGY

Human-Centered Design Approach: Modeling, Simulation, and Integration

For the last ten years, modeling and simulation technology has enabled realistic human-in-the-loop simulations (HITLS) early on during the design phase and provision of meaningful and helpful user-centered feedback based on human factors assessments (e.g., SA, workload) and user comments (Figure 19.1).

Consequently, Human-Centered Design (HCD) has been used to incrementally improve OCSIS toward an acceptable mature version (i.e., incremental prototype development, test, and modification). A scenario was developed as a sequence of normal and abnormal situations: (1) Fuel Leak, (2) Descent, (3) Approach, (4) Flaps Locked, and (5) Landing. These scenarios were used in HITLS with professional pilots for OCSIS tests. HITLS were performed on a fully functional Airbus 320 aircraft simulator at the Human-Centered Design Institute (HCDi) of Florida Institute of

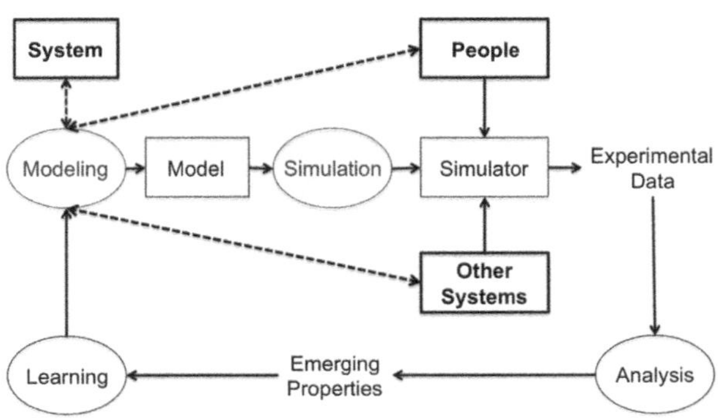

FIGURE 19.1 Human-Centered Design approach (Boy, 2014).

FIGURE 19.2 HCDi's Airbus 320 simulator.

Technology (Figure 19.2). This was the first step of OCSIS's scenario-based design (Rosson & Carroll, 2002).

Context-Sensitive Approach: Interaction Blocks

OCSIS's software was developed using a high-level procedural knowledge representation, called interactive blocks or iBlocks (Boy, 1998). Figure 19.3a shows the structure of an iBlock: it includes a set of actions, a situation pattern (triggering conditions + context pattern), and post conditions (goal + abnormal situations).

Taking the Flaps Set procedure during the approach phase (Figure 19.3b) as an example, the triggering preconditions are the Flaps position (visible on the E/WD screen) and the Flaps handle position (visible on the Flaps handle itself). When the Flaps position indicates the same value as the Flaps handle position, the goal is reached; otherwise, OCSIS informs the pilot of an abnormal situation, and the Flaps Locked abnormal procedure must be executed. Each procedure can be represented, implemented, and handled using iBlocks. Consequently, operational procedures can be described as a meaningful hyperlinked structure.

OCSIS PROTOTYPE

OCSIS's software was implemented on an iPad. It includes three types of operational information for flight, performance, and navigation. It is context-sensitive to improve safety, efficiency, and comfort in normal, abnormal, and emergency situations.

iPad-Simulator Synchronization

OCSIS design is based on the claim that context-sensitivity improves pilots' SA, decision-making, and action taking. This context-sensitivity requires an appropriate synchronization of aircraft parameters with the iPad in real time. Synchronization is performed in three steps: retrieving flight parameters from the simulator, storing them, and displaying them on the iPad.

HCDi A320 aircraft simulator is based on Prepar3D, a Lockheed Martin simulation software, which supports the control of various kinds of vehicles (e.g., marine, terrestrial, air, or space) (Prepar3D, 2014). It is based on a fully functional aircraft database and provides a realistic flight environment for professional flight crews. Typically, these parameters are retrieved through SimConnect, a library compatible with Prepar3D, and they are then stored as an XML file.

Flight parameters were transferred from the simulator to the iPad synchronously using a simple File Transfer Protocol at a sampling period of 500 milliseconds.

OCSIS Features

In principle, OCSIS includes procedures, maps, performance charts, flight plans, weather information, manuals, flight blogs, and contact information. The current

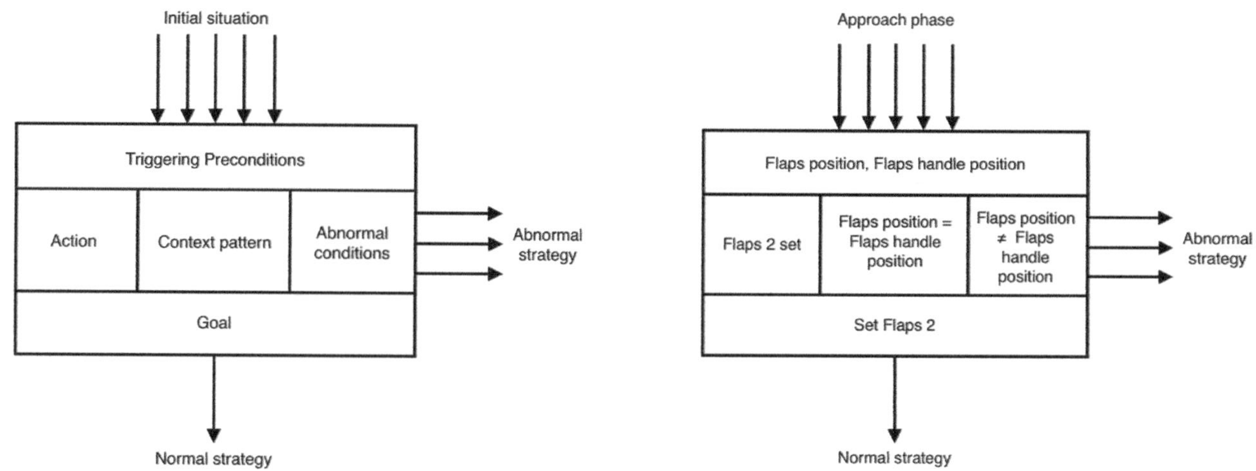

FIGURE 19.3 (a) Structure of an iBlock. (b) Application to the Flaps Set procedure.

prototype is limited to procedures, the most critical OCSIS part supporting pilots' work in the cockpit. Therefore, OCSIS is organized into three layers:

- (Level 1) Need to know or safety-critical information that pilots need to have immediately.
- (Level 2) Nice to know, or short explanations of Level 1.
- (Level 3) Technical knowledge on systems principles and troubleshooting (Blomberg et al., 2000).

Targeted normal, abnormal, and emergency procedures were developed and implemented. For example, when a pilot goes from Level 1 to Level 2, they need to know more about an appropriate action (i.e., selecting the action line leads to more explanation) (Figure 19.4).

In addition, two main features are presented in this paper: the Dynamic Color System (DCS) and the context-sensitive alert information system. The primary purpose of these features is to enhance SA. DCS is shown in Figure 19.5:

- Cyan denotes actions to be taken by the pilot.
- Amber denotes postponed actions or checks.
- Green denotes completed actions.

Implemented context patterns trigger real-time procedures in normal and abnormal situations. OCSIS- aircraft synchronization enables the visualization of pilots' physical actions, which is essential feedback to flight crews. Taking the Parking Brake procedure as an example, once the pilot releases it to OFF for taxiing, the OCSIS can detect the change of Parking Brake's status, and then the related action line turns to Green automatically. In an abnormal situation like Flaps Locked, OCSIS will immediately inform the pilot about this malfunction by displaying a pop-up information

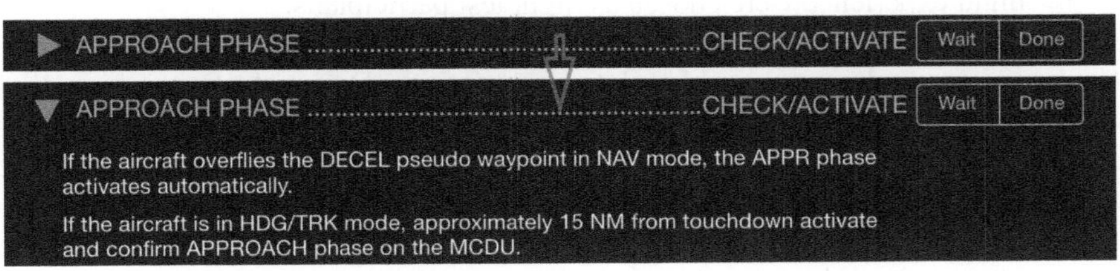

FIGURE 19.4 Information needs for flight operations: Level 1.

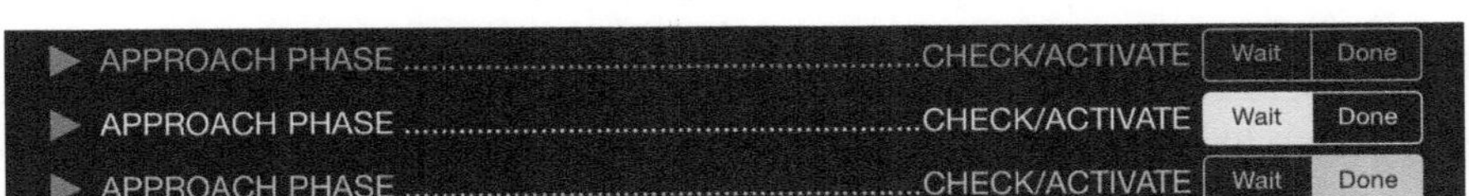

FIGURE 19.5 Dynamic Color System.

FIGURE 19.6 Abnormal situation triggering.

FIGURE 19.7 Abnormal situation reminder.

window (Figure 19.6). The flight crew is then aware of the problem and can follow the ECAM actions; consequently, they can decide to execute the additional related procedures immediately or later using OCSIS. If they choose to do it later, a reminder line will be displayed at the bottom of OCSIS, which can direct to additional Flap Locked procedures (Figure 19.7).

EXPERIMENTS

Preparation

OCSIS tests include two parts:

1. Usability and usefulness tests consist of getting feedback from pilots and checking the capability of OCSIS to assist them in their cockpit work, and
2. A location test consists of finding the best location for OCSIS. Pilots with flight experience were chosen as flight test participants.

They were required to come once each for two sessions. During the first session, they were asked to follow paper-based procedures. During the second session, they had to use OCSIS. These two sessions were a few days apart for each participant. A briefing was held before each experiment, including training, questionnaires, and surveys.

All participants performed as Pilot Not Flying (PNF) in each test and were requested to perform the following scenarios:

1. Climb, where participants could read and get used to instruments, flight parameters, system parameters, and panels.
2. Fuel Leak, where a fuel leak was triggered around 13,000 feet, required the crew to land immediately and perform related procedures. After the Fuel Leak procedure, a survey was generated for feedback, reviews, and self-assessments.
3. Approach preparation, where the Pilot Flying (PF) oversaw the Flight Management System (FMS) input, briefing for approach and other procedures, and the crew finished the Approach Checklist.

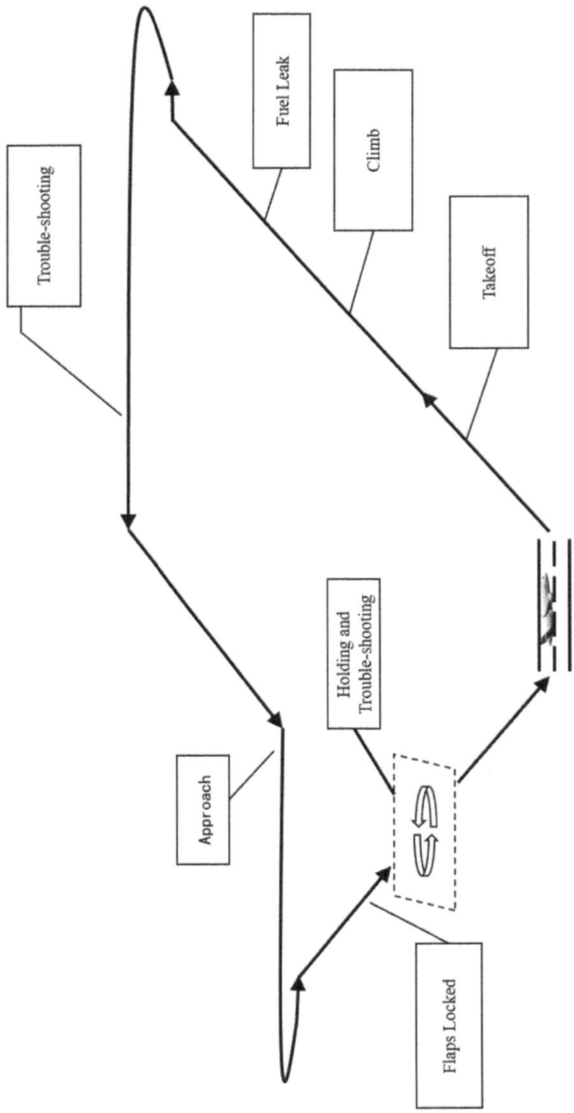

FIGURE 19.8 Experiment timeline.

4. Flaps Locked, where a Flaps Locked event was triggered when Flaps 2 was released during approach.

Experimental Protocol

Our experiments, which included questionnaires and surveys, were divided into the briefing room and the cockpit (Figure 19.8).

In the briefing room, participants were submitted to general knowledge questionnaires for the paper-based documentation test and trained to perform normal and abnormal procedures. Then, participants took a Cognitive Walkthrough test (Lewis et al., 1990) to check their understanding of the system. For the OCSIS test, another Cognitive Walkthrough test was conducted to identify difficulties encountered in using OCSIS. Consequently, a more thorough training on OCSIS was administered. Finally, pilots undertook a usability survey.

In the cockpit, the experiment was paused after the Fuel Leak malfunction test, and then participants undertook workload and SA surveys. Pilots undertook the same survey after the Flaps Locked malfunction scenario. At the end of the experiment, pilots were asked to provide their opinions on the best location of OCSIS in the cockpit.

RESULTS

Usability (Pilot–System Interaction)

OCSIS was designed following HCD principles that promote prototyping to elicit user feedback during the design phase to discover affordances, design flaws, and reformulate purposes. Cognitive Walkthrough enabled us to assess OCSIS usability. It was based on the execution of seven actions, taken without training. Unlike classical academic studies that involve a few students to satisfy statistical criteria, our first evaluation involved four experienced pilots. In HCD, specifically formative assessment, we privilege expertise and experience over statistics for efficiency and effectiveness. Summative assessment and certification will involve quantitative, statistically relevant studies. Results show that pilots can refer to the necessary information when they understand the commands. They can read actions and procedures for different flight phases and understand and use the DCS. Although the pilots suggested increasing the affordance of the title of each flight phase and the font that results from its selection, they nevertheless understood all functions after a short training.

User-Centered Assessment Methods

Workload

The pilot's workload was assessed after each malfunction during the two tests. We used NASA-TLX (Hart & Staveland, 1988) augmented with an additional visual demand criterion (Stephane, 2013). Pilot's workload was reduced for each malfunction using OCSIS (Figure 19.9).

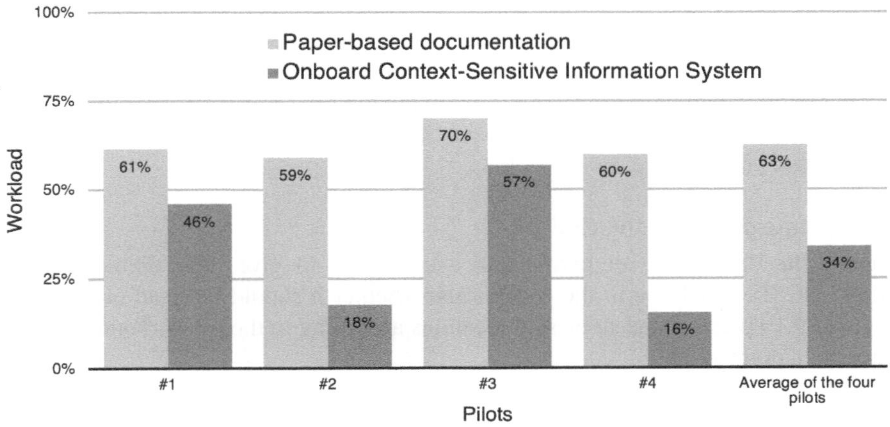

FIGURE 19.9 Average workload per pilot.

FIGURE 19.10 Different suggested positions for the OCSIS in the cockpit.

Situation Awareness

We used the Situation Awareness Global Assessment Technique (SAGAT) (Endsley, 1995). We only focused on perception and comprehension levels. At the perception level, SA assessments were performed using a scale from 0 (bad) to 10 (very good);

the average score was 9.45. Information retrieval ability, action understanding, and possible related failure on the ECAM/OCSIS were tested at the comprehension level. One of the four pilots had difficulty understanding a few actions. Still, the other three pilots performed all the actions, indicating that OCSIS can provide the correct information at the right time. The participants also assessed DCS using the same subjective scale; the average score was 9.3.

System Integration in the Cockpit

During the debriefing, all participants were asked to give their opinions about where OCSIS should be in the cockpit and whether it should be fixed or movable. Figure 19.10 presents the suggested locations according to the pilots. Four positions have been pointed out:

1. near the sidestick,
2. on a flexible arm,
3. in a sleeve near the seat, easy to handle, or
4. on the pedestal or instrument panel as a unit.

DISCUSSION: AN ITERATIVE DESIGN PROCESS

OCSIS was first designed as software that would later be used as a tangible interactive system (TIS) onboard a commercial aircraft. The main issue we had to solve was tangibility. Indeed, onboard paper-based documentation has been used since the beginning of aviation history and is tangible for pilots (i.e., natural to grasp, handle, and use). Tablets are now commonly used by the public, including pilots. Consequently, we consider tablets to be tangible objects that can support OCSIS software. The first experiments confirmed this hypothesis on a fully equipped cockpit simulator in flight operations with professional pilots. Of course, more iteration needs to be implemented to get a mature OCSIS.

This paper provides a first iteration of a participatory design of OCSIS. More generally, it shows the shift from the traditional automation approach, where additional software was added to the cockpit and induced some rigidity that sometimes results in unexpected situations, to TIS design, where tangibility must be tested against flexibility criteria. The following results came out:

1. DCS provides dynamic SA that increases flexibility in pilots' actions (e.g., when a do-list action is impossible to execute, yet the amber color reminds the pilots of later execution).
2. When using OCSIS, pilots can decide to postpone an abnormal procedure, and the system can remind them at an appropriate time.
3. Using OCSIS, pilots directly access the first layer of operational information and the others on demand (i.e., providing significant flexibility).
4. Global checks are possible, which improves flexibility in a high-pressure situation.
5. OCSIS connectivity with flight parameters provides functional affordances.

6. OCSIS knows about pilots' actions on cockpit instruments and provides redundant feedback (i.e., this additional cross-checking support is beneficial for safety and efficiency purposes).

CONCLUSION

Following context-sensitivity in operational procedure is a new feature enabled by electronic media for providing more flexibility in an aircraft cockpit. The main issue in this approach is the definition of context patterns, because when a context pattern matches the current situation, it triggers an appropriate procedure. Therefore, context patterns must be discriminated against to propose the proper procedure to the pilot. For example, in the After Engine Shut Down event, the system must know if the aircraft is on the ground or in the air to propose the appropriate procedure. It is evident that context pattern definition is a complex problem, but the current OCSIS design and positive test results encourage us to pursue this context-sensitivity approach. We plan on developing an extensive context identification effort in the next phase of this project.

HCD is about creativity and evaluation. Creativity is the integration of existing things. With the OCSIS, we integrated well-known technology and techniques. For example, we integrated a tablet context-sensitive software and appropriate connectivity with the flight simulator. Domain content and technology were realistic and enabled us to involve professional pilots. This led to very credible simulations used to test the tangible interactive OCSIS. The first formative evaluation showed that OCSIS provides much safer, efficient, and comfortable operations in the cockpit. We know this is a preliminary result, but it offered input information for an improved design of OCSIS, which we are currently developing.

This work is the first phase in the design and development of OCSIS. It is based on participatory design and agile development (i.e., at the end of each phase, the system is testable in the HITLS environment). This is now typical for the design and development of tangible interactive objects (Boy, 2014), and more generally, TISs (Boy, 2016), where the problem is no longer automation but the search for tangibility. Modeling and simulation are required to explore the possibilities and drawbacks of these TISs. The quality of both simulation capabilities and pilot participants is crucial.

NOTE

1 This chapter is based on PhD work done when the author was at the Human-Centered Design Institute (HCDi) of the Florida Institute of Technology, USA, together with Prof. Guy A. Boy (Tan, 2015).

ACKNOWLEDGMENTS

The author would like to thank Ruthvik Adloori, Sebastien Boulnois, Guy André Boy, Jarrett Clark, Vincent DeCampo, Nicholas Kasdaglis, De Vere Kiss, Pat Picornell, Pieter Rosel, and Lucas Stephane for their help during this first phase of the OCSIS project.

REFERENCES

Amalberti, R. (2012). *Flying safety* (in French: *Piloter la sécurité*), Springer, Paris.

Amalberti, R. A. (2001). *Safety–critical systems control* (in French: *La conduite de systèmes à risques*), PUF, Paris.

Blomberg, R., Boy, G. A., & Speyer, J. J. (2000). "Information needs for flight operations: Human-centered structuring of flight operations knowledge." In *Proceedings of HCI-Aero 2000, Cepadues, Toulouse, France*. ACM Digital Library.

Boy, G. A. (1998). *Cognitive function analysis* (Vol. 2). Greenwood Publishing Group.

Boy, G. A. (2014). "From automation to tangible interactive objects." *Annual Reviews in Control*, 38(1), pp. 2–10. doi: 10.1016/j.arcontrol.2014.03.001.

Boy, G. A. (2016). *Tangible interactive systems – grasping the world with computers.* Springer, UK.

Endsley, M. R. (1995). "Toward a theory of situation awareness in dynamic systems." *Human Factors: The Journal of the Human Factors and Ergonomics Society*, 37(1), 32–64.

Hart, S. G., & Staveland, L. E. (1988). "Development of NASA-TLX (Task Load Index): Results of empirical and theoretical research." *Advances in Psychology*, 52, 139–183.

Ilbert, C. G., Amalberti, R. A., Aroche, H. L., & Ariès, J. P. (2007). "Error and failures: Toward a new safety paradigm." *Journal of Risk Research*, 10, 959–975.

Lewis, C., Polson, P. G., Wharton, C., & Rieman, J. (1990, March). "Testing a walkthrough methodology for theory-based design of walk-up-and-use interfaces." In *Proceedings of the SIGCHI conference on Human factors in computing systems* (pp. 235–242). ACM Digital Library.

Prepar3d. (2014). Lockheed Martin. Accessed in March 2014. www.prepar3d.com

Ramu, J. P., Barnard, Y., Payeur, F., & Larroque, P. (2004). "Contextualized operational documentation in aviation." In D. de Waard, K. A. Brookhuis, & C. M. Weikert (Eds.), *Human factors in design* (pp. 1–12). Shaker Publishing, Maastricht, the Netherlands.

Rosson, M. B., & Carroll, J. M. (2002). "Scenario-based design." Chapter 53 in J. Jacko & A. Sears (Eds.), *The Human-Computer Interaction Handbook: Fundamentals, Evolving Technologies and Emerging Applications*, (pp. 1032–1050). Lawrence Erlbaum Associates.

Stephane, A. L. (2013). "Visual intelligence for crisis management." PhD Dissertation, Florida Institute of Technology, Proquest LLC.

Tan, W. (2014). "From commercial aircraft operational procedures to an onboard context-sensitive information system." In *Proceedings of the HCI-Aero Conference*. ACM Digital Library, Santa Clara, CA.

Tan, W. (2015). Contribution to the onboard context sensitive information system of commercial Aircraft. PhD Dissertation, Florida Institute of Technology, USA.

20 Human Acceptance of Automation and Human Systems Integration

Peter Moertl

INTRODUCTION

Automation surrounds us and promises to enhance our lives and workplaces. Automation technologies range from automated vehicles to intelligent digital assistants, seemingly becoming more numerous daily. Accordingly, automation technologies are significant economic market drivers. However, our ability to predict user acceptance of automation technologies before their introduction remains limited. For example, the widespread adoption of mobile devices such as smartphones over the past two decades was not anticipated at the time of their introduction. Similarly, with the arrival of artificial intelligence, we are facing a new type of automation that has started to change our lives and workplaces in many ways. Here, we are challenged as individuals and as a society to decide what is acceptable and what is not, and automation developers need to decide what products they can successfully introduce into the market. Therefore, introducing automation often changes the tasks and activities of the involved humans. However, it also directly and indirectly impacts the environment and organizations surrounding us in socio-technical systems. Such changes are often unpredictable as they seemingly unexpectedly emerge from human interaction and systems. Sometimes, such emergent qualities lead to unexpected technology successes, such as the internet, smartphones, or social media. However, in many cases, such unexpected emergencies can significantly hinder the uptake of automation by human users and markets, and the return on investment, such as the famous aviation accidents related to some aspects of the flight automation of the Boeing 737 Max, or the experiences with advanced automation in many new vehicles show.

Therefore, the ability to predict user acceptability of new automation should help build better products, save development costs, lead to products that people use over extended periods, and create more revenue.

This chapter reviews what is known about user acceptance by presenting well-established models. This is then set about what human factors research has been reporting about automation acceptance and summarizes problems humans have with using and interacting with different levels of automation, thereby focusing on research

DOI: 10.1201/9781003666431-24

in driving automation. This sets the role of defining how human systems integration as a discipline within systems engineering can lead toward acceptable automation.

Terminologically, the term automation refers to "[...] automation as a device or system that accomplishes (partially or fully) a function that was previously, or conceivably could be, carried out (partially or fully) by a human operator" (Parasuraman & Riley, 1997). The term "acceptability" can be differentiated from the term "acceptance" (Schade & Schlag, 2003). Both terms refer to attitudes of potential system users, where "acceptability" relates to a prospective judgment before using a system, and "acceptance" is based on experience. Furthermore, "adoption" can be differentiated from "acceptance" by referring to the actual behavior of using a system. Also, individual acceptance, which is the focus of this chapter, differs from social or societal acceptance by groups of people.

ACCEPTANCE MODELS

In their recent review, Kelly et al. (2023) identified two historically established acceptance models as the basis for subsequent research on technology acceptance for different applications: the Technology Acceptance Model (TAM) by Davis (1989) and the Unified Theory of Acceptance and Use of Technology (UTAUT) by Venkatesh et al. (2003). Both models originate from the Theory of Planned Behavior (TPB; Ajzen, 1991), which predicts and explains human behavior in specific contexts. The TPB posits that the key factor in predicting human behavior is an individual's intention to perform the behavior in question. It further theorizes that the intention or motivation to exert a behavior is influenced by three factors: the individual's attitudes toward the behavior, social norms, and the perceived behavioral control, i.e., a person's judgment about their ability to perform the behavior, i.e., the extent to which they have the required resources, such as time or skills, and opportunities. TPB also foresees weighing belief strengths and quantifying the connection between beliefs with behavioral intention and actual behavior. Although the TPB does not explicitly address automation acceptance, it can be directly applied to accepting and using specific technologies. Therefore, it is closely related to developing specific acceptance models, such as the ones by Davis and Venkatesh.

Technology Acceptance Model

The TAM by Davis (1989) posits that perceived usefulness and ease of use are the primary determinants for user acceptance. Perceived usefulness is the belief that a technical application will help them perform their job better, and ease of use is the belief about the system's difficulties. Davis assumes the human user to be a rational actor who weighs the costs and benefits of the technology to decide to use it and reports two studies investigating the model. Participants performed tasks with the computer entirely under their control (complete human control) and evaluated them with 20 questions that assessed their perceived usefulness and ease of use. The results indicated statistically significant correlations between these measures and self-reported use of computer applications. After a second study, a set of 12 items was

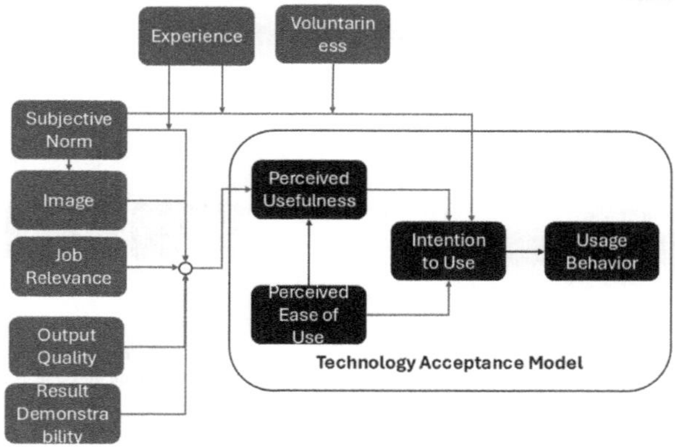

FIGURE 20.1 Automation Acceptance Model. (From Ghazizadeh et al., 2012.)

reduced, of which six assess perceived usefulness (work more quickly, job performance, increase productivity, effectiveness, and make job easier). Six assess perceived ease of use (easy to learn, controllable, clear and understandable, flexible, easy to become skillful, ease to use).

The TAM was subsequently expanded to include social influences and cognitive processes and referred to as TAM 2 (Venkatesh & Davis, 2000). Subjective norms consist of a person's beliefs about the desirability of a particular behavior from the view of people they know. Subjective norms are expected to influence mandatory behavior but not voluntary behavior. TAM 2 also considers how people make decisions concerning perceived usefulness, specifically regarding job relevance, output quality, and result demonstrability (Figure 20.1). TAM 2 was tested in four longitudinal studies that supported the postulated connections between the concepts (Venkatesh & Davis, 2000).

TAM 3 (Venkatesh & Bala, 2008) is further expanded by modeling in a more detailed way the impact of experience on various individual beliefs on technology acceptance and also validated in a longitudinal study.

Unified Theory of Acceptance and Use of Technology

The second most frequently cited user acceptance model is the Unified Theory of Acceptance and Use of Technology (UTAUT), where Venkatesh expanded existing models and combined several variables into four main predictors and moderators of user acceptance (Venkatesh et al., 2003). As Venkatesh also formulated both extensions of TAM to TAM 2 and 3, the limits between UTAUT and TAM are somewhat blurred. Venkatesh's UTAUT model (Figure 20.2) predicts "intention to use" using four dimensions.

Performance expectancy is an individual's belief that a technology will help achieve goals during job performance. "Effort expectancy" consists not only of the expected ease of use of the system but also of social influence that relates to the

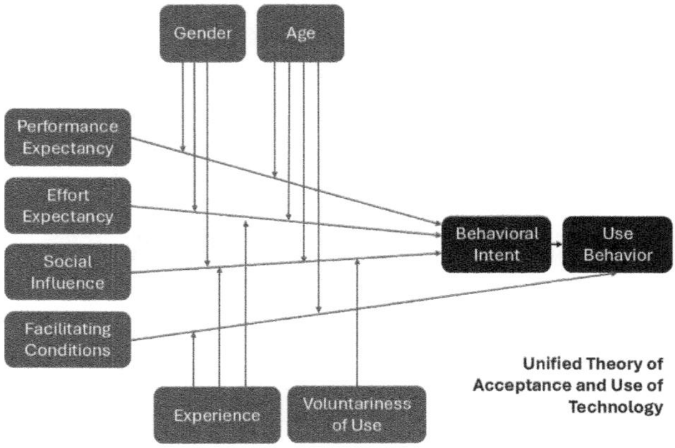

FIGURE 20.2 Unified Theory of Acceptance and Use of Technology. (By Venkatesh et al., 2003.)

perceived beliefs of others about whether to use the system or not, as well as the facilitating conditions and the individual's beliefs about whether the organizational and technical environment supports the use of the technology. The intent to use a technology directly relates to using the system, but only if facilitating conditions support it. Venkatesh's UTAUT model can be seen as a refinement of the TAM model, through a more sophisticated "rational actor" who makes decisions using a broader spectrum of information and several moderators: gender, age, experience, and voluntariness of use. The UTAUT model was initially developed to explain the acceptability of relatively simple "full-manual control" technologies such as an online meeting manager, a database application for business and banking applications, and an accounting support system (Venkatesh et al., 2003). In addition, UTAUT was extended to UTAUT 2 by adding hedonic motivation, price value, and habit (Venkatesh et al., 2012). UTAUT 3 (Farooq et al., 2017; Venkatesh, 2022) extended the UTAUT model toward AI to include specific characteristics of the human user, the technology, the environment, and training.

Specifically, the UTAUT assessment considers the following aspects:

1. *Performance Expectancy:* I find the system helpful in my job. The system enables me to accomplish tasks more quickly, increasing my productivity. If I use the system, I will increase my chances of getting a raise.
2. *Effort Expectancy:* My interaction with the system would be clear and understandable; it would be easy for me to become skillful at using the system; I would find the system easy to use; learning to operate the system is easy for me.
3. *Attitude toward Using Technology:* Using the system is a bad/good idea; the system makes work more interesting; working with the system is fun; I like working with the system.

4. *Social Influence:* People who influence my behavior think that I should use the system; people who are important to me think that I should use the system; the senior management of this business has been helpful in the use of the system; in general, the organization has supported the use of the system.
5. *Facilitating Conditions:* I have the resources necessary to use the system; I have the knowledge necessary to use the system; the system is not compatible with other systems I use; A specific person (or group) is available for assistance with system difficulties.
6. *Self-Efficacy:* I could complete a job or task using the system; if there was no one around to tell me what to do as I go; if I could call someone for help if I got stuck; if I had much time to complete the job for which the software was provided; if I had just the built-in help facility for assistance.
7. *Anxiety:* I feel apprehensive about using the system; it scares me to think that I could lose much information using the system by hitting the wrong key; I hesitate to use the system for fear of making mistakes I cannot correct; the system is somewhat intimidating to me.
8. *Behavioral Intention to Use the System:* I intend to use the system in the next <n> months; I predict I will use the system in the next <n> months; I plan to use the system in the next <n> months.

THE HUMAN FACTORS INFLUENCING THE USE OF AUTOMATION

The TAM and UTAUT models focus on information technology applications with users interacting with software and investigating their attitudes and self-reported insights. This is a somewhat different focus than the one that arises from interactions with physical automation such as washing clothes, driving a car, or flying an airplane: here some physical tasks shift from the human to automation, and new tasks may be required by the human, such as monitoring or supervising the automation (Parasuraman et al., 2000). Ghazizadeh et al. (2012) proposed the Automation Acceptance Model (AAM) to account for particularities of automation.

Automation Acceptance Model

The AAM (Ghazizadeh et al., 2012) proposes adding three factors to the TAM:

The first is trust: In most cases, the introduction of automation does not replace the human but rather forms some kind of co-agency: the human takes on new tasks as the automation takes on some of the tasks that the human previously performed. For example, people need to monitor or supervise automation. Different levels of automation need to be differentiated, depending on the tasks and responsibilities between humans and automation. Designing systems with inappropriate levels of automation can lead to brittle automation that fails in many situations or unexpectedly for the operator. Such surprises of automation need to be minimized for acceptance, which leads to the necessity of operator trust in the system (see Lee & See, 2004).

Secondly, Ghazizadeh et al. (2012) suggest that acceptance of automation systems requires compatibility between the tasks humans perform and the tasks the technology performs. Task compatibility is important because realistic automation is seen as co-agency between humans and automation. They hypothesize that task-technology compatibility also influences trust formation. Task compatibility is also emphasized by Westin et al. (2016), who differentiate compatibility of the overt behavior between human and system from the underlying processes: it is claimed that similarity in the underlying strategies and problem-solving styles between human and system is important for acceptance. This similarity is referred to as strategic conformance (Westin et al., 2016).

Thirdly, they emphasize the importance of feedback mechanisms to explain the temporal dimension of acceptance development: acceptance of a technology can change over time as trust develops and the experiences with the technology grow and give the operator a better picture of how useful and easy to use the automation is for the operator to achieve the goals.

In addition, Merat et al. (2017) point to the importance of user trust for acceptance, specifically in automated driving vehicles. Trust is "the attitude that an agent will help achieve an individual's goals in a situation characterized by uncertainty and vulnerability" (Lee & See, 2004). Similarly, the connection between automation reliability and automation reliance was investigated by Dzindolet et al. (2003), who observed that participants initially relied on automation when they had no information about its reliability. However, reliance was disrupted when they observed automation errors. When receiving explanations about the causes of system errors, their reliance increased again. Explanations about the system errors increased their reliance and trust, even when the trust was not warranted, because error rates were higher than when the participants performed the task themselves.

Extensive research investigates humans interacting with automation, as this represents a core discipline of human factors worldwide. However, automation refers to many possible constellations of human–technology interactions that differ tremendously between applications. A dishwasher is certainly an automation and an autopilot in a modern aircraft. Therefore, it is important to distinguish different levels of automation that represent the amount and quality of the automated tasks. Therefore, Parasuraman et al. (2000) introduced an important differentiation relevant to understanding and predicting acceptance of automation: the levels of automation.

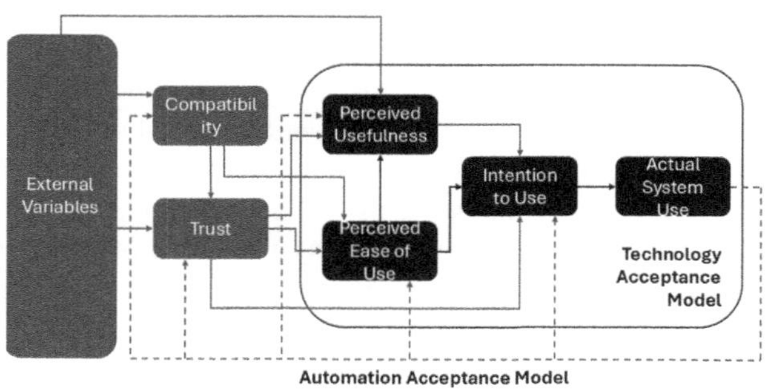

FIGURE 20.3 Automation Acceptance Model. (From Ghazizadeh et al., 2012.)

The authors propose 10 levels of automation for each of the four human information processing functions presented in Table 20.1 (left). The authors hypothesize that user acceptance relies on selecting the appropriate level of automation for each function. Differentiating these ten levels of automation seems foundational for the linkage between automation and acceptance, as it allows a more detailed look at the tasks that automation is taking over from humans. This should allow for better prediction of the amount of user acceptance that results from that shift. However, it seems not trivial to apply these levels of automation to concrete applications: for example, where does driving automation fall in this category? There, automation mostly happens on the level of action implementation, but not on the information processing level. Therefore, a different classification has been established for driving automation (SAE International, 2021) that introduces six levels of automation that differentiate the main tasks and responsibilities of the automated vehicle and human driver. This automated driving level taxonomy is widely used in the automotive industry and represents the basis for driving functionality currently sold in vehicles worldwide. These levels are summarized and contrasted on the right side of Table 20.1.

TABLE 20.1
Comparison of Levels of Automation (Parasuraman et al., 2000; SAE International, 2021)

Parasuraman Level of Automation	SAE Level of Automation for Driving Automation
L10 The computer decides everything, acts autonomously, ignoring the human	L5 Driving automation performs all tasks, there is no driver anymore
L9 Informs the human only if it, the computer, decides to L8 Informs the human only if asked	L4 Driving automation performs continuously all driving tasks in specific environments, the human take control but is not required
L7 Executes automatically, then necessarily informs the human	L3 Driving automation performs temporarily all driving tasks, the human remains fallback ready to take back control when informed by automation
L6 Allows the human a restricted time to veto before automatic execution	L2 Driving automation performs both lateral and longitudinal driving tasks, the human monitors and performs all other tasks
L5 Executes that suggestion if the human approves L4 Suggests one alternative	L1 Driving automation performs either lateral or longitudinal driving tasks, the human monitors and performs all other tasks
L3 Narrows the selection down to a few L2 The computer offers a complete set of decisions/action alternatives L1 The computer offers no assistance: humans must take all decisions and actions	L0 The driver performs all driving tasks

By comparing the two taxonomies, it becomes apparent that while Parasuraman's levels of automation increase with the cognitive aspects that are taken over by automation, the SAE's levels of driving automation increase with the driving tasks and responsibilities that are performed by driving automation. Both taxonomies stay agnostic of human acceptance, not indicating which levels would be more acceptable for users. This has been an active debate, and alternative taxonomies for driving automation have been suggested by Shi et al. (2020) and Moertl et al. (2019); they proposed an extension of the SAE taxonomy to better reflect the human role and needs, specifically the AUTOS framework (Boy, 2013). This extension is specifically targeted toward the goal of creating automation that is acceptable and useful to its human users by considering the use context more holistically and including also the tasks and situations.

While the most useful taxonomy for automation may depend on its application and differ between, for example, assisting nuclear power plant operators or allowing drivers to engage in non-driving-related activities, establishing an appropriate taxonomy should lead to a more differentiated understanding between automation and acceptance. The main factors influencing acceptance in the context of automation are described next, summarizing extensive research in human factors (Billings, 1996; McLeod, 2014; Waard & Human Factors and Ergonomics Society, 2008; Young & Stanton, 2023).

Monitoring Automation

As automation takes over tasks previously performed by humans, humans still retain responsibility for the overall outcomes. This means that tasks shift between humans and automation, and while some tasks are replaced, new tasks emerge for humans. One such common task consists of supervising automation. For example, SAE L2 driving automation takes over lateral and longitudinal vehicle control. However, it requires humans to stay responsible and intervene at any time, which requires them to supervise automation, as automation can disengage from such tasks without prior warning. Such tasks are difficult for human operators (e.g., Kaber & Endsley, 1997; Rudin-Brown & Parker, 2004; Stanton, 2019; Stanton & Young, 2005). Generally, human monitoring of automation requires the exertion of attention, which, especially over extended periods, can become monotonous and effortful and invites misuse: in 2023, Tesla had to recall virtually all their vehicles on U.S. roads because drivers did not sufficiently monitor the performance of their SAE L2 driving automation.[1]

Staying-in-the-Loop

As automation takes on tasks and performs them reliably over time, without humans having to monitor them, the role of humans changes again. Removing the monitoring tasks from humans frees them from engaging in other tasks, providing significant benefits. However, human control is often needed when automation encounters conditions it was not designed for (operational design domain or ODD) or if anomalies occur; in that case, the human must step back in and provide a safe solution. For example, during SAE L3 driving, the vehicle takes over longitudinal and lateral control. It notifies humans when to step back in case of failure or when they reach the end of the ODD. Stepping back in the loop can be difficult for humans (Endsley, 2016; Hancock, 2019; Kaber & Endsley, 1997) because it requires switching back to a task they have not been performing for a while. Therefore, bringing humans back into the loop is a new task that humans must perform with higher levels of automation.

Automation Conundrum

There seems to be an inverse relationship between the reliability of automation and the human performance interacting with it: the more reliable automation performs, the more difficult it becomes for human operators to step back into the loop and recover in situations of problems or failures. This phenomenon is called the automation conundrum and is documented in various domains (Endsley, 2019).

Automation Mode Awareness

With increasing automation, human operators are confronted with different modes of automation, which can lead to confusion (see, e.g., Degani et al., 1999; Dönmez Özkan et al., 2021), particularly regarding the currently active mode of automation. For example, when Asiana flight 214 was about to land and failed at San Francisco Airport in 2013, there were three pilots in the cockpit who all failed to detect that the airplane was in an automation mode that was intended for operation at high altitude rather than the mode they were expecting when approaching the airport. The plane crashed onto the runway as a result. As increasing automation modes are becoming available, this problem exacerbates, given that many operators are not as well trained and certified as commercial pilots.

HSI FOR DESIGNING AUTOMATION FOR HUMAN ACCEPTANCE

We can fall into many traps when designing automation that users want to use over extended periods. The reviewed TAMs and the human factors of automation have in common that designing technology for acceptance requires detailed knowledge about those who will use it, the context of use, and the goals to be achieved within the organizations in which they are used.

According to the acceptance models in Figures 20.1–20.3, many factors can influence the user's intention to use a system. Only if such information is available early on to shape the system design process do these systems have a chance to be successful.

In this way, considerations for user acceptance form the basis for HSI that attempts to facilitate engineering and research toward acceptable socio-technical systems. A primary challenge is therefore to know these acceptance factors early on and use them to shape the system design and the development process. While standards in user-centered developments of interactive systems (ISO, 2019) are available, their scope needs to be widened to build different types of systems. A car is also an interactive system, but its design must consider factors quite different from software interactions.

A critical enabler for HSI in this area is using virtual contextualization methods to assess user acceptance factors early on. If the system is not yet available, other methods than real-world data collections must be used to gather critical acceptance factors for the design process. One of these methods consists of virtual environments where participants are immersed in a world that does not yet exist to learn about user reactions and impressions and refine the system. The results are generally more valid and reliable than when participants are asked to imagine the possibilities. In automotive research, using driving simulators is essential to safely and effectively design and test human–system interactions. Similar is the case in the aviation and rail transport research and development. However, in many other fields, virtual assessment methods are still under development, especially when considering increasingly larger system scopes that include single vehicles and interactions between multiple vehicles and vulnerable road users. This requires enhanced and widened virtualization methods, as such complex information can otherwise not be directly derived from TAMs nor the existing taxonomies and human factors principles of automation.

A critical task of HSI involves planning virtual acceptance explorations early in the system design process to elicit human requirements and acceptance factors. Traditionally, such human-centered pre-design knowledge about tasks and responsibilities of humans within a to-be-designed socio-technical innovation or system is brought relatively late into the technological system development process (Pew et al., 2007). Therefore, traditionally, when such human factors knowledge enters the engineering cycle, the overall system has been designed, and functions have been allocated between humans and the system. This limitation forms an important motivation for the emergence of HSI and is observable in many fields, including automated driving developments. The belated introduction of known human factors considerations when designing the safety-critical automated driving systems causes some acceptance problems of advanced automated driving functions (CCAM, 2022). Thereby, HSI is not just about designing the interactions between humans and systems but about contributing to the design of overall automation systems that humans will accept and use.

NOTE

1 www.reuters.com/business/autos-transportation/tesla-update-software-autopilot-control-issue-2-mln- vehicles-nhtsa-2023-12-13/

REFERENCES

Ajzen, I. (1991). The theory of planned behavior. *Organizational Behavior and Human Decision Processes*, *50*(2), 179–211.

Billings, C. E. (1996). Human-centered aviation automation: Principles and guidelines. *NASA Technical Memorandum* 110381. https://ntrs.nasa.gov/search.jsp?R=19960016374

Boy, G. A. (2013). *Orchestrating Human-Centered Design*. Springer. https://link.springer.com/book/10.1007/978-1-4471-4339-0

CCAM. (2022). *CCAM Strategic Research and Innovation Agenda 2021–2027: European Leadership in safe and sustainable road transport through automation*. www.ccam.eu/wp-content/uploads/2022/05/CCAM_SRIA-report_web.pdf

Davis, F. D. (1989). Perceived usefulness, perceived ease of use, and user acceptance of information technology. *MIS Quarterly*, *13*(3), 319. https://doi.org/10.2307/249008

Degani, A., Field, M., & Kirlik, A. (1999). Modes in human–machine systems: review, classification, and application. *The International Journal of Aviation Psychology*, 9, 125–138.

Dönmez Özkan, Y., Mirnig, A. G., Meschtscherjakov, A., Demir, C., & Tscheligi, M. (2021). Mode awareness interfaces in automated vehicles, robotics, and aviation: A literature review. In *13th International Conference on Automotive User Interfaces and Interactive Vehicular Applications*, pp. 147–158. ACM Digital Library. https://doi.org/10.1145/3409118.3475125

Dzindolet, M. T., Peterson, S. A., Pomranky, R. A., Pierce, L. G., & Beck, H. P. (2003). The role of trust in automation reliance. *International Journal of Human–Computer Studies*, *58*(6), 697–718. https://doi.org/10.1016/S1071-5819(03)00038-7

Endsley, M. R. (2016). From here to autonomy: Lessons learned from human–automation research. *Human Factors*, *59*(1), 5–27.

Endsley, M. R. (2019). Situation awareness in future autonomous vehicles: Beware of the unexpected. In S. Bagnara, R. Tartaglia, S. Albolino, T. Alexander, & Y. Fujita (Hrsg.), *Proceedings of the 20th Congress of the International Ergonomics Association (IEA, 2018)* (Bd. 824, S. 303–309). Springer International Publishing. https://doi.org/10.1007/978-3-319-96071-5_32

Farooq, M. S., Salam, M., Jaafar, N., Fayolle, A., Ayupp, K., Radovic-Markovic, M., & Sajid, A. (2017). Acceptance and use of lecture capture system (LCS) in executive business studies: Extending UTAUT2. *Interactive Technology and Smart Education*, *14*(4), 329–348. https://doi.org/10.1108/ITSE-06-2016-0015

Ghazizadeh, M., Lee, J. D., & Boyle, L. N. (2012). Extending the technology acceptance model to assess automation. *Cognition, Technology & Work*, *14*(1), 39–49. https://doi.org/10.1007/s10111-011-0194-3

Hancock, P. A. (2019). Some pitfalls in the promises of automated and autonomous vehicles. *Ergonomics*, 62(4), 1–17. https://doi.org/10.1080/00140139.2018.1498136

ISO, T. 22. (2019). *Road vehicles—ergonomic aspects of transport information and control systems—calibration tasks for methods which assess driver demand due to the use of in-vehicle systems* (ISO/TS 14198:2019). International Standards Organization.

Kaber, D. B., & Endsley, M. R. (1997). Out-of-the-loop performance problems and the use of intermediate levels of automation for improved control system functioning and safety. *Process Safety Progress*, *16*(3), 126–131. https://doi.org/10.1002/prs.680160304

Kelly, S., Kaye, S.-A., & Oviedo-Trespalacios, O. (2023). What factors contribute to the acceptance of artificial intelligence? A systematic review. *Telematics and Informatics*, *77*, 101925. https://doi.org/10.1016/j.tele.2022.101925

Lee, J. D., & See, K. A. (2004). Trust in automation: Designing for appropriate reliance. *Human Factors: The Journal of the Human Factors and Ergonomics Society*, *46*(1), 50–80.

McLeod, R. (2014). *Human Factors in Highly Automated Systems: A White Paper*. https://ergonomics.org.uk/resource/human-factors-in-highly-automated-systems-white-paper.html

Merat, N., Madigan, R., & Nordhoff, S. (2017). *Human Factors, User Requirements, and User Acceptance of Ride-Sharing in Automated Vehicles* (International Transport Forum Discussion Papers 2017/10). OECD. https://doi.org/10.1787/0d3ed522-en

Moertl, P., Neuhuber, N., & Pretto, P. (2019). A human–systems integration framework around the SAE taxonomy for automated driving systems. In *Proceedings of the International Conference for Human–Systems Integration. International Conference for Human–Systems Integration 2019, Biarritz, France*. INCOSE (International Council on Systems Engineering), San Diego, CA, USA.

Parasuraman, R., & Riley, V. (1997). Humans and automation: Use, misuse, disuse, abuse. *Human Factors: The Journal of the Human Factors and Ergonomics Society*, *39*(2), 230–253.

Parasuraman, R., Sheridan, T. B., & Wickens, C. D. (2000). A model for types and levels of human interaction with automation. *IEEE Transactions on Systems, Man, and Cybernetics—Part A: Systems and Humans*, *30*(3), 286–297.

Pew, R. W., Mavor, A. S., National Research Council (U.S.), & Committee on Human-System Design Support for Changing Technology. (2007). *Human–System Integration in the System Development Process: A New Look.* National Academies Press. http://site.ebrary.com/id/10178142

Rudin-Brown, C. M., & Parker, H. A. (2004). Behavioural adaptation to adaptive cruise control (ACC): Implications for preventive strategies. *Transportation Research Part F: Traffic Psychology and Behaviour*, *7*(2), 59–76. https://doi.org/10.1016/j.trf.2004.02.001

SAE International. (2021). *Taxonomy and Definitions for Terms Related to Driving Automation Systems for On-Road Motor Vehicles: (J3016).* https://doi.org/10.4271/J3016_202104

Schade, J., & Schlag, B. (2003). Acceptability of urban transport pricing strategies. *Transportation Research Part F: Traffic Psychology and Behaviour*, *6*(1), 45–61. https://doi.org/10.1016/S1369-8478(02)00046-3

Shi, E., Gasser, T. M., Seeck, A., & Auerswald, R. (2020). The principles of operation framework: A comprehensive classification concept for automated driving functions. *SAE International Journal of Connected and Automated Vehicles*, *3*(1), 23–37. https://doi.org/10.4271/12-03-01-0003

Stanton, N. A. (2019). Thematic issue: Driving automation and autonomy. *Theoretical Issues in Ergonomics Science*, 20, 1–7. https://doi.org/10.1080/1463922X.2018.1541112

Stanton, N. A., & Young, M. S. (2005). Driver behaviour with adaptive cruise control. *Ergonomics*, *48*(10), 1294–1313. https://doi.org/10.1080/00140130500252990

Venkatesh, V. (2022). Adoption and use of AI tools: A research agenda grounded in UTAUT. *Annals of Operations Research*, *308*(1–2), 641–652. https://doi.org/10.1007/s10479-020-03918-9

Venkatesh, V., & Bala, H. (2008). Technology Acceptance Model 3 and a research agenda on interventions. *Decision Sciences*, *39*(2), 273–315. https://doi.org/10.1111/j.1540-5915.2008.00192.x

Venkatesh, V., & Davis, F. D. (2000). A theoretical extension of the technology acceptance model: Four longitudinal field studies. *Management Science*, *46*(2), 186–204. https://doi.org/10.1287/mnsc.46.2.186.11926

Venkatesh, V., Morris, M. G., Davis, G. B., & Davis, F. D. (2003). User acceptance of information technology: Toward a unified view. *MIS Quarterly*, *27*(3), 425. https://doi.org/10.2307/30036540

Venkatesh, V., Thong, J. Y. L., & Xu, X. (2012). Consumer acceptance and use of information technology: Extending the unified theory of acceptance and use of technology. *MIS Quarterly*, *36*(1), 157. https://doi.org/10.2307/41410412

Waard, D. de & Human Factors and Ergonomics Society (Hrsg.). (2008). *Human Factors for Assistance and Automation*. Shaker Publication.

Westin, C., Borst, C., & Hilburn, B. (2016). Strategic conformance: Overcoming acceptance issues of decision aiding automation? *IEEE Transactions on Human–Machine Systems*, *46*(1), 41–52. https://doi.org/10.1109/THMS.2015.2482480

Young, M. S., & Stanton, N. A. (2023). *Driving Automation: A Human Factors Perspective* (1st ed.). CRC Press.

21 Human Performance in Sociotechnical Systems

A Time-Based Perspective

Belinda S. Cham, Alexandra A. Boeing, Katrina L. Hosszu, Mark A. Griffin, and Karina Jorritsma

INTRODUCTION

Human performance is a critical concept in the design and operation of sociotechnical systems. Humans must interact with each other and with technology to perform tasks, maintain reliability, adapt to change, and anticipate problems. When human capabilities and limitations are effectively addressed in system design and development, systems can yield greater returns on investment and achieve greater safety and efficiency (Burgess-Limerick et al. 2010). On the other hand, insufficient efforts to optimize human performance as part of a wider systems engineering process can lead to systems that fail to deliver the expected efficiency gains and increase the risk of catastrophic accidents that have enduring and widespread societal consequences (Boy, 2021; Limerick & Burgess, 2010).

Conventional approaches to optimizing human performance have focused on how humans interact with a machine or an autonomous system (Boy, 2023) – for example, exploring how aspects of user-interface design impact individual cognitive performance. However, there is growing recognition that humans perform and operate within larger sociotechnical systems that involve interrelated organizational structures, business processes, and technical systems (Baxter & Sommerville, 2011). This broader lens has prompted researchers and practitioners from diverse disciplines (e.g., work psychology, information systems, and cognitive systems engineering) to explore and address how humans perform within and interact with varying aspects of sociotechnical systems.

Unfortunately, many research disciplines and domains concerned with human performance in sociotechnical systems have evolved and been practiced independently, often to achieve different goals (Clegg et al., 2000; Carayon, 2006; Baxter & Sommerville, 2011). Human performance is usually conceptualized, evaluated, and optimized in different ways depending on the goals and constraints of a specific research perspective. For example, the domains of human-factors engineering

DOI: 10.1201/9781003666431-25

and industrial/organizational (I/O) psychology both incorporate human performance concepts, yet operationalize the concept differently.

Within human factors engineering, human performance is typically viewed as a function of cognitive processes such as attention, memory, and situational awareness. By contrast, I/O psychology perspectives view human performance as a function of motivation-based processes such as engagement and job satisfaction. These different approaches can produce a partial and fragmented view of human performance. Scholars and practitioners alike have called for more integrative approaches that overcome "artificial boundaries" (Clegg et al., 2000, p. 240) between disciplines and help to navigate across diverse human performance perspectives (e.g., Baxter & Sommerville, 2011; Older et al., 1997).

We propose that variation in focal timeframes is a key reason human performance is conceptualized and operationalized differently. In other words, depending on the timeframe of the interaction between worker and system, what performance looks like may vary significantly. For example, performance over a short period of interaction with a system involves operators attending to and completing immediate work tasks. By contrast, performance over more extended periods of interaction with a system, such as over several weeks, involves operators maintaining an ongoing baseline level of energy and productivity. Unfortunately, these temporal distinctions are rarely unpacked or discussed in sociotechnical theory, methods, and approaches (Older et al., 1997). In this chapter, we explore three different timeframes of human performance (i.e., short-term, mid-term, and long-term) that help to provide a more integrated perspective across varying disciplines and bodies of research.

We describe the framework and show how the three timeframes help integrate existing literature bodies. This is followed by a discussion of how the framework can inform different aspects of system design, including how they link to system design goals and can contribute to more systematic and comprehensive approaches to determining human criteria. We conclude with directions for future research and application.

A FRAMEWORK FOR CONCEPTUALIZING HUMAN PERFORMANCE IN SOCIOTECHNICAL SYSTEMS

Below, we present a time-based framework of human performance that provides an integrative view across diverse domains of system design (see Figure 21.1). This framework is structured into three timeframes: (1) short-term performance, (2) mid-term performance, and (3) long-term performance. Each timeframe identifies the performance processes that empirical and/or theoretical literature suggests should vary and impact performance outcomes within that timeframe. The three timeframes also reflect inherent differences in the focus and methodologies of varied research domains. To this end, we provide examples of the key research domains relevant to each timeframe. It is important to note that these examples are not exhaustive, and some specific areas of inquiry within broader research domains are relevant to, and can overlap, multiple timeframes.

Human Performance Timeframes					
Short-term performance (over a task to a workday)		**Medium-term performance (over multiple days of a roster, operation, or mission)**		**Long-term performance (over multiple operations/missions to a whole career)**	
Example Processes	**Example key research domains**	**Example Processes**	**Example key research domains**	**Example Processes**	**Example key research domains**
• Situation awareness • Acute fatigue • Attention • Perception • Decision-making • Working memory • Subjective workload • Information processing	• Cognitive psychology • Human factors engineering • Human-computer interaction	• Cumulative psychological/emotional strain or fatigue • Cumulative physical strain or fatigue • Operational readiness and resilience • Motivation	• Occupational and health psychology • Industrial/ organisational psychology • Sleep science • Physicalergonomics • Macro-ergonomics • Occupational biomechanics • Industrial design	• Mental and physical health (or ill health) • Organisational commitment • Job satisfaction • Turnover intentions • Adaptability • Skill growth and development	• Strategic human resource management • Management and organisational science • Industrial/ organisational psychology • Occupational health and psychology • Macro-ergonomics
Design Goal: To limit human error at any given point in time		**Design Goal:** To support human productivity over a mission or operation		**Design Goal:** To support human sustainability over a career	

FIGURE 21.1
Time-based framework of human performance.

Short-Term Performance

The first timeframe captures "short-term" human performance processes that vary and fluctuate as workers interact with a system throughout a task or a shift (e.g., seconds to hours). Such processes include acute fatigue, subjective workload, situation awareness, working memory, attention, information processing, and decision making. These processes are commonly examined using short task-based laboratory experiments and simulation methods. In these approaches, researchers focus on how performance processes vary over short periods in response to manipulation of task-level factors like interruptions, feedback, information presentation/access, degree of automation, and number of concurrent tasks (Sharit, 2012; Wickens, 2021). For example, researchers have tested and observed how the need to complete multiple concurrent tasks can cause momentary variation in operators' attention, visual perception, and workload, which can immediately increase the likelihood of human error and missed signals (Young et al., 2015).

Both short-term performance processes and the task-level factors that can impact them tend to be the focus of research domains concerned with the interactions between humans and technical systems, such as cognitive psychology, human-factors engineering, and human–computer interaction. These domains share similar fundamental objectives – aiming to reduce human errors and increase safety in complex high-risk systems (e.g., aviation, power plants, maritime platforms) (Chapanis, 1995).

Mid-Term Performance

The second timeframe captures "mid-term" human performance processes which develop and change as workers continuously interact with a system throughout a whole mission, operation, or roster (e.g., days to weeks). Examples include psychological and physiological strain, exhaustion, resilience, job satisfaction, and engagement. These processes are commonly examined using methods that measure or track an individual's experience within a system over time, for example, through surveys, observations, interviews, and/or focus group methodologies. Often, the focus of research at this mid-term timeframe is on how work or job-level factors impact performance trajectories over a mission or operation. Key work or job-level factors include the content/organization of tasks within a job role, relationships, and responsibilities between individuals, work–life balance, shift schedules, and physical environmental factors. For example, extant research has investigated how continuous exposure to elevated levels of work demands in a job is associated with increased psychological exhaustion (Wilson et al., 2021) or how repetitive movements from constant interactions with tools or machines at work can lead to physical injury or trauma (Marras, 2012).

A broad variety of research domains examine mid-term-level processes. Psychological processes (e.g., burnout, engagement) typically fall within the remit of domains of I/O psychology and occupational health psychology (e.g., Sonnentag & Frese, 2003). In contrast, physiological processes (e.g., muscle tension/fatigue) are the focus of domains including physical ergonomics, occupational biomechanics, and industrial design (e.g., Marmaras & Nathanael, 2012). In addition, there are research domains within the broader field of occupational health and medical research that are concerned with both the build-up of psychological and physical strain, for example, sleep science research (e.g., Banks & Dinges, 2007).

Long-Term Performance

The third timeframe captures "long-term" human performance processes that evolve slowly over extended periods, such as over months to years of operations/missions within a career. Examples include mental and physical health (or ill health), organizational commitment, career satisfaction, change readiness, and turnover intentions. Long-term performance processes are investigated using similar methodologies to the medium-term (i.e., surveys, observations, interviews, and focus group methodologies); however, the scope of investigation is usually widened beyond work or job-level factors also to include the impact of organizational-level structures and factors on these long-term processes (e.g., training and career development systems, remuneration systems, organizational change processes). For example, researchers have examined how supportive career development practices may increase workers' organizational commitment and decrease voluntary turnover rates over several years (e.g., Trevor & Nyberg, 2008; Allen et al., 2003).

Key research domains focusing on long-term processes include strategic human resources management, I/O psychology, organization science, occupational health and medicine, and macro-ergonomics.

The Nested Nature of the Timeframes

It is important to note that the three timeframes are not independent, and their interactions impact human performance. While short-term performance processes contribute to immediate threats and failures, mid- and long-term processes contribute to slow system shifts towards danger and failure. For example, the likelihood/rate of attention lapses on a task-to-task basis (short-term) can increase if workers become increasingly exhausted over the multiple days of an operation (mid-term). Similarly, suppose a worker is experiencing a long-term mental health or physical health disorder. In that case, this may increase the likelihood/rate of exhaustion throughout an intensive work operation (mid-term) (for a review, see Craig & Cooper, 1992). In this way, all three timeframes must be considered for characterizing or predicting human performance in a sociotechnical system.

DESIGN IMPLICATIONS FOR SOCIOTECHNICAL SYSTEMS

This framework has important implications for the design of sociotechnical systems. A central tenet of sociotechnical systems theory is that optimal system performance relies on the joint optimization of the technical and social sub-systems (Baxter & Sommerville, 2011). In practice, this means that human performance needs to be considered alongside technological performance in the design process. The following section shows how the time-based framework introduced in this chapter enables a comprehensive and systematic approach to consider human performance in the system design process. First, we explore how the three human performance timeframes align with three distinct sociotechnical system design goals and how articulating these goals supports multidisciplinary collaboration during the design process. Second, we discuss how the timeframes can be used alongside design tools and methods to better inform human criteria for design and evaluation.

Linking Human Performance Timeframes to System Design Goals for Multidisciplinary Collaboration

As demonstrated earlier, many research domains possess diverse perspectives on human performance in complex sociotechnical systems. This multi-disciplinarity can be challenging to navigate because issues in communication and understanding can occur where one discipline does not fully understand what the other disciplines can or aim to do (Bader & Nyce, 1998; Sommerville et al., 1992). As such, there have been calls for approaches that can help to integrate across research domains by mapping diverse perspectives more usefully onto design problems (Clegg et al., 2000; Older et al., 1997). We answer this call by exploring how each of our human performance timeframes supports distinct system design goals: to design systems that (1) limit human error, (2) support human productivity, and (3) support human sustainability. These design goals are explored next.

The short-term human performance timeframe encompasses design goals to *limit human error* at any time. Human error refers to an unintended failure of a purposeful action to meet some implicit or explicit criterion (Whittingham, 2004), and the mitigation/limitation of human error in complex systems is a critical goal because of the immediate and catastrophic safety consequences that human error can have on system performance at any point in time. Many of the performance processes at the short-term timeframe are directly relevant to the mitigation/limitation of human error. This is because variability in processes such as situation awareness, acute fatigue, and attention that exceeds safe thresholds during a task can directly and immediately result in human error. For example, a momentary lapse in attention can fail to detect two converging aircraft (Dismukes & Nowinski, 2007).

The main design goal of the mid-term human performance timeframe is to *support human productivity* over a mission or operation. Key elements of human productivity include sustained performance, consistent accuracy, quality of work, and sustained motivation/morale (Cohen, 1990). This design goal goes beyond performance within safe limits at any given time, emphasizing humans remaining effective and efficient over a whole mission or operation. Ongoing human productivity is a crucial design goal because the overall performance of a sociotechnical system is often linked to achieving goals over an extended duration, for example, a whole spaceflight journey, a wartime deployment, or a rescue mission. In these situations, human workers and operators must respond effectively to routine demands and unanticipated challenges throughout the period (Cham et al., 2021; Krueger, 1989). Many mid-term processes, such as exhaustion, strain, and engagement, are directly relevant to supporting this design goal because they inform a human's capability to maintain ongoing performance.

The long-term human performance timeframe addresses design goals *to support human sustainability* over a career. Key sustainability elements include workers maintaining a state of physical and mental wellness, experiencing continued growth and development, and feeling commitment to their work or organization throughout their careers (Tortia et al., 2022, Parker, 2014). Sustainability over a career is an essential goal because many, if not all, sociotechnical systems rely on a highly skilled human workforce of a sufficient size to enable ongoing operation and sustainment. This can present challenges if a workforce has low retention rates, experiences high rates of ill health, and/or lacks necessary skills.

By linking the human performance timeframes to design goals, it is possible to see how the differing research disciplines and domains relevant to each timeframe address unique but complementary aspects of human performance for system design. This can bridge disparate perspectives and bring together the multidisciplinary expertise required to optimize different aspects of human performance within a sociotechnical system.

The Human Performance Timeframes As a Guide for Systematic and Comprehensive Human Criteria

Human criteria describe the conditions required to support human performance and provide a basis for design decisions. They are essential elements of many design tools, such as scenario tools (e.g., Boeing et al., 2020; Nadin et al., 2001; Hughes et al., 2017) and function allocation tools (e.g., Grote et al., 2000; Clegg et al., 1989). Despite their importance, human criteria are often challenging to establish in a design process because they involve diverse interpretations of human performance from varied research domains and system stakeholders, which can be challenging to reconcile (Baxter & Sommerville, 2011; Clegg, 2000).

Our time-based framework of human performance can assist with establishing an appropriate set of human criteria for system design and/or evaluation. Specifically, the three human performance timeframes can be used alongside existing design tools and methods to ensure human performance is systematically and comprehensively considered. For example, where design tools have a set of predefined human criteria, the predefined criteria can be viewed against the human performance timeframes to examine where the focus of the current criteria lies (i.e., short-term, mid-term, and/or long-term performance), and how one might expand the criteria, or pursue complementary tools to approach human performance more comprehensively. Assessing the comprehensiveness of an existing set of human criteria is essential because many design tools are developed to achieve a specific objective, such as human/machine function allocation, identifying training needs, or evaluating operator workloads. They arise from distinct research traditions and disciplines, often resulting in a narrow definition and operationalization of human performance. For example, many task allocation methods tend to specify human criteria that focus on performance throughout a task, for example, certain threshold levels of workload that impact attention, decision-making, or information processing during a task (e.g., Hoc & Debernard, 2002; Papantonopoulos & Salvendy, 2008). When viewed against the three performance timeframes, it is evident that these tools and criteria relate to short-term human performance and provide limited insight into mid-term or long-term human performance processes, such as fatigue accumulation over an operation or job satisfaction over a career. It is important to note that a narrow definition and operationalization of human performance is not a limitation of a tool or its predefined criteria per se. It is impossible to address all aspects of human performance with a single tool; however, it is essential to recognize what elements of human performance are captured and omitted. This will inform where the criteria could be expanded, or if complementary tools could be employed, to ensure human performance issues are captured comprehensively and accounted for in system design/evaluation.

Another example of how the human performance timeframes can be used alongside system design/evaluation tools is where tools rely on system stakeholder input to generate human criteria. Instead of providing predefined human criteria, some tools involve asking a group of system stakeholders, such as end-users, managers, and technical specialists, to identify and agree on relevant key criteria (e.g., Hughes et al., 2017). Several benefits of this stakeholder-driven approach are that the criteria are tailored to the work environment and facilitate end-user participation and ownership over the design and implementation of systems (Older et al., 1997; Kappelman et al., 2006). However, a challenge posed by this stakeholder-driven approach is that system stakeholders often lack the knowledge or expertise to specify a set of human criteria systematically or comprehensively without guidance. Indeed, research has shown that when non-experts are asked to design hypothetical jobs (e.g., assigning tasks to create jobs), they tend to make design decisions that prioritize efficiency in the short-term but lead to simplified and repetitive jobs, which are poor for motivation and health in the long-term (Parker et al., 2019). To address this challenge, the human performance timeframes can be used as a guiding framework to prompt stakeholder discussions about the conditions supporting human performance across all three. For example, asking end-users to describe the conditions that support or inhibit their situation awareness as they perform a task (i.e., the short-term timeframe), their energy levels and motivation as they endure a roster (i.e., the mid-term timeframe), or their mental and physical wellbeing as they sustain a long career (i.e., the long-term timeframe). In this way, the performance timeframes can inform the collection or establishment of human criteria that are more systematic and comprehensive.

Applying the Human Performance Timeframes Early in the System Lifecycle

Lastly, we recommend applying the time-based framework as a guiding approach from the earliest stages of system design. Integrating these considerations during the conceptual design stage enables a broader scope of human performance needs to be identified and captured as formal design requirements. For instance, the framework can be used to guide risk identification during Early Human Factors Analysis (EHFA) across short-term, mid-term, and long-term horizons. These insights can then be translated into design requirements, evaluation criteria, and/or human factors issues registers (HFIR) that ensure systems are immediately usable during human tasks and mitigate human error, and sustainable and supportive of longer-term human well-being, capability, and functioning.

SUMMARY

Early and careful consideration of human performance is critical to the planning and designing of sociotechnical systems to achieve overall system performance. This consideration involves operationalizing the key aspects of human performance that must be addressed in a system's life cycle. However, this can be challenging in practice because various aspects of human performance are explored across multiple, often disconnected, research perspectives. This means that how human performance is interpreted, evaluated, and optimized varies depending on what perspective is taken.

To address this challenge, we have presented a time-based framework for conceptualizing human performance in sociotechnical systems, consisting of three timeframes – short-term, mid-term, and long-term – that integrate diverse disciplines, including cognitive psychology, human factors engineering, I/O psychology, organization science, and strategic human resource management. This multidisciplinary grounding supports a more systematic, integrated, and comprehensive view of human performance across the system life cycle.

Additionally, it can inform the design process, particularly when used alongside existing methodologies and tools. By clarifying how short-, mid-, and long-term aspects of human performance are interconnected, the framework can support human systems integration practitioners in communicating the value of design decisions that account for immediate and sustained human performance. This is useful when making the case for attention to aspects of human performance that may not yield immediate benefits but are critical to long-term system safety, resilience, and effectiveness. In doing so, the framework supports a more holistic approach to system design – one that considers how human performance and functioning need to be supported across different timeframes of interaction within a system.

REFERENCES

Allen, D. G., Shore, L. M., & Griffeth, R. W. (2003). The role of perceived organizational support and supportive human resource practices in turnover. *Journal of Management*, *29*(1), pp. 99–118.

Bader, G., & Nyce, J. M. (1998). When only the self is real: Theory and practice in the development community. *ACM SIGDOC Asterisk Journal of Computer Documentation*, *22*(1), pp. 5–10.

Banks, S., & Dinges, D. F. (2007). Behavioral and physiological consequences of sleep restriction. *Journal of Clinical Sleep Medicine*, *3*(5), pp. 519–528.

Baxter, G., & Sommerville, I. (2011). Socio-technical systems: From design methods to systems engineering. *Interacting with Computers*, *23*(1), pp. 4–17.

Boeing, A. A., Jorristma, K., Griffin, M. A., & Parker, S. K. (2020). Surfacing the social factors early: A sociotechnical approach to the design of a future submarine. *Australian Journal of Management*, *45*(3), pp. 527–545.

Boy, G. A. (2021). Human–systems integration. In V. P. Glăveanu (Ed.), *The Palgrave encyclopedia of the possible* (pp. 1–11). Palgrave Macmillan. https://doi.org/10.1007/978-3-319-98390-5_13-1ff

Boy, G. A. (2023). An epistemological approach to human systems integration. *Technology in Society*, *74*, p. 102298.

Burgess-Limerick, R., Cotea, C., & Pietrzak, E. (2010). *Human systems integration is worth the money and effort! The argument for the implementation of human systems integration processes in Defence capability acquisition* (Report). Department of Defence.

Carayon, P. (2006). Human factors of complex sociotechnical systems. *Applied ergonomics*, *37*(4), 525–535.

Cham, B. S., Boeing, A. A., Wilson, M. K., Griffin, M. A., & Jorritsma, K. (2021). Endurance in extreme work environments. *Organizational Psychology Review*, *11*(4), pp. 343–364.

Chapanis, A. (1995). *Human factors in system engineering*. Wiley.

Clegg, C., Ravden, S., Corbett, M., & Johnson, G. (1989). Allocating functions in computer-integrated manufacturing: A review and a new method. *Behaviour & Information Technology*, *8*(3), pp. 175–190.

Clegg, C. W. (2000). Sociotechnical principles for system design. *Applied Ergonomics*, *31*(5), pp. 463–477.

Clegg, C. W., Gray, M. O., & Waterson, P. E. (2000). The "Charge of the Byte Brigade" and a socio-technical response. *International Journal of Human–Computer Studies*, *52*(2), pp. 235–251.

Cohen, M. M. (1990). Designing space habitats for human productivity. *SAE Transactions*, *99*(Section 1: Journal of Aerospace, Part 1), pp. 352–364.

Craig, A., & Cooper, R. E. (1992). Symptoms of acute and chronic fatigue. *Handbook of Human Performance*, *3*, pp. 289–339.

Dismukes, R. K., & Nowinski, J. (2007). Prospective memory, concurrent task management, and pilot error. *Attention: From Theory to Practice*, *4*, p. 225.

Grote, G., Ryser, C., Wāler, T., Windischer, A., & Weik, S. (2000). KOMPASS: A method for complementary function allocation in automated work systems. *International Journal of Human–Computer Studies*, *52*(2), pp. 267–287.

Hoc, J. M., & Debernard, S. (2002). Respective demands of task and function allocation on human–machine co-operation design: A psychological approach. *Connection Science*, *14*(4), pp. 283–295.

Hughes, H. P., Clegg, C. W., Bolton, L. E., & Machon, L. C. (2017). Systems scenarios: A tool for facilitating the socio-technical design of work systems. *Ergonomics*, *60*(10), pp. 1319–1335.

Kappelman, L. A., McKeeman, R., & Zhang, L. (2006). Early warning signs of IT project failure: The dominant dozen. *Information Systems Management*, *23*(4), pp. 31–36.

Krueger, G. P. (1989). Sustained work, fatigue, sleep loss, and performance: A review of the issues. *Work & Stress*, *3*(2), pp. 129–141.

Marmaras, N., & Nathanael, D. (2012). Workplace design. In G. Salvendy (Ed.), *Handbook of human factors and ergonomics*, pp. 599–615. Wiley.

Marras, W. S. (2012). Basic biomechanics and workstation design. In G. Salvendy (Ed.), *Handbook of human factors and ergonomics*, pp. 347–381. Wiley.

Nadin, S. J., Waterson, P. E., & Parker, S. K. (2001). Participation in job redesign: An evaluation of the use of a sociotechnical tool and its impact. *Human Factors and Ergonomics in Manufacturing & Service Industries*, *11*(1), pp. 53–69.

Older, M. T., Waterson, P. E., & Clegg, C. W. (1997). A critical assessment of task allocation methods and their applicability. *Ergonomics*, *40*(2), pp. 151–171.

Papantonopoulos, S., & Salvendy, G. (2008). Analytic Cognitive Task Allocation: A decision model for cognitive task allocation. *Theoretical Issues in Ergonomics Science*, *9*(2), pp. 155–185.

Parker, S. K. (2014). Beyond motivation: Job and work design for development, health, ambidexterity, and more. *Annual Review of Psychology*, *65*, pp. 661–691.

Parker, S. K., Andrei, D. M., & Van den Broeck, A. (2019). Poor work design begets poor work design: Capacity and willingness antecedents of individual work design behavior. *Journal of Applied Psychology*, *104*(7), p. 907.

Sharit, J. (2012). Human error and human reliability analysis. In In G. Salvendy (Ed.), *Handbook of human factors and ergonomics*, pp. 734–800. Wiley.

Sommerville, I., Rodden, T., Sawyer, P., & Bentley, R. (1992). Sociologists can be surprisingly useful. *People and Computers VII*, *5*, p. 341.

Sonnentag, S., & Frese, M. (2003). Stress in organizations. In W. C. Borman, D. R. Ilgen, & R. J. Klimoski (Eds.), *Comprehensive Handbook of Psychology*, vol. 12, pp. 453–491. Wiley.

Tortia, E. C., Sacchetti, S., & López-Arceiz, F. J. (2022). A human growth perspective on sustainable HRM practices, worker well-being, and organizational performance. *Sustainability*, *14*(17), 11064.

Trevor, C. O., & Nyberg, A. J. (2008). Keeping your headcount when all about you are losing theirs: Downsizing, voluntary turnover rates, and the moderating role of HR practices. *Academy of Management Journal*, *51*(2), pp. 259–276.

Whittingham, R. (2004). *The blame machine: Why human error causes accidents*. Routledge.

Wickens, C. D., Helton, W. S., Hollands, J. G., & Banbury, S. (2021). *Engineering psychology and human performance*. Routledge.

Wilson, M. K., Ballard, T., Strickland, L., Boeing, A. A., Cham, B., Griffin, M. A., & Jorritsma, K. (2021). Understanding fatigue in a naval submarine: Applying biomathematical models and workload measurement in an intensive longitudinal design. *Applied Ergonomics*, *94*, p. 103412.

Young, M. S., Brookhuis, K. A., Wickens, C. D., & Hancock, P. A. (2015). State of science: Mental workload in ergonomics. *Ergonomics*, *58*(1), pp. 1–17.

Section V

Emerging Technologies and Human Experience

22 The Metaverse and Its Effect on Human Perception and Behavior

Nirit Gavish

INTRODUCTION

Close your eyes and imagine that you can live in several worlds simultaneously and move between these worlds very quickly. Imagine that you can have robotic hands and change your skin color. Imagine there is no gravity, and you can fly or pass through walls. Imagine that a deadly spider can bite you, but you will not be injured, or you can die and return to life. Imagine that you can return to your childhood or experience your life as an older person.

Now, open your eyes. You are not imagining. All the above is happening in the metaverse.

The term "metaverse" was coined by Stephenson (2003) in his science fiction novel, Snow Crash, first published in 1992. The word is a combination of "meta" – beyond, and "verse" – "universe," denoting the next-generation internet in which users, as avatars, can interact with each other and software applications in a three-dimensional (3D) virtual space (Duan, 2021). A virtual space, or a *virtual world*, can be described as an electronic environment that visually mimics complex physical spaces, where people can interact with each other and virtual objects and are represented by animated characters (Bainbridge, 2007). The metaverse concept has attracted attention from many technological companies, such as Apple, Meta, Roblox, NVIDIA, Epic Games, and Microsoft, which are investing in R&D in metaverse-related technologies to make it more cost-effective and widely available.

Although the metaverse concept has existed for many years, the influence of performing many daily tasks in virtual worlds must be comprehensively addressed. Although these virtual environments may seem to mirror reality, one can point to several distinct characteristics that differentiate them from the real environment. These unique characteristics might affect human perception and behavior during interaction in the virtual environment. Nevertheless, perhaps more importantly, these unique characteristics also affect perception and behavior when interacting in the real world. The metaverse is a sociotechnical system by itself, and interacting with it influences the interaction with other real-world sociotechnical systems.

In the current chapter, I will present four different use cases studied in our laboratory in recent years and address these issues. The first use case will address how good we are at monitoring our abilities in the virtual environment, which can be described

DOI: 10.1201/9781003666431-27

as an environment that almost, but not entirely, reproduces the real world. The second use case will explore how we can move in the virtuality continuum to achieve better decision-making and numeracy understanding. The third use case will deal with collaboration with avatars compared to collaboration with humans remotely. The fourth and last use case will evaluate the influence of training cognitive flexibility in the real world by simultaneously being in several virtual worlds. Then, the practical implications of a human-system integration approach when interacting with the metaverse and the real world following that experience will be considered.

USE CASE #1: HOW GOOD ARE WE AT MONITORING OUR ABILITIES IN AN ENVIRONMENT THAT ALMOST, BUT NOT ENTIRELY, REPRODUCES THE REAL WORLD?

Virtual reality uses technologies to replace real-world sensory information with synthetic stimuli such as 3D visual imagery, spatialized sounds, force or tactile feedback, and more (Bowman & McMahan, 2007). The interaction in a virtual environment can almost reproduce a real-world interaction; e.g., walking during a virtual environment interaction using natural body gestures is enabled by six-degree-of-freedom input devices, etc. (Nabiyouni et al., 2015).

Virtual reality systems differ in their level of fidelity, which is the degree to which a system accurately reproduces a real-world experience and its effects (Gerathewohl, 1969). However, all virtual reality systems are in the continuum of having some fidelity to one degree or another (e.g., 3D vision only), to reproduce the real world synthetically. Nevertheless, even when there is very high fidelity, it never perfectly simulates the real world. When considering this unique character of virtual environments – partially reproducing the real world and hence being somewhere between the real world and the computerized world – several interesting questions arise regarding how virtual environments are perceived and manipulated.

In our lab, we have questioned users' ability to perform tasks in a virtual environment on their self-controlled difficulty levels in this environment (Yuviler-Gavish et al., 2021). Controlling the difficulty level of a task in a way that optimizes performance in virtual environments may deteriorate compared to our ability to do so in the real world, because people may not be aware of their capabilities in the virtual environments and of these environments' constraints. For example, the virtual reality translation of moving one's body might not correctly reflect the movement in the real world. An individual trying to grasp an object in a virtual environment may feel clumsy or awkward as they are unaware of the virtual environment's physical constraints, etc. Hence, it is possible that performing tasks in an environment that replicates the real world to a certain degree negatively affects the individual's ability to choose their preferred difficulty level of the task. The question ahead is whether, paradoxically, a fixed difficulty level when interacting with virtual reality systems will lead to better performance than letting users pick their preferred difficulty level or changing difficulty levels during the task.

FIGURE 22.1 The Toy House Boom virtual environment.

We studied this issue using a novel virtual reality game set in a Unity-3D environment, Toy House Boom, using an HTC Vive Pro head-mounted display headset with six degrees of freedom (see Figure 22.1). Gaining points in the game depends on the rate of earning points and on the length of the game. There are three levels of difficulty – easy, medium, and hard. The higher the difficulty level, the greater the point accumulation rate and the risk of terminating the game early. Forty-five participants were randomly assigned to two between-participant groups: the Experimental group, in which individual participants determined their desired level of difficulty in all five game sets, and the Controlled group, in which the level of difficulty was set to medium for all participants in all game sets.

The results demonstrated that the mean difficulty level selected by the Experimental group was lower than medium in the first two sets, and higher than medium in the last three sets. In the previous two sets, participants in this group terminated their games earlier than the allowed time (5 minutes) –2 minutes and 36 seconds for the fourth set and 2 minutes and 27 seconds for the fifth set in average, and they received lower scores – 6299 for the fourth set and 6533 for the fifth set in average, compared to the Controlled group – 3 minutes and 37 seconds with a score of 6481 in average for the fourth set and 3 minutes and 49 seconds with a score of 10,106 for the fifth set. When determining the best game strategy, we found a negative correlation between the selected level for the Experimental group and set duration, which indicates that it might have been better to select the lower levels. It seems that the high levels chosen by the Experimental group led to a deterioration in performance in sets #4 and #5.

In conclusion, letting participants choose their desired difficulty level and allowing them to change difficulty levels during the task did not lead to better performance as found in previous studies in real-world and computerized world scenarios (Alexander et al., 2013; Andrieux et al., 2012; Andrieux et al., 2016; Jaquess et al., 2020; Leiker et al., 2016; Li et al., 2019; Pathania et al., 2019). Quite the reverse – this freedom resulted in participants' poorer performance than participants assigned to the Controlled group.

These results can be explained when examining the unique characteristics of the virtual reality system. Participants have the feeling of interacting with the real world and tend to forget that the virtual environment does not perfectly replicate the real world. As a result, participants might mistakenly judge their abilities as they would when interacting in the real world, and not consider that these may be lower in this unfamiliar environment.

Consider these findings when designing virtual environments for daily use and training; perhaps it is better to limit users' freedom. We used to think that users sought control and self-management in their interactions with systems. However, given this unique characteristic of the virtual environments, more caution is needed. Computerized programs that consider the users' abilities and customize the interaction to the specific users will be more beneficial. These programs can also aid users in progressing with their difficulty level as they gain more experience.

USE CASE #2: HOW CAN WE MOVE IN THE VIRTUALITY CONTINUUM TO ACHIEVE BETTER DECISION-MAKING AND NUMERACY UNDERSTANDING?

One more unique characteristic of virtual environments is that their designers can control their virtuality levels. Different combinations of real and virtual can take place, according to the defined goals. This ability to determine which objects and interactions will be virtual and which will be real opens a spectrum of opportunities for designing the best combination of virtual and real for the tasks at hand.

In 1994, Milgram and Kishino stated, "[t]he next generation telecommunication environment is envisaged to be one which will provide an ideal virtual space with sufficient reality essential for communication." They also introduced the concept of a *Virtuality Continuum*, illustrated in Figure 22.2. *Real environments* are at the left end of the continuum, consisting solely of real objects. At the opposite (right) end of the continuum are *virtual* environments consisting solely of virtual objects, where the user interacts with a completely synthetic world. They termed their envisioned environment *Mixed Reality*. In such environments, both real and virtual objects would be part of the same scene, and the user could interact simultaneously with real and virtual

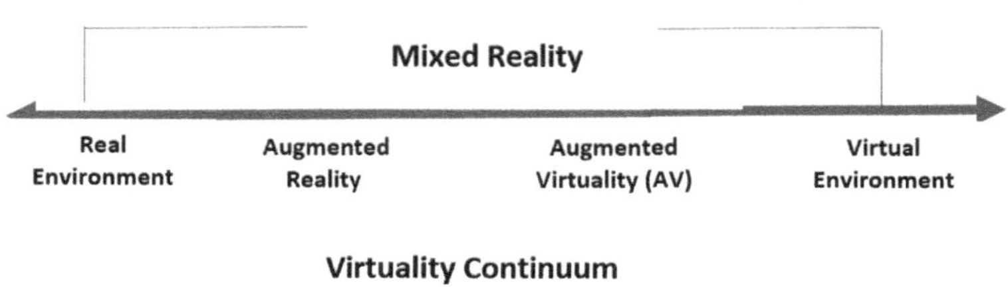

FIGURE 22.2 The virtuality continuum. (Adapted from Milgram & Kishino, 1994.)

FIGURE 22.3 The grocery store simulation with the real coin machine.

worlds. There are two discrete points along the mixed reality continuum: *Augmented Reality*, in which users are situated in the real setting and synthetic objects are added, and augmented virtuality (AV), in which the setting is virtual, and real objects are inserted into this setting.

Decades later, it seems that Milgram and Kishino's (1994) vision has come true, and many different combinations of virtual and real worlds, along with many means of interaction, are available for almost every person and every purpose.

Nonetheless, the challenge of identifying the ideal virtual space with sufficient user balance among the real world, systems, and environments is still unresolved.

In our lab, we questioned whether different levels of virtuality would affect users' decision-making processes when spending money and their numeracy understanding. In the first study (Yuviler-Gavish, Horesh et al., 2022), using a virtual grocery store simulation (see Figure 22.3), we examined whether users of an AV system, in which real coins and a real coin machine are integrated within a virtual system, will be less inclined to spend money (and, as a result, will invest less) compared to users in a virtual system, as the representation of money is more concrete in the AV system. Past research found that the move from concrete representations of money (goods and coins) to abstract representations (bank accounts) led to the general population spending more money (Schor, 1988).

Sixty-five undergraduate students were randomly assigned to two between-participant groups: a Virtual System group, where the coins were virtual, and an AV System group. The results demonstrated that participants in the AV System group invested less money, 31% of the maximum available investment, than participants in the Virtual System group, 68% of the maximum available investment.

These findings are interesting. Based on these research findings, we can cautiously speculate that adding physical objects to the task of buying will probably cause adults to be more careful in their spending behavior. In other words, we can moderate the tendency of overspending when buying with credit cards, smartphones, etc., by adding physical objects that participants will touch or hear (dropping coins into the slot), and so on, to the buying task.

In the second study (Yuviler-Gavish, Treiger et al., 2022), using the same virtual grocery store simulation, we sought ways to improve numeracy understanding among children. While adults base their numeracy understanding on symbolic representations of numbers, i.e., numbers, young children depend more on the non-symbolic representation of numbers when making decisions. Such non-symbolic representation can take several forms, e.g., visual objects, auditory cues (Bijelac et al., 1993), temporal events (Wynn, 1996), and touch (Gimbert et al., 2016). We suggested that using an AV system to represent numbers as physical objects, compared to representing them only visually on a computer screen, will improve children's numeracy decision-making. The hypothesis was evaluated by comparing children's decisions to buy more profitable versus non-profitable products, as well as their total profit, when the coins and the coin machine are real (palpable) (the AV system group), compared with a system in which the coins and the coin machine are virtual (the virtual system group).

The results supported our hypothesis. In the AV system group, the investment proportion for the profitable products, 70% of the maximum available investment, was higher than the proportion for the non-profitable products, 36% of the maximum available investment, and the difference increased as more rounds were completed. Conversely, in the virtual system group, the investment proportion for the non-profitable products, 41% of the maximum available investment, was higher than that for the profitable products, 57% of the maximum available investment. However, the difference disappeared in the later rounds.

Because of the better decision-making of the AV system group, the total profit of this group was higher than that of the virtual system group.

In conclusion, our findings from the two studies suggest the benefit of adding physical objects when interacting with money for better decision-making and a better understanding of numeracy. This was based on the unique characteristic of the mixed reality continuum, which enables a flexible move between concrete and abstract representations of money. Although going back to a concrete representation of money and abandoning the credit system is not possible, designing systems in which the level of virtuality and reality is balanced according to user needs, motivation, etc., can be achieved.

We demonstrated how to design virtual systems to improve decision-making and skills by properly balancing the virtuality and reality. This is only a starting point for addressing the challenge of identifying the ideal virtual space for each task that can be performed. However, the freedom to move along the virtuality continuum when designing the systems should be a key element in the design considerations.

USE CASE #3: IS COLLABORATION WITH AVATARS SIMILAR TO COLLABORATION WITH HUMANS REMOTELY?

In the virtual environment, users can change their physical appearance. They can choose an avatar to represent them, and this avatar can be a human or not, a male or a female, tall or short, with endless options. Users in the virtual environment interact remotely with avatars, which could be very different from real users and not with people. Does this unique characteristic of the virtual environment affect communication and interaction?

The remote communication in virtual reality platforms is being extensively developed by companies such as Apple, Meta, Roblox, NVIDIA, Epic Games, and Microsoft. Within the immense virtual reality world, focus is moving to social virtual reality (SVR). SVR comprises 3D virtual spaces in which several users can interact with one another using virtual reality head-mounted displays (McVeigh-Schultz et al., 2018; Moustafa & Steed, 2018). These virtual spaces contain immersive avatar-mediated communication that gives users a face-to-face experience, and affords a large spectrum of communication modes such as voice, gestures, eye gaze, facial expressions, etc. (Maloney et al., 2020). The aim is to increase the feeling of presence, which is defined as "a psychological state in which virtual objects are experienced as actual objects in either sensory or nonsensory ways" (Lee, 2004, p. 27). Several SVR platforms, such as Facebook Spaces, VRChat, AltspaceVR, Rec Room, and Horizon Worlds, are already used by non-professionals (Maloney et al., 2020; van Brakel et al., 2023).

While video-mediated communication (VMC), such as Zoom, Google Meet, Microsoft Teams, etc., can be interpreted as a reflection of face-to-face communication, with a reduced degree of real-world elements, SVR cannot. VMC is a 2D versus 3D interaction, where only part of the body is observed, and there is a lower resolution for facial expressions, etc. SVR environments, on the other hand, are neither reflections nor reduced-element versions of the real world. They form a different kind of world with different rules. As the rules of the SVR environment are unique and new, their effects on human behavior have been less studied, including the differences or similarities between the factors that influence communication in SVR versus VMC.

In a study performed in our lab (the reference will be updated), we focused on factors affecting the willingness to collaborate in SVR compared to VMC and, more specifically, on the effect of adding a chat option. We used a ride-sharing simulation based on the repeated prisoner's dilemma paradigm (Embrey et al., 2018; Flood, 1958; Lave, 1965; Morehous, 1966; Rapoport & Chammah, 1965). The SVR conditions included using Oculus Quest 2 headsets with six degrees of freedom and two wireless controllers for each headset. Each headset was connected to a PC with a GeForce GTX 1080 Ti graphics card. Participants saw their partner as a 3D avatar whose body movements were natural. In the VMC conditions, we used the Zoom application installed on each of the two PCs, with the voice option disabled (see Figure 22.4).

The findings were that collaboration rates were higher for both the SVR and VMC communication methods, with the chat option at 72% and 63%, respectively, compared to those without it, at 59% and 54%, respectively (see Figure 22.5). Additional t-tests performed separately on the SVR and VMC findings revealed a significant difference between individual collaboration rates with the chat option, which was only for the SVR groups, but not for the VMC groups. To conclude, when trying to improve the willingness to collaborate, both communication methods, SVR and VMC, benefit from adding a chat option, and it seems that SVR benefits more.

Is collaboration with avatars similar to collaboration with humans remotely? Although they might share some features, some considerations differentiate them. For example, chatting with avatars is crucial but might be less critical in remote collaboration with humans. It is essential to study the design factors that encourage or discourage the willingness to collaborate in the virtual environment.

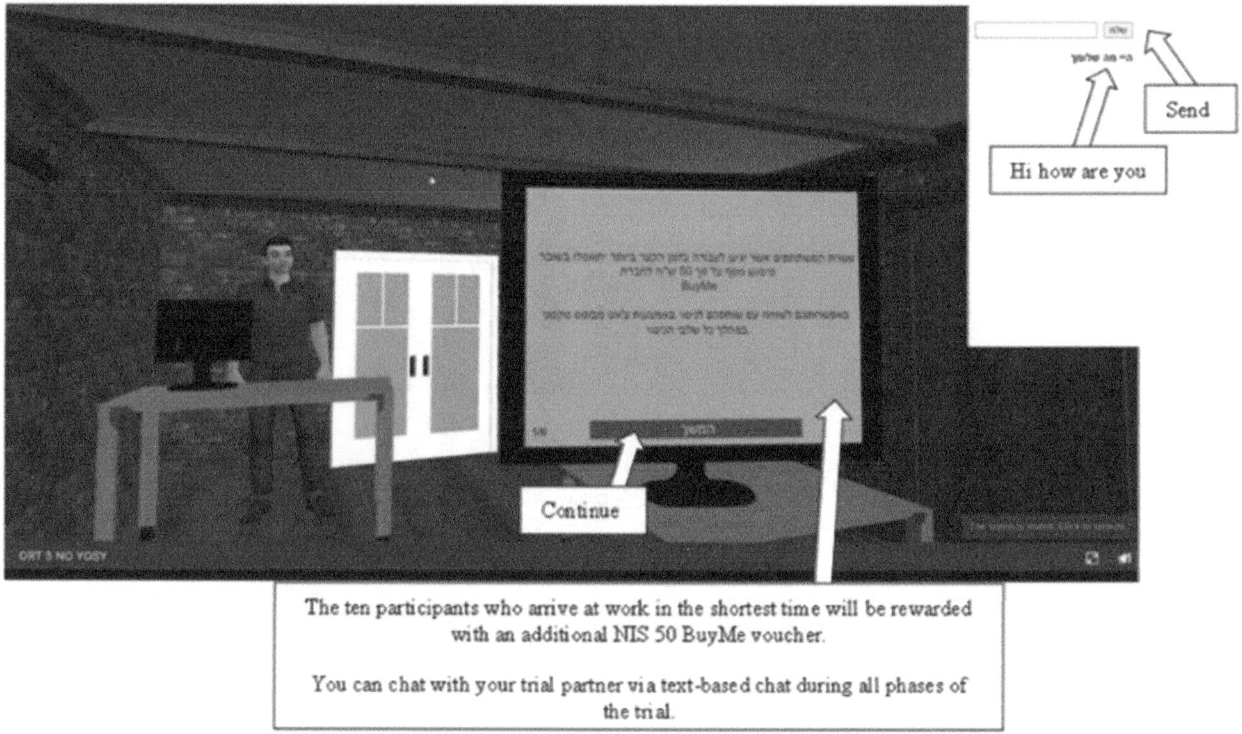

FIGURE 22.4 The ride-sharing simulation with the chat option in the SVR condition.

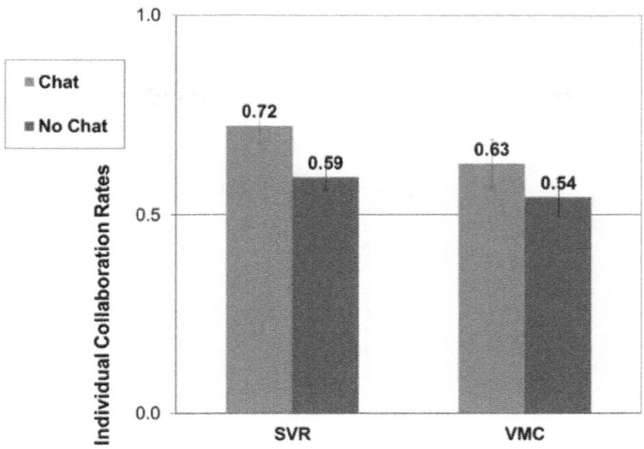

FIGURE 22.5 Mean collaboration rates for each group (with standard error bars).

USE CASE #4: CAN WE TRAIN COGNITIVE FLEXIBILITY SIMULTANEOUSLY IN SEVERAL VIRTUAL WORLDS?

In a virtual environment, humans can be in several worlds simultaneously, e.g., in a virtual world and the real world, in several virtual worlds, and can move between these worlds very quickly. In a famous TED2016 talk, Alex Kuperman introduced Microsoft HoloLens and demonstrated a "holographic teleportation." He teleported a hologram of Dr. Jeffrey Norris, who was exploring Mars in a VR environment in a NASA laboratory, to the stage where he spoke. When he asked Dr. Norris where he was, Dr. Norris replied: "*Well, I am actually in three places. I am standing in a room across the street, while I am standing on this stage with you, while I am standing on Mars, a hundred million miles away*" (TED2016 talks, 2016).

Being in several virtual worlds simultaneously is assumed to be a standard way of living for future metaverse citizens.

The ability to be in several virtual worlds simultaneously can make us perceive our environment and ourselves differently. In addition, we can use this ability to train and acquire the necessary skills.

We examined simultaneously the effect of being in several virtual worlds on cognitive flexibility training in an assembly task (Yuviler-Gavish et al., 2023). We used a novel low-fidelity virtual reality system named BLOCKS6 to train participants in three between-participants groups, 29 participants in each group: a One World group, a Two Worlds group, and a Nine Worlds group. Following the training, participants were required to assemble the model by switching between different assembly segments, and their cognitive flexibility was measured by the duration of the switching time between the segments and the mean number of errors that were made in the first step in each segment after switching to it (see Figure 22.6).

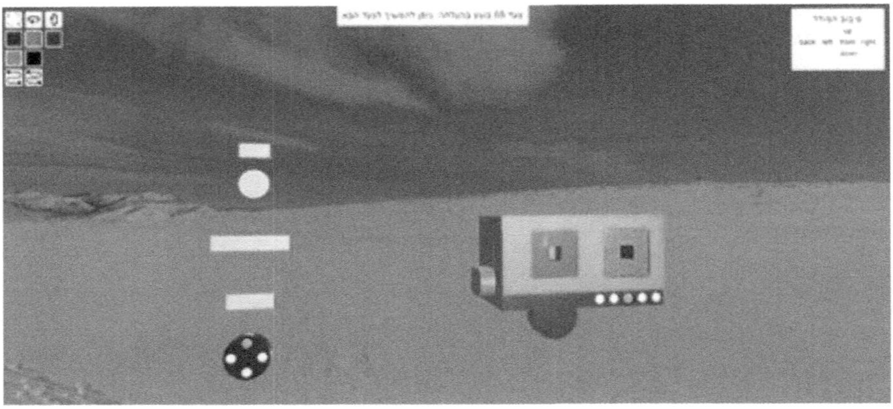

FIGURE 22.6 The virtual reality system with the assembly task.

The results demonstrated that only the Nine Worlds group had a significantly longer training time, 34 minutes and 54 seconds on average, compared to the One World group, 27 minutes and 54 seconds. This was not true for the Two Worlds group, which averaged 31 minutes and 30 seconds, versus the One World group. Hence, only the Nine Worlds group paid a price in training time for being in several virtual worlds simultaneously.

This price, it appears, was not offset by enhanced cognitive flexibility, as we saw that in the transfer phase, the Nine Worlds group did not have an advantage over the One World group in switching time, 17 seconds on average compared to 19 seconds on average, respectively. Interestingly, although the Two Worlds group did not spend more time training than the One World group, its performance in the transfer phase differed. The switching time for the Two Worlds group was shorter, 13 seconds on average, compared to the other groups.

In light of our findings, can being in several virtual worlds simultaneously and moving quickly between them improve cognitive flexibility? First, it depends on the number of virtual worlds being used. Nine worlds are too many to achieve this goal. Participants could not keep the nine worlds in their minds. At the same time, during training, they could not remember the state of the world before they left it when they came back to this world, and perceived the training phase as just randomly performing different steps without any logic. The situation was very different for the Two Worlds group. Indeed, the correct number of virtual worlds that would help improve trainees' cognitive flexibility along the continuum between two and nine should be further examined.

The unique characteristic of the virtual environment is the ability to be in several virtual worlds. It is hard to imagine how we will feel when interacting in such a manner as we are unfamiliar with it in the real world. We only started to study how it affects our perception, and whether, and how, we can use it for training and skills acquisition.

CONCLUSIONS: LESSONS LEARNED FOR DESIGNING VIRTUAL WORLDS TOWARDS A BETTER REAL WORLD

Virtual environments are not just a reflection of the real world. They have many distinct characteristics that differentiate them from the real environment. Given its unique characteristics, we can refer to the metaverse as a sociotechnical system and ask what design considerations should be applied in this system. In addition, the interaction with the virtual environment has a long-term effect on our learning, perception, and behavior. Hence, we can also ask what the design considerations of the virtual environments are to improve our lives when interacting with the real-world sociotechnical systems. Several theoretical and practical implications can be drawn from the use cases studied in our lab.

When examining the character of the virtual environment of partially reproducing the real world, we learned that users in the virtual environment are less likely to evaluate their abilities and control the necessary difficulty level efficiently.

The conclusion can be drawn for designing virtual reality sociotechnical systems. It seems that when designing such systems, it is better to limit the users' freedom to change some features of their interaction by themselves. We usually distinguish between expert users, casual users, and novices in the regular computerized sociotechnical systems. Each kind of user operates the system differently. Expert users, e.g., want an efficient and fast workflow. This freedom might not be effective for virtual reality sociotechnical systems, because users are not aware enough of their capabilities. For these systems, it might be better to leave the control of adopting the system to the user in the hands of the system, not the user.

We suggested that enabling different combinations of real and virtual opens a wide range of opportunities for changing the nature of decision-making and improving skills, as was demonstrated with the AV system for numeracy decisions. Sociotechnical systems that integrate physical components in a virtual environment or virtual components in a physical environment can gain a substantial advantage. Human-system integration experts can advise on the proper combination of virtual and real to achieve the system designers' goals, according to the pros and cons of each environment and components, virtual and real.

When studying the characteristics of the ability of users to change their physical appearance in the virtual environment, it was found that design recommendations for improving collaboration might be different for the virtual environment. Sociotechnical virtual reality systems and real-world or computerized systems may share some characteristics. Hence, human–system integration experts may suggest applying the known design recommendations when designing virtual reality systems. However, the design principles might have different effects when interacting with virtual reality systems, which should be considered.

The unique characteristic of being in several virtual worlds simultaneously enables cognitive flexibility training. Indeed, it is hard for us to imagine how we will feel when we can be in several worlds simultaneously. Will we perceive objects differently? Because objects can be duplicated and exist in many worlds. Will we perceive ourselves differently? We no longer have to decide which environment we prefer to be in at a specific moment; we can be in several. Will we gain more skills, as was demonstrated with cognitive flexibility? Or will we lose other skills, e.g., the ability to focus? The sociotechnical virtual reality environment might affect our behavior and decision-making when interacting with it. However, it might also affect our behavior, perception, skills, etc., when interacting with the real-world sociotechnical systems.

However, the wide range of the various virtual environments' characteristics and their effect on human behavior, perception, understanding, and decision-making is yet to be explored. We are just beginning a very stimulating, exciting, and amazing journey.

REFERENCES

Alexander, J. T., Sear, J., & Oikonomou, A. (2013). An investigation of the effects of game difficulty on player enjoyment. *Entertainment Computing*, *4(1)*, 53–62. https://doi.org/10.1016/j.entcom.2012.09.001

Andrieux, M., Boutin, A., & Thon, B. (2016). Self-control of task difficulty during early practice promotes motor skill learning. *Journal of Motor Behavior*, *48(1)*, 57–65. https://doi.org/10.1080/00222895.2015.1037879

Andrieux, M., Danna, J., & Thon, B. (2012). Self-control of task difficulty during training enhances motor learning of a complex coincidence-anticipation task. *Research Quarterly for Exercise and Sport*, *83(1)*, 27–35. https://doi.org/10.1080/02701367.2012.10599822

Bainbridge, W. S. (2007). The scientific research potential of virtual worlds. *Science*, *317(5837)*, 472–476. https://doi.org/10.1126/science.1146930

Bijelac, R., Bertoncini, J., & Mehler, J. (1993). How do four-day-old infants categorize multisyllabic utterances? *Developmental Psychology*, *29*, 253–269. https://doi.org/10.1037/0012-1649.29.4.711

Bowman, D. A., & McMahan, R. P. (2007). Virtual reality: How much immersion is enough? *Computer*, *40(7)*, 36–43. https://doi.org/10.1109/MC.2007.257

Duan, H., Li, J., Fan, S., Lin, Z., Wu, X., & Cai, W. (2021, October). Metaverse for social good: A university campus prototype. In *Proceedings of the 29th ACM International Conference on Multimedia* (pp. 153–161). https://doi.org/10.1145/3474085.3479238

Embrey, M., Fréchette, G. R., & Yuksel, S. (2018). Cooperation in the finitely repeated prisoner's dilemma. *The Quarterly Journal of Economics*, *133(1)*, 509–551. https://doi.org/10.1093/qje/qjx033

Flood, M. M. (1958). Some experimental games. *Management Science*, *5(1)*, 5–26. https://doi.org/10.1287/mnsc.5.1.5

Gerathewohl, S. J. (1969). *Fidelity of Simulation and Transfer of Training: A Review of the Problem*. Department of Transportation, Federal Aviation Administration, Office of Aviation Medicine, Washington, DC.

Gimbert, F., Gentaz, E., Camos, V., & Mazens, K. (2016). Children's approximate number system in haptic modality. *Perception*, *45*(1–2), 44–55. https://doi.org/10.1177/0301006615614448

Jaquess, K. J., Lu, Y., Iso-Ahola, S. E., Zhang, J., Gentili, R. J., & Hatfield, B. D. (2020). Self-controlled practice to achieve neuro-cognitive engagement: Underlying brain processes to enhance cognitive-motor learning and performance. *Journal of Motor Behavior*, *52(5)*, 544–557. https://doi.org/10.1080/00222895.2019.1651245

Lave, L. B. (1965). Factors affecting co-operation in the prisoner's dilemma. *Behavioral Science, 10(1)*, 26–38. https://doi.org/10.1002/bs.3830100104

Lee, K. M. (2004). Presence, explicated. *Communication Theory, 14(1)*, 27–50. https://doi.org/10.1111/j.1468-2885.2004.tb00302.x

Leiker, A. M., Bruzzi, A. T., Miller, M. W., Nelson, M., Wegman, R., & Lohse, K. R. (2016). The effects of autonomous difficulty selection on engagement, motivation, and learning in a motion-controlled video game task. *Human Movement Science, 49*, 326–335. https://doi.org/10.1016/j.humov.2016.08.005

Li, C., Jia, T., & Ji, L. (2019). Investigation of the effects of game difficulty on the engagement level of patients with brain injury during rehabilitation exercise. In *Tenth International Conference on Signal Processing Systems* (Vol. 11071, p. 110710X). International Society for Optics and Photonics. https://doi.org/10.1117/12.2521688

Maloney, D., Freeman, G., & Wohn, D. Y. (2020). "Talking without a Voice" Understanding non-verbal communication in social virtual reality. In *Proceedings of the ACM on Human–Computer Interaction, 4(CSCW2)*, pp. 1–25. https://doi.org/10.1145/3415246

McVeigh-Schultz, J., Márquez Segura, E., Merrill, N., & Isbister, K. (2018). What does it mean to "Be social" in VR? Mapping the social VR design ecology. In *Proceedings of the 2018 ACM Conference Companion Publication on Designing Interactive Systems* (pp. 289–294). https://doi.org/10.1145/3197391.3205451

Milgram, P., & Kishino, F. (1994). A taxonomy of mixed reality visual displays. *IEICE Transaction on Information and Systems, 77(12)*, 1321–1329.

Morehous, L. G. (1966). One-play, two-play, five-play, and ten-play runs of Prisoner's Dilemma 1. *Journal of Conflict Resolution, 10(3)*, 354–362. https://doi.org/10.1177/002200276601000307

Moustafa, F., & Steed, A. (2018). A longitudinal study of small group interaction in social virtual reality. In *Proceedings of the 24th ACM Symposium on Virtual Reality Software and Technology* (pp. 1–10). https://doi.org/10.1145/3281505.3281527

Nabiyouni, M., Saktheeswaran, A., Bowman, D. A., & Karanth, A. (2015). Comparing the performance of natural, semi-natural, and non-natural locomotion techniques in virtual reality. In *2015 IEEE Symposium on 3D User Interfaces (3DUI)* (pp. 3–10). IEEE. https://doi.org/10.1109/3DUI.2015.7131717

Pathania, A., Leiker, A. M., Euler, M., Miller, M. W., & Lohse, K. R. (2019). Challenge, motivation, and effort: Neural and behavioral correlates of self-control of difficulty during practice. *Biological Psychology, 141*, 52–63. https://doi.org/10.1016/j.biopsycho.2019.01.001

Rapoport, A., & Chammah, A. M. (1965). *Prisoner's Dilemma*. Ann Arbor: University of Michigan Press. https://doi.org/10.3998/mpub.20269

Schor, J. B. (1998). *The Overspent American: Upscaling, Downshifting, and the New Consumer*. New York: Basic Books.

Stephenson, N. (2003). *Snow Crash: A Novel*. New York, USA, Spectra.

TED2016 talks. (2016). A Futuristic Vision of the Age of Holograms. www.ted.com/talks/alex_kipman_a_futuristic_vision_of_the_age_of_holograms

van Brakel, V., Barreda-Ángeles, M., & Hartmann, T. (2023). Feelings of presence and perceived social support in social virtual reality platforms. *Computers in Human Behavior, 139*, 107523. https://doi.org/10.1016/j.chb.2022.107523

Wynn, K. (1996). Infants' individuation and enumeration of actions. *Psychological Science, 7(3)*, 164–169. https://doi.org/10.1111/j.1467-9280.1996.tb00350.x

Yuviler-Gavish, N., Horesh, E., Shamilov, E., Krisher, H., & Admoni, L. (2022). The effect of augmented virtuality on financial decision-making among adults and children. *Virtual Reality*, *26*, 1001–1008. https://doi.org/0.1007/s10055- 021-00610-6

Yuviler-Gavish, N., Muati, O., & Gabbay, B. H. (2024). How performing several task segments simultaneously affects cognitive flexibility training in an assembly task. *Interactive Learning Environments*, 7, 3181–3192. https://doi.org/10.1080/10494820.2023.2170420

Yuviler-Gavish, N., Ozery, E., & Tarazi, E. (2021). The effect of self-controlling difficulty levels in a virtual environment. *PRESENCE: Virtual and Augmented Reality*, *30*, 185–201. https://doi.org/10.1162/pres_a_00381

Yuviler-Gavish, N., Treiger, Z., Horesh, E., & Shamilov, E. (2022). Augmented virtuality systems as a tool for improving numeracy decision-making among children. *Simulation & Gaming*, *53(4)*, 317–334. https://doi.org/10.1177/10468781221099358

23 Tangible Technology for the Potential Diversity of Users

How to Reach the Goal?

Anabela Simoes

INTRODUCTION

The modern society, characterized by fast technological development, is imposing new adaptations to new technologies introduced in different contexts of our daily life. Such technologies' diversity and importance in realizing each one's communication and mobility needs require information, education, and public awareness. In some professional contexts, technological development leads to essential and fast changes, imposing new working methods, procedures, and regulations. This imposes specific training to ensure safety and reach the targeted goals. In a daily life context, it is recognized that some technologies and automation are changing our life patterns.

Such technological challenges direct every society's desired social, economic, and cultural development and should be welcomed. However, such changes are viewed, accepted, and adopted differently depending on the potential users' "*a priori*" attitudes and behaviors regarding a particular technology, such as a smartphone. In this case, the younger generations are so open to buying, trying to understand its functioning, and starting to use it that the older generations' attitude is different. On the one hand, the technological evolution is so fast that older potential users cannot keep up with such change, so they prefer to keep the old technology and extend its use for a long time. However, when holding ancient equipment approaching the end of its life cycle, they agree to buy a new one. However, they ask the new generation to help them choose easier equipment and perform the required operations until they are prepared to use it. This happens at home with technologies in daily usage, with one's car, and even when using a ticket machine to access a public transport mode.

We should look at similar situations and understand that, in each case, such technology is not as user-friendly as desired. Older people, particularly those already retired from their jobs, are not keeping up with the technological evolution and are unprepared for the fast changes. Furthermore, the technological development was not so fast during their working life, and they lived comfortably with the tools they had to use on their job, having been competent throughout their working life. Thus, we should ask ourselves about the design quality of some technological systems,

DOI: 10.1201/9781003666431-28

particularly their interfaces. Does the potential user feel that their characteristics, wills, or limits were considered in the system design?

A great variety of users should be considered when designing technologies for both professional and utilitarian users. It is known that aging is accompanied by some sensorial, perceptive, cognitive, and motor declines, and it does not start at a fixed age. Furthermore, aging is a differentiated process that depends on everyone's lifestyle, health condition evolution, and working life conditions, among other factors influencing the aging process. Another aspect to consider regarding people's diversity is focusing on the younger generation and the current educational system to prepare them for a demanding and dynamic labor market.

Still on the threshold of the third millennium, higher education faces the challenge of being prepared for its mission in a changing world. In this context, the current educational system, designed to respond to an industrial society's academic needs, is rapidly becoming obsolete. Today's society, characterized by critical demographic changes and a rapid and intensive technological development, imposes new educational needs. Furthermore, the increasing frequency of the latest threats and stresses resulting from climate change and/or human action imposes new educational perspectives, particularly in higher education programs.

FROM HUMAN CENTERED-DESIGN TO HUMAN SYSTEM INTEGRATRATION

The mission of designing and developing technological systems for human use imposes a holistic view of the entire process and the definition of the targeted functions for each system. New interfaces should be previewed, designed, developed, and tested accordingly. Meanwhile, instructions or specific training to learn and use the latest technology should be required to ensure both efficacy and efficiency, particularly in some safety-critical operational contexts.

A design approach mainly directed by the human morphology, expressed in the anthropometric variability, was imposed under the Human Factors and Ergonomics (HFE) approach. These data directed the design of physical equipment and different tools for various contexts of use for many years. However, most of these products required later intervention under the same perspective (HFE) to make the necessary corrections for a better adaptation to the human user, towards significantly improving the performance of the previewed tasks. Despite such corrections, the adopted organizational model, expressed in intensive and repetitive work, is responsible for some health issues in industrial contexts. With the technological development and the new digital era allowing for realistic modeling and simulation, a Human-Centered Design (HCD) approach to the technology design was developed, previewing the context of use and the operational system, integrating technology, organization, and people. HCD addresses the main principles and practices for designing and developing technology to meet human needs, promoting a proactive approach. Targeting effective, efficient, and sustainable systems for human use, HCD anticipates, identifies, and prevents potential adverse effects related to human health, safety, and performance using modeling and the human-in-the-loop simulation (Boy et al., 2018).

The tasks previewed and performed in every human interaction with each technology are represented by a behavior that expresses a human activity and its regulation loop from perception to the human decision and action, in coherence with the targeted goals previewed for the system. Thus, the human activity expressed in the corresponding behavior should be understood, which requires in-depth analysis to identify eventual special needs of some user groups or related to the interaction conditions in terms of organizational and/or technological issues.

According to Boy et al. (2018) and Boy (2020), "human capabilities, skills, and needs must be considered early in the design and development process and must be continuously considered throughout the system lifecycle." Thus, the integration of human needs within the system design is formalized through a Human–System Integration (HSI) program, which can reduce costs and increase efficiency, usability, and product quality. With the technological development, the increasing complexity of systems, and the emergence of systems of systems, a well-developed integration and process plan is required to deal with the challenges arising from managing multiple sub-systems and/or interacting with their automated sub-systems. Thus, the integration of human needs and user variability within the system design is formalized through an HSI program.

The interest of young students in technology is known. Still, it ends up being used as a tool of the educational process, which makes teaching monotonous today. It potentially leads to diverting students' attention to other areas of interest, in which they spend time that is not useful in the frame of the educational process. Concerning other ages, particularly older users, they avoid using new technologies designed and developed to assist the user and ease the performance of related tasks.

Nobody wins with this, not even the young generations, who are highly motivated by technology, but not sufficiently guided to discover their potential, nor by the educational system at all levels, with a growing failure resulting from its lack of adaptation to the natural motivations of young students. In what concerns the older potential users, they avoid and even refuse to use a technology that was supposed to improve safety in a particular context. It is impossible to keep our eyes closed, but there are urgent challenges to embrace and changes to make.

Human–Systems Integration

As defined by the NASA HSI Community of Practice, HSI refers to required interdisciplinary integration of the human as an element of a system to ensure that the human and both software/hardware components cooperate, coordinate, and communicate effectively to successfully perform a specific function or mission (Holbrook et al., 2021).

The ISO/IEC Joint Technical Committee defined the HSI concept as "*an approach to systems design and development that aims at making interactive systems more usable by focusing on the system's use, applying human factors/ergonomics, usability knowledge, and techniques.*" HSI emerged from the need to consider human factors during the whole life cycle of any system for human use. However, the design and development of every product started with Systems Engineering (SE), and the users had to be helped to adapt to the developed technology. This adaptation evolved progressively from designing user interfaces and operation procedures to support the human–system interaction, and later to HSI as the result of the SE and HCD merging effects (Boy & Narkevicius, 2013). Thus, HSI is typically achieved using an "*agile systems engineering*" approach, consisting of an incremental cycle that includes rapid prototyping using advanced tools, testing, analysis, and redesign. Such adaptation went progressively from the design of user interfaces and operation procedures, to support human–machine interaction (HMI), and now to HSI. HCD is now possible using very realistic modeling and simulation capabilities, allowing for running human-in-the-loop simulation (HITLS) very early, so emerging HSI properties can be identified and considered in the design, development, testing, verification, and validation. This participatory design can be achieved with professional experts from the beginning of the design and development process, adopting an incremental cycle that includes rapid prototyping using advanced tools, testing, analysis, and redesign.

The fast evolution of technology and the diversity of its applications highlight the importance of HSI concept. If, on one hand, there is an increasing diversity of application contexts, from professional to utilitarian, or leisure ones, some societal issues require to explore aging-related effects related to different application contexts; on the other hand, the aging process, particularly the individual age-related declines, should be addressed as well in the context of using recent technology in different conditions of interaction to perform a task imposing some workload resulting from the balance between everyone's resources and the task difficulty level.

Being HSI, a natural evolution of HCD, it offers research opportunities targeting the best quality systems to be designed, developed, tested, and validated in human–system interactions. As a broad approach, HSI seeks to understand how human operators interact with technology and how to design systems that enhance human performance while minimizing errors and accidents (Ahmed, 2022).

THE TECHNOLOGICAL DEVELOPMENT AND THE AUTOMATION ERA

In this era of change, the human capital needs are evolving quickly, imposing new challenges and perspectives on the new generations to be qualified in the current labor market. The most advanced industries, like the automotive and transport sectors, need to attract and keep a skilled and creative workforce, with the motivation and ability to learn throughout life, being comfortable with fast change, and able to work in an interactive and collaborative environment to design and manage highly complex sociotechnical systems. In the above-referred industrial sectors, the new challenges of this technological era call for educational areas and models that optimize and value soft skills, like cognitive flexibility, intellectual curiosity, critical thinking, adaptability to change, initiative, creativity, and efficient communication in a digital environment. These are competencies and skills for project management and problem solving, allowing for understanding human-system interaction and cooperation, communication skills in multidisciplinary teamwork, adaptability to changes, and a great motivation to keep learning throughout life.

Based on such principles, some questions arise:

1. Were the existing automated vehicles circulating on our roads developed under an HSI design and development approach?
2. Are their drivers aware of the vehicle's technology limits?
3. Have the accidents occurred been explained?
4. The system's maturity level has been discussed and referred to in explaining such accidents.

Too many issues must be explained to use the available technology appropriately, and the maturity level should be a primary requirement. Another issue requiring research regards the limits of technology and the users' or operators' awareness of such risks. Recent literature discusses the main problems related to the broad adoption of fully automated driving, particularly the vehicle technology's level of maturity (Bagloee et al., 2016) and its constraints and limits. These issues introduce new challenges regarding accuracy, reliability, and the driver's trust in advanced technology vehicles.

In-Vehicle Technology and Age-Related Attitudes and Behaviors

In-vehicle technologies (IVTs) have the potential to provide safe mobility to drivers, supporting them in navigation, hazard perception, diagnosis of the situation, and even decision-making and action. Suppose younger and experienced drivers find these devices useful, intuitive, and easy to use. In that case, older drivers do not have the same positive attitude regarding such innovation, expressing difficulties understanding its usefulness. Several studies identified the most relevant technologies to support older drivers and the benefits of their use in driving task performance. As an innovative technology, IVT requires adequate design to avoid the harmful outcomes of well-intentioned but poorly designed technology. Even though it is easy to use and somewhat intuitive, IVT might need some training, particularly for older drivers, to safely use the system and understand how to cope with technology limits. When using IVT, drivers must understand the appropriate level of trust to place in the system, as early research has shown that sometimes people place more trust in a system than it is designed to handle (Okola & Walton, 2003).

Driving a car with assistive technologies will require behavioral adaptation, which will occur with experience and progressive trust in the technology. Furthermore, integrating different technologies must ensure they work together instead of fighting for the driver's attention and giving them conflicting information. Therefore, the assistance provided by the technology should not have any negative safety consequences on other elements of the driving task or the other drivers' behavior. Unless the system interface and the previewed interactions express a HSI in their design and development, they have the potential to overload and confuse the older driver.

As a synthesis, it should be highlighted that:

- IVT has excellent potential to help older drivers perform some identified subtasks to achieve safe and automated mobility.

- The integration of several technologies in the vehicle should not overload the driver.
- A training activity is necessary to ease the required behavioral adaptation, making the technology effective, efficient, and safe.
- Some attention should be paid to the risk of individual skill loss resulting from delegating decisions and actions to the system; the assistive technologies act as a co-pilot, and even automated vehicles require an able and competent driver in the car. Age-related functional declines ease the process of skill loss if attention is not paid to this natural process.

The Advanced Driver Assistance Systems (ADAS) and the Older Users' Acceptance

In recent focus group sessions organized in the frame of a research project addressing the effects of new in-vehicle technologies on the truck drivers' activity and behavior, the older drivers expressed their disagreement about the Adaptive Cruise Control (ACC) integrated in the most recent trucks. Aiming at understanding their attitudes and behavior, a discussion was conducted to discuss their problem and how it could be resolved.

Being the ACC, a technology that assists the driver in critical and risky situations, the system acts over the vehicle in motion to keep both the previously configured safe speed and distance to the car ahead. When this one reduces the vehicle speed using a voluntary action, the distance between both cars is reduced, and the ACC acts immediately, reducing the speed to recover the configured distance to the vehicle ahead. Then, the truck driver is surprised by such fast action over the car, and the consequence is a total disagreement and the consequent decision to avoid using the system, using a permanent disconnect. They prefer to control the vehicle without any technological support that interferes with their main activity, driving it.

The ACC's fast reaction, significantly shorter than the driver's reaction time, seems to be the main factor in the expressed disagreement. Even a brief explanation of a new IVT is not enough. The older driver is not expecting to be replaced by the technology acting faster than they could. Information and training are required to ensure that the targeted effects of such technology will be reached. Thus, previous information followed by experimental sessions addressing the technology use should be provided to avoid such unexpected reactions from the system.

Complexity and Difficulty in Task Performance

Complexity and *Difficulty* are two different influences on task performance. High cognitive demands imposed by the technology versus the low resources from the human side, usually represented by the lack of knowledge or experience in using the technology, or the presence of fatigue, lead to human difficulty in interacting with the system. This means that the conditions for the task performance could represent high difficulty for the user.

If this situation occurs while driving a vehicle, the user's low resources are more evident, increasing the difficulty of performing the previewed human-system

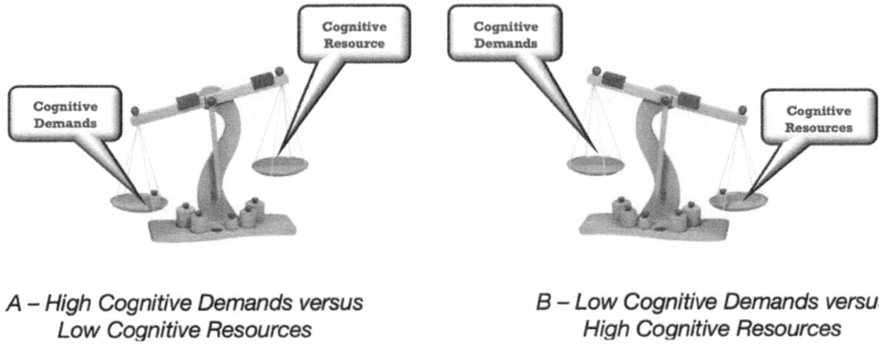

FIGURE 23.1 The balance between the user's cognitive resources and the task demands, which is expressed in Fuller's Task Capability Interface Model.

interaction. Such difficulties represent a high mental workload to humans, particularly resulting from their limited resources and lack of experience in using them. For experienced users, the available human resources are increased, and the human–system interaction is easily and successfully performed (Figure 23.1).

THE ROAD TRANSPORT SYSTEM AND DRIVING AUTOMATION

The road transport system (RTS) is dynamic and complex, integrating human–technology interaction, cooperation, and numerous social interactions in traffic. In opposition to other transport modes (e.g., railway, maritime, aviation), in which highly trained and competent professionals drive or pilot different vehicles from different transport modes, the RTS is open, accommodating an enormous variety of users: drivers, riders, and pedestrians; novice, experienced, young, and old, together with professional drivers from passengers and goods transportation areas. All of them evolve in the road environment, interacting with each other. Such a variety of users and dynamics, just covered by traffic laws and the corresponding supervision, together with the diversity of behaviors in the road environment, lead to increased dynamics, complexity, and uncertainty on the RTS. Considering the whole system as a joint cognitive system, human–technology cooperation should be viewed as being in action within a system in a perfect coupling between humans and technology, which is the basis for the system's efficiency and safety.

THE DRIVER'S ACTIVITY AND THE RELATED REGULATION LOOP

The performance of the driving task imposes on the driver a continuous activity which is developed in a regulation loop (Figure 23.2), involving the following steps (Bellet et al., 2011):

- A perceptual activity involves selecting and collecting relevant information from the road environment and the other road users' behavior.

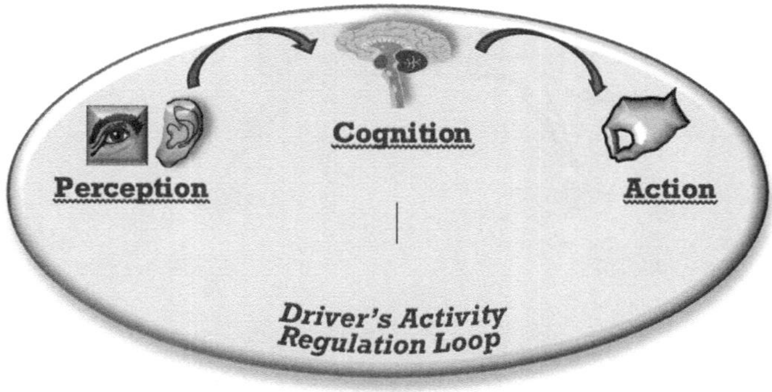

FIGURE 23.2 The Driving Activity Regulation Loop.

- A cognitive activity involving:
  - The interpretation of the collected information allows for the understanding of the current situation,
  - The decision-making process, which is based on the driver's mental representations of the driving situation,
  - The projection of cognitive simulations allows for anticipation.

- A motor activity, expressed in a body movement, to perform the actions that are required for the operational control of the vehicle.

The regulation loop of the driving activity imposes on the driver a continuous information processing requiring an essential level of attention to the driving task to select and collect the relevant information from the environment, preview the evolution of the driving situation, and promptly make the appropriate decision.

The Situation Awareness (SA) concept, introduced by Endsley (1995), plays a vital role in dynamic and complex environments. When moving in a road environment, the driver perceives the picture of each situation in the environment and understands the scenario dynamics and its evolution to project the evolution of the situation, avoiding traffic conflicts and ensuring a safe movement. These complementary phases (perception, comprehension, and projection) of the elements present and detected in the road environment allow for understanding their meaning and making the appropriate decisions to ensure safe driving.

However, some transient factors, like distraction, do not allow for paying the required attention to the dynamics of the road environment. Without paying such attention, the driver understands the situation too late and will not have time to project its evolution and make the appropriate decisions. It is essential to highlight that the SA process is not automated. If drivers are distracted, they may be surprised by an unexpected situation without time for appropriate decisions and actions. This means that SA allows road users to drive, ride, or walk, thus paying attention to the evolution of the road environment and projecting such dynamics to ensure appropriate decisions. Safe driving is so essential that it can be improved and consolidated with the driver's experience.

FINAL CONSIDERATIONS

At the end of this chapter, some final considerations regarding human systems integration in the context of driving automation complete the chapter. Regardless of the automation level, there is a need for good and efficient communication between both agents: the driver and the automation system, and the interaction with each one of the driver-support and safety-related technologies, which together act as a joint cognitive system with both elements required to cooperate in delivering safe and comfortable driving.

To finalize the chapter, some highlights about its scope as a business area, with most professionals approaching their retirement, and a new generation that should be ready to embrace the profession, but without any motivation to do so. A massive jump to the present, expecting the future, will be necessary to bridge the differences, understand them, and save the freight industry, welcoming the renewed profession, and embracing it with pleasure and a massive hope in the future. Thus, education and training throughout life represent the right ways to understand the changing present and adapt to changes, reaching the goal of using the humanized available technology to improve efficiency and safety on the job and in daily life.

BIBLIOGRAPHY

Ahmed, J. (2022). Emotional intelligence and dangerous driving behaviors (Doctoral dissertation, Montana State University-Bozeman, College of Engineering).

Ahmed, J. & Miller, E.E. (2023). Effects of traffic and road contextual language on survey design: Comparing two emotional intelligence scales. In *Proceedings of the Human Factors and Ergonomics Society Annual Meeting* (Vol. 67, No. 1, pp. 1232–1238). SAGE Publications, Los Angeles, CA.

Apesteguia, J., Ballester, M.A. & Cuhadaroglu, T. (2023). A behavioral model of adaptation. *Journal of Economic Behavior and Organization*, 207. www.sciencedirect.com/science/article/pii/S0167268123000082#:~:text=Denoteby%CE%94thesimplexofdimensionK.6Theinitialbehavioroftheindividualisdenotedbyx0%E2%88%88%CE%94,whiletheidealresponseisdenotedbyx*%E2%88%88%CE%94.Weassumepolynomialcost

Bagloee, S.A., Tavana, M., Asadi, M. & Oliver, T. (2016). Autonomous vehicles: Challenges, opportunities, and future implications for transportation policies. *Journal of Modern Transportation*, 24, 284–303.

Bellet, T., Mayenobe, P., Bornard, J.-C., Paris, J.-C., Gruyer, D. & Claverie, B. (2011). Human driver modelling and simulation in a virtual road environment. In Cacciabue, P-C., Hjalmdahl, M., Ludtke, A. & Riccioli, C. (eds.), *Human Modelling in Assisted Transportation: Models, Tools and Risk Methods* (pp. 251–262). Springer, Milan.

Boy, G.A. (2013). *Orchestrating Human-Centered Design*. Springer, London.

Boy, G.A. (2020). *Human–Systems Integration: From Virtual to Tangible*. CRC Press, T&F, New York.

Boy, G.A., Doule, O., Kiss, De Vere, M.K. & Mehta, Y. (2018). Human–systems integration verification principles for commercial space transportation. *New Space*, 53–64. https://doi.org/10.1089/space.2017.0040

Boy, G.A. & Narkevicius, J. (2013). Unifying human centered design and systems engineering for human systems integration. In Aiguier, M. (ed.), *Complex Systems Design and Management*. Springer, Cham. ISBN-13: 978-3-319-02811-8.

Carsten, O. & Martens, M.H. (2019). How can humans understand their automated cars? HMI principles, problems, and solutions. *Cognition, Technology & Work*, 21, 3–20. https://doi.org/10.1007/s10111-018-0484-0

Davidse, R.J. (2006). Older drivers and ADAS: Which systems improve road safety? *IATSS Research*, 30(1), 6–20. https://doi.org/10.1016/S0386-1112(14)60151-5

Davidse, R.J. (2007). Assisting the older driver: Intersection design and in-car devices to improve the older driver's safety. SWOV Dissertation. ISBN: 978-90-73946-02-6.

Davidse, R.J., Hagenzieker, M.P., van Wolffelaar, P.C. & Brouwer, W.H. (2009). Effects of in-car support on mental workload and driving performance of older drivers. *Human Factors*, 51(4), 463–476. doi: 10.1518/001872095779049543

Endsley, M.R. (1995). Toward a theory of situation awareness in dynamic systems. *Human Factors*, 37(1), 32–64. doi: 10.1518/001872095779049543

Endsley, M.R. (2017). From here to autonomy: Lessons learned from human–automation research. *Human Factors*, 59(1), 5–27. doi: 10.1177/0018720816681350

Holbrook, J., Novak, B., Rippy, L., Vos, G. & DeVasher, J. (2021). *Human–System Integration*. NASA. https://ntrs.nasa.gov/citations/20210010952

Kulmala, R. & Rama, P. (2013). Definition of behavioural adaptation. In Rudin-Brown, C.M. & Jamson, L. (eds.), *Behavioural Adaptation and Road Safety: Theory, Evidence and Action*. CRC Press, London.

Kyriakidis, M., Happee, R. & de Winter, J.C.F. (2015). Public opinion on automated driving: Results of an international questionnaire among 5000 respondents. *Transportation Research Part F: Traffic Psychology and Behaviour*, 32, 127–140.

Litman, T. (2018). *Autonomous Vehicles Implementation Predictions: Implications for Transport Planning*. Victoria, BC, Canada, Victoria Transport Policy Institute.

OECD (2003). *Road Safety: Impact of New Technologies*. Transport, Organization for Economic Co-Operation and Development, Paris.

Okola, A.R. & Walton, C.M. (2003). *Intelligent Transportation Systems to Improve Elderly Persons' Mobility and Decision Making within Departure Time Choice Framework*. Centre for Transportation Research, University of Texas at Austin.

Robinson, P. (2001). Task complexity, task difficulty, and task production: Exploring interactions in a componential framework. *Applied Linguistics*, 22(1), 27.

Simoes, A. (2017). Older drivers and driving automation. In Stanton, N. (ed.), *Advances in Human Aspects of Transportation. AHFE2017 Conference Proceedings*. Springer, Southampton, UK.

Stokes, G. (1992). *On Being Old: The Psychology of Later Life*. The Falmer Press, London, UK.

Suen, S.L. & Mitchell, C.G.B. (2003). Accessible transportation and mobility. In *TRB 82nd Annual Meeting*. Retrieved in 2005. https://onlinepubs.trb.org/onlinepubs/millennium/00001.pdf

Trommer, S., Kolarova, K., Fraedrich, E., Kroger, L., Kickhofer, B., Kuhnimhof, T. & Lenz, B. (2016). Autonomous driving: The impact of vehicle automation on mobility behaviour. Ifmo, Munich. http://gulliver.trb.org/publications/millennium/00001.pdf

UNESCO (1998). *World Conference on Higher Education: Towards an Agenda 21 for Higher Education*. UNESCO, Paris.

24 Applying a Systems Modeling Language Extension to Human–AI Team Design

Michael E. Miller and Michael C. Todd

INTRODUCTION

Teams are important in today's complex work environments as they include individuals who complement one another to increase cognitive capacity compared to a single individual's (Cooke et al., 2013; Salas et al., 1992).

Additionally, teams provide redundancy, which can reduce the likelihood of errors.

When discussing artificial intelligence (AI), modern literature often assumes the human replacement paradigm, where AI is advertised as replacing human activity. This paradigm ignores that humans and AI have distinct, complementary capabilities.

Further, this literature ignores the importance of integrating multiple human and AI teammates to operate in today's complex and time-sensitive environments. An alternate view of AI is to use it to augment human capabilities. This paradigm, referred to as human–AI agent teaming (Miller et al., 2020; Miller & Rusnock, 2014), human–machine symbiosis (Licklider & Clark, 1962), or augmented cognition (Bates, 2018), assumes that AI should be used to augment human cognition, further expanding the capabilities of humans. However, research and demonstrations in this field illustrate the importance of designing the AI and the interface between AI and human teammates (DeCremer & Kasparov, 2021).

We define a team as two or more agents who train and work together interdependently, with diversified knowledge, skills, and abilities to accomplish work that is complex or critical enough to require knowledge, skill, ability, capacity or redundancy that exceeds what a single agent can provide within the allowed time (Schneider et al., 2021). In Human–AI Teams (HATs), one or more agents are human and AI team members. This definition alludes to the use of teams to complete work within limited time limits, with individual agents who lack the knowledge or ability to perform the desired work, and where the individual agents tend to commit more errors than is tolerable. The demands placed on the team members must depend upon the type of team, which Sundstrom has classified into four categories, including Action, Production, Project, and Advice, based upon the diversity of the teams and their integration with the environment, where the time demands imposed on the team often depend on the degree of environmental integration (Sundstrom et al., 1990). Thus,

DOI: 10.1201/9781003666431-29

Action and Production teams, which are highly integrated with the environment, usually experience work paced by the environment, forcing rapid decision-making. Conversely, Action and Project teams typically require diverse knowledge, forcing the teams to be composed of individuals with specialized and individually unique knowledge, skills, and abilities to respond to the diversity of knowledge necessary to achieve the team's objectives.

The word interdependently within the team definition implies that two or more team members depend upon one another to achieve a goal. Neither team member completes all the tasks necessary to achieve a goal. Instead, the goal is achieved only through the agents' interaction. Interdependence can be classified as hard interdependence, where a team member cannot reach the goal without input from team members, or soft interdependence, where the interaction among agents permits them to achieve the goal more accurately, faster, or safer (Johnson et al., 2014). The type of team imposes different constraints upon this interaction. For example, for teams with high diversity, i.e., Action and Project teams, team members are more likely to experience hard dependencies, as there is little skill or knowledge overlap. When environment integration is high, the team is more likely to experience time constraints, as in Action and Project teams, the environment imposes hard dependencies to perform tasks within time constraints. Importantly, the duration and commonality of work any specific team engages in influences the work's diversity and the team members' ability to transfer their experience in coordinating work activities from one work cycle or situation to another (Sundstrom et al., 1990).

When discussing interdependence between team members, the previous paragraph describes how team members pursue common goals. Humans pursue goals by executing a perceptual loop in which they perceive the environment, assess the state of the environment against a goal, decide upon an action that they anticipate will move the environment toward their goal, and execute the selected action. Humans then iterate this process to achieve their goal. As we describe HATs, it is essential to define AI agents. An agent perceives its environment, applies this information to engage in reasoning relative to a set of goals, and applies this reasoning to drive actions in the environment (Wooldridge, 2013). Without this ability to perceive, act, and reason, the AI cannot participate in the complete perceptual loop, which is necessary to operate interdependently with humans as they complete their perceptual loops. Thus, an AI that does not fulfill the definition of an agent cannot participate as an effective teammate.

Agency also implies the ability to act without intervention, implying that the agent should be capable of autonomous action. Although the term autonomy is sometimes associated with AI agents, we can also apply this term to humans, as we do not regularly provide children with the autonomy to perform specific actions, such as driving. As suggested by Stevenson and colleagues, autonomy implies that someone gives the agent the authority to pursue a goal or take an action. Further, the agent must be available and capable of acting and can act at the desired time (Stevenson et al., 2023). The distinction between capability and capacity is essential and easily explained for humans. You or I might be fully capable of successfully controlling an automobile.

However, if we are distracted by texting, we may not have the physical and cognitive capacity to successfully control the car at the same time we are texting. In this example, because we might be devoting excessive resources to texting, we might not execute the necessary activities to control the automobile thoughtfully. Thus, we are not autonomously executing control of the car.

This introduction indicates that HATs rely on combining the unique knowledge, skills, and abilities that both human and AI agents possess. However, successful HATs require careful design of the interaction among agents within these teams. Further, the requirements for these teams will differ based on the environment and work that the team is required to perform. Thus, the interaction between team members must adapt as the demands from the environment, work, and team member capacity change to provide truly successful teams in dynamic real-world environments. These observations lead us to ask how we reliably and successfully design HATs to thrive in these unpredictable environments. This chapter aims to describe a method and model structure to support the design of HATs.

DESIGN LANGUAGE AND PROCESS FOR HATS

Developing complicated and complex systems, such as HATs, requires effective communication across diverse design and development teams. Recent innovations in Systems Engineering led to the development of Model-Based Systems Engineering (MBSE), which applies common languages, such as Systems Modeling Language (SysML), to structure system descriptions during systems design and sustainment, as well as promote multidisciplinary communication during design (Friedenthal et al., 2014). This chapter discusses Agent-Oriented Systems Modeling (AOSM), a method and extension to the SysML language to support HAT design (Miller & Rusnock, n.d.; Miller et al., 2020).

The AOSM method includes a total of six steps. This includes defining the scope of the system that the designer wishes to design. The designer then creates and models an understanding of the current or "As-Is" system. With a deep understanding of the "As-Is" system, the designer defines goals for the redesigned or "To-Be" system. The designer then designs the "To-Be" system to achieve the desired redesign. The designer then validates the design using a scenario-based design evaluation method, adjusting the "To-Be" design based on the evaluation. As the designer completes the design, they develop specifications for the AI agents, human training, procedures, and interfaces among system elements.

We apply this method to specify the shaded blocks in Figure 24.1, which shows a SysML Block Definition Diagram (BDD), depicting a metamodel of a HAT, where a system of interest includes one or more HATs.

The system also might optionally include machines that the HAT interacts with to achieve its goals within the system. The system operates in an environment and interacts with the domain, consisting of the organization and peer systems in which it resides. This diagram shows that each HAT comprises two or more agents, including at least one human and one AI agent. AI agents have one or more actuators and sensors to complete the perceptual loop. An interface between the human and AI

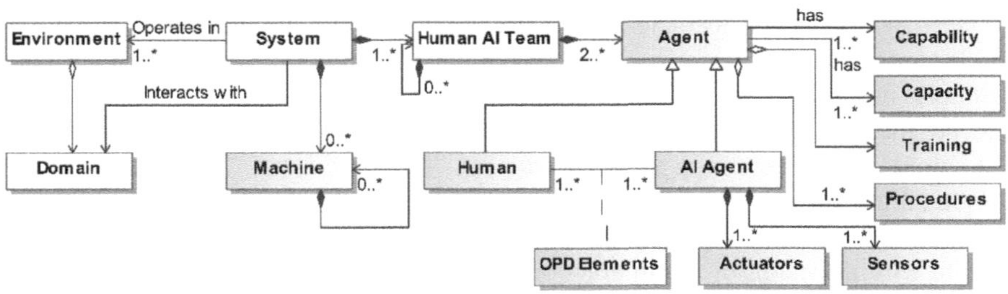

FIGURE 24.1 Structural element of a HAT in AOSM.

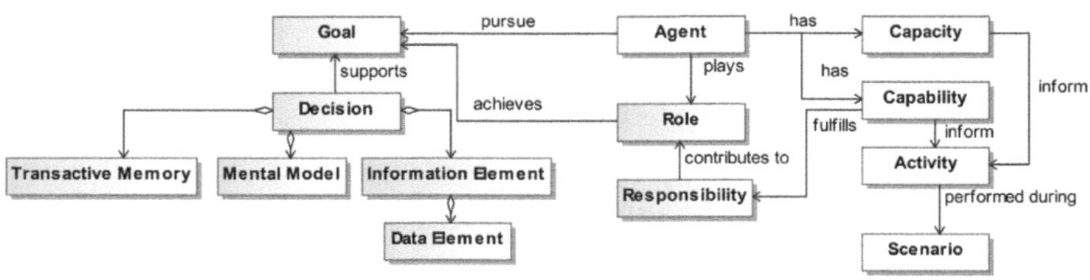

FIGURE 24.2 Behavioral elements of a HAT in AOSM.

agents permits them to observe, predict, and direct each other's behaviors. Each agent has one or more capabilities and capacities, whether human or AI. Further, they have one or more training regimens and procedures that provide the knowledge necessary to perform tasks within the environment and engage in teamwork. Each item highlighted in gray must be understood and specified to design an effective HAT.

When applying the six-step procedure, we can begin by specifying the teams, even if they are comprised of humans without AI counterparts, using this metamodel. However, specifying the existing team's physical attributes is insufficient to permit one to understand the depth of the cognitive work the team is tasked to execute. Therefore, it is essential to characterize the behavior of the HAT further. Figure 24.2 provides a metamodel of the behavioral elements that the AOSM method requires one to capture. This diagram shows that each agent pursues one or more goals as an HAT member. They pursue these goals while playing a role; each role is defined by a collection of responsibilities contributing to it. One can then understand team members' capabilities to fulfill each responsibility. To achieve goals, agents must make decisions. Each agent employs information from the environment, system, or other agents to complete a decision. Each information element might be comprised of multiple raw data elements.

Further, agents rely on their mental models of the environment, machine, and other agents within the team to support decisions. Finally, each agent will build transactive memory, which permits them to understand the capabilities of each agent within their team. The composition of the information they gather from the world and the information in their head that exists in the form of mental models and transactive memory supports decisions and actions of individual team members and the collective team.

The six-step method specifies the elements in gray with their connections to agent capabilities and capacities. It provides an in-depth understanding of the work in the environment for the "As-Is" team and a baseline for the design of the "To-Be" team.

During the design process, the behavioral elements are modified to fulfill the objectives and goals of the redesign, and these elements are used to support the redesign of the HAT. Once the HAT has been redesigned, the designer creates scenarios for applying the "To-Be" system, and the activity of the "To-Be" HAT is imagined for each scenario. This expands the specification of the responsibilities, capabilities, capacities, goals, and associated elements for the "To-Be" team to include further details for what the agent needs to sense, actuate, observe, predict, and direct among team members. That is, the scenarios further highlight and structure the interaction of the team members in the "To-Be" system. By applying the AOSM method using the structured model depicted in Figures 24.1 and 24.2, the "To-Be" HAT system is designed. The design guides the detailed specifications of each agent, as well as the training procedure and system interface.

CASE STUDY IN PROJECT TEAM

To understand the AOSM method and model, it is useful to examine a brief case study. Although the proliferation of digital devices provides many advantages, it also challenges many aspects of society, including the legal system.

Increasingly, legal evidence resides on digital devices, and acquiring this evidence requires robust and timely digital forensics. The quality of digital forensic reports is subjective, so the United States legal system requires a human to account for the evidence recovery process and explain how traces support evidence to prove or disprove an allegation. The shortage of digital forensics experts, increasing caseloads, and pressure to support the defendant's right to a speedy and public trial impose conflicting goals that can compromise the quality of investigations. Overall, these teams have attributes of both action and project teams, with team members (e.g., lawyer, caseworker, digital forensics team lead, and digital forensics expert) having diverse skills, typical of action teams, but remaining as a team with fixed goals, typical of project teams. The environment also drives the timeliness of team responses, thereby indicating tight environmental coupling typical of either team type.

Define the HAT

Like many sociotechnical teams, HATs are open, soft systems. For this reason, an initial step is to define the HAT itself and the domain in which it operates. The domain for this case study might include law enforcement and the court system. As our case will focus on the prosecution, the defense legal and forensics teams are also within the domain but external to our HAT. For this system, external perturbations typical of open systems include the insertion of high-priority cases, which impose requirements for both high quality and timely forensics, potentially to the detriment of the analysis for other cases, and novel digital systems or encryption schemes, which increase the difficulty and time required for analysis. Internal perturbations typical of soft systems include each team member's knowledge, experience, abilities, and experience working as a team. Although not shown due to limited space, the AOSM method would capture the domain members, their support systems, and a block for the HAT in a BDD. The designer captures the members of the HAT and their support systems in a second BDD. These BDDs, together with an understanding of the use cases they execute, define the scope of the HAT team that we wish to redesign.

Characterize Work

The second step is to characterize the work that the HAT performs. To accomplish this, we begin building an in-depth characterization of the work through cognitive task analysis methods to understand the goal structure of the team and related members within the domain. As is typical in systems analysis, it is essential to understand the goals of the organizations and teams that the HAT supports, as well as the HAT, as the HAT is responsible for delivering upon their goals in a way that they support the larger system, domain, or organization in which they reside (Hitchins, 1992). These goals can be captured in different formats, including hierarchical structures as illustrated in Figure 24.3 or functional abstraction networks (Duncan et al., 2023; Elm et al., 2003). As shown in Figure 24.3, goal representations can also capture the currently responsible agents or the roles associated with the goal in the "As-Is" system. While this hierarchy only shows a portion of the overall goal hierarchy, a few elements are illustrated. The HAT we are designing, referred to as Digital Forensics HAT, supports the goal of prosecuting the accused and has the overall goal of providing digital evidence. The primary responsibility of HAT is to provide digital evidence. However, other goals, as indicated by the light gray goals, including gathering allegations, documenting and gathering evidence, providing a speedy trial, and setting judge search parameters in the hierarchy, influence the HAT's ability to function. How the larger organization interprets, pursues, and executes these goals influences the definition and constraints within which the HAT must perform and is essential to recognize. Further, the HAT must communicate with the entities responsible for achieving these goals to perform the desired work.

When characterizing the work, it is essential to describe it at a high enough level that the description does not contain descriptions that constrain the ability to complete goals and the associated responsibilities, which are specified after the goal description is complete, such that specific classes of humans or current environmental artifacts, e.g., computer systems, are required. This is important, as we do not want to constrain the design to specific technical implementations as we design the "To-Be" system. It is also important to identify goals associated with the task work, such as finding digital evidence, and goals related to teamwork, such as allocating resources and maintaining team morale.

For each goal, we then consider all responsibilities the existing agents must fulfill to achieve this goal. For example, to satisfy the goal "Define Searches," the digital forensics expert might have to determine all keywords (e.g., claimant's name) and activities (e.g., logins to claimant's accounts) that might be associated with each charge. The HAT must search the digital device for content referencing these

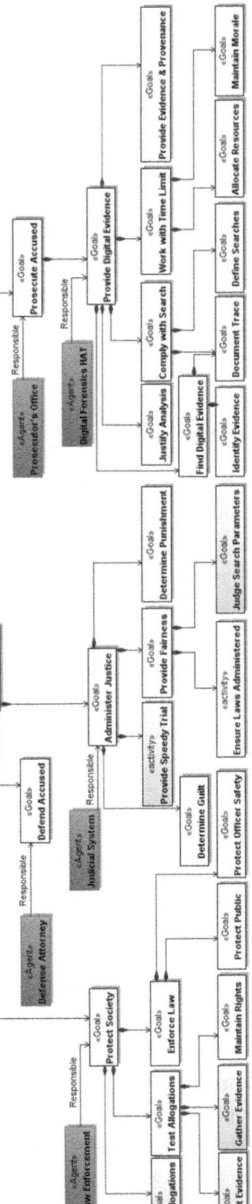

FIGURE 24.3 An example goal hierarchy for the "As-Is" system.

keywords or accounts to identify evidence. Document traces require the team to track the progress of the search, track any hits to understand activity, and log any relevant evidence or counterevidence associated with each search. Other goals, such as justify analysis, which is necessary to demonstrate that the search is consistent with the allowances of any search warrant, require a rationale for analysis steps to be associated with a rationale.

As we consider responsibilities, it is essential to consider them as residing in a perceptual loop where each goal should have responsibilities associated with sensing or perceiving information, interpreting that information, forming a decision, and taking an action. During analysis of the "As-Is" system, the designer can evaluate the responsibilities associated with each goal to ensure that they capture responsibilities encompassing each step in the perceptual loop necessary to fulfill each goal. In a human team, which is often present in today's systems, each of these responsibilities is fulfilled by a human, with the assistance of simple tools, such as databases or manual search tools.

Establish "To-Be" System Goals

Having characterized the system, the next step is to establish development goals for the "To-Be" system. As we have already noted, forensics experts are often the bottleneck in the process, as they are in high demand and usually need to complete a large amount of work in a limited time. Further, analysis of this system provides insight that the case worker and the team lead have goals that are associated with reducing the communication burden on the attorney and the forensics expert, thus it is reasonable to establish design goals related to reducing the work of the forensics expert and enhancing communication among the team, but particularly for the attorney and the forensics expert.

Design "To-Be" System

To achieve these design goals, the designer analyzes each of the goals and responsibilities to determine if the goals and responsibilities need to be modified or augmented. Then the designer develops the underlying capabilities that an entity must have to complete each responsibility. Once all capabilities have been determined, the designer assesses each capability to determine if it is reasonable to require a human or an AI agent to execute each responsibility. We then group the capabilities into a two-dimensional grid with roles, which are collections of capabilities requiring similar knowledge in rows, and the capabilities arranged in three columns indicating whether humans, AI agents, or both humans and AI agents can possess these capabilities. The designer then creates arrangements of AI agents and humans that might fulfill the capabilities, indicating information flows. An example of a single agent arrangement for a single role is provided in Figure 24.4, which shows capabilities on the left, a depiction of the agent arrangement and communication in the "As-Is" system, and a potential "To-Be" system that includes modified descriptions of the human agents, added AI agents, and supporting infrastructure. Once completed, the designer synthesizes collages of these arrangements for each role into a larger team diagram.

We then capture which agent (human, AI, or both) in the final design must be able to support each capability. Further, by using associations between capabilities

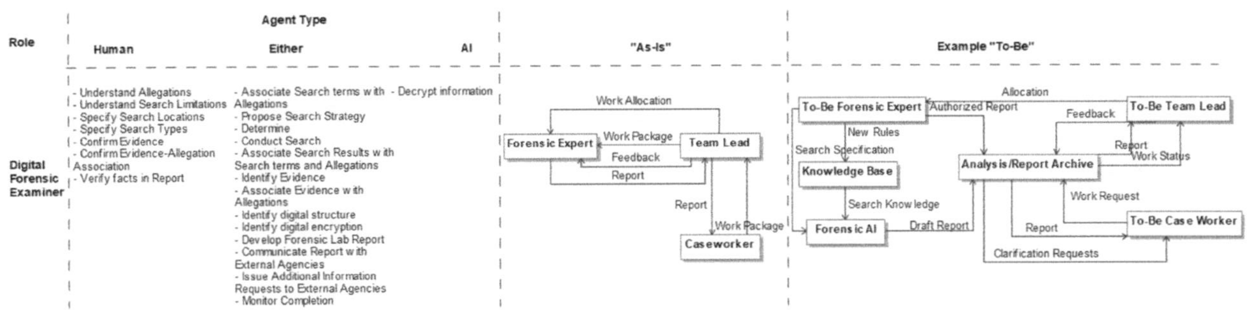

FIGURE 24.4 An example diagram used to brainstorm agent configurations to support roles or collections of capabilities.

and responsibilities, we identify which agents are associated with each responsibility and then analyze which agent is responsible, accountable, supporting, consulted, or informed for each responsibility to form an initial team design.

Analyze "To-Be" System

We can then analyze this team design to describe further the team members' interaction using scenario-based, coactive design techniques, such as Interdependence

Manual Forensic Analysis Report		**Primary Agent**		
Scenario	**Tasks**	**Capabilities**	**Agent Name**	**Perf. Rating**
T1	Receive lab request	Ability to communicate with customer	TL	1
	Assign acoropriata DFE	Ability to see work request hacking Ability to chango work focus and priorities Ability to urcorstand workflow process Ability to communicate with DFE	TL	1
	Im mocie Exras, tela from cigital	Ability to produce code for software solution Ability to extract knowledge from source Ability to acc new knowlecge base entry Ability to make technical design decisions Ability to research new tech	DFE	1
	Build a sroмeнцebase for platform	Ability to produce code for salware solution Ability to extract knowledge from source Ability to aco now knowlocge base ontry Ability to make technical design decisions Ability to research new tech	DFE	1
	Qually check report	Ability to communicate with DFE Ability to communicate with Ml. Developer Ability in comm.rinale wilt Worker Ability to assess automation perfomanc Ability to chango work focus and priorities	QAM	1
	Deliver rezort to external customor	Ability to communicate with customer Ability to communicate with DFE Ability to assess automation performance	TL	1

FIGURE 24.5 An example interdependence table with tasks, capabilities, agents, and OPDs.

Analysis (Johnson et al., 2014). Through this approach, we can further understand and describe the prescribed behavior of the team, including interaction among team members by characterizing what each team member must be able to Observe, Predict, or Direct (OPDs) for another team member to ensure adequate coordination. Figure 24.5 presents a table that illustrates the interdependence in a typical scenario described as a 'manual forensic analysis report.' Here we see tasks assigned to specific agents, such as Team Lead (TL), Digital Forensic Examiner (DFE), System

Secondary Agent			**OPDs**
Capabilities	**Agent Name**	**Team Rating**	**Observe, Predict, Direct**
Ability to communicate with customer	DFE	1	TL and DFE should observe communication with external agency and soo report request in backlog.
Ability to understand workflow process Ability to prioritize quouod jobs Ability to communicato with customer Ability to understand and change dev pipeline.	DFE	1	DFE should be ready for a leaking and coordinate with SysDev il requirod.
Abilily to produce code for sotheare solution Ability to add now knowledge base entry Ability to make lachnical thesign docisions Ability to research new tech	Sys Dev	2	DEF must understand current media platforms and be ready to research unseen devices; SysClev must be notified first thing when new model required.
Ability to produce onde for software solution Ability to add new knowledge base entry Ability to make lachnical design decisions Ability to research new tech	Sys Dev	2	DFE aliserves extracted media to determine whal conditions are lappropriate for ML model.
Ability to communicate with Ivil Developer Ability to communicate willı Al Agent Ability to compare automated vs. manual reports	DFE	2	QAM observes cutput from DFE report and manual resort to verify accuracy and quality
Ability to communicate with customer Ability to compare automated ve, manual reports	DFE	1	TL or DFE provides report la santernal agency and receives foodback

Developer (SysDev), and Quality Assurance Manager (QAM). Additionally, the table shows these assignments based on two criteria: performance and team rating, to assess the capabilities of each agent in the scenario. This approach encourages the designer to refine the existing HAT by incorporating a secondary or backup agent and relevant OPDs to support effective completion, coordination, and teamwork for each task. Primary agents are typically secondary agents, while secondary agents are supporting or accountable agents based on the earlier team description. Both primary and secondary agents can contribute to or complete the task. The team's overall performance rating decreases if either agent's capabilities are not well aligned with the task or the designer assesses that the agent cannot perform the task effectively due to a lack of information, overload, or other limitations. Throughout this analysis step, adding new agents, capabilities, and tasks or eliminating existing ones within the design is common. This iterative process explores opportunities to create a more efficient, error-tolerant, and redundant system. Further, the designer conducts this analysis for various scenarios, including standard or typical and less common, anomalous scenarios. While the earlier responsibility analysis aims to determine responsibilities that the HAT must fulfill, the designer does not consider the temporal arrangement of the responsibilities. The scenario-based analysis shown here includes tasks that might be associated with responsibilities.

However, the designer analyzes the tasks in a time-sequential fashion, which provides insight into capacity limitations that are difficult to consider when discussing the more abstract responsibilities.

Develop Specifications for "To-Be" System

Finally, the capabilities assigned to each agent provide functional requirements for each human and AI agent. These capabilities indicate autonomous behaviors that the AI agents must support and create an initial understanding of their sensing and actuator requirements. The designer can use this description to define procedures and training requirements for humans. The designer uses the OPDs and the interdependence analysis to define additional functional and interface requirements for the agents and software interfaces to support human–AI agent interaction. Finally, the designer uses this information to define data and information structures necessary for record keeping and communication among the team.

Assessing Human–AI Agent Teams

Generally, the HAT design should improve the productivity of the digital forensics expert, requiring fewer of these individuals to perform an equal number of examinations with equal quality. However, experience with this method in multiple application domains has shown that the approach rarely decreases the number of classes of individuals. In fact, in most systems, it is common to see the need to include data scientists or AI developers who are closely involved with or embedded in the team to permit the team to adapt to changes in real-world circumstances in near real time.

Without this ability to adapt to the system, the system will not evolve with changes in real-world circumstances, as adversaries, e.g., criminal elements, continually adapt to obfuscate incriminating information within their digital systems.

Additionally, the approach typically leads to developing multiple AI agents to support the system entirely. As such, the altered systems usually require the construction of integrated digital systems with human interfaces and multiple AI agents. The creation of such systems is likely to require tightly integrated design teams. The ability to specify the requirements for the overall system, the human–systems interfaces, and the AI agents can permit the simultaneous development of these system components. This design method is unlikely to lead to the development of error-free requirements or user stories. However, it can bootstrap the development of a highly integrated system that the designer can evolve through agile and human-centered design processes.

To evaluate these system designs, it is essential to continue to ask whether we are truly designing for teaming. That is, are we incorporating interdependent work, enabling the sharing of work, enabling backup behaviors for times where an agent might not be capable of performing the work the designer designed them to accomplish, as well as enabling effective communication and the construction of robust mental models of the work among agents. This evaluation involves several aspects. First, it assesses if the system allows the team to perform mission analysis, define goals, and plan actions for standard and exceptional conditions before work execution. Further, it tests whether the team can use implicit and explicit communication and coordination methods while conducting work. Lastly, it determines if the system enables team members to review each other's work, assist in focusing on critical activities, and redistribute workload and responsibilities if a member is unable to complete their work (Dorneich et al., 2023; Marks et al., 2001). Finally, we can ask whether the system permits one to monitor the team's state, detect off-nominal conditions, and adapt the work in response. These attributes are essential to achieve a successful HAT design.

SUMMARY

As humans, we are constrained by limited capabilities. However, we routinely "make sense" of the world around us and apply our understanding to innovate solutions to unforeseen circumstances and to make judgments that lead to acceptable outcomes. AI rapidly evolves to consider exact facts, collected across numerous dimensions, to simulate future outcomes. While recent advances, particularly in machine learning, have been rapid, allowing AI to outperform humans in constrained environments, they cannot truly perform "sense making" to enable innovation and judgment in the same fashion as humans. Further, these systems are augmented by sensors and actuators, permitting them to participate in the full perceptual loop, enabling so-called "autonomous" systems, which can function well in constrained, expected circumstances. However, these systems are also prone to failures in situations that appear trivial to humans (Gordon, 2023).

Importantly, our cognitive constraints often lead us to use simplified models of the world around us. For example, we might predict future outcomes using a model that includes a linear chain of causal effects. At the same time, much of the world can be explained more accurately as a series of interconnected feedback loops. Our simplified models often lead us to view the world as chaotic, with unexpected behaviors and outcomes. To reduce the world's complexity and enable timely decisions, we often simplify the world, considering only localized effects when making decisions. However, our decisions will be more effective as we develop tools that enable us to better model and understand effects at a more global level. In the end, it is this ability to consider larger contexts, improving the quality and effectiveness of decision-making, that a HAT has the potential to provide. Systems that improve decisions by linking the diverse knowledge and expertise of multiple humans with specialized knowledge and AI agents with the ability to estimate outcomes using complex, rigorous models were envisioned as Licklider and Clark proposed human–machine symbiosis more than six decades ago (Licklider & Clark, 1962).

While this vision is compelling and technologies such as the World Wide Web and graphical interfaces have helped to enable portions of this vision, we lack the tools and methods necessary to reliably design and integrate robust teams of humans and AI agents to fulfill significant tenets of this vision. For example, AI agents are rarely constructed to participate as team members with humans, although concepts such as methods to enable AI to support joint work (Vignati et al., 2022), methods for human intent recognition (Schneider et al., 2023), and methods to support AI mental model construction (Lapso et al., 2023), which are necessary to enable true human–AI teaming, are in early stages of development.

The design approach discussed in this chapter provides a method for understanding and characterizing teamwork in existing human teams to design future HATs that leverage and enhance the capabilities of these agents. We can contrast this design approach with AI-focused development, which too often aims to replace humans with AI, at least in limited circumstances. This design goal ignores the complexity of the world and effective real-world work systems, which must not only deal with real-world complexity but also continually adapt to changes in the world.

These work systems are inherently open, soft systems situated worldwide. They must continually adapt to changes in the real world and rely on human operators, who vary in skills and abilities, to succeed. The design of these systems is inherently more complex than the closed, complicated systems that system designers typically construct. While the method described in this chapter and elsewhere (Miller & Rusnock, n.d.; Miller et al., 2020) provides a valuable method and construct for describing and designing these systems, we expect to see these methods evolve to enable successful, integrated teams that include multiple humans and AI agents that can couple together to better respond to objective world complexity.

DISCLAIMER

The views expressed are those of the authors and do not reflect the official guidance or position of the United States Government, the Department of Defense, the United States Air Force, or the United States Space Force.

REFERENCES

Bates, A. (2018). *Augmented Mind: AI, Humans and the Superhuman Revolution*. San Diego, CA: Neocortex Ventures LLC.

Cooke, N.J., Gorman, J.C., Myers, C.W., & Duran, J.L. (2013). Interactive Team Cognition. *Cognitive Science* 37 (2): 255–85. https://doi.org/10.1111/cogs.12009

DeCremer, D., & Kasparov, G. (2021). AI Should Augment Human Intelligence, Not Replace It. *Harvard Business Review* 18: 1–8. https://hbr.org/2021/03/ai-should-augment-human-intelligence-not-replace-it

Dorneich, M.C., Gilbert, S., Francis, R.F., Talyat, M., & Didic, E. (2023). Team Skill Metrics That Span Human–Human and Human–Agent Teams: An Initial Assessment. Proceedings of the Human Factors and Ergonomics Society Annual Meeting, 67(1), 123–130. https://doi.org/10.1177/21695067231192892

Duncan, M.C., Miller, M.E., & Borghetti, B.J. (2023). Analysis and Requirement Generation for Defense Intelligence Search: Addressing Data Overload through Human–AI Agent System Design for Ambient Awareness. *MDPI Systems* 11: 561. https://doi.org/https://doi.org/10.3390/systems11120561

Elm, W.C., Potter, S.S., Gualtieri, J., Easter, J., & Roth, E.M. (2003). Applied Cognitive Work Analysis: A Pragmatic Methodology for Designing Revolutionary Cognitive Affordances. In *Handbook of Cognitive Task Design*, edited by E. Hollnagel, 357–82. Boca Raton, FL, CRC Press.

Friedenthal, S., Moore, A., & Steiner, R. (2014). *A Practical Guide to SysML*. Waltham, MA: Elsevier Science and Technology.

Gordon, A. (2023). Week of Cone: Activist Group Is Protesting Driverless Cars by Disabling Them With Traffic Cones. *Motherboard*. 2023. www.vice.com/en/article/bvjv48/week-of-cone-activist-group-is-protesting- driverless-cars-by-disabling-them-with-traffic-cones

Hitchins, D.K. (1992). *Putting Systems to Work*. New York, NY: Wiley.

Johnson, M., Bradshaw, J.M. Feltovich, P.J. Jonker, C.M. Van Riemsdijk, M.B., & Sierhuis, M. (2014). Coactive Design: Designing Support for Interdependence in Joint Activity. *Journal of Human–Robot Interaction* 3 (1): 43. https://doi.org/10.5898/JHRI.3.1.Johnson

Lapso, J.A., Peterson, G.L., & Miller, M.E. (2023). A Hybrid Cognitive Model for Machine Agents. *Cognitive Systems Research* 81 (February): 1–10. https://doi.org/10.1016/j.cogsys.2023.02.007

Licklider, J.C.R., & Clark, W.E. (1962). Online Man–Computer Communication. *AFIPS Proceedings – 1962 Spring Joint Computer Conference, AFIPS 1962*, pp. 113–28. https://doi.org/10.1145/1460833.1460847

Marks, M.A., Mathieu, J.E., & Zaccaro, S.J. (2001). Remember Me: A Temporally Based Framework and Taxonomy of Team Processes. *Academy of Management Review* 26 (3): 356–76.

Miller, M.E., McGuirl, J.M., Schneider, M. F., & Ford, T.C. (2020). Systems Modeling Language Extension to Support Modeling of Human–Agent Teams. *Systems Engineering* 23 (5): 519–33. https://doi.org/10.1002/sys.21546.

Miller, M.E., & Rusnock, C.F. (2014). *Integrating Artificial and Human Intelligence through Agent-Oriented Systems Modeling*. Boca Raton, FL: CRC Press, Taylor & Francis Group.

Salas, E., Dickinson, T.L., Converse, S A., & Tannenbaum, S. I. (1992). Toward an Understanding of Team Performance and Training. In *Teams: Their Training and Performance*, edited by R. W. Swezey & E. Salas, pp. 3–29. Norwood, NJ: Ablex.

Schneider, M.F., Miller, M.E., Jacques, D.R., Peterson, G.L., & Ford, T.C. (2021). Exploring the Impact of Coordination in Human–Agent Teams. *Cognitive Engineering and Decision Making* 15 (2–3): 97–115. https://doi.org/https://doi.org/10.1177/15553434211010573

Schneider, M.F., Miller, M.E., & McGuirl, J.M. (2023). Assessing Quality Goal Rankings as a Method for Communicating Operator Intent. *Journal of Cognitive Engineering and Decision Making* 17 (1): 26–48. https://doi.org/https://doi.org/10.1177/1555343422 1131665

Stevenson, J.A., McGuirl, J.M., & Miller, M.E. (2023). Interface Design for Collaboration with Semi-Autonomous Agents from an Airborne Aircraft. In the *22nd International Symposium on Aviation Psychology*, pp. 1–6. Rochester, NY, Wright State University, Dayton, OH.

Sundstrom, E., De Meuse, K.P., & Futrell, D. (1990). Work Teams: Applications and Effectiveness. *American Psychologist* 45 (2): 120–33.

Vignati, M., Johnson, M., Bunch, L., Carff, J., & Duran, D. (2022). Interdependence Design Principles in Practice. *Frontiers in Physics* 10 (December): 1–18. https://doi.org/10.3389/fphy.2022.969544

Wooldridge, M. (2013). Intelligent Agents. In *Multiagent Systems*, edited by Gerhard Weiss, Second edition, pp. 3–51. Cambridge, MA: The MIT Press.

25 PF-HCPS

Extending Problem Frames for Supporting Human Cyber Physical System Collaboration

Yilong Yang, Bingjie Zeng, Zhi Ching Chen, and Juntao Gao

INTRODUCTION

Cyber–physical system (CPS) is an intelligent system that integrates computational and communication technologies with the physical world to achieve real-time perception, intelligent control, and adaptivity (Baheti et al., 2011). As a natural extension of CPS, the human–cyber–physical system (HCPS) (Jackson et al., 2000) considers human interaction and cooperation with information systems and the physical world, aiming to achieve user goals through human–cyber–physical collaboration. Overall, HCPS adds the human concept compared to CPS. For HCPS, humans are not only users of the system but also a part of the system, contained within the system boundaries and supporting the realization of the system's functions.

HCPS has found widespread use across various domains, notably in autopilot systems. However, a series of safety accidents involving Tesla Autopilot in recent years has undoubtedly dealt a massive blow to the development of HCPS. Erroneous behavior can lead to fatal accidents. The complex human–cyber–physical collaboration patterns in HCPS lead to the very complex behavior of such systems. Ensuring the safety of HCPS has become the focus of recent research. There are two challenges to modeling requirements in HCPS as follows.

The dissimilar ways humans and machines interact with their surroundings in diverse environments present a formidable challenge in accurately capturing and modeling the behaviors within HCPSs while ensuring their adherence to the stipulated requirements.

In addition, the uncertain and dynamic nature of the environment creates an additional layer of complexity in the HCPS requirements. This uncertainty poses a significant challenge that requires careful consideration and planning.

Problem frames (PF) [2] provides a way to describe the domain problem to help developers understand the environments in which the system is related. PF considers the effect of the software system on the natural world as the source of the software problem, and a structured analysis of the system's impact on the real world is the entry point for requirements analysis [3]. Compared with other approaches, PF emphasizes

DOI: 10.1201/9781003666431-30

the portrayal of the natural world in which the software system will act and refers to the meaning of requirements to a description of the relevant domain of the real world. However, the existing PF model cannot zoom in on the details of the problem diagram to describe the human–machine collaboration pattern in HCPS, which is the core of HCPS functional implementation. We propose a PF extension model, PF-HCPS, for HCPS concerning the meta-model proposed by Hatebur et al. (Hatebur, Heisel & Schmidt, 2008). The main contributions of this paper are as follows:

- *Extending Domain Elements of the PF Model:* To help developers better understand how the system interacts with the external environment, we set the system boundary and categorize the domains into system and environment domains.
- *Extending State Machines to Represent Human–Machine Collaboration:* We extend the state machine to model human–machine collaboration patterns in HCPS. This allows us to zoom in on the requirements for human–machine collaboration in HCPS to facilitate developer understanding.
- *Improving the Usability of the PF Model:* We introduced "human" and common concepts in HCPS into PF to support PF's description of human behavior in HCPS, lowering the threshold for PF's use.

The remaining parts of the paper are structured as follows. Second section presents related work and compares other work with the PF-HCPS meta-model. Third section shows the details of the PF-HCPS. Fourth section verifies the usability of our technique through a practical problem. Finally, fifth section concludes the paper and proposes future work.

RELATED WORK

In the last decade, many efforts have started to combine model-driven engineering (MDE) with PF and apply it in various domains. The first formal PF meta-model was proposed and used in security engineering by Hatebur et al. (Hatebur, Heisel & Schmidt, 2008). They divided PF into four parts: domain, interface, requirements, and requirements reference. Colombo et al. [6] studied the meta-model and ontology of PF and proposed a PF model for problem decomposition (Colombo et al., 2009). Yu et al. (Tun et al., 2009; Yu et al., 2011) constructed a PF meta-model based on the PF theory and implemented the OpenPF [9] tool using EMF to address interaction problems. Xiao et al. [10] proposed an extended meta-model of the PF approach to support modeling causal behaviors representing the CPS context (Wei, Li, Yang & Xiao, 2021). Wei et al. [11] allow users to zoom in on the details of the problem diagram by adding UML state machine diagrams and SysML block definition diagrams to the domain description. Wei et al. allow users to zoom in on the details of the problem diagram by adding UML state machine diagrams and SysML block definition diagrams to the domain description.

Table 25.1 compares the models in related work and the model proposed in this paper. We compare the meta-models in ten aspects: problem domain, system boundary, and domain properties. The table shows that although the first four models

TABLE 25.1
Comparison of Meta-Models in Related Work

	Hatebur' 2008	Lavazza' 2009	OpenPF	PF-CPS	Zoom4PF	PF-HCPS
Problem domain	√	√	√	√	√	√
System boundary	×	×	×	×	×	√
Human domain	×	×	×	×	×	√
Machine domain	√	√	√	√	√	√
Thing domain	√	√	√	√	√	√
Domain property	×	√	×	√	×	√
Requirement	√	√	√	√	√	√
Interface	√	√	√	√	√	√
State machine	×	×	×	×	√	√
Collaboration pattern	×	×	×	×	×	√

can represent the concepts in PF, they cannot describe the details in the problem domain. Zoom4PF solves this problem by introducing the state machine in PF, but it fails to represent the human–machine collaboration pattern, which does not apply to HCPS. The broader and more precise representation of PF in PF-HCPS applies more to describing the problem domains and requirements in HCPS.

PF-HCPS

To represent requirements in HCPS, we extend the PF model proposed by Hatebur et al. [4] using a model-driven approach. This chapter presents PF-HCPS in three parts: model overview, meta-model, and graphical notations.

Overview

PF-HCPS serves as an extension of PF tailored specifically for HCPS. It is structured into three fundamental elements: the meta-model, the graphical notations, and the diagrams. As shown in Figure 25.1, the meta-model of PF-HCPS has three parts: the initial PF, the PF extension, and the state machine extension. The model consists of nodes and edges. It contains two diagrams: the problem diagram and the human–machine collaboration state machine diagram. The problem diagram can describe the problem domain in HCPS, and the human–machine collaboration state machine diagram can explain the human–machine collaboration patterns in HCPS requirements.

Meta-Model

Figure 25.2 presents the meta-model of PF-HCPS, which consists of three parts: the original PF meta-model (marked with a yellow background), the extension of the PF meta-model (marked with a blue background), and the state machine extension

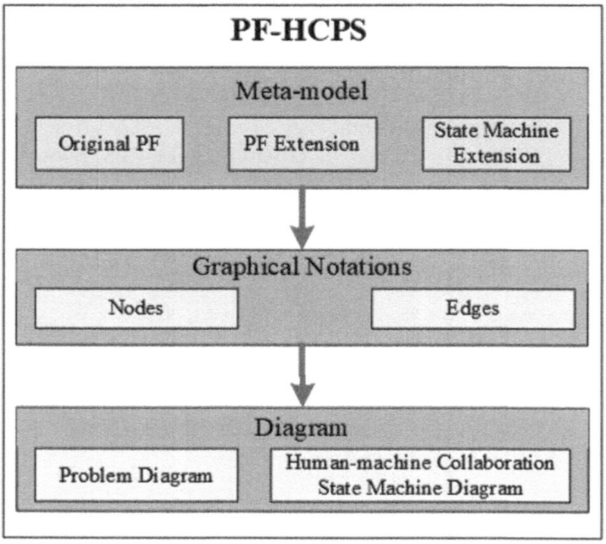

FIGURE 25.1 Overview of PF-HCPS.

(marked with a red meta-class background). However, we cannot provide detailed information about all the elements of the original PF meta-model marked in yellow due to space limitations.

Extension of PF Meta-Model

In Figure 25.3, we can see the enhancements to PF for the characteristics of HCPS. As the HCPS is closely linked to the environment in which it operates, we divided the domains into the system and environment domains. The system boundary element defines the system boundary, where the system domain is contained within the system boundary, and the environment domain is outside the system boundary. The system domain is further divided into four elements: *Machine*, *Internal Biddable Domain*, *Internal Causal Domain*, and *Lexical Domain*. The machine element represents the software that needs to be developed, which is the "Cyber" part of HCPS. The *Internal Biddable Domain* represents the system user, the "Human" part of HCPS. The *Internal Causal* and *Lexical Domain* represents the hardware part of the "Cyber" of HCPS. The *Internal Causal Domain* represents the hardware related to the realization of software functions in the system, such as sensors and actuators. In contrast, the Lexical Domain represents the devices responsible for I/O operations, such as control panels.

The environment domain is a part of the HCPS that represents the physical aspect. It can be further divided into the *External Biddable* and the *External Causal Domain*. The *External Biddable Domain* represents the environment's uncertain, uncontrollable, or unpredictable aspects, while the *External Causal Domain* represents the controllable or predictable aspects of the environment. The environment domains vary

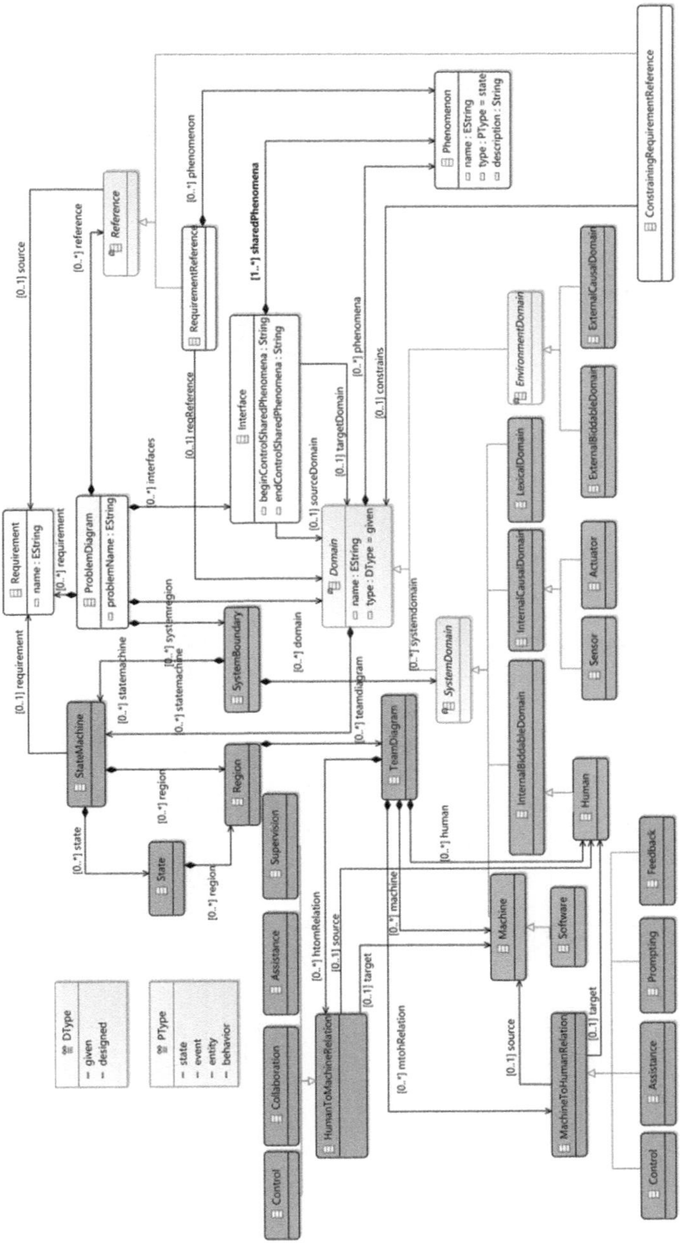

FIGURE 25.2 Meta-model of PF-HCPS.

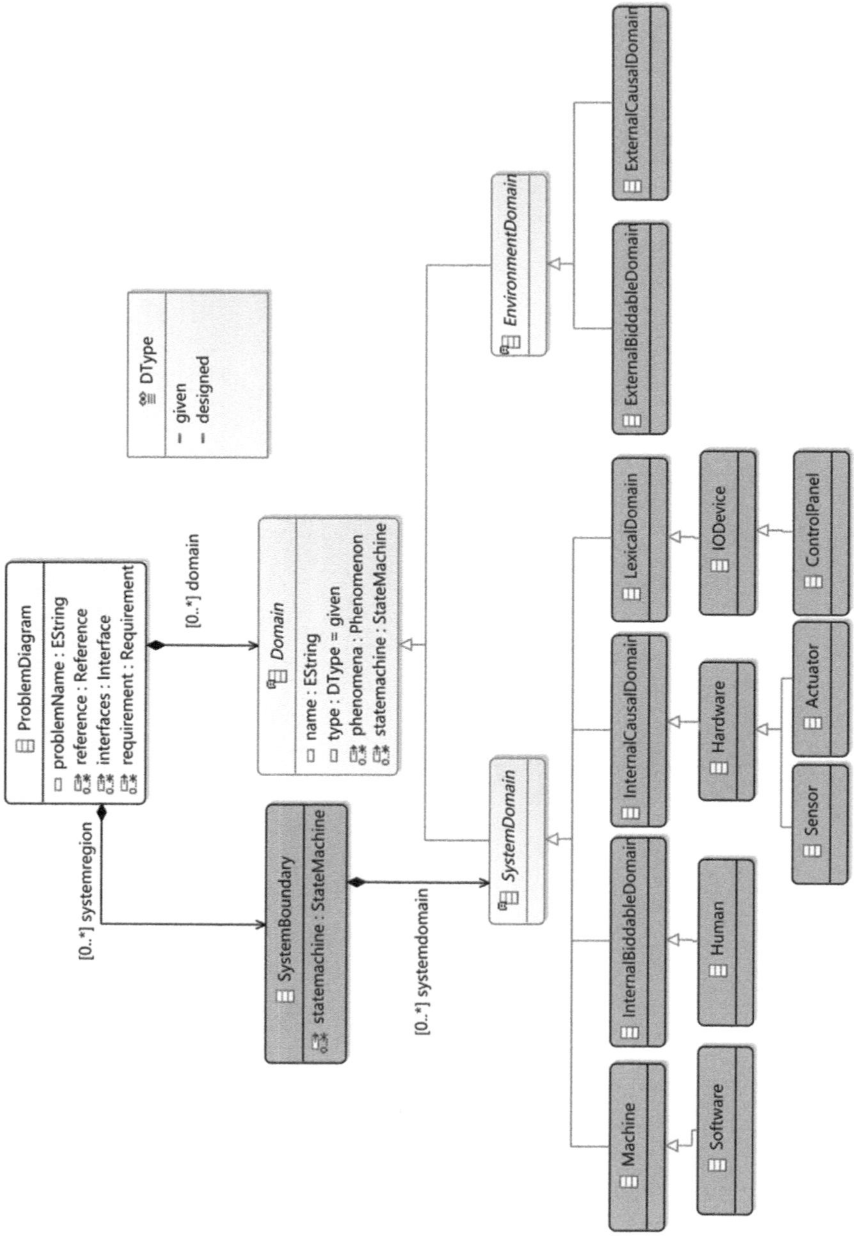

FIGURE 25.3 Extension of the PF meta-model.

across different problem domains due to the varying environments of other systems. For example, uncertain domains in Tesla Autopilot include pedestrians and external vehicles, while causal domains include lanes and traffic signs.

Extension of State Machine

To represent the complex human–machine collaboration patterns in HCPS requirements, we have built upon the work of Zoom4PF (Wei, Li, Yang & Xiao, 2021) and extended the state machine. Figure 25.4 presents a detailed view of our extension to the state machine. To clearly illustrate the various aspects of the system's state machine, we have associated it with the system boundary, domain, and requirement. This enables us to represent the state machine of each domain and the system's state machine under a given requirement constraint.

To depict the human–machine collaboration, we have introduced a team diagram in the region of the state machine. Within the collaboration diagram, we have added two elements, human and machine, to serve as discussion objects. This provides a clear illustration of how the human–machine collaboration works within the context of the system. We believe that this approach will help to improve the overall effectiveness of the HCPS requirements process. We are confident that this work will interest those involved in developing HCPS requirements, and we look forward to sharing our findings with the broader community.

To represent the human and machine collaborative relationship, we configure the *HumanToMachineRelation* and *MachineToHumanRelation* for human and machine elements. We categorized and ranked the human-to-machine relationships according to the degree of human dominance over the machine's behavior: *control*, *collaboration*, *assistance*, and *supervision*. In the control relationship, the human has direct control over the behavior of the machine and can directly direct and determine the actions of the machine; in the collaboration relationship, the human and the machine are jointly involved in problem-solving and task execution; in the assistance relationship, the machine dominates the execution of the task, and the human assists the machine by providing information and decision-making support, etc.; in the supervision relationship, the machine is responsible for the execution of the task, and the human supervises the machine's behavior to ensure that the machine works according to expectations. We categorize and rank the machine-to-human relationships according to the degree of dominance of the machine over human behavior: *control*, *assistance*, *prompting*, and *feedback*. In the control relationship, the machine can determine human actions and decisions directly; in the assistance relationship, the machine dominates the execution of the task, while the human plays an auxiliary and supportive role in providing information or decision-making in the necessary moments; in the prompting relationship, the machine guides the human to carry out a specific action by providing prompts or reminders, but the human maintains a certain degree of initiative during the execution process; in the feedback relationship, the machine transmits the machine's working condition and the results of the execution to the human by providing real-time feedback to enable the human to carry out a specific action based on such information.

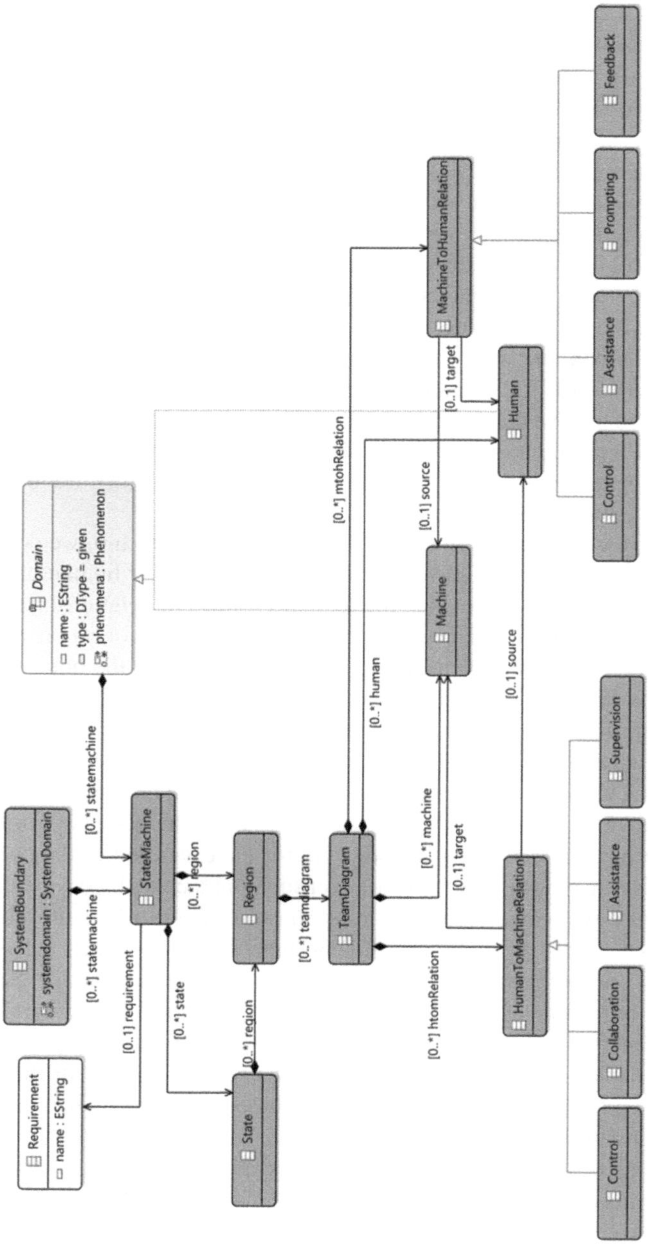

FIGURE 25.4 Extension of state machine.

To establish the nature of interactions between human and machine, we have formulated two relation categories: *HumanToMachineRelation* and *MachineToHumanRelation*. Additionally, we have categorized the degree of human control over machine behavior into four groups: *control*, *collaboration*, *assistance*, and *supervision*. In a *control* relationship, the human governs the machine's actions and behavior, with the ability to directly determine its actions. In a *collaborative* relationship, humans and machines work cohesively to accomplish goals and solve problems. In an *assistance* relationship, the machine oversees the task while the human provides guidance or support. Lastly, in a *supervision* relationship, the machine operates independently while still being monitored and supervised by the human.

Graphical Notations of PF-HCPS

We have used Sirius to implement our model graphically in this paper. Our model comprises nodes and edges, representing the relationships between the nodes. The meta-model in Figure 25.2 has been mapped to the nodes and edges. Table 25.2 illustrates how the meta-classes in the meta-model are mapped to the model nodes. Similarly, Table 25.3 shows how the meta-classes in the meta-model are mapped to the model's edges.

TABLE 25.2
Graphical Notations of PF-HCPS

Nodes	Graphical Notations	Meta-Classes	Diagrams
Machine	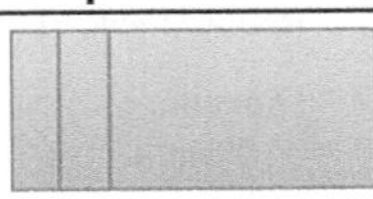	Machine	Problem diagram and collaboration diagram
Human		Human	
State Machine		StateMachine	
System Boundary		SystemBoundary	Problem diagram
InternalCausalDomain		InternalCausalDomain	
LexicalDomain		LexicalDomain	

(continued)

TABLE 25.2 (Continued)
Graphical Notations of PF-HCPS

Nodes	Graphical Notations	Meta-Classes	Diagrams
ExternalCausalDomain		ExternalCausalDomain	
ExternalBiddableDomain		ExternalBiddableDomain	
Requirement	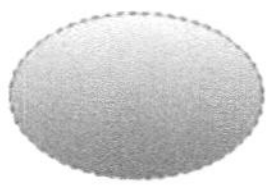	Requirement	
TeamDiagram		TeamDiagram	Collaboration diagram

TABLE 25.3
Graphical Edges of PF-HCPS

Edges	Graphical Notations	Source	Target	Diagrams
Interface		Domain	Domain	Problem diagram
ReqReference		Requirement	Domain	
ReqConstraint		Requirement	Domain	
statemachine+		Domain, System Region	StateMachine	
HumanToMachineRelation		Human	Machine	Collaboration diagram
MachineToHumanRelation		Machine	Human	

CASE STUDY: TESLA AUTOPILOT

This paper utilizes the Tesla Autopilot system as a case study to evaluate the expressive capability of PF-HCPS. The Tesla Autopilot system is a widely used HCPS that involves a typical human–machine–object collaboration problem. It comprises 17 autopilot-related function points that require collaboration between the human component (i.e., the driver), the network component (i.e., the software controller that needs to be developed), the complex physical component (such as the hardware of the vehicle and various obstacles in the environment), and the human–machine collaboration pattern during the system's operation. The collaboration between the driver and the software controller is a crucial aspect of ensuring the safe transportation of the vehicle during the system's operation.

The PF methodology uses a problem diagram as a graphical model of the software problem's scope and structure. In the case of the Tesla Autopilot system, Figure 25.5

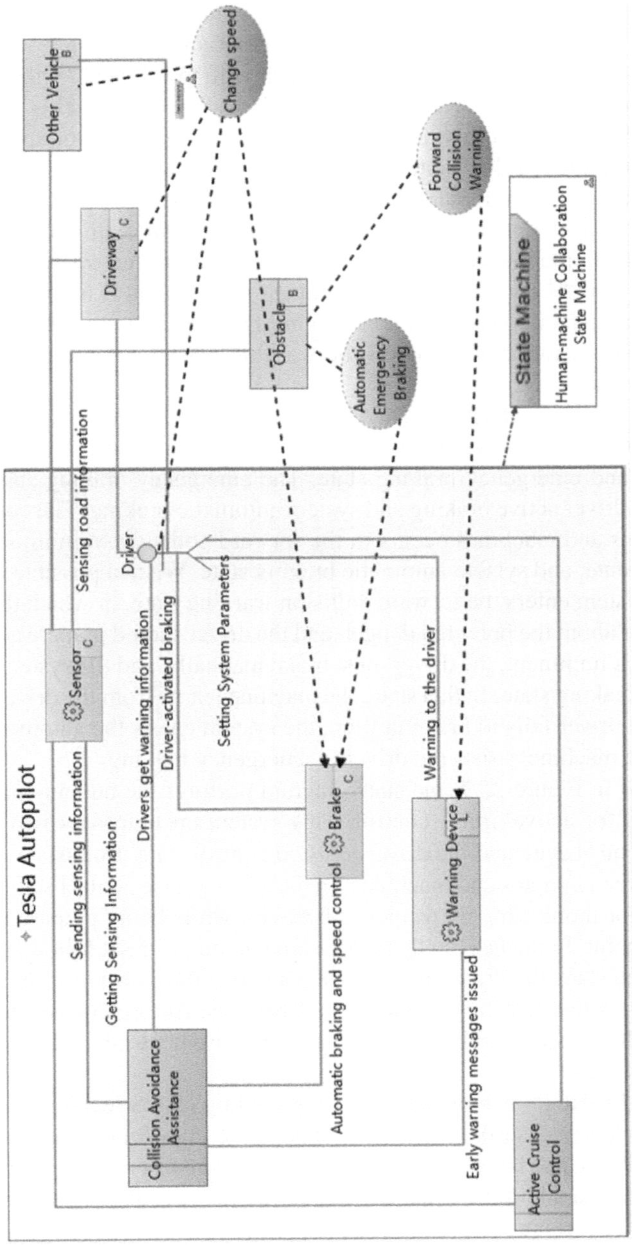

FIGURE 25.5 Problem diagram of Tesla Autopilot.

shows a partial problem diagram highlighting issues in two function points: active cruise control and collision avoidance assistance. The whole system, including software and hardware, is delineated, and system boundaries are identified. Three requirements were identified based on what the system needs to accomplish: change speed, forward collision warning, and automatic emergency braking.

The "Change Speed" requirement is associated with the "Lane," "Other Vehicles," and "Driver" domains, which constrain the "Brake" domain. This requirement indicates that, during active cruise control, the software will control the speed based on information about other vehicles, the current lane, and user actions. The "Automatic Emergency Braking" requirement is associated with the "Obstacle" domain and constrains the "Brake" domain. It indicates that the software will control the brakes based on information about obstacles outside the system to realize automatic emergency braking. The "Forward Collision Warning" requirement is associated with the "Obstacle" domain and constrains the "Warning Device" domain. This requirement indicates that the software will control the warning device based on out-of-system obstacle information.

Figure 25.6 shows the state diagram of human–machine collaboration for collision avoidance assistance. The system has three states: driving state, forward collision warning state, and emergency braking state. The emergency braking state includes two sub-states: driver active braking and system automatic braking. The collaboration between humans and machines occurs in the forward collision warning state, driver active braking state, and system automatic braking state. When a possible collision is detected, the system enters the forward collision warning state, in which the machine warns the driver about the potential danger, and the driver should prepare accordingly. If the collision is imminent, the driver must brake manually, and the system enters the driver's active braking state. In this state, the machine learns from the driver's braking behavior. If the driver fails to brake in time, the system enters the automatic braking state, where the machine assists the driver in emergency braking.

As presented in Figure 25.7, the state diagram portrays the human–machine collaboration state for active cruise control. This system encompasses two states: the active cruise control state and the driver-assisted control state. Both states entail collaboration between humans and machines. In the active cruise control state, the driver is responsible for monitoring the vehicle's behavior while being prepared to resume control at any point. In the next state, the machine maintains a set following distance. In the overtaking state, the driver initiates the maneuver through a turn signal, and the machine regulates the vehicle's acceleration. The Tesla Autopilot case study results show that PF-HCPS can completely describe the requirements and problem domains in HCPS.

Meanwhile, PF-HCPS is more suitable for modeling HCPS requirements because PF-HCPS can clearly describe the human–machine collaboration pattern in HCPS. However, PF-HCPS also has some weaknesses. It cannot describe the scenario information in detail. In addition, when the scenarios become numerous or complex, the scenario transformation relationships represented by PF-HCPS become very complicated.

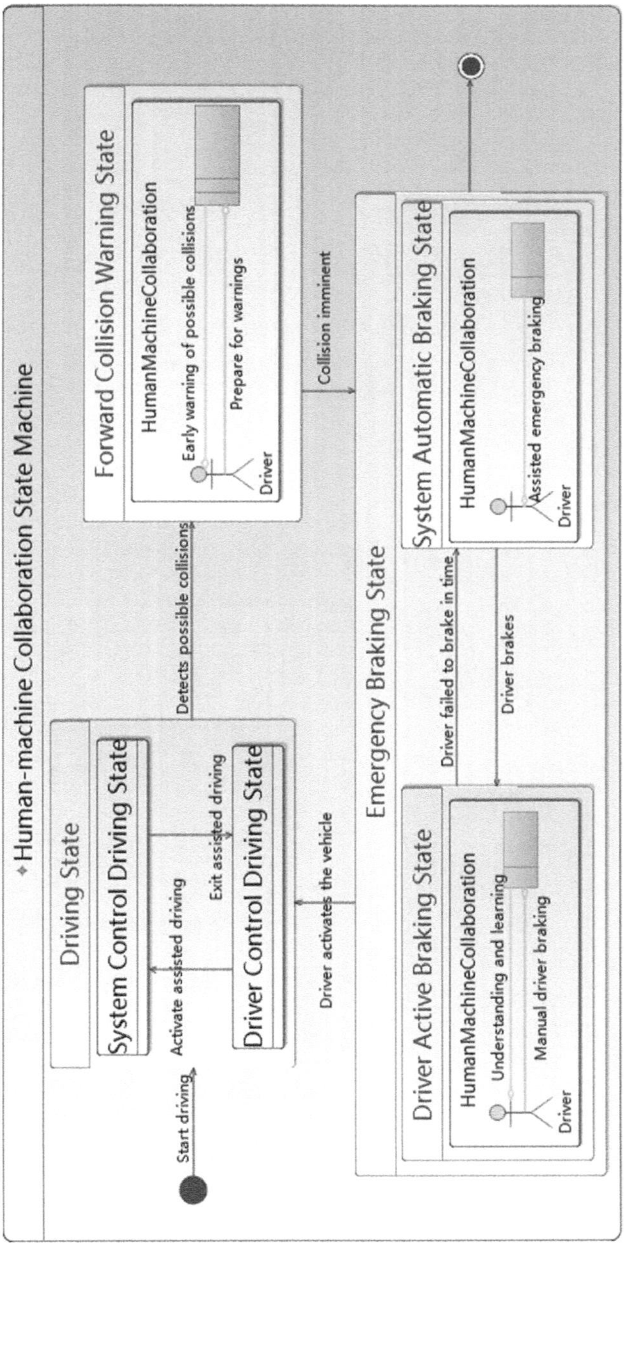

FIGURE 25.6 Human–machine collaboration state machine diagram (collision avoidance assistance).

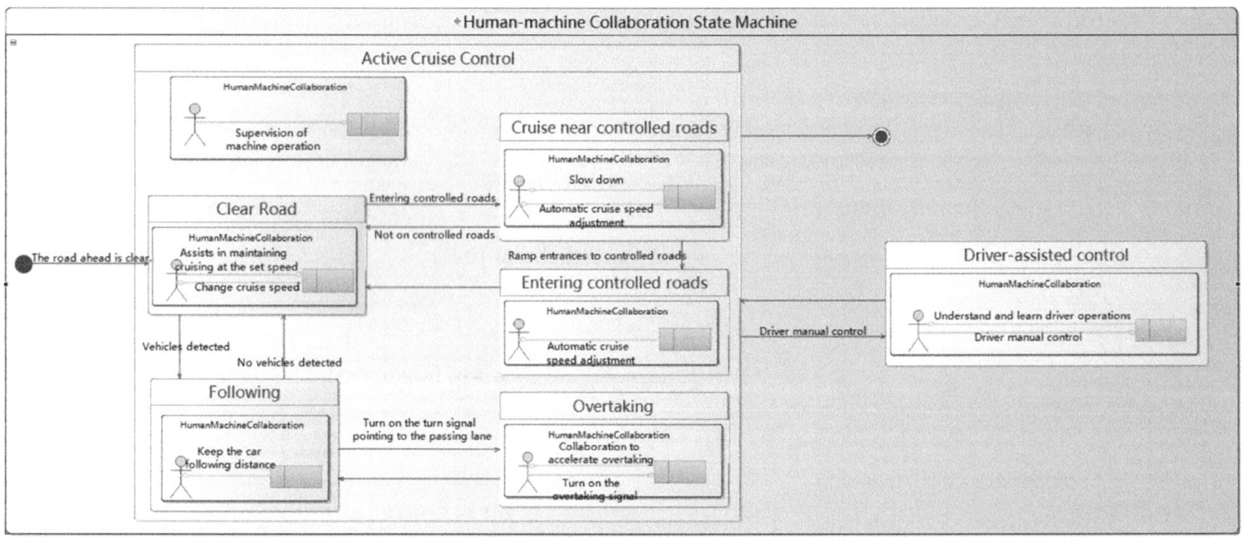

FIGURE 25.7 Human–machine collaboration state machine diagram (active cruise control).

CONCLUSION AND FUTURE WORK

This paper focuses on the criticality of requirements engineering within the development process of HCPS. To represent problem domains and human–machine collaboration patterns in HCPS, the extension of the problem framework and state machine has resulted in the proposal of PF-HCPS. Utilizing EMF and Sirius, a modeling tool for PF-HCPS has been developed. The expressiveness of PF-HCPS has been evaluated through the case study of the Tesla Autopilot system, and the findings demonstrate that PF-HCPS has a strong ability to express the requirements inherent in HCPS.

As we improve the PF-HCPS model, we have identified several critical areas that require our attention. Firstly, we aim to incorporate the description of nonfunctional requirements within the problem frame, specifically concerning human–machine collaboration patterns. Additionally, we are committed to integrating the PF-HCPS model with other models such as SysML/UML. Moving forward, our research will explore how to leverage the models generated by PF-HCPS to develop system prototypes for validating requirements.

REFERENCES

Baheti, R., & Gill, H. (2011). Cyber-physical systems. *The Impact of Control Technology*, 12(1), 161–166.

Colombo, P., Lavazza, L., Coen-Porisini, A., & Del Bianco, V. (2009, September). Towards a meta-model for problem frames: Conceptual issues and tool building support. In *2009 Fourth International Conference on Software Engineering Advances* (pp. 339–345). IEEE.

Hatebur, D., Heisel, M., & Schmidt, H. (2006, June). Security engineering using problem frames. In *International Conference on Emerging Trends in Information and Communication Security* (pp. 238–253). Springer Berlin Heidelberg.

Hatebur, D., Heisel, M., & Schmidt, H. (2008, September). A formal metamodel for problem frames. In *International Conference on Model Driven Engineering Languages and Systems* (pp. 68–82). Springer Berlin Heidelberg.

Jackson, M. (2000). *Problem Frames: Analyzing and Structuring Software Development Problems*. Addison-Wesley Longman Publishing Co., Inc.

Jackson, M. (2015). System behaviours and problem frames: Concepts, concerns and the role of formalisms in developing cyber-physical systems. In *Dependable Software Systems Engineering* (pp. 79–104). IOS Press.

ProblemFrames. (2021). OpenPF Website (open-source). Retrieved on August 1, 2021, from https://github.com/problem-frames/openpf

Tun, T. T., Yu, Y., Laney, R., & Nuseibeh, B. (2009). Early identification of problem interactions: A tool-supported approach. In *Requirements Engineering: Foundation for Software Quality: 15th International Working Conference, REFSQ 2009, Amsterdam, The Netherlands, June 8–9, 2009 Proceedings 15* (pp. 74–88). Springer Berlin Heidelberg.

Wei, S., Li, Z., Yang, Y., & Xiao, H. (2021, September). Zoom4PF: A tool for refining static and dynamic domain descriptions in problem frames. In *2021, IEEE 29th International Requirements Engineering Conference (RE)* (pp. 414–415). IEEE.

Xiao, H., Li, Z., Yang, Y., Deng, J., & Wei, S. (2021, September). An extended meta-model of problem frames for enriching environmental descriptions. In *2021, IEEE 29th International Requirements Engineering Conference Workshops (REW)* (pp. 428–434). IEEE.

Yu, Y., Tun, T. T., Tedeschi, A., Franqueira, V. N., & Nuseibeh, B. (2011, August). OpenArgue: Supporting argumentation to evolve secure software systems. In *2011, IEEE 19th International Requirements Engineering Conference* (pp. 351–352). IEEE.

26 Ingescape, a Hands-On Solution for Model-Based HSI, from Early Design to Post-Deployment

Stéphane Valès

INTRODUCTION

Modern industrial systems are often defined as sociotechnical and cyber–physical systems, consisting of heterogeneous, distributed systems with humans, artificial intelligence (AI), simulations, digital twins, and hardware in the loop at several levels.

The lifecycle of a sociotechnical system (STS) may be seen as a multiplayer puzzle game, where the puzzle pieces and the assembling change continuously during the game. Moreover, our puzzle game players do not all share the same vision, objectives, and methodologies.

This paper addresses and illustrates the challenges and solutions to provide a thorough "puzzle board" for STSs, containing the tools to continuously *interoperate, observe, measure,* and *verify* both the pieces and the assembling, with multidisciplinary and iterative collaboration in mind, orchestrated by a model-based human–system integration (MBHSI) methodology (Boy, 2023).

The proposed solution consists of a breakthrough software tool suite, Ingescape, enabling the creation and exploitation of model-based software and hardware platforms dedicated to all the relevant multidisciplinary HSI activities in real-world STSs (Boy & Kennedy, 2023). This paper covers all the currently identified key enablers relevant to model-based HSI methodologies, so that project teams of all sizes get efficient support for the iterative *design, simulation, prototyping, integration, assessment,* and *validation* of such systems.

The Ingescape tool suite is structured in several parts, handling both systems' and users' sides. Most parts are flexible and used whenever a project requires them. All the parts are linked around a unified model-based approach, consistently combining industry standards, such as SysML and business process model notation (BPMN), and other open standards and models from the system engineering, business processes modeling, software engineering, user eXperience (UX)/user interface (UI), human factors, and networking domains.

DOI: 10.1201/9781003666431-31

The main Ingescape breakthrough is its ability to provide models serving as *continuous interactive blueprints* for the whole lifecycle STSs, i.e., models making the actual systems observable, actionable, measurable, and verifiable by all involved stakeholders relying on their different backgrounds, and exposing all necessary types of organizational, operational, functional, logical, and physical information.

SOLVE THE INTEROPERABILITY PROBLEM

Modern systems are, in fact, systems of systems (SoSs) (Popper et al., 2004), composed of heterogeneous and distributed hardware and software technologies that must reactively and asynchronously interoperate. Technologies in a real-world project shall be picked and mixed as freely as possible to ensure the best fit between identified needs and the most efficient solutions. Proper performance, control, and scalability of complex systems require various complementary communication patterns that cannot be limited to a single restrictive one. Interface contracts between system components shall be specified precisely so that the resulting STS can work while minimizing integration and verification efforts.

All these issues are addressed by the open-source Ingescape library,[1] offering a unique interoperability approach *through interactive model-based architecture and communications*, making STSs directly observable and actionable through their models.

The core concept used in Ingescape to embed pieces of a SoS is called an *agent*, in Minsky's (1986) sense. An *agent* is a piece of connected software or hardware that can be developed in any language and run on any operating system. Humans are integrated into the system through UI applications and various connected sensors. Agents communicate using two main communication patterns offered by Ingescape: *data flows* and *services* (Figure 26.1).

Data flows are created by connecting inputs and outputs between agents (Figure 26.2). Following this pattern, an agent receives and handles information through its inputs, computes it, and delivers it through its outputs. This creates loosely coupled communications that bring remarkable advantages in SoSs:

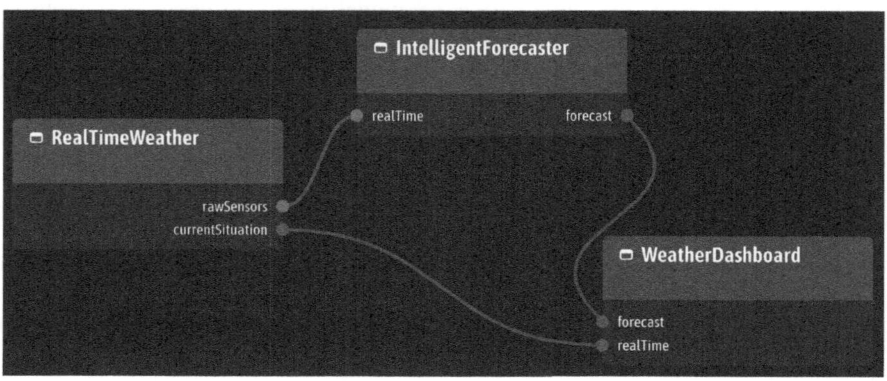

FIGURE 26.1 Typical Ingescape diagrams.

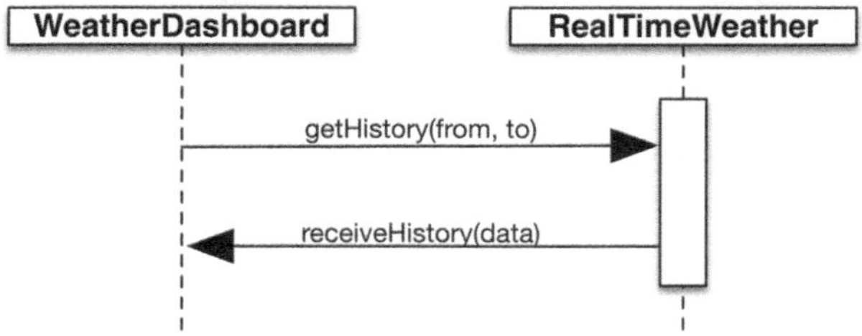

FIGURE 26.2 Data flows in a simple example.

- Adding (or removing) an input/output to an agent only affects the inputs/outputs mapped to it. The rest of the system is unaffected.
- Only the data it publishes on its outputs is affected when an agent is missing or failing. The rest of the system still behaves normally.
- Data exchanges on computer networks are high-speed with actual point-to-point exchanges, removing the need for central servers or brokers.

Services in Ingescape enable a pervasive communication pattern, also known as request/reply, client/server communication, or remote procedure/function calls (RPC). Using this pattern, an agent offers one or several services that can be called by other agents with specific arguments. The callee may optionally offer replies, which are also services, called the other way around, from the callee to the initial caller.

Data flows and services shall not be opposed. When designing a specific data exchange in each system, picking one or the other requires a sound analysis of the necessary coupling between sources and destinations for the information.

In our experience, data flows are applicable in 80% of the cases. Systems should be designed as loosely coupled using data flows, with some exceptions for using services. As a design rationale, agents shall preferably be conceived as entities transforming information from their inputs to their outputs. In contrast, services handle specific individual queries from other agents. For instance, an agent offering authentication features can use services to receive individual queries with usernames and passwords and respond to its callers with replies indicating the success or failure of the authentication, possibly attaching an authentication token in case of success.

Systems-Wide Model-Based Logical Architecture and Interface Contracts

Ingescape agents form the software layer of an STS. Each agent contains functions, which can, in turn, also contain sub-functions. Functions also provide inputs, outputs, and services, providing a consistent hierarchical representation at all logical levels.

The definition of agents and systems functions with their inputs, outputs, and services, combined with a definition of the input/output detailed specifications and sequence diagrams for service calls, enables a reliable and complete description of a model-based logical architecture and the information exchanges inside an SoS.

This model-based approach to interoperability opens the way to unique ways of interacting with information systems through their models, opening an extended approach to model-based system engineering (MBSE) (Madni et al., 2023) and making room for HSI-related challenges. These topics are explored in the following sections of this paper.

Software Licenses Are Essential for Interoperability

The Ingescape library and its dependencies are developed under the Mozilla Public License version 2 (MPLv2). With this license, any developer can create Ingescape agents embedding the Ingescape library without any constraint on the ability to apply commercial licenses to the developed agents. This makes collaborations and partnerships much easier in large HSI projects.

Some tools on top of the Ingescape library, aggregated in the Ingescape™ Circle application, are available under commercial licenses. These tools deliver productivity features that affect neither the MPLv2 license on the agents themselves nor the openness of the STSs.

Opening the Data and the Models

The Ingescape library and Ingescape™ Circle manipulate data models composed of systems architecture, data flows, services, and interface contracts. The entire model is expressed in javascript object notation (JSON) and is suitable for human-readable inspection.

This open-format approach, combined with careful work on the simplicity and expressivity of the concepts, guarantees that the Ingescape models can be generated and edited by any third-party actor. This has proven useful when handling complex systems with tens of thousands of inputs and outputs, such as aeronautics simulation systems. Despite the large legacy models in such systems, exposing them and making them instantly usable in agents has been very easy.

Gathering Remote Platforms (Cloud, Edge, and Hybrid)

In SoSs, components may be located on different local area networks (LANs) and geographical locations, requiring distant communications to integrate them, using dedicated infrastructure or the Internet. Similarly, Ingescape agents may run in web browsers or mobile devices and may require connecting to a given infrastructure using virtual private network (VPN) tunnels or the Internet.

For these situations, Ingescape proxies provide a unique and secure ability to manage the geography of distributed systems. In minutes and compliance with state-of-the-art cybersecurity requirements, system administrators can design and install connections between remote Ingescape platforms to project agents from their physical platforms to distant platforms where they are needed. The resulting flexibility is straightforward and controllable, reaching and often exceeding the abilities of microservices architectures and web technologies.

Cybersecurity for Fully Decentralized Systems

Open, fully decentralized systems can be as secure as classic client/server or broker-based infrastructures. Ingescape uses state-of-the-art asymmetric encryption mechanisms provided by its ZeroMQ layer using simple authentication and security layer (SASL) and elliptic curve cryptography (ECC) using Curve25519, also known as Ed25519. These mechanisms ensure sender and receiver authentication, message signatures from senders, and message encryption for receivers. As for all other asymmetric encryption types, this is achieved by sets of public and private keys. The use of Curve25519, compared, for example, to RSA, enables the use of shorter encryption keys and offers equivalent if not better security with reduced computational efforts. Ed25519 is becoming the default choice in SSL/TLS and SSH exchanges, progressively replacing RSA.

Even with its fully decentralized design, Ingescape complies with the most advanced state-of-the-art practices regarding cybersecurity and lets the cybersecurity teams decide the strategies to deploy for advanced control of program execution and network communications.

MAKE SYSTEMS OF SYSTEMS OBSERVABLE, ACTIONABLE, MEASURABLE, AND VERIFIABLE

The complexity and heterogeneity of SoSs raise a specific stake around the ability to observe, capture, and interact with such systems. This includes ways to measure if a SoS is properly working and its internal activities, together with ways to interact with the different parts of the SoS at the unit, partial, and global levels, for simulation, integration, verification, and validation purposes.

Our innovative approach introduces interactive model-based representations above actual SoSs, so that visual tools are combined to model elements to make SoS **observable, actionable, measurable, and verifiable** through their models.

This approach is made possible thanks to the Ingescape software library's specific ability to embed each agent's model and by the capacity of all the meshed agents to synchronize their models with one another. This is precisely how the models become interactive and can be used to observe and control SoSs (Figure 26.3).

Observation of SoSs includes the following facets:

- Knowing if and where parts of the system are running or not,
- Get a statement communicated by all executing parties,
- Monitor and inspect data exchanges between all parts of the system,
- Inspect logging data for all systems.

The Ingescape library provides these abilities for all Ingescape agents on a given platform. Moreover, Ingescape™ Circle provides the no-code visual tools to take advantage of all the information made available by the Ingescape agents, primarily an *output observer* and a *service observer*.

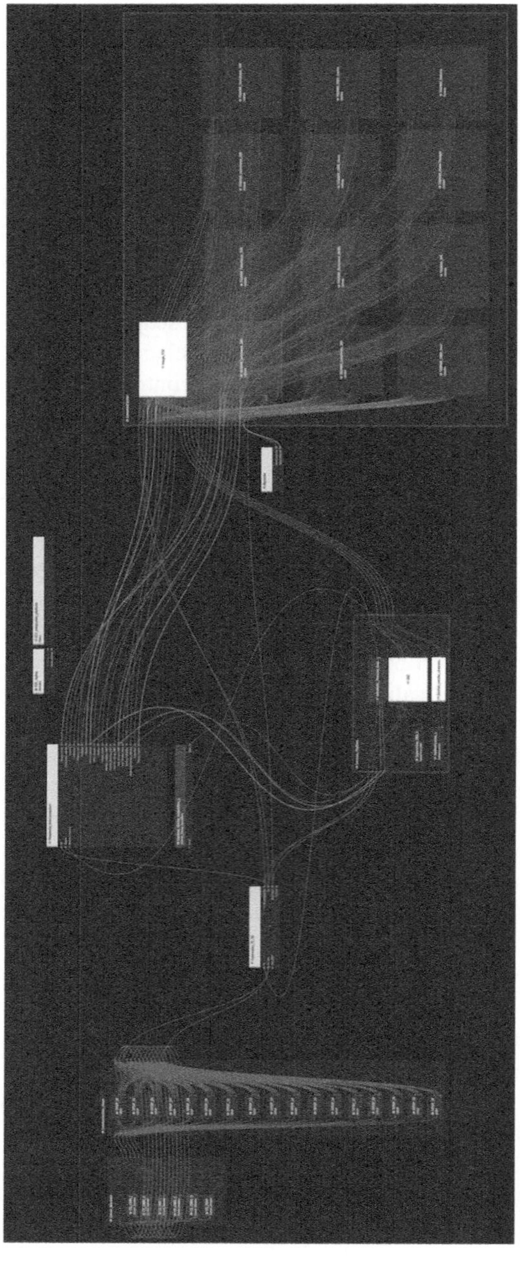

FIGURE 26.3 Example of a system of systems representation on Ingescape.

Actionability is obtained with Ingescape agents, who also expose their inputs to be written to and services to be called. Moreover, Ingescape™ Circle provides both an *input writer* and a *service caller,* enabling all user profiles to inject data or trigger services in the SoS on any accessible agent.

The first use case for **Observation** and **Actionability** is the unit testing of a single agent, where developers can combine the *writer/caller* and *observer* tools to inject data into an agent and then check the logs of the agent and the published outputs and service replies. This can be achieved in real time, while the developers are coding, compiling, and iteratively testing one or more functions inside an agent. With Ingescape™ Circle, unit and integration testing become no-code, interactive, and model-based.

No-Code Record/Replay/Export

Ingescape platforms can record systems-wide activity, including the outputs and service calls for any agent inside a platform. Such records can then be replayed and exported, for example, for external statistical analysis.

Replay capabilities are beneficial in various situations:

- During projects, teams collect external data from external systems and develop the new system by injecting this realistic data.
- During exploitation, recorded data is collected on operational platforms and replayed on integration or test platforms for deeper analysis outside inaccessible infrastructures.
- Records are also a nice way to execute training scenarios in realistic contexts.

Record and replay do not require any change in the code or configuration of the Ingescape agents. It is a transversal feature, which requires a data lake for records storage, but does not exert constraints on other agents.

Systems Scripting for Automation, Verification, and Validation

Ingescape has a domain-specific language (DSL) dedicated to creating scripts, allowing for the injection and checking of data inside any Ingescape platform. Data injection consists of writing data to agent inputs and calling agent services. This DSL also includes features to start and stop agents, create timers, execute system commands, ask for user actions, emulate mouse, touch, and keyboard events, etc.

The DSL syntax has been designed to be accessible to non-developers. The usage of the DSL is a natural extension of the visual tools inside Ingescape platforms in a more compact and scalable manner. Scripts are easy to generate from existing routines or test logic to transpose into an Ingescape model. Scripts have been used to create test data, help in the learning phases of AI engines, set systems in particular states for testing or simulation of failures, etc.

FEDERATE HSI TEAMS AROUND RELEVANT MODELS AND STANDARDS

Ingescape enables interactions with SoSs using models. In this section, we cover relevant models and tools to be engaged in HSI activities. The ability to unify or combine these models is a key enabler for new, unique opportunities around *simulation, integration, measurement, verification, and validation* of STSs.

Activity Modeling and User-Centered Scenarios

Ingescape™ Circle comes with support for SysML Activity Diagrams, SysMLv2 Actions, and BPMN diagrams, combined with timeline-based modeling. These standards and tools enable the description and the execution of system-wide workflows with opportunities to go into details and test alternative paths—such workflows position tasks and functions at different levels of depth. Tasks are split into tasks or functions to be allocated in an SoS, attached to hardware, software, and human actors.

Workflows can be modeled and executed at any stage of maturity during a project (Figure 26.4), providing qualitative insights about the behavior of an STS from early phases to advanced validation campaigns, and during the actual exploitation of the STS. *Hybrid simulations* involving tasks, simulators, software, hardware, and humans in the loop can be freely designed and executed, keeping consistent representations and immediate ways to run workflows inside the STS.

Metrics and Resources Computation During STS Simulations

With Ingescape, tasks and functions involved in workflow executions may be attached to the calculation of system-wide metrics and resources, so that the execution also generates quantified results for all involved performance indicators.

In addition, optimization and simulation engines, scenario execution engines, and any other software involving actual quantitative and possibly realistic data are seamlessly mixed inside Ingescape platforms, offering new opportunities for dynamic performance measurements in STSs.

MBSE Mixed with Live Executable Agents

The innovative approach of Ingescape with tangible, observable, actionable, measurable, and verifiable models has been successfully extended to MBSE. Activity

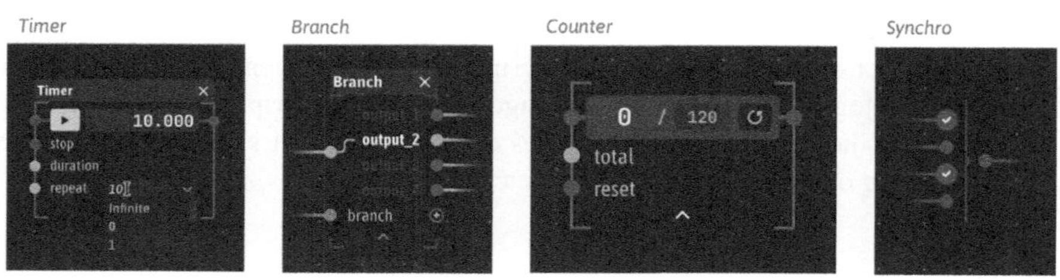

FIGURE 26.4 Examples of workflows and connectivity of Ingescape.

diagrams and State diagrams can be visually executed and combined with all the other Ingescape tools and agents. Sequence diagrams can be declared theoretically, checked, and even discovered when running a simulation or an actual system inside Ingescape™ Circle. SysML parametric, calculations, and actions can be directly included and executed inside Ingescape™ Circle.

The Systems Engineering models built in this context can be executed, visualized, interacted with, and verified in live conditions, relying on Ingescape platforms. The benefits of iterative MBSE and collaboration between *system architects, software developers, system integrators, and system validators*, all using shared visual and interactive representations, are immediate and significant. This approach guarantees that the modeled workflows and sequences remain useful, relevant, and up to date during the STS lifecycle.

Requirements and Automated Verification and Validation

Model-based integration, verification, validation, and qualification (IVVQ) become possible when model-based approaches are continuously propagated and preserved through software development.

In this context, requirements may be expressed in executable scripts, using the Ingescape scripting DSL. Such scripts, when executed on an STS at any stage of maturity, enable the generation of reports embedding the traceability and compliance data for the STS requirements. This is achieved at the systems level, without any constraint on the underlying software code in the systems, and with automatic execution supported by DevOps toolchains. Resulting traceability and compliance reports are processed by application/product lifecycle management (ALM/PML) solutions, offering a complete digital continuity on requirements engineering, and thus an opportunity to iteratively improve and refine such requirements.

ASSEMBLE ALL THE PIECES FOR CONTINUOUS MODEL-BASED DESIGN, SIMULATION, DEVELOPMENT, AND IVVQ

With all that precedes, we can iteratively create and combine models, both for the systems' and users' sides of things, and make the SoS observable, actionable, measurable, and verifiable interactively through their models. All these activities are orchestrated inside a consistent HSI process, continuously supported by Ingescape™ Circle. This process occurs inside global MBSE processes and seamlessly fits inside them at all stages by actively supporting multidisciplinary collaboration and digital continuity.

This HSI process is structured into five steps covering the various involved HSI activities (Figure 26.5). Step 5 materializes the pivot to the next iteration in the process. In early phases, Steps 1, 2, and 5 are iteratively engaged to support the emergence of context, challenges, and solutions. Then, Steps 3 and 4 are activated to provide more tangible and mature instances of system components and more advanced ways to assess and qualify them while enabling the iterative refinement of context, requirements, models, and developed solutions.

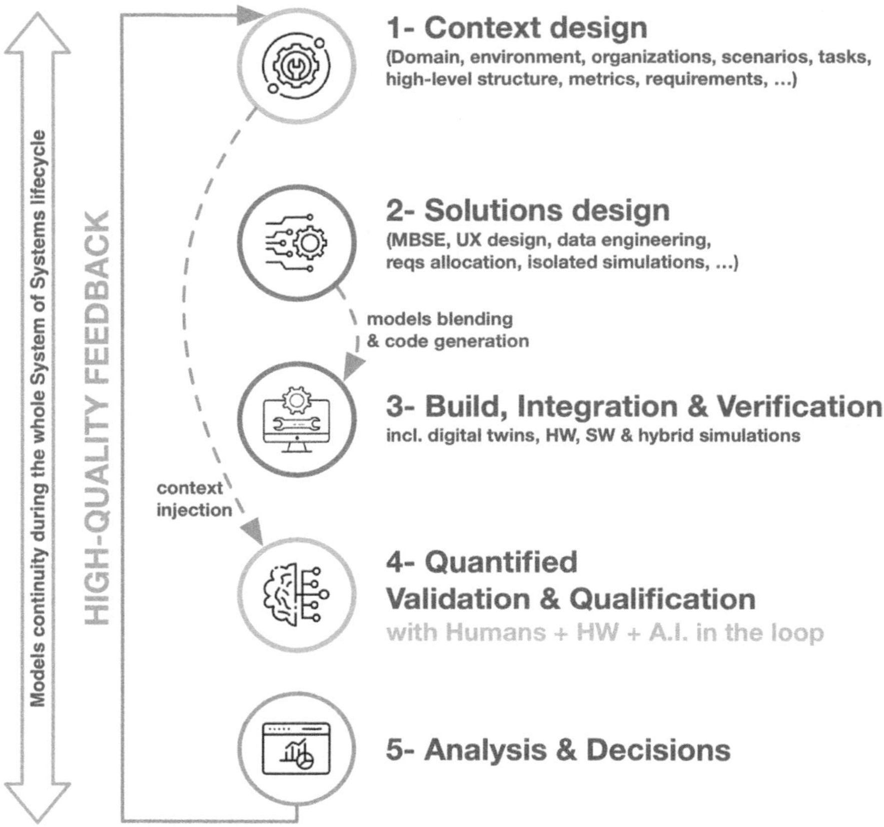

FIGURE 26.5 HSI process using Ingescape.

The underlying activities and tools are all iterative and multidisciplinary, with a unified model-based approach, offering an advanced digital continuity between models, behaviors, code, and assessment capabilities across iterations.

Strategies for Human-in-the-Loop Simulation

Simulation environments in the aeronautics and defense domains with high throughput, low latency, model-based data exchanges have been at the core of Ingescape since its inception, to enable such simulations to communicate with very heterogeneous and highly scalable software and hardware platforms.

Human-in-the-loop simulation (HITLS) is achieved within Ingescape distributed architectures, where dedicated agents serve as mediators between humans and the rest of the system, and where tasks, workflows, and scenarios are executed in dedicated tools above the models of the systems. These interactive agents range from encapsulations of physical devices and sensors to graphical and natural user interfaces, fully integrated with other systems.

Here are the most relevant strategies that we have developed to guarantee HITLS capabilities in various conditions, even when the simulators themselves are not available:

1. In modernization projects, when a system is already deployed, Ingescape enables the early record and replay of field activities, so that the project team can **continuously rely on realistic operational data and contexts**.
2. Ingescape actively supports **hybrid simulation** scenarios, integrating any simulation technologies. The flexible approach of Ingescape toward time management and real-time constraints significantly eases the time synchronization requirements at the system level for such activities.
3. Ingescape proxies are a powerful solution to **remotely access simulations and external systems from their dedicated infrastructure**, without moving the HITLS environment close to them.
4. **Mixing simulations, existing operational systems, replays, and basic emulators** is seamlessly possible on Ingescape platforms. This means that even for operational systems, already in a production stage, Ingescape creates a space for **realistic experimentation for new features, without risking breaking or destabilizing the operational system itself**.

Scientific Assessments in Sociotechnical Systems

Ingescape comes with a state-of-the-art extension dedicated to scientific assessment. This extension enables the definition of scientific protocols with independent and dependent variables, hypotheses, and scenarios compatible with all the other Ingescape concepts and tools. Protocols and scenarios can be executed on physical Ingescape platforms in various environmental conditions. The resulting datasets can then be replayed and analyzed with statistical computations.

MEASURED INGESCAPE BENEFITS IN A REAL-WORLD HSI PROJECT

Ingescape has been deployed in ambitious and complex real-world projects ranging from exploratory R&D platforms in defense and aeronautics to critical systems in transportation and energy. These projects have served to analyze the needs of their multidisciplinary teams and to iteratively introduce and improve solutions, while measuring the performance gains offered by Ingescape in the various stages of these projects. This section focuses on a transportation project consisting of the modernization of the command-and-control center for a réseau express régional (RER) train line in Paris (Figure 26.6), conducted from start to finish through a consistent use of Ingescape technologies and tools. We used an iterative HSI approach, performed by a multidisciplinary team, including system engineers, software architects, UX/UI designers, human factors specialists, integrators, and verification and validation (V&V) engineers.

FIGURE 26.6 Command-and-Control Center for an RER train line in Paris.

The illustration below shows a before/after result on the command-and-control operators' workstations, moving from electromechanical systems combined with data on paper, to a 100% digitalized system providing much improved situation awareness and decision-making, especially during a crisis.

Such complex projects necessarily rely on MBSE. Ingescape has enabled the importation of Capella models so that they can be collectively improved by systems engineers, software architects, and requirements specialists. The gain for these activities could not be rigorously quantified. However, we have observed a significant improvement in collaboration, shared vision, and the ability to progressively improve the models, requirements, and systems along the project's multiple iterations. Moreover, we broke the usual silos, engaging all the actors in an actual iterative and multidisciplinary process.

This project consisted of more than 60 systems to integrate at its core, with connections to several external systems. Integration was achieved on dedicated platforms, first for teams in charge of each system, then with larger testing and integration platforms between teams and partners, finally providing the client with a validation platform and, of course, the deployed system. All integration activities were conducted using Ingescape live models to enforce and verify interface contracts, with a massive simulation and injection of realistic data recorded in the field.

V&V relied on the execution of requirements scripts through the dedicated Ingescape DSL mentioned earlier, and the automation of this execution into DevOps pipelines, which were triggered every time a new version of a system component was pushed. This massive automation **reduced time and effort for V&V activities by 80%**, compared to initial estimations.

The continuous model-based integration approach was combined with the formal expression of the requirements through system scripts, the DevOps-automated and repeatable V&V activities, and the consistent use of simulation relying on realistic data. Compared to initial estimations, it enabled the **division by 5 of the integration and verification time and efforts** on the dedicated platforms.

In such large projects, engaging critical systems on a functioning train line, the on-site validation and qualification shall be conducted very efficiently in small time slots, generally at night during the weekends, when field work is minimal, and the line is closed to passengers. Because of this, any delay in the validation and qualification campaigns may have huge impacts on project planning and delivery dates. With the described approach, such campaigns could confidently rely on all the work already achieved in the various integration platforms, already engaging trusted context and field data, and an already mature system. Validation and qualification campaigns involved the customer, who benefited from the visual and interactive models to observe and qualify the system. **On-site validation and qualification time and efforts were divided by 8**, with no regression or unexpected behavior observed during their execution.

This global modernization project on one of the largest urban train lines in the world took **less than 3 years, when usual deployments of command and control for simpler metro lines take 4–5 years**. This demonstrates that model-based HSI projects with iterative approaches and multidisciplinary teams are relevant and that Ingescape can be considered a serious solution for conducting them.

CONCLUSION—INDUSTRIAL METHODOLOGIES WITH INGESCAPE

The previous sections demonstrate that, by its very nature, Ingescape empowers and actively supports multidisciplinary and iterative organizations in industrial projects. The tangible model-based approach enabled by Ingescape is also a unique asset for continuous lifecycle management of complex industrial projects through HSI (Boy, 2020), including requirements, models, and code, from the early design stages to deploying a verified and qualified system.

Ingescape is compatible with and improves the efficiency of all agile methodologies, such as rapid application development (RAD), Scrum, eXtreme Programming (XP), and all lean, feature-driven, and behavior-driven methodologies.

With its highly supervised and yet fully decentralized features, Ingescape is also suitable for the scaled agile framework (SAFe), intended for agile practices at the enterprise level. Ingescape provides absolute freedom for multidisciplinary teams regarding their organizational and technological choices and solid contracts and meeting/synchronization points between these teams for design, development, integration, and validation activities.

MBSE methodologies such as Arcadia and the many SysML-based methods are compatible with Ingescape, naturally extending them toward more iterative and multidisciplinary approaches. Model importation from Capella and several proprietary and open-source SysML tools has been successfully achieved.

On the users' side of things, user-centered design (UCD) methodologies, but also design thinking and more generally UX and HCI processes, usefully rely on the task modeling, HITLS, and scientific assessment tools provided by Ingescape.

The unique combination and (re)conciliation of user-centric and system-centric methodologies through Ingescape ultimately offers new grounds for long-awaited holistic MBHSI practices.

NOTE

1 https://github.com/zeromq/ingescape

REFERENCES

Boy, G.A. (2020). *Human Systems Integration: From Virtual to Tangible*. CRC Press–Taylor & Francis Group, Boca Raton, FL.

Boy, G.A. (2023). Model-Based Human Systems Integration. In *Handbook of Model-Based Systems Engineering*, A.M. Madni & N. Augustine (Eds.). Springer, USA, pp. 471–499. https://doi.org/10.1007/978-3-030-27486-3_28-1

Boy, G.A. & Kennedy, G. (Eds.). (2023). *Human Systems Integration Primer Volume One*. INCOSE, San Diego, CA. ISBN 978-1-937076-12-2.

Madni, A.M., Augustine, N. & Sievers, M. (2023). *Handbook of Model-Based Systems Engineering*. Springer Nature, USA.

Minsky, M.L. (1986). *The Society of Mind*. Simon and Schuster, New York. ISBN 978-0-671-60740-1.

Popper, S., Bankes, S., Callaway, R. & DeLaurentis, D. (2004). *System-of-Systems Symposium: Report on a Summer Conversation Archived 2011-09-28 at the Wayback Machine, July 21–22*. Potomac Institute for Policy Studies, Arlington, VA.

Section VI

Safety, Security, and Operational Contexts

27 Optimizing Overall System Performance and Human Well-Being in High-Technology Manufacturing

Tero Sotamaa, Ari Teppo, Arto Reiman, and Elina Parviainen

INTRODUCTION

Companies worldwide have become increasingly aware of the need to deepen their understanding of sociotechnical systems (STS) complexities to maintain and develop competitiveness sustainably (Polojärvi et al., 2023). Leveraging STS understanding is a significant challenge in today's turbulent and rapidly evolving business environments, which rely on intense competition for market leadership. In this regard, companies tend to lean on business- and technology-oriented management paradigms that they have typically relied on but do not always allow full consideration of the human work perspective in this complexity. Service orientation, another general trend in manufacturing (e.g., Jiang et al., 2022; Lahiri et al., 2022), adds more complexity from the STS perspective. Companies need to consider the human work perspective internally and how outsourcing and location-free work tasks affect STS performance (e.g., Kolade & Owoseni, 2022).

In manufacturing, business development has strongly leaned on technological development, often associated with Industry 4.0, which promotes digitalization as the key to manufacturing performance development (Dolgui et al., 2022). In this regard, the human work perspective is often overlooked (e.g., Neumann et al., 2021; Reiman et al., 2021). Although there is proof that these new technologies have reduced physical labor in manufacturing, evidence also shows they have introduced new risks for productivity, health, and safety. The literature associates these risks with cognitive and psychosocial domains (Min et al., 2019) related to interactions between humans and the system (Shamsi & Aris, 2021). These interactions often stem from process or resource deficiencies within a company and are influenced by external business factors (Grondys et al., 2021). In this regard, the profoundness, applicability, and suitability of risk management (RM) practices and processes are crucial to companies (e.g., Zare et al., 2016). Manufacturing discussions often center on holistic

DOI: 10.1201/9781003666431-33

RM, encompassing human factors (HFs), social aspects, occupational health, safety, technology, manufacturing processes, stakeholders, services, performance, products, and quality (Cheng et al., 2021).

As a design-oriented discipline, human factors and ergonomics (HF/E) play a pivotal role in STS design and related RM (Karwowski, 2012). From the STS perspective, HF/E theories, principles, data, and methods, combined with other information and principles of the system, serve to achieve simultaneous development in terms of human well-being and overall system performance. Regarding systems engineering (SE), a central focus in this textbook, HF/E also adopts a system-oriented approach. When examined from the systems ergonomics perspective, HF/E can be approached from the micro and macro domains. Microergonomics is interested in individual-level work systems, whereas macroergonomics focuses on broader STS complexities, such as manufacturing lines and their optimization (Wilson, 2014). This chapter explores how SE, HF/E, and RM can be approached in sociotechnical systems engineering (STSE) through system modeling and utilizing information and data from failure mode and effects analysis (FMEA) reports. This integrative approach allows for a dual emphasis on the technical and social aspects of the system, hence offering a link and bridging a gap in data-driven system design and development (Baxter & Sommerville, 2011).

This chapter presents the complexities of optimizing STSs using a system modeling approach supplemented with HF/E and RM knowledge. We point out that modeling alone does not guarantee optimization outcomes, but it provides support when analyzing and designing system structures and processes, system capabilities, outputs, and overall performance. The modeling example in this chapter resembles real manufacturing processes from Nokia Mobile Networks' radio prototype production. With its ambiguous vision of operating a "conscious factory," the company can be considered a forerunner in adopting new Industry 4.0 technology, making it an interesting example for analysis. Through modeling, we aim to identify the human–system interfaces in the production system and understand how they impact the human phases of the manufacturing process and the system's overall performance. We analyze the sociotechnical complexity of these interfaces at both the micro and macro levels to understand how they impact the overall performance of the manufacturing process as a system through various human phases. To add an RM perspective to the examination, we utilize real data and information from the process produced by the company's FMEA process. FMEA helps prioritize the system's failure modes that should be addressed, considering their severity, likelihood of occurrence, and detectability (e.g., Stamatis, 2019). To achieve the above STS modeling objective, we aim to answer the following research questions (RQs):

1. What critical elements must be considered in STSE when modeling STS at the micro and macro levels?
2. How can STSE be harnessed to enhance manufacturing performance and human well-being?
3. How do RM and FMEA data support system performance optimization within a socio-technical manufacturing environment?

BACKGROUND

STS Engineering

In this chapter, we approach SE from the STS perspective. We refer to an STS as a complex human–machine system combining technology, organizations, and people (Boy, 2023). SE can be considered an integrative approach to help teams collaborate to understand, design, and manage systems and their complexity and to deliver successful systems (INCOSE, 2023a). We adhere to the principles of SE to prevent system failures often caused by unidentified complexities, communication breakdowns, and a lack of awareness (Holt, 2023). However, we are also willing to highlight the sociotechnical aspect and incorporate the human element into the discussion (Polojärvi et al., 2023). Incorporating SE activities and comprehending operator characteristics, objectives, and needs together with organizational objectives are key activities in the SE lifecycle when defining HF requirements and considering performance (INCOSE, 2023b). In this regard, successful SE emphasizes communication as a key element. This scope includes communication within information technologies, service-based, and technical systems. Another critical aspect of SE involves comprehending all the stakeholders connected to the system (Holt, 2023). These stakeholders may be internal (e.g., employees, supervisors, and management) or external (e.g., customers, service providers, and suppliers) (see, e.g., Dul & Neumann, 2009).

SE can be applied to various classifications and taxonomies of systems, such as natural, designed abstract, transcendental, designed physical, and human activity systems (Holt, 2023, p. 3). Each system type is defined by a standard set of characteristics, allowing it to be understood and developed. A system consists of elements that interact, eventually forming its overall structure (Holt, 2023, p. 4). In the STSE perspective, the significance of human activity systems and designed physical systems is emphasized because they typically involve technological objects and human interactions to achieve a common objective or target (Holt, 2023, pp. 3–4). In this regard, planning and control are critical in designs that aim to optimize system performance and, simultaneously, human well-being concerning the objectives of HF/E. Planning determines what and how something is intended to happen, while control makes the necessary adjustments to ensure that the operation achieves its objectives (Slack et al., 2022, p. 481).

In manufacturing, a common concern in process optimization is ensuring that machines and personnel are logically organized to optimize production efficiency. The technology must suit its production purpose at the micro level, necessitating well-trained personnel to operate the equipment. Continuous enhancement of design and processes is vital to meet quality or customer requirements (Collier & Evans, 2021, pp. 84–85). Seamless alignment of the manufacturing process with macro-level commitments, such as obligations to various stakeholders, including customers, distributors, suppliers, owners, lenders, employees, and the community, is crucial (Heyzer et al., 2017, p. 57).

When optimizing their manufacturing processes, companies must confront operational, technological, and performance-related risk factors from internal and external sources that may influence their operations (Cheng et al., 2021; ISO 31000, 2018). When considering these risks, a company should also pay attention to those that stem from the human work perspective (e.g., Reiman et al., 2021), including aspects of occupational health and safety. RM should be guided by an understanding of STS and serve as an integral part of STSE. From a systems lifecycle perspective, RM in STSE should address organizational readiness for change, such as training and recruitment, and focus on improving overall performance. To enhance well-being, it must also consider operator aging and its effects on human performance and provide support and prepare operators for retirement and broader sociotechnical changes (INCOSE, 2023b). By anticipating potential risks and taking proactive steps to minimize their impact, RM enables organizations to maintain their performance and capitalize on opportunities (COSO, 2017, pp. 3–4). When developing sustainable processes that are efficient, effective, and productive, the recognized challenges lie in producing responsible, safe, high-quality products, motivating employees, creating a safe workplace, and honoring all possible (internal and external) stakeholder commitments (Heyzer et al., 2017, p. 57).

Performance Management in STS

Performance management varies according to the organization's context (Ashdown, 2014), and it can be viewed from diverse perspectives, including strategy, organizational behavior, operations management, economics, accounting, and human resource management (Hutchinson, 2013, p. 26). Performance is driven by the interaction of people, planning, and processes, supported by a foundation of commitment, culture, and communication (Oakland et al., 2021, p. 117). In performance management, human behavior is viewed in three dimensions: (1) task performance involves duties contributing to production, (2) citizenship includes actions impacting the psychological environment, and (3) Counterproductivity involves actions that are potentially harmful to the organization. Performance developed using a holistic approach should ensure that employees excel in more than one or two dimensions. Key factors include skills, competencies, and motivation, in which motivation results from the interplay between an individual and a situation (Robbins et al., 2017, pp. 148, 501). With the complexities of STS, performance management should acknowledge that internal stakeholders may have diverse perspectives and interests. This highlights the importance of considering individuals outside the organization (Ashdown, 2014, p. 12).

To support overall performance, capacity adjustments must address micro-level considerations involving the design of production processes, technology allocation, and various HFs, including personnel and subcontractors within the broader macro-level context (Heyzer et al., 2017, p. 347). The core aim of production technology is to enhance production and productivity by seamlessly integrating micro-level considerations into macro-level systems. Their integration should meet diverse macro-level stakeholder demands, such as customer expectations, supplier dynamics, and innovations by producers (Heyzer et al., 2017, p. 332). At the micro level, performance gaps tend to be significantly influenced by information, motivation, technology, and tools within the system. These elements stand out as predominant factors impacting performance. In managing performance, it is crucial to assess the process using a holistic approach and identify performance gaps between the current and desired states (Stefaniak, 2021, pp. 4, 75).

MICRO AND MACRO PERSPECTIVES ON STS OPTIMIZATION

In this chapter, the system of interest (SoI) is the manufacturing process in Nokia Mobile Networks' radio prototype production. The manufacturing process represents a macroergonomic STS. As a macroergonomic system, it cannot be fully optimized without addressing microergonomic considerations in human–technology interfaces. In response to RQ 1, at the micro level, the primary focus was on the roles and tasks of the assembly operators, who are critical system components. The operators are influenced by immediate (work) environmental factors at the assembly line, such as noise, lighting, temperature, cleanliness, and order. These factors may impact operator well-being, safety, and productivity, affecting overall system performance. From the operational RM perspective, these microergonomic factors should be identified and optimized through management practices and processes, as they directly influence production efficiency, effectiveness, and product quality. Consequently, performance monitoring through product quality metrics is crucial, as deficiencies can negatively affect external stakeholders, including customers, and lead to warranty issues or rework, ultimately impacting overall system performance (see Figures 27.1 and 27.2).

Macroergonomic considerations must incorporate microergonomic insights and vice versa; thus, micro-level, human-centric perspectives require awareness of broader, macro-level organizational and STS considerations. As highlighted by Holt (2023), communication, encompassing information processing at both micro and macro levels, is vital given the complexity of STS. In the Nokia Mobile Networks production plant, assembly operators regularly communicate and interact with each other and operate through human–technology interfaces. At the same time, operators are embedded in their physical work environment, influencing how tasks are carried out on the assembly line. These micro-level interactions include revising instructions, monitoring equipment, making real-time process adjustments, and ensuring the quality of assembled products.

At the broader macro level, the perspective expands beyond individual workstations to encompass broader organizational factors that shape manufacturing operations, such as external stakeholder commitments (e.g., to customers, service providers, suppliers) and organizational policies and decisions (Dul & Neumann, 2009; Heyzer et al., 2017). As shown in Figure 27.2, macro-level organizational factors such as decision-making quality, communication strategies, and disruption management practices directly shape the social environment and influence micro-level conditions like assembly cell dimensions and physical arrangements. For example, at Nokia Mobile Networks' radio prototype production plant, managerial decisions on product quality, required safety protocols, and output objectives directly affect the ergonomic setup of assembly cells (i.e., determine the cells' ergonomic design at the micro level, thus influencing operator well-being, workload, and the risk of operational failures.

From an operational RM perspective, the scope of risk identification should extend beyond internal process deficiencies (Grondys et al., 2021) to encompass the entire manufacturing system. This includes all interconnected stakeholders (Holt, 2023), technological components, product quality, and overall performance outcomes. Crucially, it must also address HFs, the social environment, and occupational health and safety. Particular attention should be paid to maintaining an organized and orderly work environment, ensuring cleanliness, and safeguarding both the physical and the cognitive well-being of workers (Ashby, 2022, pp. 197–200; Palumbo et al., 2022).

As Holt (2023) highlighted, a key aspect of external stakeholder relations is the customer interface, which serves as a two-way communication link between manufacturing operations and customers through the supply interface. Viewed as a macro-level, interorganizational perspective, this interface underscores how customer demands and feedback influence not only the manufacturing process but also the roles of suppliers, supply chains, and end customers (Holt, 2023, p. 241; Slack et al., 2022). Planning and control are essential for maintaining system efficiency and operating through demand and supply interfaces. The demand interface focuses on understanding customer demands and market trends, while the supply interface involves managing production resources and logistics. At the macro level, core planning and control activities, such as scheduling, loading, sequencing, monitoring, and control, are central to optimizing production and resource allocation (Slack et al., 2022). These activities are critical in production efficiency, resource allocation, and manufacturing performance. Addressing these planning and control strategies from an HF/E perspective requires a solid understanding of the theoretical principles of RM and performance management and HF/E's potential to support the development of an STS in manufacturing.

MODELING THE MANUFACTURING PROCESS

In our modeling example, we applied model-based systems engineering (MBSE), in which all the information is included in the model (Holt, 2021). Using a Cameo Systems Modeler tool and the Systems Modeling Language (SysML), we created a visualization of our SoI. The logical model diagrams presented in this chapter consist of SysML block definition diagrams (bdd) and SysML use case (uc) diagrams. Visualization through modeling eases understanding of the SoI more than a plain textual description. It also allows the system information to be stored in the database used by the modeling tool. This database enables the use of model information in simulation tools and system data for automated document generation. In our macroergonomics-oriented bdd model (see Figure 27.1), the overall hierarchy and process dependencies of our SoI can be observed. The SoI is presented in this chapter in a simplified form. Blocks within the diagram represent different system elements and illustrate their relationships and the overall performance flow. A block represents a conceptual element that forms part of the system, while relationships establish connections between these blocks within the system. In a simplified form, internal and external stakeholders (visualized as human characters in Figure 27.1) associated with the system characterize who or what has an interest in the system (Holt, 2021).

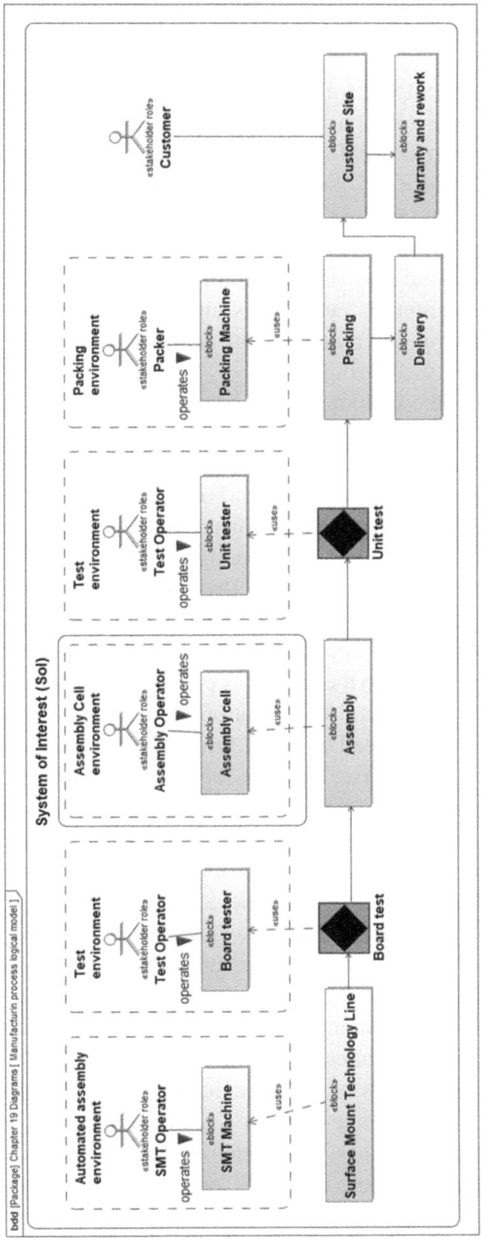

FIGURE 27.1 Model diagram of the SoI: a manufacturing process in prototype production.

Modeling Human Work as Part of STS

To shed light on the dynamic interactions, interfaces, and their relationships within our SoI, we deepened the analysis by focusing on a distinct part of the system. We concentrated on the assembly cell, where the final assembly of the radio product prototype was conducted.

In response to RQ 2, in STSE, comprehensive design contributes to overall system performance by promoting a holistic understanding of the system. Integrating design principles into decision-making fosters communication and planning, facilitates iterative development, prioritizes the user experience, and ensures systems' adaptability for the development process or changing conditions. From a human-centric perspective, the microergonomic work system is integral to consider how organizational factors shape the social environment, influencing interactions among operators, tools, instructions, jigs, materials, and the assembly environment. Process functionality must be balanced with key requirements designed to enhance system performance, operator safety, and well-being.

Figure 27.2 visualizes the system from a microergonomic perspective. This micro-level approach allows system presentation from the individual assembly operator's perspective. The diagram in Figure 27.2 divides the process into input and output components. On the input side, the social and organizational factors influencing the manufacturing process and system performance are defined. These organizational influences include management practices, company policies, workload expectations, and compensation structures, which shape the social environment and potentially impact operator performance and well-being. The manufacturing process block specifies the process requirements, technical criteria, quality standards, and safety protocols from the manufacturing system to the assembly cell. The *assembly cell* block indicates the specific location of the assembly operation. This block is characterized by its physical attributes and is influenced by *(work) environmental* factors, including workspace conditions, cleanliness, noise, lighting, temperature, and occupational hygiene. In the assembly cell, the *assembly operator* can be considered a critical stakeholder whose performance is influenced by competencies, skills, physical capabilities, and cognitive abilities. Physical and cognitive attributes, influenced by social and organizational factors, emerge as workload, fatigue, stress, motivation level, and job satisfaction, each significantly shaping operator well-being and performance.

To perform work tasks, the operator relies on instructions, tools, production jigs, and materials provided within the assembly environment. The diagram illustrates interactions between these elements and thereby identifies their interfaces and relationships. Understanding connector interfaces is crucial in determining whether the interaction occurs both ways or solely unidirectionally. To provide insight into human well-being and safety issues within the manufacturing process, we identified physical factors within the assembly environment block that are related to the operator's health and performance, including occupational hygiene, noise, lighting, temperature, workload, fatigue, stress, motivation level, and job satisfaction. For instance, poor occupational hygiene can potentially lead to health issues, while a hygienic process environment contributes to the health and well-being of the assembly operator. Poor occupational hygiene or inadequate environmental conditions (e.g.,

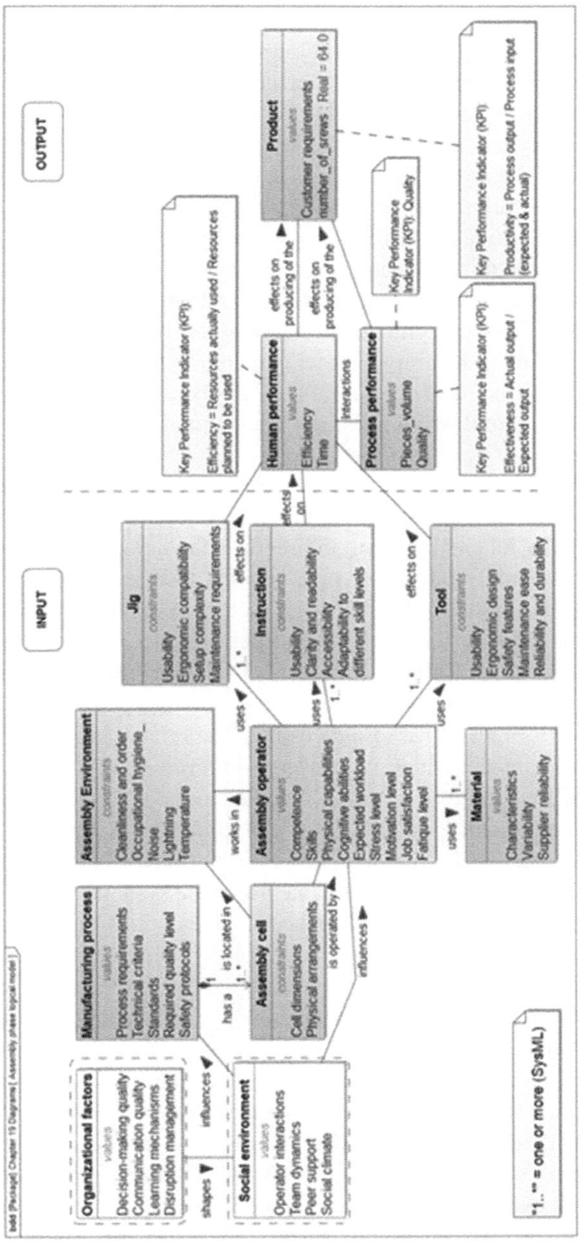

FIGURE 27.2 Elements and indicators of overall performance in high-tech manufacturing.

noise, lighting, temperature) can lead to increased operator fatigue and stress, lower motivation, decreased job satisfaction, and potentially to health issues. In contrast, optimal conditions in these areas positively contribute to operator health and well-being, ultimately enhancing overall performance.

An organized and orderly assembly environment is presumed to reduce stress for the assembly operator by providing a safer and more comfortable working environment that positively impacts well-being. In addition, operator competencies play a crucial role in safety, as trained and skilled operators are more likely to perform tasks more safely, which reduces the risk of accidents or injuries. Adequate training and continuous skill development promote operator safety and well-being. However, we identified several critical issues related to the operator's physical and cognitive aspects that required attention. When designing assembly tasks, operators must consider their physical capabilities to ensure that job and physical demands align closely with their abilities; a close alignment promotes well-being and prevents physical strain or injury. Aligning task complexity with operators' cognitive abilities is equally crucial for mental well-being. Tasks optimized to operator capabilities, with appropriate workloads and complexity levels, mitigate stress and reduce the likelihood of errors and injuries.

Furthermore, organizational factors and social conditions, such as company policies, compensation, workload management, and overall organizational culture, significantly shape the social environment, indirectly influencing manufacturing process performance and ultimately impacting operator well-being and safety. Our insights highlight that considering human and process factors in manufacturing system modeling contributes to safety and overall system performance. An integrated approach ensures a holistic focus that aligns RM and STSE principles by comprehensively addressing operator well-being and system safety.

The output side in Figure 27.2 visualizes how the relationships on the input side impact human performance, which is measured in terms of efficiency, based on the utilization of human resources over time. When examined from a technological perspective, process performance is defined by piece volume and piece quality, which can be considered the attributes that include key performance indicators (KPIs). KPIs such as timings and specific objectives are integral to the system's input–output scope. Considering human and process performance, it is essential to account for factors influencing the process and those placing strain on it, including potential constraints. The output side results in the product, specifically the produced radio product prototype. Ensuring this *product* block meets customer requirements without a warranty or rework is important.

When examined from the HF/E perspective, our diagram shows, in a simplified and visualized form, how HFs influencing system performance can have various complexity issues. The diagram in Figure 27.2 is structured to illustrate the interactions and contributions of various elements during the assembly phase that ultimately impact the system's overall performance and the final output (i.e., the product being produced). Human-specific constraints, such as the usability of tools, instructions, materials, and jigs, directly influence the performance of both the human operator and the process. This has implications for the quality and volume of the product, aligning with customer requirements. In this regard, modeling diagrams are helpful in MBSE for visualizing complex systems and their interactions. Modeling involves integrating work planning, organizing, time management, and the efficient execution of human-operated work processes by each individual. Such integration allows for a better comprehension of overall performance.

Additionally, models can enhance understanding of relationships and communication among blocks and stakeholders, aiding in analysis and decision-making processes within an STS (Polojärvi et al., 2023). Referring to Reiman et al. (2023), we ask whether and how companies organize, orchestrate, and manage the above-described human-centric STSE process in practice. Reiman et al. (2023) provide evidence that even in manufacturing companies considered forerunners in adopting new production technologies, organizational capabilities for STSE remain limited. A key challenge is that HF/E expertise is often applied by occupational health and safety professionals, whose focus tends to be restricted to personnel RM rather than the broader development of human work performance.

FMEA Considerations for the SoI

Considering design-driven solutions that address process and HFs, RM becomes an integral element of STSE when optimizing overall system performance (Grondys et al., 2021). In response to RQ 3, in our example, four process phases influencing system performance were analyzed using Nokia Mobile Networks' process FMEA information and data. The process phases were volume line, mechanical parts preparation, final assembly, and packing. Assembly tasks and their potential failure modes in each phase were evaluated to assess their impact on process, product, and customer. The analysis aimed to identify failure modes requiring attention by evaluating factors such as severity, occurrence, and detectability, as well as the risk priority number (RPN), following the general principles of process FMEA (Stamatis, 2019). In the RM process, each factor was ranked on a 1-to-10 scale, where higher values indicate greater severity or probability (Stamatis, 2019, pp. 66–70). RPN was calculated by multiplying the severity, occurrence, and detectability ranks, serving as a metric for risk prioritization. In our dataset, six final assembly tasks with potential failure modes were identified among 69 assembly tasks, according to the process FMEA report (see Figure 27.3).

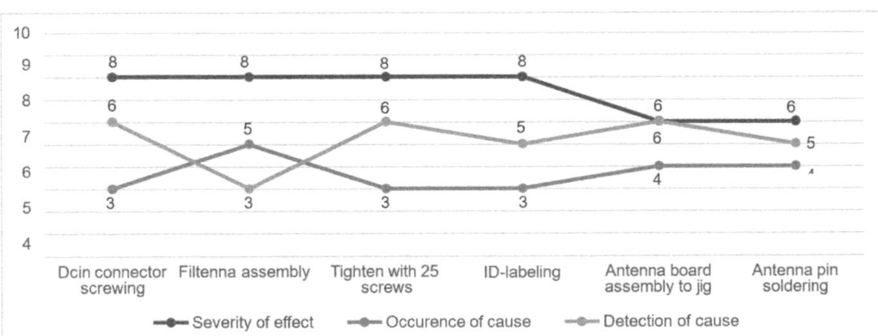

FIGURE 27.3 Failure rankings according to assembly task.

These six tasks exceeded RPN 120, indicating critical risks where human error could significantly affect process performance. In the final assembly phase, failures with a severity rank of 6–8, primarily caused by operator mistakes, had been identified. In the severity scale, ranks 4–6 indicate a moderate impact, potentially causing some customer dissatisfaction, while 7–8 suggest a higher potential for dissatisfaction and disruptions on subsequent processes or services, ultimately affecting manufacturing process performance. The failure occurrence was rated 3–5, indicating that these assembly tasks are often associated with failed processes (see Figure 27.3). These values suggest a low to moderate probability of failure, which is typically related to occasional or isolated incidents. Half of the identified failures had a detection rank of 6, indicating a relatively low detectability, while others ranked at 5 or 3, suggesting a moderate (5) or high (3) detectability. It is important to note that higher detection ranks (e.g., 6) pose challenges for controllers in identifying and addressing failures. High RPN scores indicate bottlenecks and inefficiencies affecting operational performance. Effective planning and control strategies must address technical issues while promoting sustainable quality and performance.

In the final assembly phase, nearly all 69 assembly tasks, including those with RPN values exceeding 120, were associated with potential failures caused by operator errors, such as missing components or incorrect assembly. However, if the analysis is left at that stage, it may be misleading and insufficient to identify the failures' root causes. A closer examination of the process environment and communication processes between human interfaces could reveal underlying issues or HFs contributing to the observed failures. Identifying operator errors as a common cause of potential failures underscores the importance of HF/E in manufacturing (Wilson, 2014). Internal (e.g., technicians, quality control personnel, and maintenance staff) and external stakeholders (e.g., technology developers) may also have an interest and capabilities in finding solutions and helping identify root causes. Therefore, a comprehensive understanding of macroergonomics, including the complex interfaces and interactions among these stakeholders, is required for a thorough STS design analysis. From the STSE perspective, development efforts should focus on eliminating root causes of failures in the assembly tasks to prevent their recurrence.

The probability rankings in process FMEA data revealed frequent occurrences of specific effects of failure (e.g., functional or reliability issues), suggesting systemic weaknesses in the process or possible gaps in operator training and competence. Despite moderate (5) and high (3) detection rankings, existing controls may have lacked sufficient preventive efficacy for improvement, particularly in early process stages. These insights highlight the need for a comprehensive STS analysis addressing microergonomics (operator–equipment interactions) and macroergonomics (design processes and procedures for risk mitigation). Shamsi and Aris (2021) noted that risks often stem from HFs within the system. In Nokia Mobile Networks' manufacturing system design, the process FMEA insights also revealed that deficiencies in process resources should be comprehensively considered through system modeling. As human and technological components are critical system elements, system modeling must integrate them in the assembly process to meet stakeholder demands and customer requirements at the micro and macro levels. This integration is necessary to achieve performance and quality objectives while mitigating inefficiencies that may affect operational performance (Cheng et al., 2021; Mourtzis, 2020).

Future Research

Future research in STSE should explore integrating micro-level considerations, such as work-phase design, process requirements, technology allocation, and HFs, with macro-level strategies involving stakeholder engagement. Particular attention should be devoted to effectively incorporating RM and HF/E practices. Investigating the integration of RM into system design and capacity adjustments could provide valuable insights into enhancing manufacturing resilience and performance in a dynamic manufacturing environment. Future research should also address the impacts of outsourcing work tasks on overall manufacturing performance and examine specific HFs critical to system risk assessment, including their potential quantitative impacts on system performance. Moreover, future research should incorporate real-life information and data with stakeholder input into models to enable realistic simulations supporting process design from an STSE perspective.

CONCLUSION

The theoretical perspectives presented in this chapter enhance understanding of STSs from an SE perspective. STSE is a key concept in integrating HF/E and RM into STS thinking. The chapter emphasizes the interplay between people (the social system) and the manufacturing process (the technical system) within an STSE framework. Alongside STSE, our visualization through modeling facilitates comprehension of the SoI beyond a textual description. Comprehending key concepts, such as HFs, RM, and sociotechnical interactions, is critical when modeling an STS. When aiming to develop and optimize manufacturing performance, process visualizations, in the form of SE diagrams, can provide a basis for understanding the interfaces and interconnections within the manufacturing process. In the context of STSE, emphasis should be on identifying interfaces, understanding their impact on human phases at the SoI, and enhancing overall system performance. Social and technical aspects must be integrated to prevent failures. This requires technical improvements and a focus on the critical human element in an STS to enhance understanding of the dynamic interactions, interfaces, and relationships at the SoI. Integrating design principles into decision-making can promote effective communication, support iterative development, prioritize the user experience, and ensure adaptability for evolving or dynamic conditions during development.

From an RM standpoint, the FMEA analysis underscores the need for proactive measures and an integrated approach to addressing high-severity issues in the final assembly phase. These measures should focus on detecting, mitigating, and preventing failures, particularly those associated with potential human errors. The central role of the assembly operator and the influence of immediate environmental factors highlight the importance of considering both social and technical aspects for an effective RM strategy in a manufacturing process. RM principles should guide the prioritization and mitigation of issues within STSE. A proper understanding of HF/E informs understanding of how changes in process design can reduce human error and enhance overall human performance and resource efficiency. In this regard, operator diversity and (for instance) aging should be considered critical factors impacting human performance and require careful attention. Thus, considerations related to operators throughout the SE lifecycle should extend beyond the process perspective to encompass broader organizational changes. These perspectives align to improve overall performance, ensuring that the system is effective, efficient, productive, and capable of meeting both customer production requirements and the needs of its human operators.

In adopting the synergy of HF/E and RM-driven design strategies, manufacturing organizations can proactively reduce cognitive and physical strain on human operators, thereby minimizing the likelihood of failures. A strategic focus on high-risk areas provides operational benefits for well-being and safety and delivers advantages such as decreased waste and enhanced product quality. This holistic approach fosters sustainable improvement in operational performance and workplace well-being, creating an environment where excellence in operations is continually pursued.

REFERENCES

Ashby, S. (2022). *Fundamentals of operational risk management: Understanding and implementing effective tools, policies, and frameworks*. Kogan Page Limited.

Ashdown, L. (2014). *Performance management*. Kogan Page.

Baxter, G., & Sommerville, I. (2011). Socio-technical systems: From design methods to systems engineering. *Interacting With Computers*, 23(1), 4–17. https://doi.org/10.1016/j.intcom.2010.07.003

Boy, G. A. (2023). An epistemological approach to human systems integration. *Technology in Society,* 74, Article 102298. https://doi.org/10.1016/j.techsoc.2023.102298

Cheng, S., Jianfu, S., Alrasheedi, M., Saeidi, P., Mishra, A. R., & Rani, P. (2021). A new extended VIKOR approach using q-rung orthopair fuzzy sets for sustainable enterprise risk management assessment in manufacturing small and medium-sized enterprises. *International Journal of Fuzzy Systems*, 23, 1347–1369. https://doi.org/10.1007/s40815-020-01024-3

Collier, D. A., & Evans. J. R. (2021). *Operations and supply chain management* (2nd ed.). Cengage Learning Inc.

Committee of Sponsoring Organizations of the Treadway Commission (COSO). (2017). *Enterprise risk management integrating with strategy and performance*. COSO eBook Publications. www.aicpa-cima.com/cpe-learning/publication/coso-enterprise-risk- management-integrating-with-strategy-and-performance

Dolgui, A., Sgarbossa, F., & Simonetto, M. (2022). Design and management of assembly systems 4.0: Systematic literature review and research agenda. *International Journal of Production Research*, 60(1), 184–210. https://doi.org/10.1080/00207543.2021.1990433

Dul, J., & Neumann, W. P. (2009). Ergonomics contributions to company strategies. *Applied Ergonomics*, 40(4), 745–752. https://doi.org/10.1016/j.apergo.2008.07.001

Grondys, K., Targalski, J. and Piechocki, R. (2021), "The role of finance in small and medium-sized enterprises' business continuity: evidence from Poland", Entrepreneurial Business and Economics Review, 9(1), 119-135.

Heyzer, J., Render, B., & Munson, C. (2017). *Operations management: Sustainability and supply chain management* (12th ed.). Pearson Education Ltd.

Holt, J. (2021). Systems engineering demystified: A practitioner's handbook for developing complex systems using a model-based approach. Packt Publishing.

Holt, J. (2023). *Systems engineering demystified: Apply modern, model-based systems engineering techniques to build complex systems*. Packt Publishing.

Hutchinson, S. (2013). *Performance management: Theory and practice*. Chartered Institute of Personnel and Development.

International Council on Systems Engineering (INCOSE). (2023a). *The human systems integration (HSI) primer: Volume one*. Wiley.

International Council on Systems Engineering (INCOSE). (2023b). *Systems engineering handbook: A guide for system life cycle processes and activities* (5th ed.). Wiley.

ISO 31000. (2018). *Risk management – guidelines*. International Standardization Organization.

Jiang, Z. Z., Feng, G., Yi, Z., & Guo, X. (2022). Service-oriented manufacturing: A literature review and future research directions. *Frontiers of Engineering Management*, 9, 71–88. https://doi.org/10.1007/s42524-021-0171-3

Karwowski, W. (2012). A review of human factors challenges of complex adaptive systems. *Human Factors*, 54(6), 983–995. https://doi.org/10.1177/0018720812467459

Kolade, O., & Owoseni, A. (2022). Employment 5.0: The work of the future and the future of work. *Technology in Society*, 71, 102086. https://doi.org/10.1016/j.techsoc.2022.102086

Lahiri, S., Karna, A., Kalubandi, S. C., & Edacherian, S. (2022). Performance implications of outsourcing: A meta-analysis. *Journal of Business Research*, 139, 1303–1316, https://doi.org/10.1016/j.jbusres.2021.10.061

Min, J., Kim, Y., Lee, S., Jang, T.-W., Kim, I., Song, J. (2019). The Fourth Industrial Revolution and its impact on occupational health and safety, workers' compensation, and labor conditions. *Safety and Health at Work*, 10(4), 400–408. https://doi.org/10.1016/j.shaw.2019.09.005

Mourtzis, D. (2020). Simulation in the design and operation of manufacturing systems: State of the art and new trends. *International Journal of Production Research*, 58(7), 1927–1949. https://doi.org/10.1080/00207543.2019.1636321

Neumann, W. P., Winkelhaus, S., Grosse, E. H., & Glock, C. H. (2021). Industry 4.0 and the human factor – a systems framework and analysis methodology for successful development. *International Journal of Production Economics*, 233, Article 107992. https://doi.org/10.1016/j.ijpe.2020.107992

Oakland, J. S., Oakland, R. J., & Turner, M. A. (2021). *Total quality management and operational excellence*. Routledge.

Palumbo, R., Casprini, E., & Montera, R. (2022). Making digitalization work: Unveiling digitalization's implications on psycho-social risks at work. *Total Quality Management & Business Excellence*, 1–22. https://doi.org/10.1080/14783363.2022.2055458

Polojärvi, D., Palmer, E., & Dunford, C. (2023). A systematic literature review of sociotechnical systems in systems engineering. *Systems Engineering*, 26, 482–504. https://doi.org/10.1002/sys.21664

Reiman, A., Kaivo-oja, J., Parviainen, E., Lauraeus, T., & Takala, E-P. (2023). Human work in the shift to industry 4.0: A road map to managing technological changes in manufacturing. *International Journal of Production Research*, 62(16), 5613–5630. http://dx.doi.org/10.1080/00207543.2023.2291814

Reiman, A., Kaivo-oja, J., Parviainen, E., Takala, E-P., & Lauraéus, T. (2021). Human factors and ergonomics in manufacturing in the Industry 4.0 context – a scoping review. *Technology in Society*, 65, Article 101572. https://doi.org/10.1016/j.techsoc.2021.101572

Robbins, S. P., Judge, T. A., & Campbell, T. T. (2017). *Organizational behaviour* (2nd ed.). Pearson Education Ltd.

Shamsi, N. S. N. A., & Aris, A. (2021). Ranking of risk management factors influencing SMEs' organizational performance. *International Journal of Sustainable Construction Engineering and Technology*, 12(5), 180–188. https://doi.org/10.30880/ijscet.2021.12.05.018

Slack, N., Brandon-Jones, A., & Burgess, N. (2022). *Operations management* (10th ed.). Pearson Education Ltd.

Stamatis, D. H. (2019). *Risk management using failure mode and effect analysis (FMEA)*. ASQ Quality Press.

Stefaniak, J. E. (2021). Needs assessment for learning and performance: Theory, process, and practice. Routledge/Taylor & Francis Group.

Wilson, J. R. (2014). Fundamentals of systems ergonomics/human factors. *Applied Ergonomics*, 45(1), 5–13. https://doi.org/10.1016/j.apergo.2013.03.021

Zare, M., Croq, M., Hossein-Arabi, F., Brunet, R., & Roquelaure, Y. (2016). Does ergonomics improve product quality and reduce costs? A review article. *Human Factors and Ergonomics in Manufacturing & Service Industries*, 26(2), 205–223. https://doi.org/10.1002/hfm.20623

28 Safety, Security, and Health

Considerations on Individual and Organizational Professionalism

Alexander Rudolph

INTRODUCTION

Safety, security, and health considerations play an essential role throughout the lifecycle of a system. What they have in common is the need to understand and ultimately to accept the interactions between the system-of-interest (SoI) and the humans that use it, as well as the associated risks, making Human System Interaction (HSI) a key element for all of them.

Process-Oriented Systems Engineering (INCOSE, 2023) aims to embed safety, security, health engineering, and HSI into an overarching concept. Still, it remains vague about a generalized interaction between the system of interest and any specific human or organization. While the valid definition of the SoI is a fundamental activity of Systems Engineering, an equivalent definition of a human-of-interest (HoI), e.g., of a particular user type with a vital competency, is not laid out in the standards.

As one consequence, today's standards on safety engineering, both in the military (US Department of Defense, 2012) and in the civilian (ISO/IEC, 2014) context, as well as on security engineering (ISO/IEC, 2022), are widely technical without direct reference to HSI. Health or healthcare engineering is still an emerging engineering discipline. The term has been used for decades, but the definition and the scope of healthcare engineering remain ambiguous (Chyu et al., 2015). HSI techniques can and should complement the existing frameworks with complex and controversial aspects, such as human intention, determination, and professionalism.

Consider, e.g., the photo illustrated in Figure 28.1, which, to point to some elements, is intentionally without a technical system. The torero is undoubtedly not in a healthy situation. However, would the situation itself be rather unsafe or insecure? Or both? Or none of both? The answers will ultimately refer to the professionalism of the torero, being exposed to a risk he had decided to accept.

This paper aims to establish a top-level view on professionalism, both by human individuals and by the organizations composed of them. Some implications on the notion and the meaning of HSI are further derived and discussed, as well as their incorporation in Systems Engineering.

DOI: 10.1201/9781003666431-34

FIGURE 28.1 An unsafe, insecure, and unhealthy situation (© Northrop Grumman Corporation 2007).

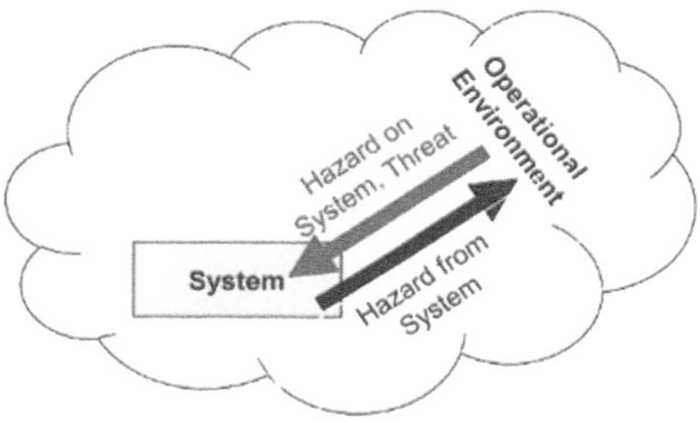

FIGURE 28.2 Safety and security at the system boundary.

SAFETY, SECURITY, AND HEALTH

Traditionally, safety and security are defined at the boundary of a system. The presence of safety is preoccupied with an acceptable level of risk from the hazard potential that the system, by itself, introduces to its operational environment. Security, the other way around, is considered present if the level of risk emanating from environmental threats is acceptable. So-called security systems intended to reduce the environmental risk need a careful trade-off between their intended use and the new risk they will introduce simply by their existence (Figure 28.2).

Health is special as it is a state of the human itself, not a system. The relationships between a specific human and the system of interest and other humans must be analyzed to understand the factors that influence the health of this human.

Engineering View

Table 28.1 gives an overview of safety, security, and health. From an engineering perspective, there are some relevant implications:

TABLE 28.1
Safety, Security, and Health

Aspect	Definition	Meaning	Examples
Safety	State of system risk	Absence of accidental harm caused by system	System failure, misuse
Security	State of environment risk	Absence of natural and intentional harm from the environment	Abuse/sabotage, acts of God, violent aggression
Health	State of a single human or of a group of humans	Presence of well-being	Physical, mental, and social dimensions

1. Improving security typically comes with lowered safety margins. An environmental control system ensuring human survival,[1] e.g., introduces new failure modes.
2. A safe system, e.g., a zero-accident system, would still be prone to environmental risks. It is therefore not possible to build it.
3. A system is deemed fit-for-purpose if the environmental risk, including the accidental risk the system introduces to the environment, is acceptable.

A trade-off must balance the fitness for purpose of the system of interest, ultimately based on utilitarian analyses of its overall risks and benefits.

For this trade-off, we must discriminate between a "safe system" and an "effective system." In the trade-off between system risk and environmental risk, the security effect of the second may justify reduced safety margins for the first.

The Human Element

The considerations described in the above section do not, in general, apply to any specific human or a group of humans with unknown relationships. Any system declared fit-for-purpose does not, by principle, increase or decrease the health of humans or any individual human in its context.

The special relationship between the system and a specific human need needs to be analyzed. The notion of the HoI as an equivalence to the SoI is a useful denotation. Conventional Systems Engineering (INCOSE, 2023) merely subsumes users as stakeholders, which is known to lead to oversimplifications initially. The notion of a HoI allows us to consider specific motivations.

The obvious HSI implication is the need for a basic taxonomy of those humans who are somehow expected to interact with the system. These interactions by principle include emerging properties of the system, e.g., properties that are not known during the design time of the system. Amongst them may also be a misuse or even an abuse of the system, e.g., by fundamentally repurposing it.

This leads to a key statement of this paper: HSI as a discipline needs the following:

1. An alignment with the intended users of the system, as dedicated HoI, and
2. An alignment also with the unintended users of the system regarding misuse and abuse, as a dedicated HoI.

One necessary consequence of the two alignments is the need to consider human intention and professionalism. HSI in the sense of human systems integration must primarily focus on integrating professional humans in reasonable belief, and also on defending against all other humans.

A SIMPLE TAXONOMY OF HUMANS

Unfortunately, the health of the human depicted in the center of Figure 28.3 is ultimately a limited entity for some obvious reasons.

First, nature will eventually decrease individual health and life, intrinsically through disease or infirmity, and extrinsically through natural danger. There is an additional health effect from societal pressure, such as that caused by a reference group of humans rather than one specific individual human.

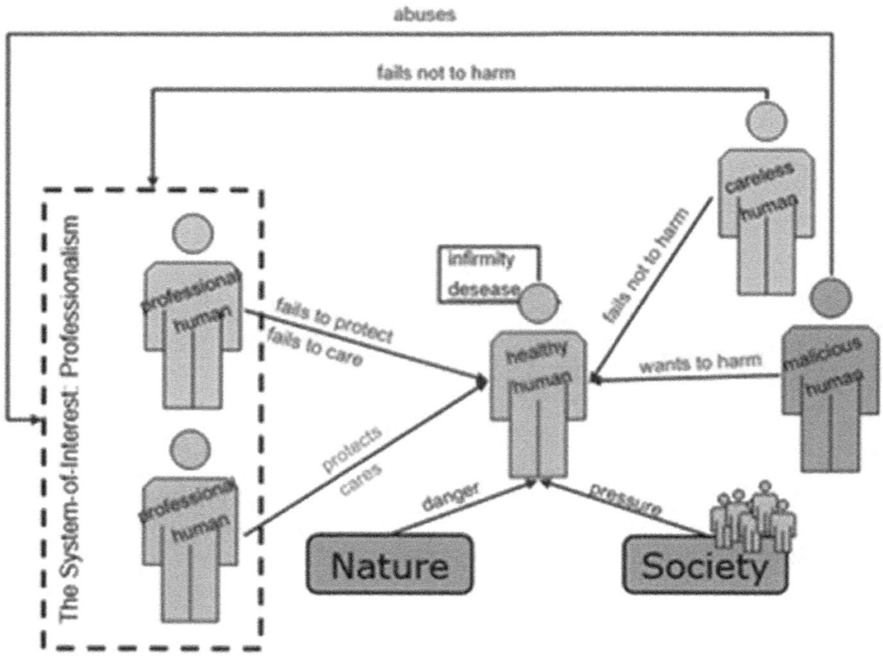

FIGURE 28.3 The "healthy human" is in the center.

Second, humans who reasonably believe they positively impact human health, e.g., by providing either care or protection, may fail to do so. Nevertheless, with their intention and professionalism, humans of this kind are the only real source of any positive health effect for the healthy human.

Third, two types of humans have or may hurt the healthy human:

- A malicious human deliberately harming a healthy human. This may imply the abuse of the system of interest.
- A careless human accidentally fails to harm a healthy human. This may imply the misuse of the system of interest.

Preventing abuse and reasonably foreseeable misuse is vital to security engineering and safety engineering. This typically includes societal and legal aspects, as the criminalization of abuse and, depending on the specific case and the legal and ethical duties of care, also of misuse.

Individual Professionalism

As mentioned in the section above, a professional human in good faith is the only source of a positive health impact. This paper aims to acknowledge this bare fact for the common HSI understanding. It is therefore proposed

1. to refer to the SoI as the system that is used by these HoIs, who are professional humans in good faith,
2. to refer to human-system integration as the engineering discipline
   a. of empowering these HoI, who are professional humans in reasonable belief, to use the SoI as intended.
   b. we must prevent these HoI, not professional humans with reasonable belief, from misusing and abusing the SoI.

This second reference to HSI may seem controversial at first sight, as it means the opposite of integrating humans with the system. However, there is no paradox behind it, but the necessity to incorporate the right humans into the system, only those who will use the SoI as intended, and only as intended.

Domain Examples

In summary, the domain examples in Table 28.2 back the pattern described in Figure 28.3. The systems that mitigate environmental risk (e.g., contribute to security) simultaneously introduce systemic risk, just by their bare existence.

A healthy human has the support of a professional human with reasonable belief, in each domain underpinned by a legal or ethical framework. This may include personal oaths or a legal need to bear a license.

TABLE 28.2
Domain Examples for Human Profession

Domain	Environment Risk ("Security")	System of Interest	System Risk ("Safety")	Professional Human	Healthy Human	Careless Human	Malicious Human	Particularities
Defense and military	Hostility	Weapon system	Weapon failure	Soldier	Civilian		Enemy soldier	Oath of allegiance, martial/ international law
Commercial aerospace	In-humane conditions	Aircraft	Aircraft failure	Pilot, flight crew members	Passenger	Drunken passenger	Hi-Jacker	ICAO, international and national law, licensing
	Separation Violation	Air traffic control	ATC failure	Air traffic controller				
Automotive	Bad road conditions	Vehicle	Vehicle failure	Driver	Passenger	Drunken passenger	Hacker	International and national law, licensing
Healthcare and medicine	Natural danger, infirmity, disease	Medical drug	Adverse drug effects	Medical doctor	Citizen with reduced health	Public smoker	Violent offender	Hippocratic oath, laws of emergency
Manned space flight	In-humane conditions	Spaceship incl. S/G interface	Space system failure	Astronaut	Space tourist	Impudent space tourist	Hi-Jacker	Very high level of professionalism
	Lost spacecraft	Ground control	Transmission loss	Spacecraft operating crew				
Public security	Catastrophes, riots	Early warning system (EWS)	EWS false warnings	Policeman	Citizen	Undutiful person	Criminal	Official oath, national civil law with duty of care

SOCIETIES AND ORGANIZATIONS

System Acceptance

Accepting an SoI, including its risks and adverse effects, is a human privilege and obligation. From this perspective, the entire discipline of Systems Engineering, as represented by INCOSE (2023) and standardized by ISO (ISO/IEC/IEEE, 2023), can ultimately be considered as a means of enabling acceptance from the early ideation phase of the system through the utilization phase of the useful life of the system until its final retirement phase.

Figure 28.4 gives an overview of an acceptance system, composed of three components:

1. If one, the socio-technical component provides the acceptance scheme, e.g., the process, the criteria, and the acceptance authority.
2. The domain-specific component provides assumptions and justifications available from domain wisdom and may be difficult to understand outside the domain.
3. The engineering domain provides proven engineering lifecycle practice. Typically, these practices include specified processes to inform decision-making attributed to acceptance by giving arguments and evidence.

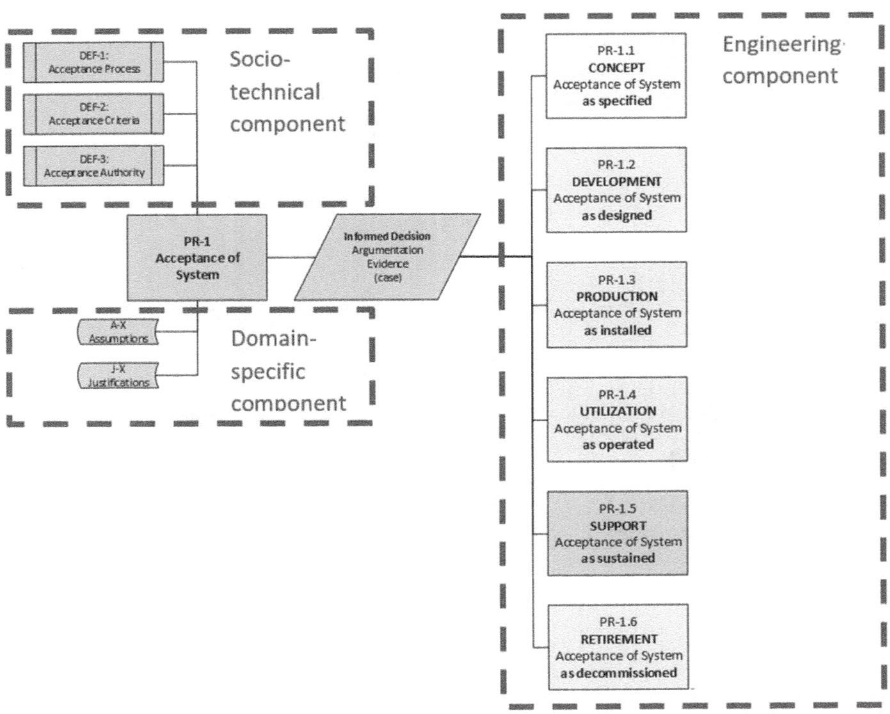

FIGURE 28.4 System acceptance.

Informed decision-making results in the unconditional or conditional acceptance of the system for the next lifecycle stage.

The rigor and the formalism of the individual acceptance processes depend on the society and the domain. There are fully controlled ecosystems, e.g., for large commercial aircraft, weapons of war, medical devices, and certain pharmaceuticals. The usage of these systems is restricted to registered or authorized professionals.

The most critical acceptance scheme is for operational use (PR-1.4 in Figure 28.4). Reality is in this phase created by the humans who use the system, as shaped by the humans who have accepted it before.

The Organizational Triangle

As depicted in Figure 28.5 for the example of safety, the risks of a system can be viewed, judged, and ultimately accepted from essentially three different perspectives:

1. System and product safety focuses on the design and implementation[2] of the system itself, aiming at minimizing the number and the severity of hazardous conditions, both during nominal operation and in the presence of system failure. A designated design authority is usually in charge, together with a dedicated conformance officer for production.

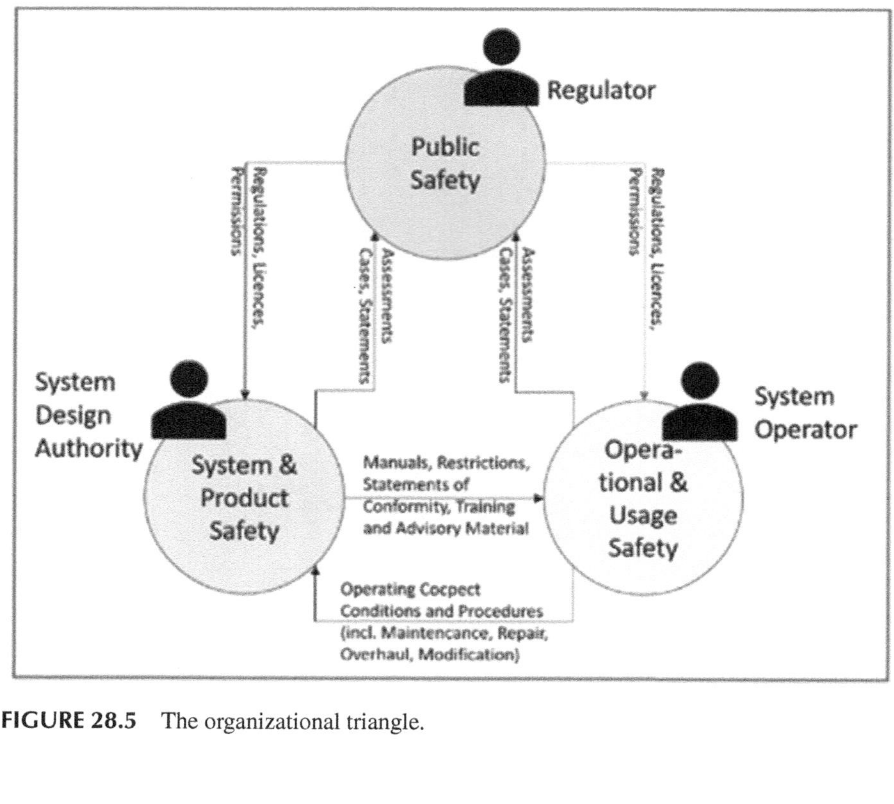

FIGURE 28.5 The organizational triangle.

2. Public safety refers to the interaction of the system and its users with the public. It aims at overall hazard control, both proactively and reactively, as a response to incidents or accidents, and includes regulation, licensing, operation, and usage permissions. A regulator typically acts on behalf of public safety.
3. Operational safety refers to normal, abnormal, and emergency operating conditions and the readiness of the users to manage them appropriately. It aims to ensure the system's and its users' continued worthiness, including maintenance and renewal. The system operator or a delegated person is responsible for operational safety.

The organizational triangle in Figure 28.5 illustrates these three corners where a system's risks and benefits are negotiated, questioned, and accepted.

ORGANIZATIONAL PROFESSIONALISM

HSI is conventionally attributed to operational and usage safety, being a means to identify nominal and safe operating conditions and the human activity needed to establish and to retain them. Organizational professionalism goes beyond and aims at a continued trade-off within the organizational triangle. This may include a temporary decrease in public safety, e.g., introducing a new technology or changing the mode of operation. Organizational professionalism is thereby a paramount enabler of actual health risk reduction. It may somehow be referred to as civilization itself.

THE SYSTEM ENVIRONMENT

To develop a general notion of organizational professionalism with safety, security, and health risks, it is helpful to think of a bare model of the environment in which the system is used, misused, and abused by the HoIs, as shown in Figure 28.6.

Any human involved is of a type introduced in Figure 28.3. Rules, as well as a certain level of rule enforcement, may exist under the assertiveness of the professionals among these humans. However, there are no relationships, and the separation between use, misuse, and abuse is primarily arbitrary and reactive.

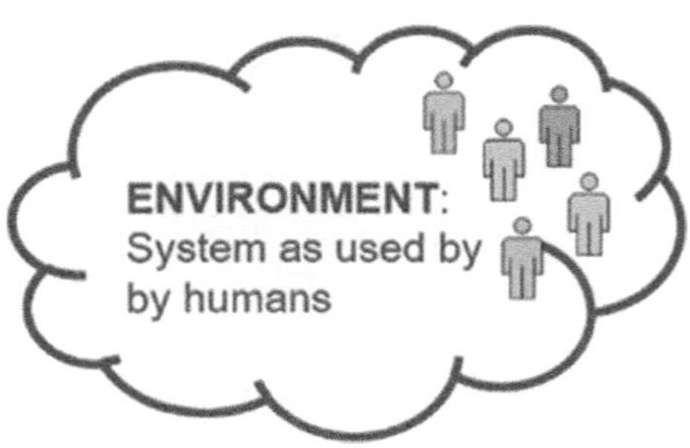

FIGURE 28.6 The system environment.

Using Processes and Measures

The processes introduced in Figure 28.4 are arranged around the system environment according to Figure 28.7, as a feedforward means to establish

1. Access rules for systems about to enter the environment, e.g., due to the need to accept their concept, their design, and their installation.
2. Continuation rules for retaining systems in the environment, e.g., due to the need to periodically accept the current modes of operation or the licenses of the users.
3. Exit rules for systems about to leave the environment, e.g., due to the need for complete retirement or the management of obsolescence.

Rule-making requires respective bodies that have both the capability and the authority. In addition to the rules, measures of effectiveness (MoEs) for relevant safety, security, and health properties may be defined at the environmental level. These could, e.g., be a general life expectancy above 75 years or an average probability of fatal road accidents below 1e-6/hr.

However, in the absence of organizational supervision, there is neither any credibility for the defined MoEs nor an association between these MoEs and the rules represented by engineering processes. The rules and processes are not negotiated amongst themselves, and their practical contribution is difficult to determine.

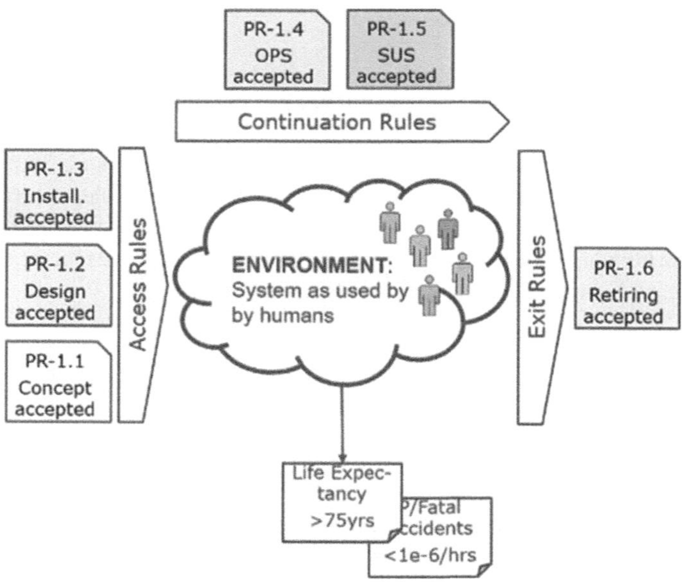

FIGURE 28.7 Engineering processes controlling environment access, continuation, and exit.

Closing the Loop

Considering the processes in closed-loop control of determined MoEs against targets is a significant extension. As shown in Figure 28.8, this, so to speak, socio-technical control loop enables:

1. to derive from the obtained indications to applicable rules, and
2. to deduce from hypothetical rules to their potential effectiveness.

The MoE definition and its structured determination from credible measurement depend on organizational authority and empowerment. This may be referred to as organizational professionalism and as societal maturity.

The derivation of adequate rules and processes requires a continuous trade-off of the systemic, operational, and environmental benefits and risks. Imagining a single entity capable of conducting this trade-off is difficult. As the corner points of the organizational triangle, the regulator, the operator, and the design authority described in Figure 28.5 may be considered the generic constituents of the overarching environmental control system.

Organizational professionalism may be seen as the deliberate and active provision of a triangle between a system design authority, a regulator, and a system operator for a continuous and balanced management of safety, security, and health issues.

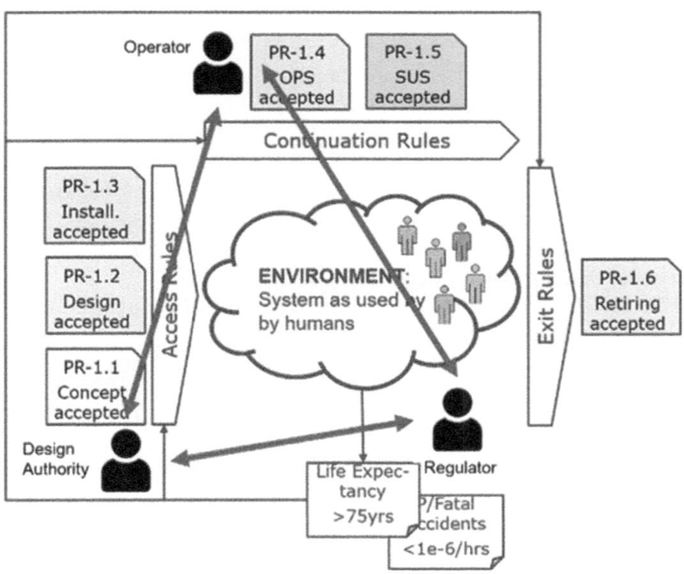

FIGURE 28.8 Closing the loop with the organizational triangle.

SUMMARY AND CONCLUSION

Based on a simplified human taxonomy, this paper attempts to develop a new HSI perspective focusing on individual and organizational professionalism for managing safety, security, and health risks. As three central concepts, the paper proposes:

1. The term HoI is equivalent to SoI.
2. The goal is to associate the "integration" in HSI primarily with the intended use by individual professionals in reasonable belief and to put secondary emphasis on the separation from misuse or abuse.
3. The organizational triangle for socio-technical concerns is a valuable model for the organizational professionalism behind balanced solutions based on defined MoEs and their structured feedback.

On one hand, focusing on individual and organizational professionalism helps define HSI objectives in Systems Engineering. On the other hand, it may also represent a practical aid for the HSI application in specific ventures and projects.

NOTES

1 For example, the cabin pressure control system of a passenger aircraft.
2 Implementation here includes the serial production and assembly of the system in conformity to the previously accepted design.

REFERENCES

Chyu, M.-C., Austin, T., Calisir, F., Chanjaplammootil, S., Davis, M. J., Favela, J., … Haddas, R. (2015). Healthcare Engineering Defined: A White Paper. *Journal of Healthcare Engineering*, *6*(4), 635–648.

INCOSE. (2023). *Systems Engineering Handbook.* San Diego, CA: Wiley.

ISO/IEC. (2014). *Guide 51:2014 – Safety Aspects: Guidelines for Their Inclusion in Standards.* International Organization for Standardization.

ISO/IEC. (2022). *27001:2022 – Information Security, Cybersecurity and Privacy Protection.* International Organization for Standardization.

ISO/IEC/IEEE. (2023). *15288:2023 – Systems and Software Engineering.* International Organization for Standardization.

US Department of Defence. (2012). *MIL-STD-882E Standard Practice System Safety.* US Department of Defence.

29 Human Systems Integration (HSI) in Safety Management Systems

Putting the Pieces in Sync ...

Jean-Jacques Speyer

THE IMPLICATION

The concept of zero accidents has led airline customers to expect this as a guarantee. While we are nearing this goal, it is not yet assured. This paper describes the journey towards achieving it. Safety Management Systems (SMSs) are not exclusive to aviation; they also play crucial roles in various industries such as healthcare, food production, nuclear safety, chemical and oil industries, civil construction, and finance. These systems help improve relationships between integrated systems in organizations, aiming to better correct, prevent, or predict failures through operational safety control and dynamic risk management. Human System Integration (HSI) is leading the final steps.

The implementation of SMS was formally established by ICAO's Safety Recommended Practices (SARPs) under Annexe 19 in 2006. This initiative led to considering technical specifications for implementation at both the state and operator levels. This dual approach has gradually evolved since the first version of the Global Aviation Safety Plan (GASP) in 1997, which formalized a series of conclusions and recommendations developed by the Air Navigation Bureau and industry.

In the 2006 Global Strategy for Aviation Safety, the Conference of Directors General of Civil Aviation endorsed the development of a Global Aviation Safety Roadmap. It is recommended that ICAO develop an integrated approach to safety management. This approach provides two distinct but complementary frameworks for coordinating worldwide safety policies and initiatives (ICAO, 2013):

- The implementation and maintenance of a State's Safety Program (SSP).
- Implementing and maintaining a service provider's safety management system (SMS).

As a management program, SMSs must be an integral part of every level of the entire structure of any company. An interactive and proactive SMS process requires that

DOI: 10.1201/9781003666431-35

a CEO be directly involved in its deployment for effective safety assurance. This is where pieces here are not necessarily in sync, whether at the state, airline, or manufacturer's level.

On one hand, we prioritize safety risk management by analyzing data to identify hazards and assess risks to mitigate their consequences (Speyer, 2023). Conversely, we monitor performance to manage change and continuously enhance the SMS.

We must safeguard safety information about aircraft design, manufacture, and operation by local legal regulations.

Although ICAO member states have agreed to collaborate to achieve uniformity, this does not imply compliance; variations may exist between SARPs of individual countries and the practices of local operators.

Our principal objective is to illustrate that HSI can address these challenges by highlighting safety themes and issues recognized as significant beyond subjective interpretation. This approach enables early identification and prompt safety, regularity, and efficiency actions.

EXTRACTING LESSONS LEARNED

Let us review the journey that helped extract lessons, eventually using HSI.

The US Commitment

In 2017, the Aviation Rulemaking Advisory Committee (ARAC) was created to achieve stakeholders' consensus on FAA rulemaking issues (Figure 29.1). This set

FIGURE 29.1 SMS FAA: Resolution for Safety Risk Management and Safety Assurance.

the stage for sharing vital safety information proactively, allowing stakeholders to detect hazards, risks, and weak signals and develop industry solutions. ICAO's distinction between reactive, proactive, and predictive safety measures is becoming a reality in mature SMSs.

We must remain vigilant about emerging trends and precursors despite lower aircraft accident rates due to increased flights. At the core of an SMS, Safety Risk Management and Safety Assurance play essential roles (Speyer, 2023).

The European Commitment

The European Human Factors Advisory Group, now known as the Collaborative Action Group (CAG), developed a framework addressing European systemic, operational, and emerging aviation issues. EASA Advisory Bodies, CAGs, and the European Network of Analysts work together to assess these issues (Figure 29.2). CAGs analyze European aviation safety, including human factors, with members from industry, academia, and partners. They regularly review safety information, conduct analyses, identify emerging issues, monitor safety performance, and provide feedback on actions taken.

They assist EASA in establishing and maintaining safety risk portfolios using a standard risk classification for national authorities. The European Plan for Aviation Safety (EPAS) aligns with ICAO's Global Aviation Safety Plan every three years. EPAS Volume I sets strategic priorities for this period, with a mid-term review for relevance. Volume II (EPAS actions) and Volume III (Safety Risk Portfolios) are updated yearly (EASA, January 2024).

FIGURE 29.2 EASA: Safety Risk Management Process.

The Systems Approach with Models and Methods

We will analyze data to extract safety knowledge using a systems approach. Air transport employees, being human, will make errors; this is not a moral issue but rather due to latent conditions within the system (pathogens). Those at the "sharp end" often inherit problems rather than create them. Therefore, remedial actions should improve defenses and eliminate error traps by addressing systemic causes. Errors are indicators of deeper failures, and safety improvements require identifying and managing these underlying issues. Initially developed by Professor Elwyn Edwards, the SHEL model was later revised to the SHELL model by Captain Paul Hawkins in the 1980s. When an error source cannot be managed within one of the components, it is possible to adjust the interfaces with the other components.

- Errors can occur within any model component.
- Errors can also happen at the interfaces between components.

The "Environment" element helps the model account for work situations and their variability over time. Reason's "Swiss Cheese" model, developed in the 1990s and 2000s, describes how pathogens in a system can cause risks under certain conditions after incubation. Risk levels are influenced by the number of pathogenic agents, often masked by opaque systems. Factors can be weighted positively or negatively based on their impact.

Reason's model was incorporated into ICAO's Human Factors accident investigation manual in 1993, based on the idea that organizational elements must work together for safe and efficient operations. Latent and active failures can compromise system integrity, making it susceptible to hazards. These failures create "holes" in layers of an "Emmenthal Cheese," including organizational influences, unsafe supervision, preconditions for unsafe acts, and unsafe acts themselves. Absent defenses can lead to hazardous breaches, causing irreversible events.

The Human Factors and Classification System (HFACS) framework was developed to define latent and active failures, serving as a safety investigation and analysis tool. HFACS emerged from the study of numerous accident reports and human causes, detailing four levels of latent and active failures:

- Unsafe Acts.
- Preconditions for Unsafe Acts.
- Unsafe Supervision.
- Organizational Influences.

The BowTie risk management methodology is a sequential way to represent an event's development. It provides a diagrammatic illustration of the hazardous event resulting from release, the trigger hazards/threats causing the undesirable top event, its potential consequences, and the risk controls that should be implemented to minimize the identified risks through prevention or mitigation. The BowTie method has been widely adopted since the 1990s to describe and capture an overview of risk management practices. Originating in the oil and gas industry, its use has expanded to other sectors such as aviation, mining, maritime, railroads, the chemical industry, banking and finance, health care (including COVID-19), and social welfare.

DATA MINING AND ARTIFICIAL INTELLIGENCE TECHNIQUES

Focusing on aviation safety, we now discuss EASA's position on artificial intelligence (AI) implications.

EASA Roadmap

As AI advances, it rapidly permeates aviation. Although AI has been conceptualized since the 1950s, its development has significantly accelerated in the past decade due to three concurrent factors:

- The capacity to collect and store massive amounts of data.
- A dramatic increase in computational power.
- The rise of powerful algorithms and architectures.

As an airworthiness authority, EASA has proactively developed a roadmap (EASA, 2024) to promote the safe integration of new and emerging technologies in aviation. Besides numerous potential AI applications in design and operations, aircraft production and maintenance, air traffic management, drones and urban air mobility, cybersecurity, and environmental issues, safety risk management stands out for its significant data-mining potential to extract safety information from diverse sources. In this context, EASA's role includes monitoring progress in investigations, processing and following up on recommendations, and being aware of and disseminating safety issues and corrective actions, which are central to SMSs.

In the context of EASA safety intelligence and management, AI is poised to become a key enabler for the detection of emerging risks, risk classification of occurrences, safety risk portfolio design, and prioritization of issues. Specifically, looking at the application of AI to the EASA Data 4 Safety (D4S) project, machine learning (ML) could provide practical solutions to manage D4S data (i.e., significant sources of operational data like flight data, safety reports, weather data) and traffic data (which include huge volumes, variable and complex data silos, and numerous potential quality issues) during both the collection/gathering/preparation of the datasets and the analysis/resolution steps (Figure 29.3).

AI can enhance knowledge by:

- Understanding data through performance measurement and risk modeling with large historical datasets.
- Identifying hidden correlations across different data silos using data fusion.
- Detecting issues and anomalies during the incremental data flow analysis in Big Data architecture, identifying threats, vulnerabilities, anomalies, incidents, and human errors.

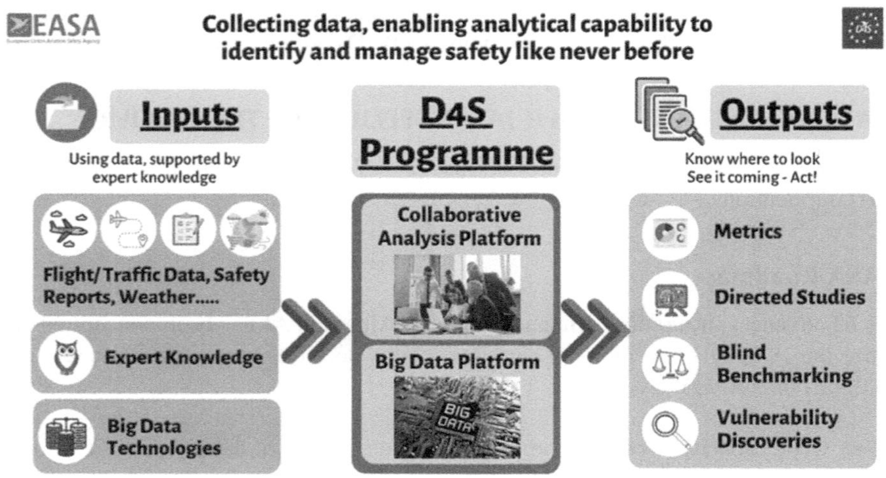

FIGURE 29.3 EASA: the essence of the D4S Programme.

A concept paper reissued in April 2024 (EASA, April 2024), following the initial 2020 release (EASA, February 2020), outlines EASA's future guidance for ML applications approval via a W-shaped process. This process highlights research areas that must be established to establish compliance methods for AI trustworthiness. The MLEAP project focuses on three main tasks regarding data and models (Figure 29.4):

- Ensuring completeness and representativeness,
- Guaranteeing ML generalization,
- Ensuring algorithm robustness and stability.

This task involves guiding, approving, and certifying AI/ML applications:

Level 1 AI/ML (Assistance to Human): starting with routine assistance (1A), then reinforced assistance (1B);

Level 2 AI/ML (Human/Machine Teaming): humans perform and machines monitor (2A), then machines perform and humans monitor (2B);

Level 3 (Advanced Automation) by 2035: machines perform without human intervention in operations, with humans involved in design and oversight.

AI Applications Ahead in the Coming Decade

A review of AI in aviation safety literature drew attention to a recent article published in December 2023 (Ziakkas & Pechlivanis, 2023). This article compared human subject matter experts (SMEs) with AI's ChatGPT to revisit high-profile aviation accidents using the systems-based HFACS model. The investigation aimed to determine:

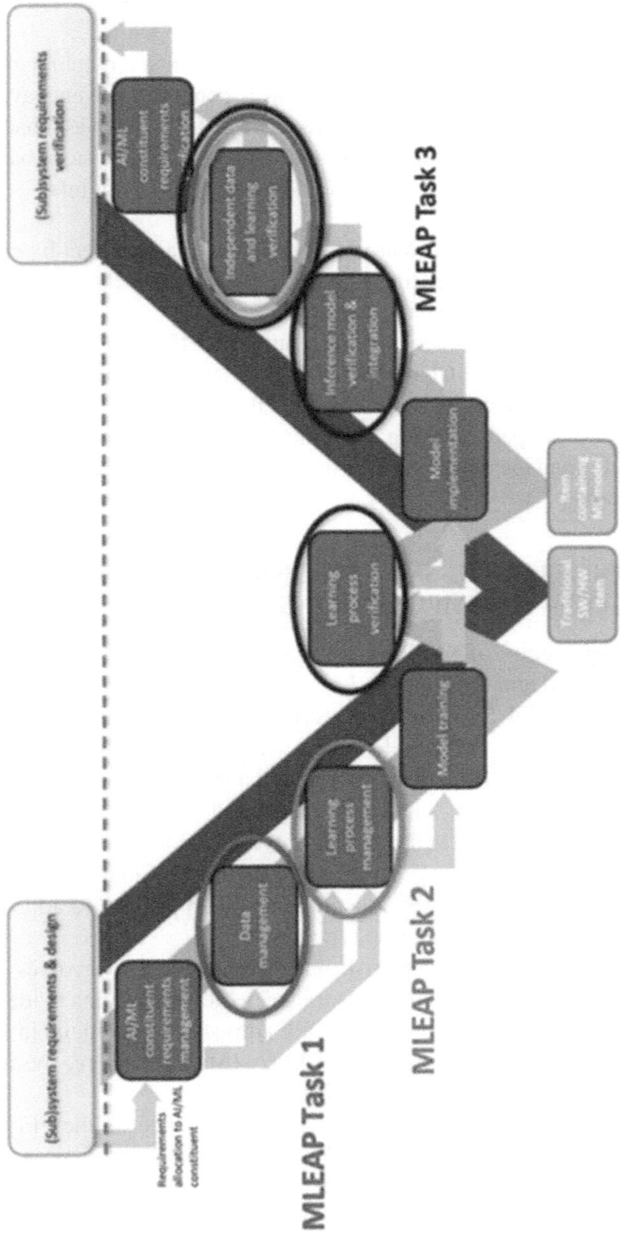

FIGURE 29.4 The MLEAP Development Assurance Process.

- Whether the ChatGPT AI agent could conduct accident analysis and compare the results with the raters?
- Whether it is possible to verify AI's learning abilities?
- Whether it could address other systems-based models?

AI applications are already integrated into the aviation sociotechnical ecosystem and are expected to expand further, embodying a genuine HSI endeavor. Ziakkas and Pechlivanis demonstrated that open-source AI/ML tools, such as ChatGPT, could effectively conduct accident classification and analysis using system-based models for more efficient training. Their findings highlight several key points:

- Contributing to continuous learning and change management by creating new knowledge from successes and failures, utilizing more appropriate analytical methods and tools for decision-making.
- Achieving "fair intelligence" between human SMEs and AI analysts.
- Performing better with ChatGPT on linear models like BowTie compared to systemic accident causation models like HFACS.
- Attaining W-shaped learning assurance (EASA), relying on multidisciplinary data management teams comprising human factors, IT, and accident investigators.

CONCLUSION: SHARING SMS LESSONS LEARNED WITH SIE AND AI – TUI – HSI IMPLICATIONS

Achieving synchronization of SMS systems has been a lengthy process, but we are progressing as the need and the requisite knowledge have been identified. This author emphasized that HSI and Systems Engineering are closely interconnected during INCOSE's HSI 2021 Conference held on November 16–19, 2021 (Speyer, 2021). Both disciplines require well-timed and finely tuned information-exchange processes to support SMSs for effective "change management" and "operational learning." Safety improvements have reached a point where individual airlines cannot gather sufficient information from their operational data alone. This underscores the importance of D4S and explains why EASA has initiated an AI process to address this particular challenge using various database types.

Central to SMS systems is safety risk management, which involves identifying hazards and mitigating risks, and safety assurance, which tracks data to identify emerging hazards. Safety Information Exchange (SIE) processes are crucial in this context and are facilitated by organizations such as IATA (Steades). Here, the information must be concrete and precise to inform decisions on SIE.

A pertinent example illustrating our current stage concerns the "FLCH" mode on a non-Airbus twin wide-body aircraft. While most pilots certified on this aircraft understand the mode well, some failed to realize that when using this mode with the flight director, with the autopilot disconnected but auto thrust armed, the latter would not respond upon reaching the new altitude. This misunderstanding contributed to the crash of Flight 214 at SFO. The FLCH issue had evaded detection through SIE. Ziakkas & Pechlivanis, 2023 suggests that open-source AI/ML tools (such as ChatGPT) could improve procedural comprehensibility and address potential cognitive gaps with significant safety implications by processing knowledge resources like operational or training manuals. As Eric Hollnagel says, "Can we ever imagine how work is done?" (Hollnagel, 2017).

Automated systems are not fully designed to take over when the aircraft leaves its flight envelope. According to Boy (2020), tangibility refers to the goal of enabling a human's ability to properly grasp and manipulate human–machine interfaces as tangible user interfaces (TUI) in the fullest operational sense. Therefore, tangibility needs to be a part of the design process. Boy et al. aim for systems safety by considering systems as sociotechnical, which is achievable but requires numerous iterations. This underscores the need to specify HSI certification requirements to avoid untreated risks. The current level is 1B, which is required to reach level 2 for HSI.

Through scenario simulations, HSI supports IT, Operations, Systems Engineering, and Human Factors and Ergonomics to prevent and predict occurrences. AI will be integrated into aircraft certification tests and processes, while safety management systems for risk management will soon be operationally synchronized. These elements will gradually align and become a practical reality, putting the pieces in sync.

REFERENCES

Boy, G.A. (2020). *Human–Systems Integration: From Virtual to Tangible*. CRC Press, ISBN: 9780429351686, https://doi.org/10.1201/9780429351686

EASA (2024, April). *Artificial Intelligence Roadmap 2.0: Guidance for Level 1 & 2 Machine Learning Applications, a Deliverable of the EASA AI Roadmap*, www.easa.europa.eu/en/domains/research-innovation/ai

EASA (2020, February). *Artificial Intelligence Roadmap 1.0: A Human-Centric Approach to AI in Aviation*, www.easa.europa.eu/en/domains/research-innovation/ai

EASA (2024, January). *European Plan for Aviation Safety (EPAS)*, 13th Edition, www.easa.europa.eu/en/downloads/137466/en www.easa.europa.eu/en/downloads/141520/en www.easa.europa.eu/en/downloads/141521/en

Hollnagel, E. (2017). *Can We Imagine How Work Is Done? Hindsight 25, Eurocontrol*, www.eurocontrol.int › portal › skybrary

ICAO (2013). *ICAO Safety Management Manual*, Third Edition, AN/474, Doc 9859, www.icao.int/SAM/Documents/2017-SSP-GUY/

Speyer, J.J. (2021). HSI and Safety: An Eye Opener from Aviation and Aeronautics Experience. In *Presented at INCOSE, IW 2021*. www.incose.org/flyer_hsi2021-flyer_final

Speyer, J.J. (2023). *Rulemaking in Aeronautics, Safety Methods and Tools*. KUL University, Belgium, Bruges.

Ziakkas, D. & Pechlivanis, K. (2023). "Artificial Intelligence Applications in Aviation Accident Classification: A Preliminary Exploratory Study," *Elsevier Decision Analytics Journal* s9, www.researchgate.net/publication/375356936_Artificial_intelligence_applications_in_aviation_accident_classification_A_preliminary_exploratory_study

30 Considering Context in Air Traffic Control System Design

Alexandre Disdier

INTRODUCTION

Today, air traffic control (ATC) mainly relies on physical towers near an airfield. Air Traffic Controllers (which we often abbreviate ATCos) are human operators working within the tower to ensure a safe and efficient aircraft movement. ATC towers are part of a larger system of systems, including, but not limited to: ATC towers, En-route centers, communication satellites, radars, and controlled planes. Altogether, these systems cooperate to ensure proper air traffic services, like maintaining an orderly and fluid aircraft traffic, and preventing collisions between them.

An ATC tower is already a complex system, as it is made of many interacting parts that together exhibit emergent behaviors, which are difficult to anticipate when the system is not in operation. It is even more than that: many of these parts are human ATCos with different roles, interacting with other teams of human operators (e.g., crash teams and fire brigades) and both hardware and software devices (e.g., radars and software modules for managing flight information). We will further justify the socio-technical nature of control towers in the section "ATC Systems Are Complex Socio-Technical Systems."

Studies of alternatives to physical towers have been carried out by industrials and Air Navigation Service Providers (ANSPs) since the end of the 1990s. The most researched system of interest in this respect is the Remote and Virtual Tower (RVT), which consists of moving ATCos from the physical tower to a remote center, thus replacing the traditional out-the-window view with screens surrounding the ATCos. The section "Why Context Matters: The Case of RVTs" will outline the evolution of the developed RVT prototypes and their limitations regarding the Human Systems Integration (HSI) of people, organizations, and technologies. In particular, we will discover that such systems are susceptible to their *operational context*.

The notion of context is crucial in Systems Engineering (SE), as its intricacies make it hard to predict the behavior of a system before it has been integrated and operationalized. Context seems like a rather intuitive notion, yet studying the literature about it shows that it is difficult to grasp and define, as we will observe in the section "Can We Define System Context?"

DOI: 10.1201/9781003666431-36

ATC SYSTEMS ARE COMPLEX SOCIO-TECHNICAL SYSTEMS

An ATC tower is not only a matter of technology. It is not something that involves only ATCos, either. The HSI principles tell us that when we design such a system, we should hear the concerns from a broad spectrum of people (NASA, 2021). When an ATC tower operates, these people are primarily the ATCos. However, they also include (but are not limited to) maintenance personnel, technicians, ground support personnel, fire brigades, weather forecast services, rescue teams, crash teams, and pilots. Considering the tower lifecycle, the people involved are designers, developers, engineers, testers, managers, project supervisors, training teams, manufacturers, suppliers, qualifiers, and regulators. All these stakeholders, whether humans or organizations, along with technical components, must cooperate and coordinate throughout the system lifecycle, from concept to retirement.[1] Therefore, we should treat ATC towers as complex, life-critical socio-technical systems of systems. They are life-critical because an unanticipated or unexpected event may lead to critical consequences regarding the life or integrity of human beings. They are socio-technical because their technological aspect is only a subpart of their nature, human, and organizational dimensions. Finally, they are systems of systems because we can decompose them into more granular independent subsystems, which have their functions and purpose.[2]

You should be convinced by the reality of human and organizational complexities of ATC towers just by looking at Figure 30.1. This diagram represents the structural organization of a military ATC tower (which may differ in civil airports and may depend on the country, but the basic roles are the same. We can divide the control tower into the cabin (or cab) and the approach. The cabin supervises the traffic on the track and around it. The approach room does not offer a direct view of the airfield. However, it gathers the ATCos who guide aircraft beyond the responsibility area of the cab (i.e., from about 20 km from the tower) to the responsibility area of the en-route ATCos (i.e., from about 90 km from the tower). The en-route ATCos are generally located in separate en-route centers, which guide aircraft coming from and heading to other regional airfields. Note how even among the ATCos themselves, there are many different roles, which mainly determine the nature of their mission and the geographical area under their responsibility (e.g., ground ATCos supervise what happens between the tracks and the apron, departing ATCos guide departing aircraft, radar ATCos use radar information to guide landing aircraft).

A glance at Figure 30.1 helps us understand that the tower complexity is not solely technological. There is a diversity of human and organizational stakeholders involved in the system during operations. Their behavior is highly unpredictable and may differ from the behavior that designers predicted they would have when the system was being designed. Therefore, the system is likely to exhibit emergent properties that may not have been expected. This situation creates what systems engineers call *emergence*. Checkland (1999) characterizes emergence as "*the principle that whole entities exhibit properties which are meaningful only when attributed to the whole, not to its parts.*" Therefore, anticipating emergent properties before the system is integrated and operationalized is problematic. This is one reason why socio-technical system design is so hard.

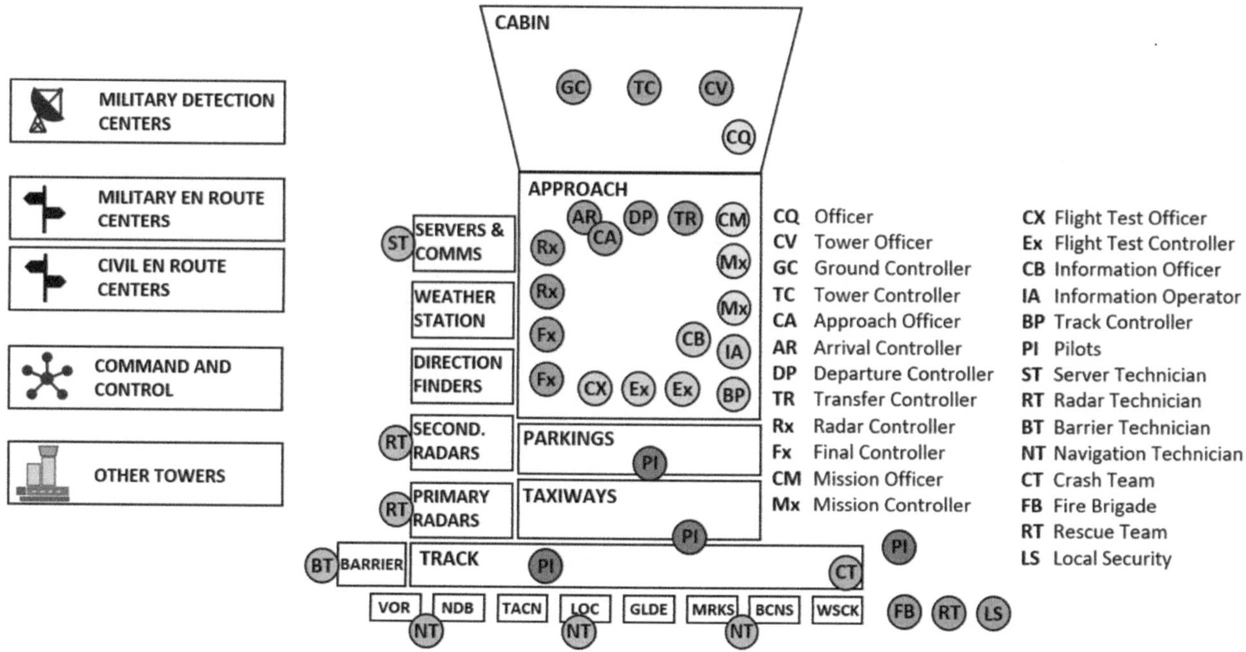

FIGURE 30.1 An ATC tower is a complex socio-technical system.

If we dive further into the literature, we find that there is a need for more synergy between technology and social concerns during design (Whitworth, 2009). Ruault et al. (2012) explain that socio-technical systems are prone to disruptions due to the complexity of their interdependent parts. Thus, the challenge is to learn the appropriate rules to maintain system stability and performance under unforeseen conditions. Disruptions may stem from unexpected external events (e.g., natural disasters) or internal faults (e.g., miscommunication). They are difficult to anticipate *a priori* and are the keystone of the Resilience Engineering domain (Hollnagel et al., 2006). Resilience Engineering itself has connections to the Cognitive Engineering and Human Error fields, whose most influential ideas are represented by works from Rasmussen (1997), Hollnagel et al. (2006), Woods and Roth (1988), Leveson (2020), and Dekker and Nyce (2012).

WHY CONTEXT MATTERS: THE CASE OF RVTS

The primary function of ATCos is to establish a comprehensive air situation around the tower. We have seen that to achieve this end, they must interact with technological elements, organizations, and other ATCos whose roles may differ. Since the late 1990s, researchers and industrials have been working on conceptualizing alternative and disruptive ATC paradigms (Fürstenau, 2022). Schaik et al. (2014) tried to remove and replace the physical tower with an RVT, located several kilometers from the supervised airfield. In an RVT, a wall of screens has replaced the out-of-view window. The screens stream a real-time video signal from cameras around the distant airport (Figure 30.2). Separating the controlling center from the controlled airfield like this presents at least two advantages. On the one hand, there are no longer

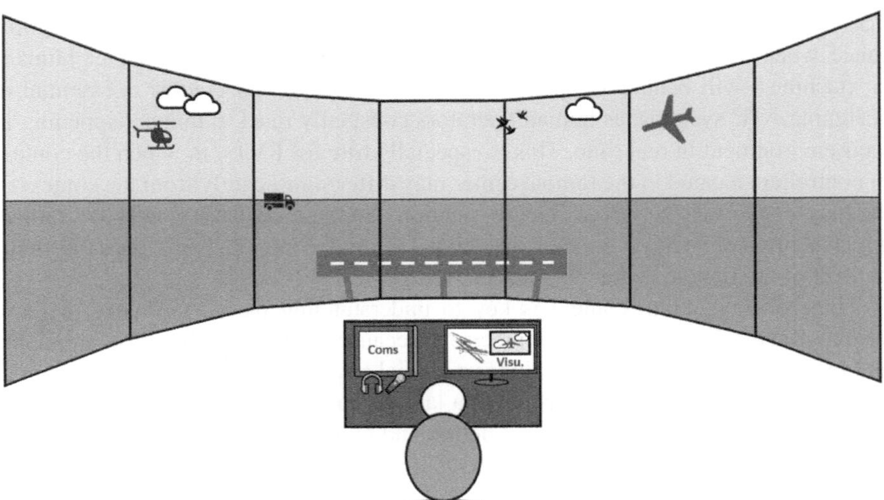

FIGURE 30.2 A schematic representation of an RVT.

any costs associated with the construction and maintenance of the tower, which can amount to several million dollars.[3] On the other hand, ANSPs may pool their human and technical resources from several airfields with little traffic into one RVT. The RVT is then called a facility enabling Multiple Remote Tower Operations, or MRTO (Papenfuss & Friedrich, 2016).

HungaroControl, Avinor, and ENAV are examples of national ANSPs having already demonstrated the technological feasibility of single and multiple remote tower operations instead of conventional towers (Kearney & Li, 2018). However, it turns out that camera-based RVTs like the one shown in Figure 30.2 have at least two drawbacks. First, technological constraints and costs limit its applicability, as transmitting a video signal from the airfield to the remote center requires high-bandwidth communication facilities along with screens of high enough resolutions to allow for visual detection, recognition, and identification of entities over the airfield (Ellis & Liston, 2022). All this cumbersome equipment may make the deployment of a remote center lengthy and tedious. The second drawback of camera-based RVTs is that the performance of most of these prototypes is primarily assessed on technological criteria. Still, the role of the human element and its issues have not been reconsidered. These issues may include trust in the system, focus, situation awareness, comfort, and fatigue. In particular, the ATCos' workload in RVTs has not improved as the ATCos still need to scrutinize the track and its surroundings continuously. Moreover, stress and trust in the system may be altered because the sole restitution of the air situation around the airfield onto screens is insufficient, as it has been shown that ATCos are also sensitive to non-visual cues such as sounds and vibrations (Reynal et al., 2019).

What properties do such systems possess that make them difficult to design? The INCOSE's Complexity Primer for Systems Engineers (INCOSE, 2016) lists 14 characteristics often found in complex systems, including system and unexpected emergence, intricate environmental interactions, and *contextual influences*. It seems that environment and context are two interrelated notions that carry deep significance when understanding how the system and its components (whether humans or machines) will behave at operation time. The *operational context* is essential in designing ATC systems, as human operators constantly react to things happening in their environment in real time. This is especially true for RVTs, in which the context of controllers located in the remote center may differ significantly from the context of the local center they supervise. Not to mention that these controllers must be trained, which is often done in yet another specialized training center that simulates the environment of the remote center (Figure 30.3).

Thus, understanding context is key to understanding the performance of these systems. Their design is complex precisely because context is difficult to anticipate and model early on, even though this is a crucial requirement if one does not want to face negative emergent behavior due to a lack of consideration of the context of the system. Consequently, we must first define and characterize further the operational context of socio-technical systems.

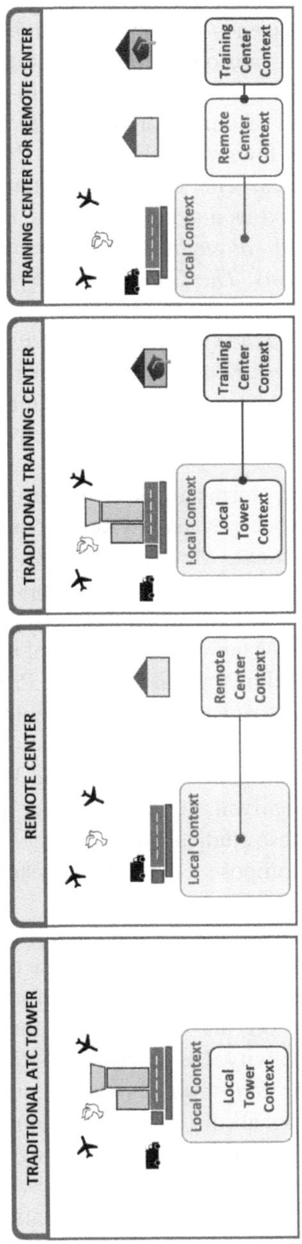

FIGURE 30.3 Remote tower operations and training introduce multiple layers of contextual complexity.

CAN WE DEFINE SYSTEM CONTEXT?

There is no consensus on the definition of the word *context*. Bazire and Brézillon (2005) establishes a comparative study of context definitions in the literature and demonstrates that context transcends multiple research and industrial areas. As complex socio-technical systems are our subject, a natural reflex is to look for the definitions of context provided by primary references in SE and HSI. By doing so, we find that SE and HSI-related resources often refer to context, but seldom define it. The NASA Human Systems Integration Handbook (NASA, 2021) mentions the *operational* and *mission contexts*. The INCOSE Systems Engineering Handbook (INCOSE, 2021) refers to context as a synonym for the *operating environment* of a system, which also includes a *threat environment*, a *resource environment*, and *collaborating and competing systems. The Guide to the Systems Engineering Book of Knowledge* (SEBoK Editorial Board, 2023) references Flood's definition (Flood & Carson, 1993): context "*describes the system relationships and environment, resolved around a selected system-of-interest.*" Context is described as a "*diagram defining the highest-level view of a system in its environment.*" Hence, system context is often reduced to a static representation of anything in the environment, i.e., outside the system boundaries.

Context research takes many forms in other scientific and engineering fields, spanning from linguistics to cyber–physical systems. Some studies attempt to provide a formal definition of context. McCarthy (1993), for instance, constructs a theory of contexts that revolves around symbolic propositional sentences such as *ist(c, p)*, meaning that proposition *p* holds in context *c* (e.g., *ist("fog around airport," "ATCo's visibility is reduced")*). Many works in artificial intelligence were built upon McCarthy's results, as Akman did in Natural Language Processing (Akman & Surav, 1997). Context is often studied in software engineering fields, like context-aware computing and distributed systems (see the survey from Koç (2014)). Areas of interest in context also include ambient intelligent systems (Kofod-Petersen & Cassens, 2007), human–computer interaction (Agarwal, 2018), and cyber–physical systems (Daun & Tenbergen, 2022). In most of these studies, context is loosely defined as some object surrounding the elements that compose a distributed system, whose role is to gather and process data from multiple sensors.

Context research has also been carried out concerning human behavior. Contextual Graphs (CxG) and Context-Based Reasoning (CxBR) are two examples of developing representations of context relative to the activity performed by human operators. Brézillon (2007) introduces (CxG) as a formalism that helps represent context when reasoning about human activity. He describes context as the partition of three sets: the *external knowledge*, the *contextual knowledge,* and the *proceduralized context*. External knowledge includes all the contextual elements that have no influence whatsoever on executing a task performed by a human agent. Conversely, contextual knowledge comprises all the information that somehow affects the execution of a task. Proceduralized context is the subset of this information extracted and processed by an agent when performing the task. An important point is that these three subsets are not static: they evolve through time and differ depending on the current *focus*, meaning they vary according to which task is being processed. As such, context is dynamic.

On the other hand, CxBR (Stensrud et al., 2004) provides an object-oriented terminology for modeling tactical situations that require some agents to make real-time decisions. CxBR defines terms such as *missions*, *goals*, *constraints*, and *context.* One of the key ideas of CxBR is that at a given time, only a fraction of the context (called the *active context*) is relevant to understanding how an agent may perform a task. The active context changes over time depending on environmental conditions and agent stimuli. CxBR has roots in Context-Mediated Behavior (Turner, 2017) and Case-Based Reasoning (Watson & Marir, 1994).

When we analyze the different definitions provided by the literature, we can detect patterns and common properties of context. We have already seen that context is not static: it changes through time (therefore, it is "*transient*"). Furthermore, CxBR tells us that only a fraction of our knowledge can be classified as being part of the context of a system, as most information is irrelevant to understanding the system's behavior. Hence, context must be "*curated*" to contain only the necessary information. By aggregating all the different definitions and properties from the literature, we came up with the six properties of context listed in Table 30.1.

Besides the properties, we propose our context definition tailored to the system ontologies and frameworks defined in the HSI literature. Our proposed definition is: "T*he operational context of a socio-technical system is* a *historical sequence of situations, triggered by events, that influence and help explain the behavior of a focus*." Let us break down this definition: a *situation* is the set of contextual elements relevant to a focus at a given time. The *focus* is the system or subsystem under study,

TABLE 30.1
The Six Properties of Context

Context Property	Symbology	Short Description
Specific	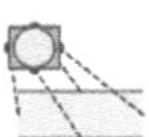	Context is always relative to some focus, defined as a couple *(structure, function)*
Curated	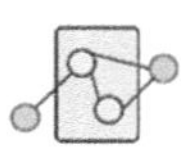	Only a few contextual elements have a real relevance to the system and its behavior
Holistic		The whole system's context is more than the individual contexts of its subsystems
Transient	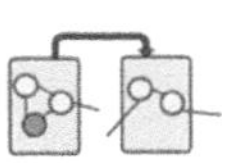	Context is not static and definitive, but changes through time
Entangled	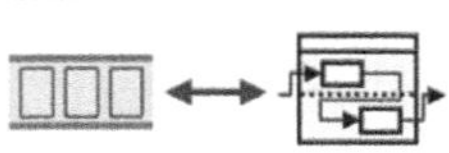	Context affects the system's resources, and the system's behavior affects context
Persistent	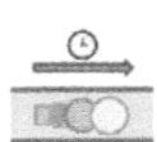	A contextual element's former value can still have a relevance to the current situation

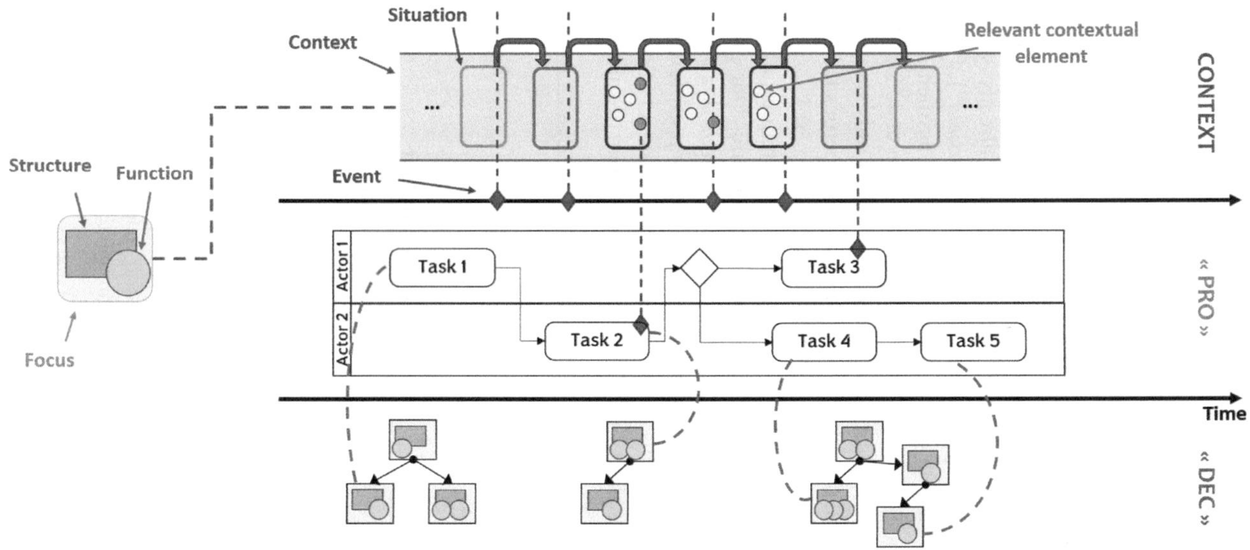

FIGURE 30.4 Our framework for the operational context of a socio-technical system.

performing some function.[4] The relevant contextual elements that compose a given situation are the variables whose values influence the function's execution. When the values of contextual elements change because some event has occurred, we transition from one situation to another. The resulting sequence of situations is the context of the focus. This definition is visually illustrated in the upper layer of Figure 30.4. The events that affect contextual element values can be external to the system (e.g., a storm is an external event as it does not result from the behavior of the system or one of its components) or they can be internal (i.e., the closure of a track is a consequence of one ATCo's decision). The system's behavior is described by the middle and bottom lanes, representing the procedural knowledge (what the system does) and the declarative knowledge (what the system is made of). This synergy between procedural and declarative knowledge is further developed by the PRODEC research effort (Boy et al., 2024).

CONCLUSION

In this chapter, we took the case study of ATC systems to understand what makes such systems complex socio-technical systems. What stands out from this ATC example is that a complex socio-technical system has a deep relationship with its operational context. When we design a system, we create new contexts for the human and machine agents that compose it. Understanding early on how a system and its context relate to each other enables flexibility in system design. It helps ensure system robustness and adaptability, as stakeholders can better anticipate how a system will react to its surrounding and evolving environment. The context of a system is not its environment, nor is it what resides outside its boundaries. It is more than that: the operational context is an intricate, dynamic, evolving object that the system and its constituents must probe continuously when it is in operation. Based on the SE and HSI literature, we proposed a simple definition of the operational context of complex socio-technical systems. We complemented this definition by providing six immutable properties of context.

We want the definition and the properties to form the basis for long-term research that aims to better integrate contextual information about a system in early system models. Indeed, there is still a lack of expressiveness in modeling notations and languages regarding integrating contextual information into system, user, task, and scenario models. A research effort is needed to combine better contextual descriptions of one system's environment with traditional procedural and declarative descriptions of this system, usually modeled in languages such as SysML, UML, Capella, or BPMN. It is not an easy endeavor: context, as we have discussed, is a fuzzy, hard-to-grasp concept. Moreover, it is not clear yet which information is to be categorized as context: depending on the boundaries of a system, an element may be included or excluded from what SE practitioners call system context. Weather, for instance, is outside the boundaries of a control tower. Does this mean it should not be considered in the tower design and ATCo's operation design? Of course not, because the impact of weather on ATC procedures is immense. Our point is that socio-technical design by nature calls for adopting a holistic approach to design that considers non-technical, societal, natural, and political impacts, even when those lie beyond the system boundaries. Engineers alone cannot carry this burden; this effort calls for specialized HSI practitioners and appropriate feedback from subject matter experts.

NOTES

1 In the sense of ISO/IEC/IEE 15288:2023, see www.iso.org/standard/81702.html *(last accessed by author in March 2025).*
2 There are arguments in the engineering community about the precise definition of "systems of systems," which is a term we use here to describe what others may call "integrated systems" or just "systems" depending on the managerial and operational interdependence of these systems; however, the HSI literature often assumes that as long as a system has at least two parts, it ought to be considered a system of systems (Boy et al., 2024).
3 *See April 2022 article "FAA Commissions New Air Traffic Control Tower at Charlotte Douglas International Airport" at* www.faa.gov/newsroom/faa-commissions-new-air-traffic-control-tower-charlotte-douglas-international-airport *(last accessed by author in March 2025).*
4 Hitchins (2007) defines a function from a SE viewpoint as "*an action, a task, or an activity performed to achieve a desired outcome*."

REFERENCES

Agarwal, N. (2018). What do we mean when we talk about context? In *IEEE 34th International Conference on Data Engineering Workshops* (pp. 29–32). Paris. doi:10.1109/ICDEW.2018.00012

Akman, V., & Surav, M. (1997). The use of situation theory in context modeling. *Computational Intelligence, 13*(3), 427–438. doi:10.1111/0824-7935.00046

Bazire, M., & Brézillon, P. (2005). Understanding context before using it. In A. Dey, B. Kokinov, D. Leake, & R. Turner (Eds.), *Modeling and Using Context* (pp. 29–40). Springer.

Boy, G., Masson, D., Durnerin, E., & Morel, C. (2024). PRODEC for human systems integration of increasingly autonomous systems. *Systems Engineering, 27*(5), 805–826. doi:10.1002/sys.21751

Brézillon, P. (2007). Context modeling: Task model and practice model. In Boicho Kokinov, Daniel C. Richardson, Thomas R. Roth-Berghofer, and Laure Vieu (Eds.), *Modeling and Using Context* (pp. 122–135). Springer.

Checkland, P. (1999). *Systems Thinking, Systems Practice.* Wiley.

Daun, M., & Tenbergen, B. (2022). Context modeling for cyber–physical systems. *Journal of Software: Evolution and Process.* Wiley. doi:10.1002/smr.2451

Dekker, S., & Nyce, J. (2012). Cognitive engineering and the moral theology and witchcraft of cause. *Cognition, Technology & Work, 14*, 207–212.

Ellis, S., & Liston, D. (2022). Visual features used by airport tower controllers: Some implications for the design of remote or virtual towers. In N. Fürstenau (Ed.), *Virtual and Remote Control Tower – Research, Design, Development and Validation.* Springer, Cham. doi:10.1007/978-3-030-93650-1_2

Flood, R., & Carson, E. (1993). *Dealing with Complexity: An Introduction to the Theory and Application of Systems Science.* Springer.

Fürstenau, N. (2022). *Virtual and Remote Control Tower: Research, Design, Development, Validation and Implementation* (2nd ed.). Springer. doi:10.1007/978-3-030-93650-1

Hitchins, D. (2007). *Systems Engineering: A 21st Century Systems Methodology.* Wiley.

Hollnagel, E., Woods, D., & Leveson, N. (2006). *Resilience Engineering: Concepts and Precepts.* Ashgate Publishing.

INCOSE. (2016). *A Complexity Primer for Systems Engineers.* TP-2016-001-01.0.

INCOSE. (2021). *Systems Engineering Handbook: A Guide for Systems Life Cycle Processes and Activities* (4th ed.). Wiley.

Kearney, P., & Li, W. (2018). Multiple remote towers for a single European sky: The evolution from initial operational concept to regulatory approved implementation. *Transportation Research Part A: Policy and Practice, 116*, 15–30. doi:10.1016/j.tra.2018.06.005

Koç, H., Hennig, E., Jastram, S., & Starke, C. (2014). State of the art in context modelling – a systematic literature review. In *Advanced Information Systems Engineering Workshops* (pp. 53–64). Springer.

Kofod-Petersen, A., & Cassens, J. (2007). Explanations and context in ambient intelligent systems. In Boicho Kokinov, Daniel C. Richardson, Thomas R. Roth-Berghofer, and Laure Vieu (Eds.), *Modeling and Using Context* (pp. 303–316). Springer.

Leveson, N. (2020). Safety III: A systems approach to safety and resilience. *Mit Eng. Syst. Lab.*

McCarthy, J. (1993). Notes on formalizing context. In *13th International Joint Conference on Artificial Intelligence* (pp. 555–560). San Francisco: Morgan Kaufmann.

NASA. (2021). *Human Systems Integration Handbook. (*NASA/SP-20210010952).

Papenfuss, A., & Friedrich, M. (2016). Head up only – a design concept to enable multiple remote tower operations. In *35th Digital Avionics Systems Conference* (pp. 1–10). Institute of Electrical and Electronics Engineers (IEEE). doi:10.1109/DASC.2016.7777948

Rasmussen, J. (1997). Risk management in a dynamic society: A modelling problem. *Safety Science*, 27, 183–213. doi:10.1016/S0925-7535(97)00052-0

Reynal, M., Imbert, J., Aricò, P., Toupillier, J., Borghini, G., & Hurter, C. (2019). Audio focus: Interactive spatial sound coupled with haptics to improve sound source location in poor visibility. *International Journal of Human–Computer Studies, 129*, 116–128. doi:10.1016/j.ijhcs.2019.04.001

Ruault, J.-R., Vanderhaegen, F., & Kolski, C. (2012). Sociotechnical systems resilience. In *INCOSE International Symposium*, vol. *22*. International Council on Systems Engineering (INCOSE) & Wiley.

Schaik, F., Roessingh, J., Lindqvist, G., & Falt, K. (2014). Detection and recognition for remote tower operations. In N. Fürstenau (Ed.), *Virtual and Remote Control Tower: Research, Design, Development, Validation and Implementation* (pp. 53–65). Springer, Cham. doi:10.1007/978-3-319-28719-5_3

SEBoK Editorial Board. (2023). *The Guide to the Systems Engineering Body of Knowledge.* (N. Hutchinson, Ed.). The Stevens Institute of Technology Systems Engineering Research Center, the International Council on Systems Engineering, and the Institute of Electrical and Electronics Engineers Systems Council.

Stensrud, B., Barrett, G., Trinh, C., & Gonzalez, A. (2004). Context-based reasoning: A revised specification. In *Seventeenth International Florida Artificial Intelligence Research Society Conference.*

Turner, R. (2017). Context-mediated behavior. In Patrick Brézillon, Roy Turner and Carlo Penco **(Eds.),** *Modeling and Using Context.* Springer. doi:10.21494/ISTE.OP.2017.0131

Watson, I., & Marir, F. (1994). Case-based reasoning: A review. *The Knowledge Engineering Review, 9*(4), 327–354. doi:10.1017/S0269888900007098

Whitworth, B. (2009). *Handbook of Research on Socio-Technical Design and Social Networking Systems.* IGI Global.

Woods, D., & Roth, E. (1988). Cognitive engineering: Human problem solving with tools. *Human Factors, 30*(4), 415–430.

31 Human–Autonomy Teaming for Future Commercial Aviation Concepts of Operations

Benjamin Berton, Elodie Bouzekri, Amandeep Singh, and Philippe Doyon-Poulin

INTRODUCTION

Single-Pilot Operations

Commercial aviation has experienced a "de-crewing" trend, gradually reducing the number of crew members in the cockpit (Harris, 2007). Over time, the traditional five-member flight crew has been scaled down to two, i.e., the captain (CPT) and first officer (FO). The reduction in crew size has been made possible by the advancement in flight technology, which has replaced the need for flight engineers, navigators, and radio operators.

Currently, the Federal Aviation Administration (FAA), under *14 CFR Part 121 Subpart M – Airman and Crewmember Requirements* (n.d.), mandates the minimum requirement of two pilots in the cockpit. However, ongoing technological advancement suggests a future possibility for commercial airliners' single-pilot operations (SPO). SPO refers to operating a commercial aircraft with only one pilot in the cockpit, augmented by advanced onboard technology and potential support from ground operators to assist piloting operations (Bilimoria et al., 2014). The rationale behind SPO is based on the necessity for future aviation to increase both profit margins and operational efficiency. Airlines are under pressure to reduce the operating costs of flights further. Crew costs, including benefits, training, and salaries, account for a significant portion of an airline's operating budget, ranging from 19% to 35% of total operating expenses, depending primarily on the aircraft size (Stanton et al., 2016). Another challenge for airlines is the expected shortage of qualified pilots to match the industry's growth. Despite the temporary downturn during the COVID-19 pandemic, long-term forecasts by Boeing and Airbus have remained consistent, projecting the need for over 550,000 new pilots by 2040. The global aviation industry's recovery further substantiates these projections, with 2024 air traffic surpassing pre-pandemic levels (Aviation Business News, 2024), highlighting the ongoing demand for qualified pilots (Li et al., 2022; Ung et al., 2021). The traditional solution of recruiting and training pilots may raise airline operating costs. With the transition to SPO, airlines

 DOI: 10.1201/9781003666431-37

can operate more aircraft with the same number of pilots, thus solving the pilot shortage issue. Ultimately, this transition will enhance airlines' operational efficiency and profitability in the long run.

However, transitioning from a two-pilot flight deck to a single-pilot setup presents more significant challenges than previous crew reductions (Matessa et al., 2017). Human factors emerge as an essential concern in adopting commercial SPO. Reducing the crew to one removes a redundancy layer from the flight deck. The fundamental requirement of SPO is to ensure safety levels equivalent to or higher than those of the current two-pilot operation (Boy, 2014). There is ongoing debate on compensatory measures for the loss of redundancy to maintain (or enhance) the current level of safety, as in two-crew operation (Li et al., 2022). A recent survey on commercial SPO highlighted a preference for a concept of operations centered on onboard systems with increasing levels of autonomy, referred to hereafter as "Autonomy" (Pechlivanis & Harris, 2025). For more information on the various operational concepts proposed for SPO, readers are encouraged to consult the work of Puca and Guglieri (2024).

The effectiveness of an SPO concept depends on the teaming between the pilot and the onboard autonomy, known as human–autonomy teaming (HAT). This teamwork approach must integrate the strengths of a human pilot and automated technologies and lead to safer, more efficient, and overall better performance in single-pilot flights.

Human–Autonomy Teaming

HAT concept builds upon the progression from automation to autonomy. Automation is the technology that processes data, makes decisions, and controls processes (Hoff & Bashir, 2015). In other words, automation follows predetermined steps to accomplish specific outcomes (Hancock, 2017). It can perform functions that were previously or could be performed by a human operator, which can range from fully manual to fully automated (Parasuraman et al., 2000). Autonomy, on the other hand, pertains to the capabilities of a system to independently take on tasks with minimal human intervention over a prolonged duration (Endsley, 2017; Holbrook et al., 2020). The progression from lower levels of automation to more autonomous systems illustrates a fundamental shift in human–automation relationships, moving from simple interactions to teamwork (McNeese & McNeese, 2020). While increased automation can enhance safety, it also introduces new challenges such as "automation complacency," where over-reliance on automated systems may lead to vigilance decline and failure to respond effectively when needed (Lee & Seppelt, 2023). In SPO, it is imperative to design automated systems that promote collaboration between the pilot and onboard systems while avoiding falling into this automation conundrum.

This underscores the importance of drawing upon key insights from HAT research, particularly the foundational tenets, interaction patterns, metaphors, and teaming schemes that guide the design of effective collaboration. For instance, Flemisch et al. (2003) introduced influential teaming metaphors that conceptualize human–machine relationships; Johnson and Vera (2019) emphasized the critical role of teaming intelligence in autonomous systems; and comprehensive overviews by Lyons et al. (2021) and O'Neill et al. (2022) synthesize the research on HAT. Due to the limited scope of this chapter, we do not detail each of these contributions but instead direct interested readers to the works. The remainder of the chapter focuses on a practical use case to illustrate how these theoretical foundations can be translated into a design process for SPO contexts.

HAT FOR SPO: A CASE STUDY

We aim to design an autonomous system to team with a single pilot during takeoff. We are particularly interested in the function allocation and human–autonomy interface that supports HAT. The scenario consists of a takeoff and initial climb phase with an engine failure due to a bird or drone strike. Takeoff with an engine failure is a challenging scenario for single-pilot operations because it is highly dynamic and requires fast and precise decision-making from the captain. Bird and drone strikes will be an issue for the foreseeable future of commercial aviation and are therefore a scenario of particular interest in studying HAT in the future SPO flight deck. We will see through the case study how a design process can be used to address and evaluate the teaming dynamics between the onboard captain and the autonomy. Figure 31.1 presents an overview of the HAT framework we used. Each step is described next.

STEP 1: DEFINING THE OPERATIONAL CONSTRAINTS

The initial step of this approach involves making assumptions about the operational domain to reduce uncertainty during later stages. A clearly defined ConOps, based on existing research, offers a structured foundation for generating ideas and innovating within set operational boundaries. Similarly, reviewing autonomous systems in the market or research will help the design team understand the state-of-the-art technology's capabilities, reliability, and efficiency. This overview provides an early understanding of the system's potential role and abilities, guiding the focus of subsequent task and function analysis.

Application to the Case Study

We chose to define the autonomy capabilities and the teaming characteristics to align with the dynamics of the current two-pilot crew, which has already been established with training standardization, such as Crew Resource Management. This operational constraint respects the HAT tenets requiring interdependency between the autonomy and the human teammate while having a degree of agency and proactivity for the autonomy (Lyons et al., 2019; O'Neill et al., 2022). In our case, we assume that autonomy can self-initiate a task but must inform its human teammate of the nature of the action and the goal currently pursued and allow the human to cancel or modify the execution. Regarding technological capabilities, we assume that the autonomy is integrated into the aircraft's avionics suite. It can read instrument values, interact with the Flight Management System (FMS) and autopilot, and execute specific commands such as shutting down an engine. These assumptions reflect current and emerging capabilities in state-of-the-art autonomous systems under research and early deployment. Finally, consistent with current regulatory frameworks, we assume that take-offs must

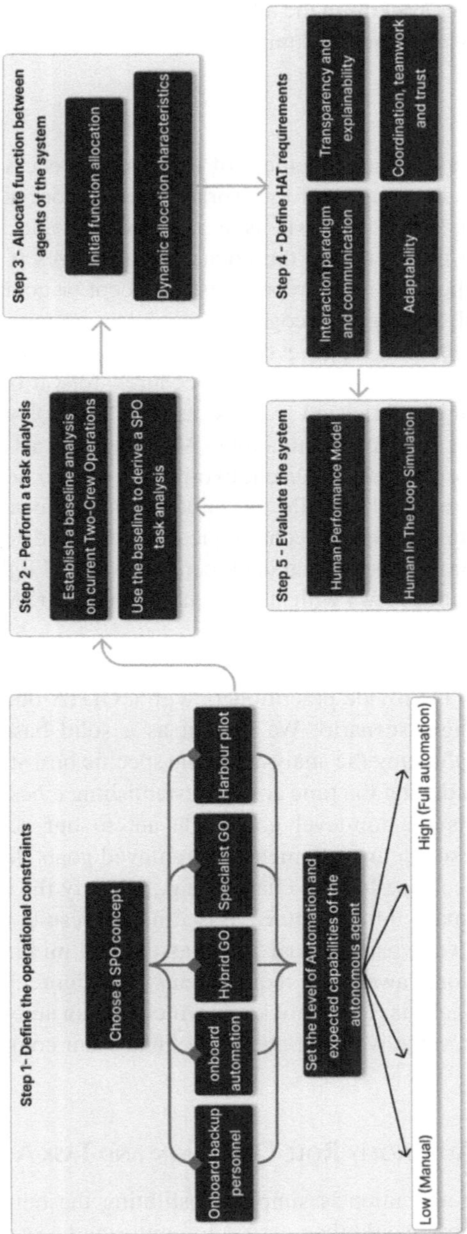

FIGURE 31.1 Five steps of the HAT framework.

be conducted under the pilot's manual control, which is treated as a non-negotiable function allocation (Harris et al., 2015).

Note that the case study assumes a cockpit-based ConOps where the onboard pilot interacts with autonomous systems in the aircraft. However, this work could be adapted to ground-based SPO as it depends on similar advanced autonomy (Brandt et al., 2015).

Step 2: Task Analysis

Task analysis emerges as a crucial element of the design process that will help the analyst build a solid understanding of the work domain before allocating functions between system agents. Task analysis is a multifaceted and versatile methodology. When searching the literature, the novice soon realizes it is not confined to a single, definitive method but rather represents a concept or goal: comprehensively understanding physical actions and cognitive processes required by operators in achieving specific system states. However, how can one perform a task analysis of a yet-to-be-built system? Unlike established procedures, forward-looking SPOs lack existing frameworks and experienced experts, and the technology envisioned may not yet be available for practical simulations. Methods like a Goal-Directed Task Analysis (GDTA) performed on the current two-pilot crew can help the design team to list the information required to build the pilot's situation awareness (SA) of the scenario without relying on specific technological means (Endsley et al., 2003). To do so, analysts must consult operational documentation (e.g., Flight Crew Operating Manual) and engage in interviews with subject-matter experts (SMEs).

Application to the Case Study

Endsley and Jones (2011) provide practitioners with a GDTA output for a complete commercial aviation flight scenario. We used it as a solid baseline for engaging SMEs in refining and validating the analysis for the specific bird strike–engine failure takeoff scenario, thus reducing the time cost for establishing a baseline task analysis.

Figure 31.2 presents the top-level goals relevant to our scenario: navigation planning, flying, and resource management. We removed goals that were not safety-related or outside of the scope for this chapter (e.g., "Satisfy the customer" or goals related to ATC and airspace infrastructure). Based on the results we found from our interviews with SME, we expanded goal 2.5 to assess and minimize the bird strike situation with its situational awareness requirements (see Figure 31.2).

At this point in the analysis, we do not know which human autonomy, or both, will require the information since we have not yet defined team composition and roles. This is done next.

Step 3: Human and Autonomy Role Definition and Task Allocation

One approach for task allocation is simply substituting the human copilot with a virtual one. The autonomy would then perform monitoring tasks previously handled by the copilot, since the human pilot is mandated to fly manually during the takeoff. These tasks may include cross-checks, system and environmental monitoring, and keeping track of the human pilot's status. However, the taskwork that makes a

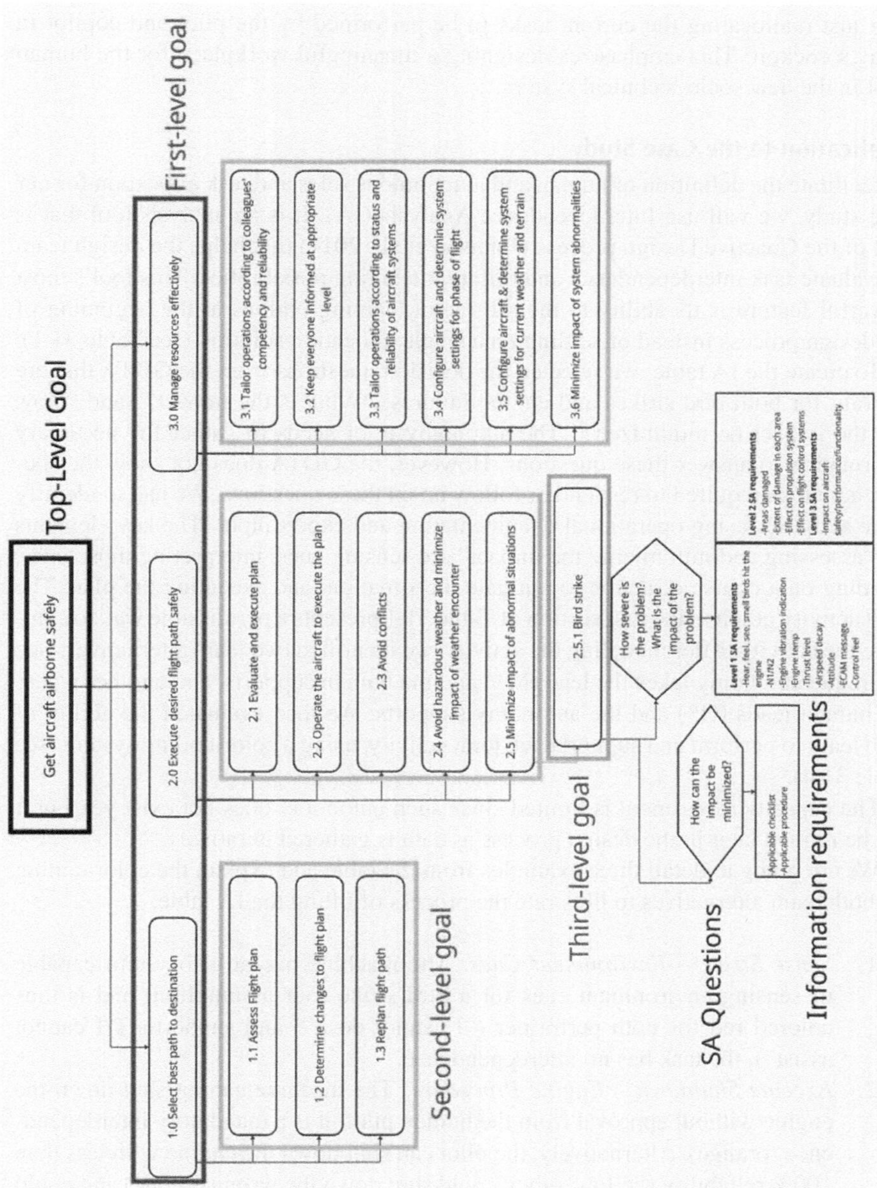

FIGURE 31.2 Excerpt of GDTA output for the takeoff phase.

human–human flight crew perform well might be difficult to translate with a silicon-based teammate. Another approach considers autonomy as a "complement" to human capabilities. The complemation approach (Schutte et al., 2017) focuses on defining the new role of the single pilot and how the autonomous system can support it, rather than just reallocating the current tasks to be performed by the pilot and copilot in today's cockpit. This emphasizes designing a meaningful workplace for the human pilot in the new socio-technical system.

Application to the Case Study

To facilitate the definition of human and autonomous roles and task allocation for our case study, we will use Interdependence Analysis (IA). It is an analysis tool that is part of the Coactive Design process (Johnson et al., 2014) that helps the design team to evaluate task interdependence and different teaming possibilities. This tool's most powerful feature is its ability to map different teaming options at the beginning of the design process instead of settling on a single allocation solution (see Table 31.1).

To create the IA table, we selected the decision questions from the GDTA that are relevant for both bird strikes and engine failures: "What is the impact?" and "How can the impact be minimized?" The autonomy pilot needs to gather the necessary information to answer these questions. However, the GDTA does not show the specific activities required to respond or follow up on these questions. We had to identify these activities using operational documentation and expert input. The key elements for "assessing and minimizing the impact" are sensing cues, interpreting those cues, deciding on a course of action to mitigate the situation, and executing the plan. The joint activity graph in the first section of Table 31.1 presents a permissible way to complete the activity. After modeling the activity, we identified two team alternatives: one where the autonomy takes the lead (M*) and the human supports, and another where the human leads (H*) and the autonomy supports. We then evaluated the ability of each team to perform and support each joint activity, using a color-coding system (see Table 31.4).

The capacity assessment is limited since such autonomy does not exist yet, but it can be refined later in the design process as data is gathered iteratively.

We are going to detail three examples from the table and explain the color-coding for both team alternatives to illustrate the process of filling the IA table:

1. *Sense Strike – Environment Cues:* The machine has no subsystem capable of sensing environment cues for a bird strike (our assumption) and is thus colored red for both performer ("I cannot do it") and supporter ("I cannot assist"); the task has no interdependence.
2. *Execute Shutdown – Engine Procedure:* The machine cannot shut down the engine without approval from the human pilot; it is a mandatory interdependence (orange). Alternatively, the pilot can shut down the engine with less than 100% reliability (yellow – they could shut down the wrong engine) and could benefit from the automation for efficiency (green – e.g., cross-check).
3. *Decide – Engine-Out Climb Procedure:* The autonomy can decide to load the navigation for the airport-specific engine-out Standard Instrument Departure reliably (green), but it could benefit from the pilot to improve reliability (yellow – e.g., the pilot ensures that there is no traffic).

TABLE 31.1
Interdependence Analysis Table with Three Main Sections Labeled

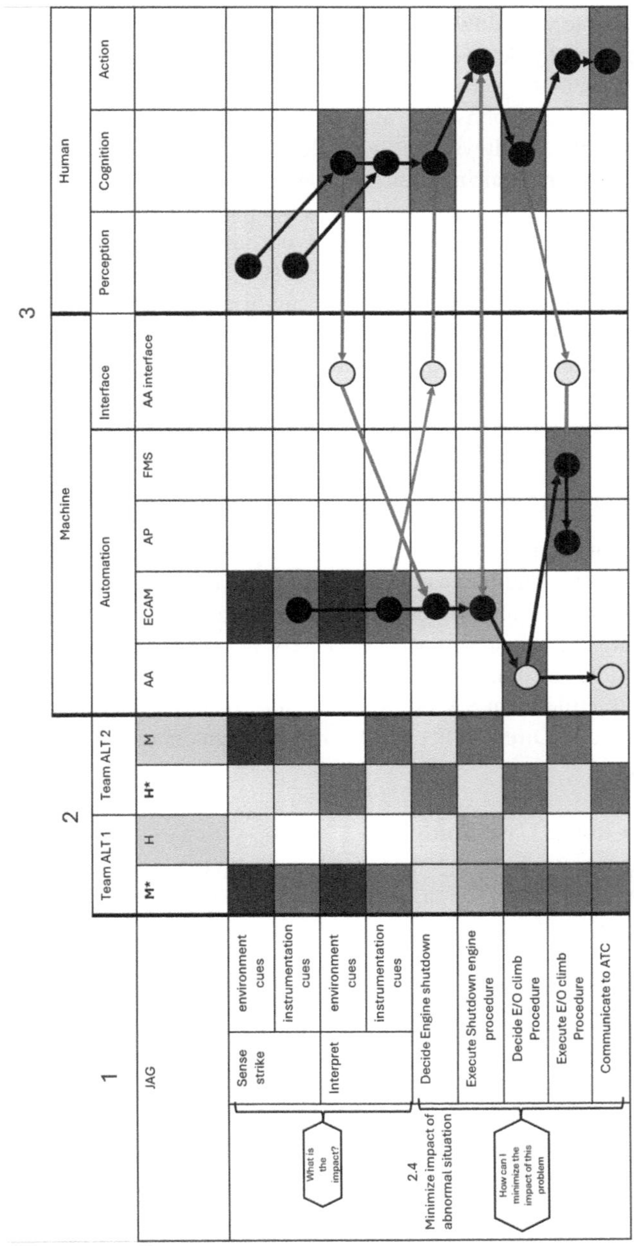

[1] Model of the joint activity, 2: potential interdependence in the activity, 3: potential workflows in the human–machine system.

The final step in the IA involves defining workflows to understand how the teaming alternatives will be physically instantiated. We expanded the team to detail relevant subsystems, interfaces, and human cognitive processes. A dot was placed to indicate which component has a role in the activity, and connections between the dots represent possible workflows to achieve the goal. The outcome, presented in the third section (right-hand side of Table 31.1), is "an enumeration of the possible ways a team can complete the joint activity, providing a *solution space* illustrating the flexibility of the system" (Johnson et al., 2018). The third section of the IA table shows various potential workflows. One workflow is primarily carried out by automated systems, except for requiring the pilot's approval to shut down the engine. Another workflow is entirely manual, with the pilot performing all tasks independently, without support from the autonomy. Both workflows have limited reliability in certain activities, suggesting they may not be the best solutions. Considering the supporting team member, we can see how teaming might enhance reliability or efficiency in alternative workflows, as illustrated by the blue arrows. For instance, the pilot could increase the reliability of the engine shutdown decision by providing the system with interpretations of environmental cues through the autonomy's interface. The yellow dot represents the future autonomy system and its interface.

Step 4: Define HAT Requirements

At this stage of the process, researchers should have a clearer understanding of the operational domain and the teaming possibilities. This knowledge will help establish requirements for the autonomous system that align with HAT principles. Flying a plane is a complex human activity; conversely, designing an autonomy that will act as a virtual copilot is a complex engineering task. The GDTA helps identify the information needed to build team SA. At the same time, the IA will determine Observability, Predictability, and Directability (OPD) requirements necessary for each joint task's interactions.

Application to the Case Study

First, we use the IA table of the previous step to systematically identify the OPD requirements across every interdependent task for each team's alternatives. To elicit the OPD requirements, we examine each row of the table and, for both team alternatives, answer these three questions:

- Who needs to observe what from whom?
- Who needs to be able to predict what from whom?
- Who needs to direct what from whom? Table 31.2 presents the resulting OPD requirements.

The GDTA goal hierarchy and its information requirements help determine the critical information the interface needs to show to the pilot and communicate to the autonomy. The GDTA also helps organize information presentation, aligning it with specific goals to support SA directly. Table 31.3 lists examples corresponding to activities in the scenario.

TABLE 31.2
Bird Strike and Engine Failure at Takeoff: Interdependence Relationships and OPD Requirements

Required capacities	Team ALT 1		Team ALT 2		Interpretation and OPD requirements
	M*	H	H*	M	
Sense environment cues					Only the pilot can sense the environment cues, there is an **observability** requirement to let the pilot feed its input to the machine (e.g. abnormal vibration)
Sense instrumentation cues					The ECAM keep sensing instrumentation cues and automatically raise an alarm when detecting a problem. The pilot may not be 100% reliable and thus the system must concurrently perform the task and make **observable** when sensing a discrepancy
Interpret environment cues					Only the pilot can interpret the environment cues, there is an **observability** requirement to let the pilot feed an additional interpretation to the machine (e.g. fire suspicion), he should be able to **direct** the automation to perform a procedure associated with its interpretation
Interpret instrumentation cues					The ECAM is reliable enough to interpret instrumentation cues and select the relevant checklist in case of a system failure, it should be able to make **observable** its recommendation and reasoning, it should be able to **direct** the pilot to perform the appropriate procedure or checklist.
Decide engine shutdown					Either can decide to perform the engine shutdown procedure, the automation must be **directable** by the pilot to make the decision, the automation must make **observable** the status of the current impacted system, the automation must **direct** the pilot to crosscheck and agree or not with the decision
Execute engine shutdown					Shutting down the engine is a critical task that the automation cannot do without the approval of the pilot. The automation must make **observable** the specific action it is performing, the pilot must be able to **predict** the next actions that the automation will make. The pilot must be able to **direct** the automation to modify or cancel the execution of the procedure. The automation must be able to **direct** the pilot into performing the correct actions
Decide E/O climb procedure					Either can decide to perform the airport specific engine-out climb procedure, the automation should make **observable** the relevant parameter for the navigation, and should be **directed** by the pilot to make the decision and should **direct** the pilot to crosscheck and agree or not with the decision
Execute E/O climb procedure					The automation must be **directable** to execute to take on performing the task at any time, the pilot must be able to **predict** the next actions of the automation during the procedure, the automation must make **observable** the state and relevant navigation parameters entered in the autopilot as well as current and expected performance in navigating the procedure
Communicate to ATC					Either can perform the task of communicating to ATC, the automation should make the past, current, and next messages **observable** beforehand and allow the pilot to **direct** who will communicate it

TABLE 31.3
Information Requirements for Each Activity

JAG			Level 1 SA Requirements	Higher Level SA Requirements	Teamwork Information Requirement
Goal number Minimize impact of abnormal situation	Sense strike	Environment cues	• Hear, feel, see, smell birds hit the engine		• Environmental parameters list
		Instrumentation cues	• N1/N2 • Engine vibration indication • Thrust level • Engine temp		• Sensed parameters • Discrepancies detected
	Interpret severity	Environment cues	• Hear, feel, see, smell birds hit the engine • Airspeed decay • Attitude	• Impact on flight safety/performance • Aircraft's ability to maintained controlled flight • Structural integrity • Impact on flight path • Immediate risk • Area of damage • Extent/severity of damage in each area	• Environmental input entered

Instrumentation cues	• N1/N2 • Engine vibration indication • Thrust level • Engine temp • Airspeed decay • Attitude	• Effect on propulsion system • Effect on flight control systems • Engine usable • Area of damage • Extent/severity of damage in each area	• Summary of sensor data and possible damaged area • Trend of critical parameters and possible outcomes • Prioritized areas to monitor
Decide engine shutdown	• Applicable checklist • Time to restart • phase of flight • ECAM message	• Engine usable • Aircraft configuration • Prioritization of procedures	• Possible course of actions • Outcome of course of actions • Decision status
Execute shutdown engine procedure	• Applicable checklist	• Aircraft response to shut down • Flight envelope affected by reduced thrust • Flight path affected by reduced thrust	• Checklist step completion • Step division between teammates • Future checklist steps • Aircraft status update • Confirmation and cross-checks
Decide E/O climb procedure	• Distance to airport • Applicable procedure • Time to airport • Airport facilities • Traffic • Approach available at airport	• Ability to reach airport • Ability to execute the procedure given the affected flight performance	• Options and associated risk • Optimal path • Decision status • Agreements and adjustments made

(continued)

TABLE 31.3 (Continued)
Information Requirements for Each Activity

JAG	Level 1 SA Requirements	Higher Level SA Requirements	Teamwork Information Requirement
Execute E/O climb procedure	• Applicable procedure • Aircraft system status • Flight parameters	• Climb performance • Deviation from the climb path	• Flight system status • Task allocation between teammates • Task completion • Aircraft performance update • Aircraft configuration
Communicate to ATC	• Situation report • ATC instructions	• Aircraft state • Navigation intentions • Assistance requirements	• History of communication • Communication allocation between teammates • Next message • Clearance • Emergency status

Step 5: Evaluate the System

Based on the previously defined requirements, we developed an initial prototype of the autonomy and evaluated it using traditional human factors and ergonomics methods. HAT can be assessed at various levels: from the individual perspective, team performance, or the overall air traffic system (Strybel et al., 2018). We focus on evaluating both individual and team performance, particularly team processes and team SA. For team processes, we assess coordination between team members, such as the time spent on communication and interactions like cross-checking, modifying, or canceling tasks (Directability). Team SA is reflected in the SA of individual members (Endsley et al., 2003). Since the pilot and autonomy form a closely integrated team, they share a significant amount of SA, which needs to be evaluated to ensure common ground. Besides assessing the SA of the current situation, we also evaluate each team member's understanding of the other's status. We aim for each agent to be aware of the other's current activities and progress (Observability) and future actions (Predictability).

We propose two complementary approaches to evaluate the autonomy: cognitive modeling and human-in-the-loop simulations (HITLS). Mental models help with early system evaluation (Li et al., 2023). As simulations are fast, teaming parameters such as allocation scheme and contextual elements can quickly be adjusted to find hidden design flaws and choose a relevant prototype. HITLS allows further exploration of the design and identification of the emergent team behavior. Using both methods provides a thorough evaluation while saving time and costs (Wang et al., 2023).

Application to the Case Study

We are developing a working prototype of the autonomous system that meets the previously defined HAT requirements in a flight simulation environment. The prototype is a tablet app connected to the simulated aircraft and displayed in the cockpit. We now aim to test the system's design and interface to ensure it effectively supports the pilot handling a bird strike scenario shortly after takeoff.

For the cognitive model, we chose a QN-ACTR-SEEV program (Cao & Liu, 2012; Rehman, 2020) that simulates a human pilot's cognitive processes, such as mental workload and SA, while enabling real-time interaction with the autonomy and flight simulator. We can then simulate each workflow from the right-hand side of the IA table and automatically compute:

- Mental workload predictions for the pilot allow us to investigate whether workflows with greater teamwork interdependence impose a "coordination cost" on the pilot's task load.
- We can also explore whether increased interdependence improves the pilot's SA by having the model answer Situation Awareness Global Assessment Technique (SAGAT) queries during the scenario and analyzing the model's eye movements.

One advantage of the cognitive model is that it provides continuous and dynamic assessments of both SA and mental workload (MWL) at each simulation step. This helps us understand the causes of workload peaks and SA breakdowns, allowing us to refine the interface and teaming requirements. After iterative refinements, when the prototype is satisfactory, we can recruit airline pilots to participate in simulator studies using the same scenario with autonomy. This HITLS serves two primary purposes:

- *First*, to validate and refine the cognitive model. Behavioral data from pilots will help adjust model parameters, such as how long it takes to physically press a button on the autonomy's interface. By having pilots "think aloud" during the scenario, we can verify whether the model's mental steps are cognitively plausible. We can also check for correlation between the model's computed workload and the pilot's subjective workload using standardized NASA-TLX questionnaire scores. We can also compare SAGAT responses and gaze patterns between the model and the human pilots (e.g., time spent dwelling on certain displays).
- *Second*, to overcome the limitations of the cognitive model and explore other aspects by collecting subjective feedback on the autonomy and its interface, including usability and trust. HITLS is particularly useful for identifying emergent properties of the system (Boy & Morel, 2022), which could be added to the computational model.

CONCLUSION

This case study outlined a structured five-step approach to enhance HAT in SPO, using an illustrative example of a bird or drone strike scenario during takeoff. The process began by defining operational constraints and understanding the tasks involved, moving toward a detailed analysis of roles, functions, and interdependence between the human pilot and the autonomy. By exploring these tasks through tools like GDTA and IA, we gained valuable insights into the teaming dynamics needed for effective collaboration. As the design progressed, specific requirements emerged, guiding the development of an interface that contributes to team SA and supports the interdependent tasks. Finally, the evaluation phase integrates cognitive modeling and HITLS to refine and validate the system, ensuring it supports team performance and pilot workload management. The framework provides a practical guide for designing and evaluating HAT systems in future flight decks.

CHALLENGES FOR HAT IN SPO

The transition to SPO introduces various technical, human, and regulatory challenges that must be addressed to ensure safe and effective implementation. First, communication between the single pilot and the autonomous system presents a critical challenge. Effective decision-making and system management rely heavily on the timely and accurate exchange of information between the pilot and the autonomy. This is particularly crucial during high workload or emergency situations. Second, training enables pilots to work effectively with autonomous systems. New training programs necessitate a re-evaluation of Crew Resource Management principles.

Adapting CRM to SPO contexts means reinforcing skills such as communication, task delegation, mutual monitoring, and trust, not between two humans, but between the pilot and the autonomy. Another critical challenge lies in the acceptance of SPO by both pilots and passengers. Pilot unions and professional associations have voiced strong opposition to the deployment of SPO, citing safety, workload, and ethical concerns (ALPA, 2019). Similarly, passenger trust in flights operated by a single pilot with autonomous support remains uncertain, and public perception could significantly influence regulatory and commercial adoption. Finally, the integration of autonomy raises substantial challenges for regulatory bodies. Autonomous systems must meet stringent reliability, integrity, and redundancy requirements. However, many operational rules and certification standards for autonomy remain underdeveloped. While technical standards for airworthiness are beginning to address these issues, further regulatory innovation is required to ensure safe and certifiable deployment Tambon et al. (2022).

While this chapter focused on HAT as a central lens to explore SPO design in commercial aviation, it is essential to acknowledge the broader implications that extend beyond the scope of our contribution. We did not address the organizational, regulatory, and sociological dimensions that would inevitably be affected by introducing such a paradigm shift. A comprehensive approach to the development and deployment of SPO should incorporate Human Systems Integration (HSI), which considers not only cockpit technologies and human–machine interaction but also the alignment of technical, human, and organizational elements across the system's full life cycle (Boy, 2024). While our case study illustrates how HAT principles can support collaborative interaction between a single pilot and intelligent systems, the successful implementation of SPO will ultimately require integrating HSI practices to ensure system-wide safety, usability, and operational resilience. Addressing these challenges will support the transition to real-world HSI with autonomous systems.

REFERENCES

14 CFR Part 121 Subpart M – Airman and Crewmember Requirements. (n.d.). Retrieved April 15, 2025, from www.ecfr.gov/current/title-14/part-121/subpart-M

Air Line Pilots Association, International. (2019). *The dangers of single-pilot operations* (White paper). Air Line Pilots Association, International. www.alpa.org/-/media/Project/ALPA/Alpa/Files/White-Papers/white-paper-single-pilot-operations.pdf

Aviation Business News. (2024). *IATA Declares the Airline Industry has "Fully Recovered" from the COVID Pandemic*. Aviation Business News. www.aviationbusinessnews.com/industry-news/iata-declares-the-airline-industry-has-fully-recovered-from-the-covid-pandemic/

Bilimoria, K. D., Johnson, W. W. & Schutte, P. C. (2014). Conceptual Framework for Single Pilot Operations. In *Proceedings of the International Conference on Human–Computer Interaction in Aerospace* (pp. 1–8). Association for Computing Machinery. https://doi.org/10.1145/2669592.2669647

Boy, G. A. (2014, January 1)). Requirements for single pilot operations in commercial aviation: A first high-level cognitive function analysis. In F. Boulanger, D. Krob, G. Morel, & J.-C. Roussel (Eds.), *Proceedings of the 5th International Conference on Complex Systems Design & Management (CSD&M 2014)* (pp. 227–234). Springer.

Brandt, S. L., Lachter, J., Battiste, V. & Johnson, W. (2015). Pilot Situation Awareness and Its Implications for Single Pilot Operations: Analysis of a Human-in-the-Loop Study. *Procedia Manufacturing*, *3*, 3017–3024. https://doi.org/10.1016/j.promfg.2015.07.846

Cao, S. & Liu, Y. (2012). QN-ACTR Modeling of Multitask Performance of Dynamic and Complex Cognitive Tasks. *Proceedings of the Human Factors and Ergonomics Society Annual Meeting*, *56*(1), 1015–1019. https://doi.org/10.1177/1071181312561212

Endsley, M. R. (2017). From Here to Autonomy: Lessons Learned From Human–Automation Research. *Human Factors: The Journal of the Human Factors and Ergonomics Society*, *59*(1), 5–27. https://doi.org/10.1177/0018720816681350

Endsley, M. R., Bolte, B. & Jones, D. G. (2003). *Designing for Situation Awareness: An Approach to User-Centered Design*. CRC Press. https://doi.org/10.1201/9780203485088

Endsley, M. R. & Jones, D. (2011). *Designing for Situation Awareness: An Approach to User-Centered Design* (2nd ed.). CRC Press.

Flemisch, F., Adams, C., Conway, S., Goodrich, K., Palmer, M. & Schutte, P. (2003). *The H-Metaphor as a Guideline for Vehicle Automation and Interaction*. (NASA TM-2003-212672). National Aeronautics and Space Administration, NASA Langley Research Center. http://ntrs.nasa.gov/citations/20040031835

Hancock, P. A. (2017). Imposing Limits on Autonomous Systems. *Ergonomics*, *60*(2), 284–291. https://doi.org/10.1080/00140139.2016.1190035

Harris, D. (2007). A Human-Centred Design Agenda for the Development of Single Crew-Operated Commercial Aircraft. *Aircraft Engineering and Aerospace Technology*, *79*(5), 518–526. https://doi.org/10.1108/00022660710780650

Harris, D., Stanton, N. A. & Starr, A. (2015). Spot the Difference: Operational Event Sequence Diagrams Are a Formal Method for Work Allocation in the Development of Single-Pilot Operations for Commercial Aircraft. *Ergonomics*, *58*(11), 1773–1791. https://doi.org/10.1080/00140139.2015.1044574

Hoff, K. A. & Bashir, M. (2015). Trust in Automation: Integrating Empirical Evidence on Factors That Influence Trust. *Human Factors: The Journal of the Human Factors and Ergonomics Society*, *57*, 407–434.

Holbrook, J., Prinzel, L. J., Chancey, E. T., Shively, R. J., Feary, M., Dao, Q., Ballin, M. G. & Teubert, C. (2020, June 15). Enabling Urban Air Mobility: Human–Autonomy Teaming Research Challenges and Recommendations. In *AIAA Aviation 2020 Forum. AIAA Aviation 2020 Forum*. (AIAA Paper 2020-3250). American Institute of Aeronautics and Astronautics.. https://doi.org/10.2514/6.2020-3250

Johnson, M., Bradshaw, J. M., Feltovich, P. J., Jonker, C. M., van Riemsdijk, M. B. & Sierhuis, M. (2014). Coactive Design: Designing Support for Interdependence in Joint Activity. *Journal of Human–Robot Interaction*, *3*(1), 43–69. https://doi.org/10.5898/JHRI.3.1.Johnson

Johnson, M. & Vera, A. (2019). No AI Is an Island: The Case for Teaming Intelligence. *AI Magazine*, *40*(1), Article 1. https://doi.org/10.1609/aimag.v40i1.2842

Johnson, M., Vignati, M. & Duran, D. (2018). Understanding Human–Autonomy Teaming through Interdependence Analysis. In *Symposium on Human Autonomy Teaming* (pp. 1–20).

Lee, J. D. & Seppelt, B. D. (2023). Design for Human–Automation and Human–Autonomous Systems. In S. Y. Nof (Ed.), *Springer Handbook of Automation* (pp. 433–455). Springer International Publishing. https://doi.org/10.1007/978-3-030-96729-1_19

Li, M., Wang, M., Ding, D. & Wang, G. (2022). Development and Evaluation of Single Pilot Operations with the Human-Centered Design Approach. *Aerospace*, *9*(10), Article 10. https://doi.org/10.3390/aerospace9100601

Li, M., Wang, M., Wang, G., Chen, Y. & Zhong, K. (2023). Predictive Mental Workload Modeling Methodology for Single-Pilot Operations System Design. *Journal of Aerospace Information Systems*, 20, 207–230. https://doi.org/10.2514/1.I011314

Lyons, S. K., Lewis, M. & Capiola, A. (2021). Human–Autonomy Teaming: Definitions, Debates, and Directions. *Frontiers in Psychology*, *12,* 101–116. www.frontiersin.org/articles/10.3389/fpsyg.2021.589585

Lyons, W., Wynne, K. T., Mahoney, S. & Roebke, M. A. (2019). Chapter 6 – Trust and Human–Machine Teaming: A Qualitative Study. In W. Lawless, R. Mittu, D. Sofge, I. S. Moskowitz & S. Russell (Eds.), *Artificial Intelligence for the Internet of Everything* (pp. 101–116). Academic Press. https://doi.org/10.1016/B978-0-12-817 636-8.00006-5

Matessa, M., Strybel, T., Vu, K., Battiste, V. & Schnell, T. (2017). *Concept of Operations for RCO/SPO* (20170007262). NASA Ames. https://ntrs.nasa.gov/citations/20170007262

McNeese, M. D. & McNeese, N. J. (2020). Chapter 9 – Humans Interacting with Intelligent Machines: At the Crossroads of Symbiotic Teamwork. In R. Pak, E. J. de Visser, & E. Rovira (Eds.), *Living with Robots* (pp. 165–197). Academic Press. ISBN: 9780128153673. https://doi.org/10.1016/B978-0-12-815367-3.00009-8

O'Neill, T., McNeese, N., Barron, A. & Schelble, B. (2022). Human–Autonomy Teaming: A Review and Analysis of the Empirical Literature. *Human Factors*, *64*(5), 904– 938. https://doi.org/10.1177/0018720820960865

Parasuraman, R., Sheridan, T. B. & Wickens, C. D. (2000). A Model for Types and Levels of Human Interaction with Automation. *IEEE Transactions on Systems, Man, and Cybernetics – Part A: Systems and Humans*, *30*(3), 286–297. https://doi.org/10.1109/3468.844354

Pechlivanis, K. & Harris, D. (2025). Single Pilot Concept of Operations: Hazard Identification and Mitigation Measures. *The International Journal of Aerospace Psychology*, 35(4), 1–24. https://doi.org/10.1080/24721840.2025.2481880

Puca, N. & Guglieri, G. (2024). Enabling Civil Single-Pilot Operations: A State-of-the-Art Review. *Aerotecnica Missili & Spazio*, 104, 187–212. https://doi.org/10.1007/s42 496-024-00223-7

Rehman, U. (2020). Modelling Level 1 Situation Awareness in Driving: A Cognitive Architecture Approach. *Transportation Research Part C: Emerging Technologies*, *165*, 104737.

Schutte, P., Goodrich, K. & Williams, R. (2017). Synergistic Allocation of Flight Expertise on the Flight Deck (SAFEdeck): A Design Concept to Combat Mode Confusion, Complacency, and Skill Loss in the Flight Deck. In N. A. Stanton, S. Landry, G. Di Bucchianico, & A. Vallicelli (Eds.), *Advances in Human Aspects of Transportation* (Vol. 484, pp. 899–911). Springer International Publishing. https://doi.org/10.1007/978-3-319-41682-3_74

Stanton, H. D. & Starr, A. (2016). The Future Flight Deck: Modelling Dual, Single and Distributed Crewing Options. *Applied Ergonomics*, *53*, 331–342. https://doi.org/10.1016/j.apergo.2015.06.019

Strybel, T. Z., Keeler, J., Mattoon, N., Alvarez, A., Barakezyan, V., Barraza, E., Park, J., Vu, K.-P. L. & Battiste, V. (2018). Measuring the Effectiveness of Human Autonomy Teaming. In C. Baldwin (Ed.), *Advances in Neuroergonomics and Cognitive Engineering* (Vol. 586, pp. 23–33). Springer International Publishing. https://doi.org/10.1007/978-3-319-60642-2_3

Tambon, F., Laberge, G., An, L., Nikanjam, A., Mindom, P. S. N., Pequignot, Y., Khomh, F., Antoniol, G., Merlo, E. & Laviolette, F. (2022). How to Certify Machine Learning

Based Safety-Critical Systems? A Systematic Literature Review. *Automated Software Engineering*, *29*(2), 38. https://doi.org/10.1007/s10515-022-00337-x

Ung, K., Doyon-Poulin, P. & Robert, J.-M. (2021). Exploiting Forward-Looking Data in Prospective Ergonomics: The Case of Aviation. In N. L. Black, W. P. Neumann & I. Noy (Eds.), *Proceedings of the 21st Congress of the International Ergonomics Association (IEA, 2021)* (Vol. 219, pp. 805–812). Springer International Publishing. https://doi.org/10.1007/978-3-030-74602-5_110

Wang, G., Li, M., Wang, M. & Ding, D. (2023). A Systematic Literature Review of the Human-Centered Design Approach in Single Pilot Operations. *Chinese Journal of Aeronautics*, *36*(11), 1–23. https://doi.org/10.1016/j.cja.2023.07.026

APPENDIX

See Table 31.4.

TABLE 31.4
Interdependence analysis color scheme

Color Key for Team Member Role Alternative Capability Assessment

Performer	Supporting Team Members
I can do it all	My assistance could improve efficiency
I can do it all but my reliability is less than 100%	My assistance could improve reliability
I can contribute but need assistance	My assistance is required
I cannot do it	I cannot provide assistance
Not applicable/not significant	Not applicable/not significant

Source: Johnson et al. (2014).

Note: The "Performer" column has a different meaning than the "Supporting Team Member" column.

Section VII

Assessment, Measurement, and Management

32 Maturity Assessments

TRL (Technical Readiness Levels), HRL (Human Readiness Levels), ORL (Organizational Readiness Levels), and Other Maturity Levels

Paul Lambertson

INTRODUCTION

In some form or another, maturity levels have been around as far back as the ancient Greeks and Romans. They had their methods for design and construction, which could be seen as a predecessor to modern design maturity assessments. However, these assessments are not as formal as modern-day assessments.

The Greeks made significant contributions to the history of building technology. They developed innovative building methods and required considerable resources to construct monuments. The Romans perfected several of these innovations, such as techniques for lifting heavy loads. The Greek building technology and methods include the material remains from ancient buildings, detailed financial accounts that the Greeks kept for major building projects, and the Latin and Greek works of writers such as Vitruvius, Pliny, and Theophrastus.

A Roman architect and engineer, Vitruvius wrote "De Architectura Libri Decem," the oldest extant architectural treatise available today. This work provides a way of understanding how mathematics was used by the architects of the Classical Greek and Roman worlds to both solve practical problems and to create buildings that conformed to the highest aesthetic aspirations. Although these methods might not be directly comparable to a modern design maturity assessment, they do show that the Greeks and Romans had a systematic approach to design and construction, with a focus on aesthetics, functionality, and the use of resources. This could be seen as an early form of assessing design maturity. However, it is essential to note that the concept of design maturity, as we understand it today, did not exist in the same form as it did in ancient times. The focus was more on the practical aspects of construction and the aesthetic outcome than on the era's formalized process of continuous improvement and maturity. Human factors from the safety perspective were not a focus during ancient times.

Today, we can conduct detailed assessments to see if a design is ready to move forward in stages for eventual release to the public. It includes a perspective on

DOI: 10.1201/9781003666431-39

human readiness and how an organization could be prepared to advance in a project or a design. Consequently, technical readiness levels (TRLs), human readiness levels (HRLs), and organizational readiness levels (ORLs) introduced by Boy (2021), and general maturity levels will be defined; examples will be discussed, and thoughts on the future of maturity assessments will be provided.

TECHNICAL READINESS LEVELS

The definition of TRLs: a method used to measure and assess the maturity of a particular technology. They allow engineers, managers, and project personnel a consistent reference point for understanding how a technology evolves. The purpose of TRLs is to measure a system's maturity of technology components. This measurement lets project personnel know how much development a particular technology needs before being utilized. A TRL rating helps in measuring the progress of a project. TRLs were initially developed by NASA in the 1970s for technologies in space exploration. TRLs are based on a scale from 1 to 9, with 9 being the most mature technology. Similar to our ancient Romans or Greeks, who focused on usability, the TRL focuses on setting up the technology to be ready for use.

The TRLs are described in Table 32.1 based on NASA TRL Best Practices guide 2021. The steps are in a logical progression of how a technology should be prepared

TABLE 32.1
Technical readiness levels

TRL	Description
1	*Basic Principles Observed:* This is the research stage of TRL. During this stage, the research can include the technology's basic properties, paper studies. It will need supporting information that can be published research for identifying the principles of technology.
2	*Technology Concept and Application Formulated:* After a thorough observation of basic technical principles, you can go to the next level for inventing the practical application.
3	*Experimental Proof of Concept:* TRL 3 means the initiation of active research and development of technology concepts. Besides analytical studies, this process will also include laboratory studies.
4	*Technology Validated in Lab:* During this stage, all the basic technological components will be integrated.
5	*Technology Validated in Relevant Environment:* At this stage, the components of the technology concept are representative or integrated.
6	*Prototype Technology Demonstrated in Relevant Environment:* At this stage, a prototype of the technology is demonstrated in a relevant environment.
7	*System Model or Prototype Demonstration in Operational Environment:* At this stage, a prototype system is demonstrated in an operational environment.
8	*System Complete and Qualified:* At this stage, the system is complete and qualified.
9	*Actual System Proven in Operational Environment:* At this stage, the technology has been proven to work in its final form and under expected conditions.

to be ready for use. These assessments help to identify requirements, new technology, critical elements of the technology, and the level of integration for testing. Each component of the system will have a technology readiness assessment. Identifying whether the technology performs a new or novel function or is an existing one is essential. This will allow for a determination of whether the technology requires readiness or is an existing technology. A technical readiness report is created for each level, comprehensively explaining the assessed TRL level.

HUMAN READINESS LEVELS

HRLs are a nine-level scale to evaluate and communicate the readiness of a system for human use. They are intended to address all of the Human Systems Integration (HSI) domains: workforce, personnel, training, human factors engineering, safety and occupational health, survivability, and environment and habitability. HRLs can complement and supplement the TRL scale. They help understand the readiness of technology to operate within human capabilities and limitations. HRLs provide an overview of the human interfaces and complement the TRLs. HRLs facilitate understanding of the advantages of integrating the human perspective and how to overcome some of its limitations for the interface to the item or technology being evaluated. The Romans and Greeks focused on basic human usability and function; the HRLs have a more focused approach on the human-to-system interface. The table is a list of HRLs (Table 32.2) based.

A Simple Example of an HRL

A company is developing a new display software interface intended to be used for diverse groups of users. To assess the HRL of the display software interface, user testing would

TABLE 32.2
Human Readiness Levels

HRL	Description
1	Relevant human capabilities, limitations, and basic human performance issues and risks identified.
2	Human-focused concept of operations is defined, and human performance design principles established.
3	Requirements for supporting human performance established.
4	Modeling, part-task testing, and trade studies of human systems design concepts completed.
5	User evaluation of prototypes in mission-relevant simulations completed to inform design.
6	System/subsystem model or prototype demonstration in a relevant environment
7	System prototype demonstration in an operational environment.
8	Actual system completed and "qualified" through test and demonstration.
9	Actual system through successful mission operations.

be conducted with a diverse group of users representing the demographic intended for the display software interface. This will be used to identify areas where improvements could be made. The testing could ask users to perform specific tasks using the display software interface and observe their interactions with the application. Data on usability, safety, and human performance would be gathered from the users. Based on the results of the user testing, areas where improvements could be made to increase the usability of the display interface software could be identified. For example, if users have difficulty navigating the display interface, changes could be made to make it more intuitive.

Conducting user testing is an effective way to assess the HRL of a technology and can identify areas where improvements could be made.

ORGANIZATIONAL READINESS LEVELS

Organizational readiness takes a proactive approach to change. By operating from a holistic mindset for end-to-end transformation, organizational readiness establishes early leadership alignment, builds understanding and desire across the organization, and drives optimization through a commitment to continuous improvement. An organizational readiness assessment is an official measurement of a company's preparedness to undergo a major change or take on a significant new project. It provides the knowledge and assurance that the company's proposed endeavor can succeed if it decides to go ahead and do it. It can also save a company's reputation by allowing it to avoid a potentially high-profile failure for engaging in a project it was not ready to complete. ORL levels are defined as in Table 32.3.

TABLE 32.3
Organizational Readiness Levels

ORL	Description
0	Exploring the first principles for the organizational model.
1	Goal-oriented research making choices from first principles to full practical digital model.
2	Proof of principle development and active research and development is started in a virtual environment.
3	Virtual agile organizational prototype development and first HITLS (virtual HCD).
4	Proof of organizational concept development using concrete scenario-based design from full virtual to more tangible environments.
5	Assessing organizational capabilities in terms of authority sharing (responsibility, accountability, and control plus other items such as trust, coordination, and collaboration).
6	Real-world use-case tests in a wider variety of situations – tangibilization continues.
7	Practical integration with respect to criteria such as safety, efficiency, and comfort at various levels of granularity of the organization – tangibilization continues.
8	Readiness for effective implementation on a real site (fully tangible) based on feedback for deployment approval.
9	Deployment involving both personal and real machines.

Simplified Example

A company is starting to consider a transition to a fully digital environment. They begin by considering how the transformation would fit into the company from a people, organizational, and technology perspective (ORL 0). The company begins researching various aspects of digital transformation and starts to create the goal-oriented principles. They consider the impacts on the organization and set some goals, such as improved work accuracy and traceability of the information (ORL 1). The team developed proof of the principles and created a virtual model to inform the organization of its advantages and disadvantages (ORL 2). Creating a virtual prototype that will be used with humans in the loop. This will provide data that can be used to improve the processes and enhance the system for the users (ORL 3). The prime scenarios for the organization are created, and more than just a virtual environment is started. This testing phase will prove the organization concept (ORL 4). The team needs to assess the organization's capabilities and understand where items are located, such as how the final authority is executed and how the system can be trusted to perform its function. In a digital environment, the system may be used for the final signoff, and tracking the changes could provide a way for people to trust the system (ORL 5). The team starts to test in real-world test-case situations. This could include using the digital environment and the process it is replacing side by side for a comparison (ORL 6). The team starts practical integration and looks at the digital environment from the perspective of all levels of the organization. This is a method to continue to analyze the impact on the organization (ORL 7). The team now assesses the readiness to roll out the digital integration at a real site and uses a review to gain approval for the rollout (ORL 8). The team moves to a full rollout of the digital environment into the real company system and documents lessons learned (ORL 9).

GATE REVIEWS AND HOW COMPANIES ADDRESS READINESS LEVELS

Many companies have methods to review if a project is progressing, to understand how it is progressing against milestones. The milestones can include items from ORLs, HRLs, TRLs, timing, and financial status. It varies by company on how integrated TRLs, HRL, and ORLs are into a company's gate or readiness reviews. What they tend to have in common is a method to assess if a project is on time and its financial status. These reviews are often the focus of the project management team.

While the other reviews tend to be more technically driven, key items for gate reviews are as follows:

1. *Define the Scope of the Project:* What is the project intended to accomplish, and what are its goals and objectives? The scope could be what functions are intended to be released at launch from a project to when the product is designed for launch.

2. *Assess the Current State of the Project:* This can be done with the assistance of TRLs, HRLs, and ORLs. This is typically performed during several intervals of the project. An example of something to assess is where the technology is in terms of readiness level and its relationship to the planned release schedule.
3. *Set Targets:* This should be for the project's overall scope, and HRLs, TRLs, and ORLs can all add helpful information. Knowing the technology is ready to use and its interfaces are evaluated for the human in the loop provides critical feedback to provide information on the project's targets. Cost targets are currently one of the most commonly tracked targets.
4. *Develop a Plan:* The plan should be detailed enough to achieve the project objectives. This should include technical activities, human factors, and organizational needs. One of the key items in a plan is the timeline for when key items are intended to be completed. This would include TRLs and HRLs. The intended times for those reviews would be developed and documented.
5. *Implement the Plan:* This would include any relationships to HRLs, TRLs, and ORLs. The organization needs to understand how to implement and work on its TRL and HRL plans. They need to work on items such as demonstrating the technology and the human interfaces, with cost and delivery dates in mind.
6. Use gate reviews at the key points of the project that are determined, such as when human and technological interfaces are critical, to ensure that timing and cost are on track.
7. *Review and Lessons Learned:* Learning from past mistakes and their successes is a key to understanding where issues can occur and how they were solved in the past. This could include items such as those that did not pass or caused issues at past reviews, and understand if any items in the TRLs, HRL, or ORLs point to items that could be improved.

CONCLUSION

The ability to relate ORL, TRL, and HRLs together can take a company to the next level. HFES has completed some work to join TRL and HRLs together (Human Factors and Ergonomics Society, Table 32.4). The ability to coordinate human readiness with technical readiness provides the ability to allow for technology and human usability to be coordinated with each other. This coordination will allow for a system with technology that is more integrated into how humans use the system. This integration will allow for better user-centered designs and improvement of the system. If an organization can prepare for the tasks to be completed by merging the TRLs and HRLs in advance, it will help the company to be ready for the work that needs to be completed. These projects will allow an organization to be more prepared to support the setup of the technology and people. The backbone for the organization will be how ready it is to help with the integration of TRLs and HRLs with its gate reviews.

TABLE 32.4
HRLs and TRLs

Level	HRL	TRL
1	Basic principles for human characteristics performance, and behavior observed and reported	Basic principles observed and reported
2	Human-centered concepts, application, and guidelines defined	Technology concept and/or application formulated
3	Human-centered requirements to support human performance and human technology interactions established	Analytical and experimental critical function and/or characteristic proof of concept
4	Modeling, part-task testing, and trade studies of human systems design concepts and application completed	Component and/or characteristic proof of concept
5	Human-centered evaluation of prototypes in mission- relevant part-task simulations completed to inform design	Component and/or breadboard type validation in relevant environment
6	Human systems design fully matured as influenced by human performance analysis, metrics, prototyping and high-fidelity simulations	System/subsystem model or prototype demonstration in relevant environment
7	Human systems design fully tested and verified in operational environment with systems hardware and software and representative users	System prototype demonstration in an operational environment
8	Total human–system performance fully tested, validated, and approved in mission operations, using completed system hardware and software and representative users	Actual system completed and qualified through test and demonstration
9	System successfully used in operation across the operational envelope with systematic monitoring of human–system performance	Actual system proven through successful mission operations

The benefits of integrating readiness levels with TRLs, HRLs, and ORLs include:

1. Creating coordination between the people, organization, and technology.
2. Such that ORLs, TRL, and HRLs work together to achieve and launch a design solution.
3. Gate reviews can be strengthened through the coordination of the readiness levels. This can give an organization a better understanding of the cost, timing, and ability to incorporate the technology into a human environment.

4. The ability to have a well-integrated plan that knows the tasks that must be completed for gate reviews. This can help to understand the timing required for a high-quality product.
5. The information from the readiness levels allows data-informed decisions for the gate reviews. This will enable an organization to respond better to issues and challenges found during the project.

Using readiness reviews, a company can understand the progress of its technology and human readiness so that it can plan for the organizational needs of a product with the needed technology and support the user interface.

REFERENCES

Boy, G.A. (2021). Socioergonomics: A Few Clarifications on the Technology–Organizations–People Tryptic. In *Proceedings of INCOSE HSI2021 International Conference*. Wiley Online Library (Open Access).

NASA. (2021). Technical Readiness Assessment Best Practices Guide SP-20205003605 ANSI/HFES 400-2021 Human Readiness Level Scale in the System Development Process The North American Standards Institute, ANSI/HFES.

33 Design and Measurement for Socio-Technical Change Using OSCAR-TI

Grace A. L. Kennedy, Robyn Coman, Leanne Treadwell, and John Paul Kennedy

INTRODUCTION

Traditionally, Systems Engineering (SE) has focused on developing the technological system of interest. With the introduction of human systems integration (HSI), the human element has been considered an integral part of the system and a necessary consideration within the systems engineering processes. As we embark on the digitalization of both the system and the SE processes, the digitalization of humans and the broader modelling of that human within the socio-technical system are also progressing.

While the human factors/ergonomic (HF/E) domains of physical and cognitive—the capabilities, behavior, and performance of the user—have been integrated within SE, the integration of the organizational domain is seemingly not as mature. For many, one of the first SE activities within the system lifecycle that focuses on the system from a human perspective is the creation of the operational concepts, which describe the envisaged system. This activity strategically addresses the end-user's perspective, which becomes the impetus for activities that define user requirements—ideally in consultation with the stakeholders as per human-centered design principles.

However, the end-user is not the only human involved in the system. Firstly, users do not exist in isolation; they exist within complex adaptive organizational systems that span across one or more systems in operation. Secondly, many human stakeholders of the system are not necessarily direct users. For example, imagine a technology system validated as fit-for-purpose that "ticks all the boxes" for a user to perform and reach their goal effectively, efficiently, and in an enjoyable manner. Suppose the organization's decision makers are not supportive of the system or have not appropriately planned for its introduction, operations, or maintenance. In that case, the technology might be rejected before the user can interact with the system.

As organizations are complex adaptive systems, achieving change is not as simple as developing or designing a precise, static plan for the future organization. It is necessary to iteratively plan, regularly monitor, and actively review the organizational system's maturity and readiness for the envisaged change. This chapter describes

DOI: 10.1201/9781003666431-40

efforts to incorporate measures of organizational readiness into systems engineering practices that will enable HSI practitioners to ensure that the organizational system design and organizational change are adequately considered alongside technology and human readiness and ultimately reduce the risk that an organization will be unable to adopt a new technology effectively.

Organizational Systems, Change, and Readiness

A *system* is an *interacting* combination of *system elements* to accomplish a defined *objective*(s). The system interacts with its *environment*, including other systems, users, and the natural environment (INCOSE, 2021). This definition seems innocuously simple. However, when organizations are viewed as systems, there is no single definition of the system as a whole because the inclusion of and/or interactions between the system elements often depend on the perspective of the viewer/observer. When viewing an organization as a system, there can be many ways to decompose the system into its subsystems (or entities) (Bednar & Welch, 2020). For example, Carmichael (2018) suggests that an organization is a group of at least three people working toward some common purpose, and that the elements of the organizational system are thus the people. Other researchers consider the elements of the system to be the sub-units (e.g., teams or departments) within the broader organization (Urban, 2017). The traditional solution to this problem has been to use multiple levels of decomposition to describe the organizational system (individuals, groups, and organization) (Robbins & Judge, 2007; Schein, 2010), and this approach has been applied in studies on organizational learning and evolution (Annosi et al., 2020). Ainsworth and Feyerherm (2016) build on this three-level notion by adding a fourth supersystem level, the "trans-organizational" system, which is composed of multiple organizations.

The Viable System Model (VSM) (Beer, 1984) forms the basis of much of the existing literature and incorporates many of the ideas above. The VSM views an organizational system as being composed of five interacting systems: (1) operations—the viable system itself and its recursions, (2) coordination, (3) delivery, (4) development, and (5) policy. Hoverstadt (2020) focuses on the applications of the VSM to human activity systems, particularly in using the model to diagnose gaps in real-world organizations and subsequent design efforts to improve the situation. Donaldson (2017) suggests that because an organization's environment is constantly changing, designing adaptation in response to change that has already occurred may be a futile endeavor. Instead, he proposes that organizational systems should be designed and changed over time to adapt to environmental changes as they happen. This dynamic, adapting view of viable systems informs the work's core.

This research lies firmly in the HSI field, so recent progress in Human Readiness Levels (HRLs) is utilized as a starting point for developing Organizational Readiness Levels (ORLs). ANSI-HFES 400-2021 outlines the HRLs and guides how the levels are assessed (ANSI/HFES, 2021). Aligned work on models that include the Human View (HV) within SE has evolved over the past 15 years (Handley, 2022; see also Chapter 5), with human elements receiving greater consideration within architecture frameworks and model-based systems engineering. As the organization is one of three domains of HF/E, any extensions to organizational aspects should fit within the HRL and HV work schema, rather than compete with it. Therefore, the Technology Readiness Levels (TRLs) (Mankins, 1995) and HRLs are appropriate for building our ORLs.

Academic databases (Scopus, Google Scholar, ABI Inform complete, ScienceDirect, IEEE Explore, and Web of Science) were searched using the search string <organizational readiness level> from 2013 to 2023. A total of 74 results were found, of which 30 were excluded as repetitions/unretrievable, and 11 were excluded as they mentioned the keywords only in passing. This left 33 results in the set of interest, and of these, 15 used the term "organizational readiness level" as part of the prose to denote an unspecified quantity of sufficiency. The remaining 18 results used the term ORL in one of the following five ways: (1) factors affecting organizational readiness, (2) change to paradigm or process, (3) change to a technology, (4) ORL scales, or (5) readiness for change/instruments for assessment (see Table A1.1).

Delving further, an initial literature review showed only three published scales for ORL, coincidentally all published between 2020 and 2022 (see Appendix 1). The Vik et al. (2021) scale is focused mainly on the work processes and integration with existing technologies that organizations are already utilizing. It is reassuring that the scale emphasizes that a technology must be adapted in response to organizational change. However, this adaptation does not happen until relatively late (ORL8). The Boy (2022) scale looks at the concept of an organization maturing from a virtual perspective, through levels of increasing tangibility before eventually entering into the real world (ORL6–9). Boy's conception is that an organization comprises multiple agents that can be tested and simulated alongside a first virtual, then increasingly tangible, technological design. In many ways, this scale offers that elusive structure in which organizational and technological design can be tested together with change driven from either perspective. Finally, the Bruno et al. (2020) scale focuses on system maturity from an organizational design view, considering the specific aspects of an organizational system (e.g., roles, processes, structures) that HF/E practitioners would be most familiar with.

Each of the ORL scales described is slightly different and provides insight into different aspects of ORL, and thus, each scale has different benefits for an assessor. The Vik et al. (2021) schema does not align readily to the TRLs and HRLs and was therefore excluded from further comparison. Before ORL6, the Boy (2022) and Bruno et al. (2020) scales have different foci. Boy (2022) considers organizational design with the future operational organization in mind, seeking to refine models of the future organization. Bruno et al. (2020) approach these early levels through an organizational change perspective, assessing first the impact of a proposed technology on the needs of the existing organization before beginning the transformation of the organization. From ORL6 onwards, both scales focus on real-world testing in the operational environment. It is clear, therefore, that two dimensions to organizational readiness need to be considered during system development: (1) the maturing of the system design of the future organization (ORL-SD) and (2) the change readiness of the actual organization in its current state (ORL-CR).

A FRAMEWORK FOR ORGANIZATIONAL SYSTEMS CHANGE AND READINESS FOR TECHNOLOGY INTRODUCTION

In Kennedy (2024), the lead author proposed a framework to provide a simple structure through which HSI practitioners might map out and evaluate both organizational system design and planned organizational change in a way aligned with the technological system development. This framework did not establish the methods or mechanisms for the decisions made around the actual design or planned change, but instead gave a set of constructs around which activities could be performed. Kennedy (2024) depicted the framework as a cube (Figure 33.1) with three dimensions: (1) aligned TRL/HRLs (assuming them to be matched or in sync), (2) maturity of the organizational system design (ORL-SD), and (3) organizational change readiness (ORL-CR).

To introduce a new technology, this framework suggests that system developers will seek to move from the bottom left of the cube [*coordinates* 0,0,0] to the top correct diagonal of the cube (marked with a star) [*coordinates* 9,9,9]. On reaching this goal, the system will have TRL/HRL9, the designed organizational system will have matured to use the technology (ORL-SD9), and the organization will be ready for the change (ORL-CR9).

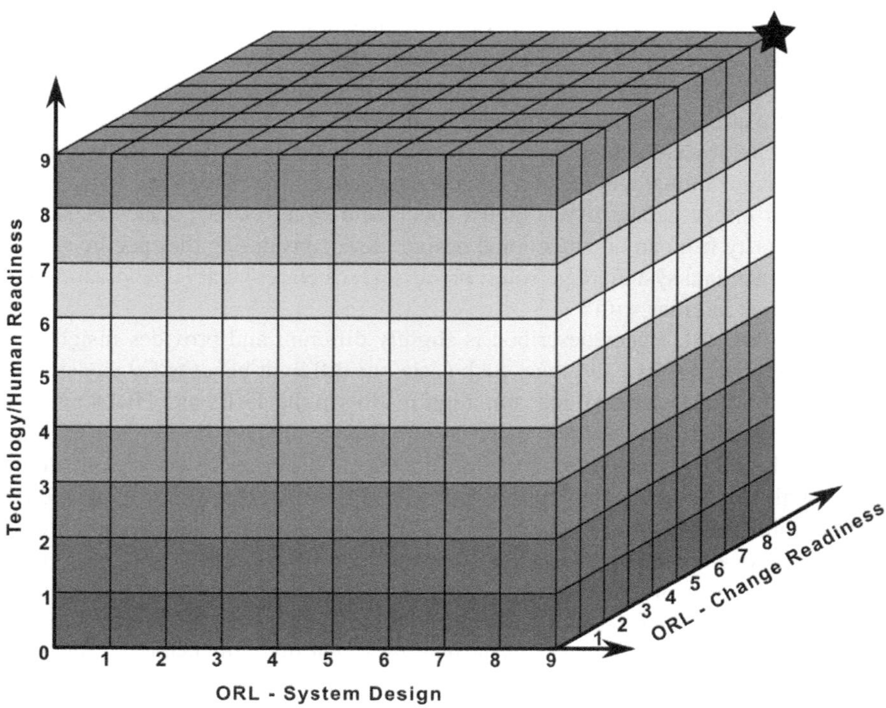

FIGURE 33.1 Framework for organizational systems change and readiness.

By way of illustration, consider three hypothetical scenarios:

1. *A technology has been developed in isolation from any organizational planning.* Here, the framework suggests that a narrow column is built at the front left of the cube and then assumes a step change that aims to fill in the rest of the cube in one go. The organization is not ready for the change, and/or the technology is not fit for the organization's purpose. There is inertia or resistance to change, and confusion in the organization exists around operations, as in essence, the organization must now fit around the new technology. Decision makers are surprised that their technology investment is not producing the expected gains.
2. *The organization is not considered until the technology has reached a tangible prototype (TRL6+).* As mentioned before, a narrow column is built at the front left of the cube until TRL/HRL6 is achieved. The two ORL dimensions are considered, and the column tries to grow outward in those directions. Although the users' needs have been designed for, the organization is not positioned to use the technology, and the organization's constraints for change and adoption become clear. It is now that a redesign may be required.
3. *The technology and the organization develop at a similar maturity rate.* By evolving the tangibility of technology and organization together (Boy, 2022), the organizational system has seen small step changes that bring the actual organization and its readiness for change to full maturity just in time for technology adoption.

All three scenarios fit well within the framework, with pros and cons. Scenario 1 involves no upfront investment in time or development funds from the organization. However, it will likely involve an extended transition period, teething troubles, or even rejection of the technology altogether. Scenario 2 is the most common in contemporary technology adoption.

Organizations are unlikely to contemplate adopting technological solutions until they have reached a certain level of tangibility, as technological development failure rates pre-TRL6 are known to be high, so investing in an organization at these early stages is seen as risky. Scenario 3 is the costliest upfront, but it may reduce overall development costs as design changes can be made whilst they are still flexible. As with all SE practices, while scenario 3 derisks the development and transition, due care must be taken not to overburden the organization with change initiatives. It is important to stress here that there is no implied correct situation. Organizations must decide on the strategy that best fits their needs and resources while maintaining risk at an acceptable level.

While useful, there are some practical limitations to viewing systems change using the cube visualization. Therefore, we propose an alternative radar-style plot to show the same four measures on four separate scales. Note that while the cube model was described as having matched TRL and HRL, we chose to represent these two measures separately, as in practice, these may differ. Figure 33.2a shows an example plot with balanced levels (Lvl 6) across all four measures. The shaded elliptical area shows the best fit involving the four points. In this case, as the measures are equal, the centroid of the plot (marked with a diamond) is aligned with the center.

Representing the previous scenarios on similar plots (Figure 33.2b–d) shows the change in ellipse position under different situations. Scenario 1 (TRL9, HRL9, ORL-SD1, ORL-CR1) shows the centroid skewed the furthest to the top right. Scenario 2 (TRL6, HRL6, ORL-SD1, ORL-CR1) shows the centroid skewed toward the top right, but not as far. While some effort is still required to bring the centroid closer to the balanced situation in this scenario, the effort required is not as much as in scenario 1. Scenario 3 (TRL5, HRL5, ORL-SD5, ORL-CR5) shows balanced development, with each level in sync.

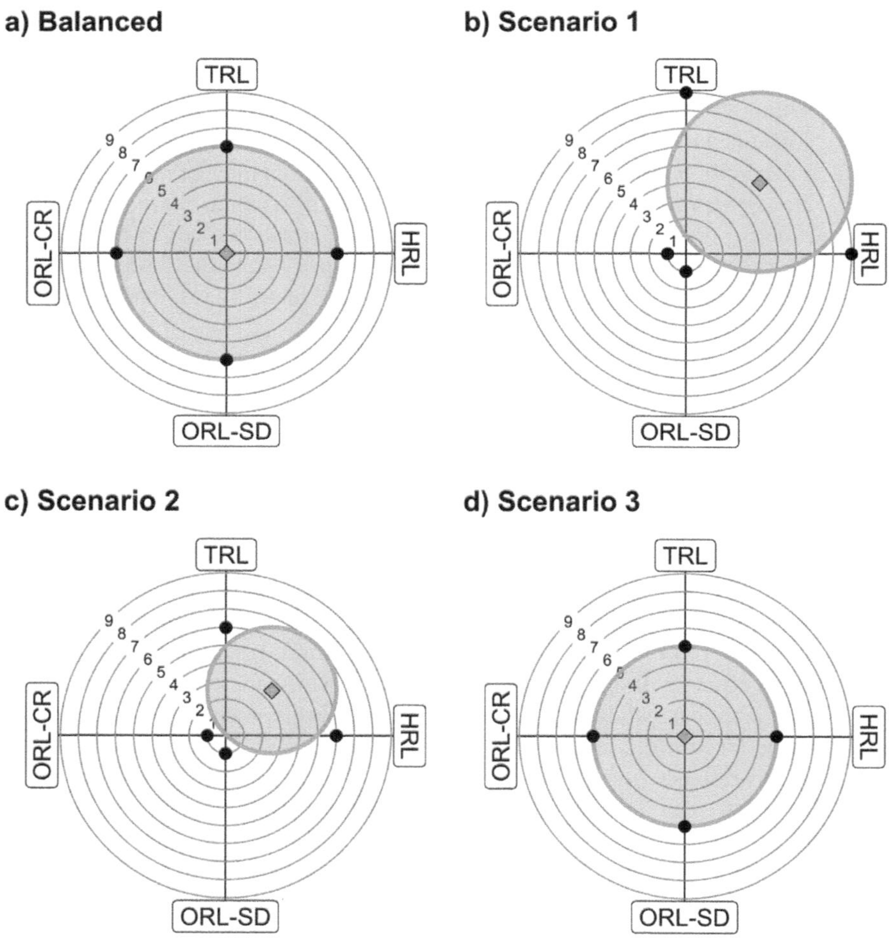

FIGURE 33.2 Measuring organizational systems change and readiness within the OSCAR-TI framework. *Note:* Framework diagrams for (a) a balanced system development, (b–d) scenarios 1–3.

Now, let us assume that systems engineers, as part of the development process, are concerned about the project's risk. During the planning stages, they have predetermined the anticipated, acceptable levels at various stages of the development process. They have determined that maintaining a balanced development process is desirable in this hypothetical situation. However, the planned readiness levels do not necessarily have to be balanced (i.e., the developers may have opted to accept justified project risk in favor of staggered development), which would then yield eccentric plots. Figure 33.3 shows the actual (assessed) level at a point in time, overlaid on the planned level. Generally, the larger the non-intersecting pink area, the greater the potential project risk:

$$\textit{potential project risk} = \textit{planned} - (\textit{planned} \cap \textit{actual})$$

It can be easily distinguished that scenario 1 shows the largest pink area and therefore represents the highest potential project risk. Scenario 2 shows a smaller area of project risk, whilst scenario 3, being balanced, shows no potential project risk due to a mismatch in development as the actual and planned ellipses overlap.

USING THE ORGANIZATIONAL SYSTEMS CHANGE AND READINESS FOR TECHNOLOGY INTRODUCTION FRAMEWORK FOR THE INTRODUCTION OF VIRTUAL REALITY FOR MEDICAL TRAINING: A BRIEF CASE STUDY

Virtual reality (VR) technologies are emerging as a feasible mechanism for enhancing the training of individuals within medical education systems (e.g., university medical schools, hospital-based skills training, and clinician refresher practice). However, the adoption of such technologies is not yet widespread, with various factors such as quality of the training, quality of the technology, cost, resource scarcity, proven learning outcomes, usability, and "buy-in" affecting both an individual's and organization's decisions (Glegg & Levac, 2018; Pedram & Kennedy, 2024). A significant factor in successful VR adoption is the alignment of the VR solution development and the positive reception of end users and their organizations. A human-centered design approach to developing the wider educational system, within which the VR will be introduced, mitigates this risk of misalignment. However, ensuring the appropriate maturity level for each technological, human, and organizational aspect within suitable timescales remains an issue. Measures such as the TRLs initially developed by NASA (Hirshorn & Jefferies, 2016; Mankins, 1995) and standardized as ISO 16290:2013 (International Organization for Standardization [ISO], 2013) have been broadly used in industry to ascertain the maturity of technological systems. However, there has been less focus on measuring system maturity through HRLs and ORLs. HRLs are intended to align with TRLs and ensure that the human has been adequately considered through human system integration activities at the appropriate level of maturity. In an ideal case, we assert that ORL measures should align similarly with HRLs and TRLs.

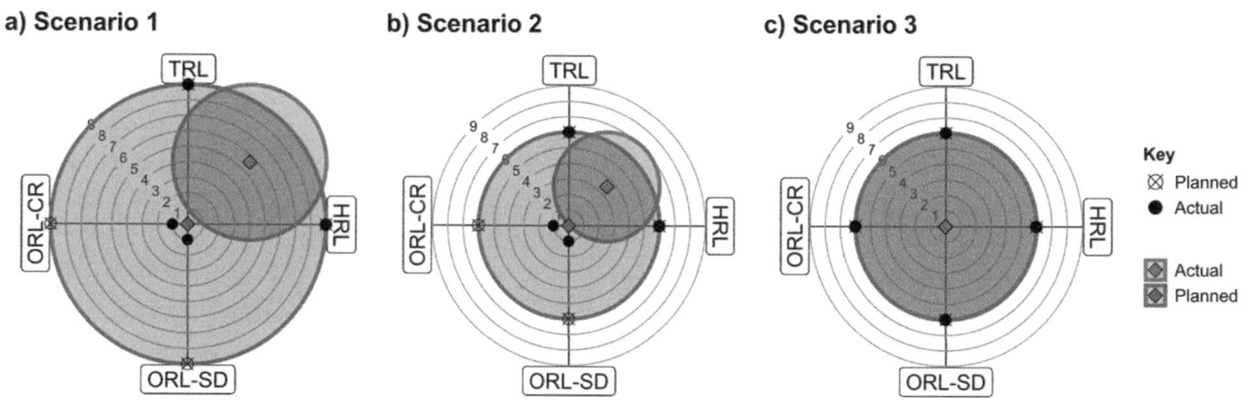

FIGURE 33.3 Potential project risk as visualized for scenarios 1–3.

A scenario-based approach demonstrated how the Organizational Systems Change and Readiness for Technology Introduction (OSCAR-TI) framework and readiness level measurements can be applied. The scenario involves the introduction of a VR head-mounted display (HMD) that provides individual users with simulated tasks and psychomotor skills training to perform clinical procedures. In this case, the system of interest is not simply the VR user (the learner) and the VR technology, but the overall training system in which the learner exists with their organizational system around them. The VR training is not intended to be used as a standalone activity, but as an integrated part of a broader medical education and training program. Hence, it is appropriate to consider the micro-level human–computer interaction and the meso-level organizational training system (Figure 33.4). The key actors are the Learner, their Lecturer and Trainer/Tutor (together the learning team), the resource coordinators at the place of learning, the hospital settings where the training may take place, and the VR developers. Current learning activities involve traditional formal learning (e.g., lectures, demonstrations, book work), a practical that involves physical simulation (e.g., an anatomical and/or physiological manikin), and a practical assessment of learning outcomes. The VR-HMD training system extends the traditional learning use cases by providing an additional VR simulation of the procedure for the learner. This also involves the lecturer providing guidance to ensure that the skills being taught are correct and assisting in the ongoing development of the VR training system in partnership with the VR developer.

The following five-step method was used to explore how the OSCAR-TI framework could be used for assessment (Figure 33.5).

Step 1: With the development team's support, ascertain the technological development's maturity using the TRL, then assess the HRL for the corresponding TRL using ANSI/HFES 400-2021.

VR simulations of clinical procedures are becoming more common, with increasingly higher fidelity and/or complicated procedures being made available for training. While some commercial solutions approaching TRL9 are on the market, most are sub-TRL7. Semi-structured interviews were conducted with the VR developers. This successful MedTech start-up has commercialized a suite of clinical procedures already being used at TRL9 (*Actual system proven in operational environment*).

However, this study focuses explicitly on introducing Intravenous Cannulation (IVC) with a Learning Assistant. The IVC procedure has been available to learners for many months, and the developers have been evaluating the fit-for-purpose of the system. They have also been developing an in-VR learning assistant that demonstrates the physical movements to provide corrective and explanatory feedback when the learner enlists the assistant's aid via a controller button. This functionality is new for the software and is in the prototyping stage, so the system is assessed to be at **TRL6** (*Prototype technically demonstrated in a relevant environment*).

ANSI/HFES 400:2021 was used to assess the level of HSI activities that had been performed during the development of the system. The standard provides sets

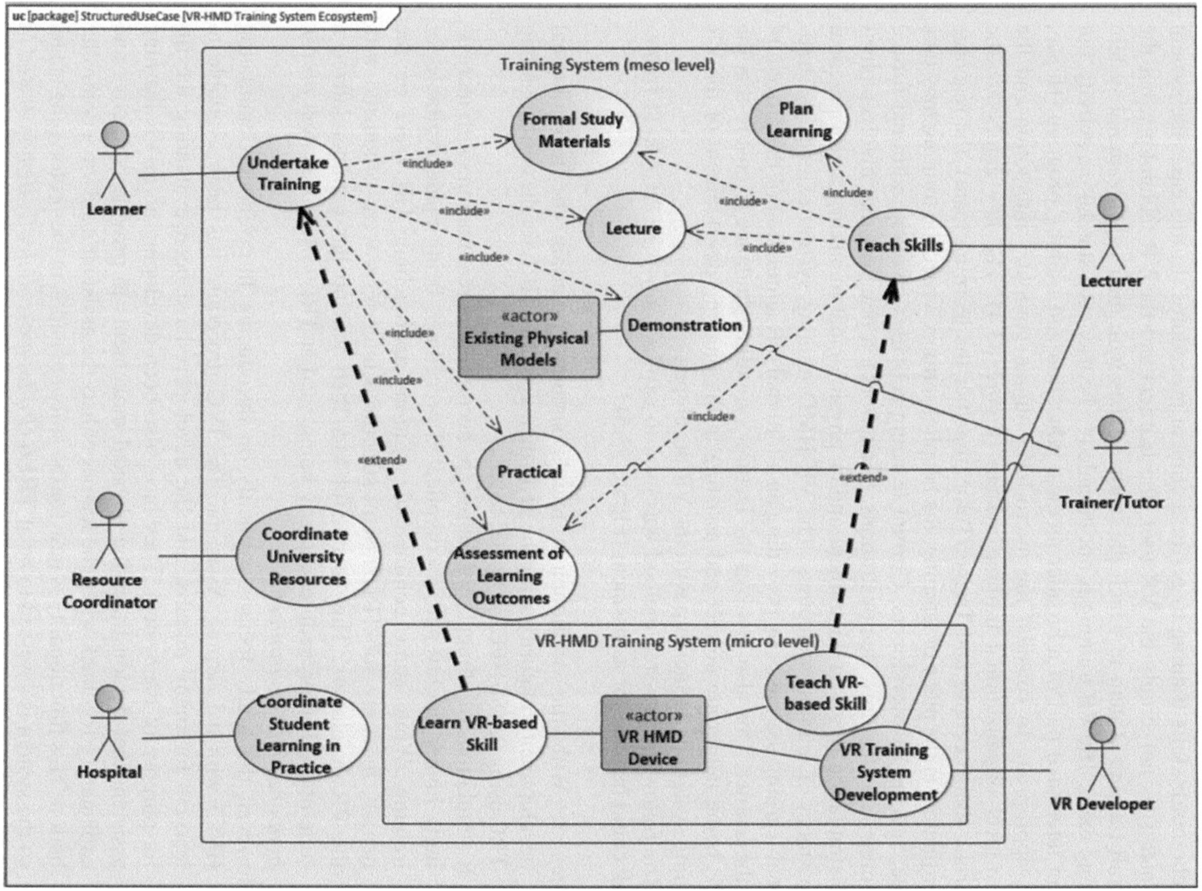

FIGURE 33.4 The training system operational concept in use-case notation.

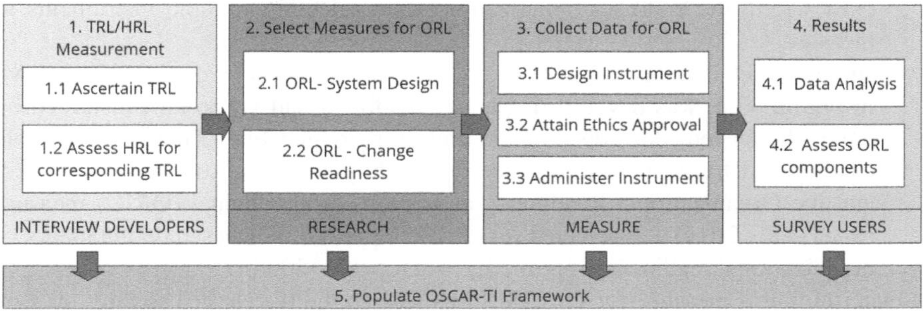

FIGURE 33.5 OSCAR-TI methodology for VR case study.

of questions and descriptors that aid the assessor in identifying the extent to which the appropriate level of HSI activities has been performed. Appendix 2 summarizes the HRL assessment, provides the HRL questions, and also provides complete semi-structured interview responses. However, ANSI/HFES 400:2021 does not provide a method to combine the individual ratings into a single global scale. If each applicable part of the assessed system does not reach a given HRL (e.g., HRL6), should it indicate that the system has not attained that level, but where should the HRL be scored? For this case study, the highest HRL achieved was HRL6 in seven sub-headings. For each sub-heading not at HRL6, the HRL descriptors were used to ascertain the highest prior level, resulting in one HRL3, two HRL4s, and two HRL5s. If we were to assess the global HRL as the maximum level that *all* sub-questions reached—in this case, HRL3—then it could be easily argued that we would not be a fair assessment of the overall HRL. To place the HRL quantitatively into the OSCAR-TI framework, we propose a ratio-based mechanism that considers the differing HRLs and provides a global single score based on all the components HRL_{global}. This represents HRL_{global} the fraction of individual HRL measures that attained the target level.

$$HRL_{global} = \frac{\sum HRL_{assessed}}{\sum HRL_{max}} \times HRL_{target}$$

$$HRL_{global} = \frac{7\times6+2\times5+2\times4+1\times3}{12\times6}\times6 = \frac{63}{72}\times6 = 5.25$$

The system in this case study is therefore assessed at $\boldsymbol{HRL_{global}} = \mathbf{5.25}$.

Step 2: Decide appropriate measures for the ORL components: the Organizational System Design Level (ORL-SD) and the Organizational Change Readiness Level (ORL-CR). Assessors should engage with the VR developers and end-user organizations to consider which organizational system design models fit the system development approach and what information gathering mechanisms exist within the end-user organization when selecting suitable instruments to measure.

A survey targeted at a cross-section of potential stakeholders identified, as shown in Figure 33.4, was developed. Change readiness had yet to be considered by either the system developer or the end-user organization. Therefore, suitable candidate instruments from the literature highlighted previously were identified, and the Organizational Readiness for Implementing Change (ORIC) measure (Shea et al., 2014) was selected for ORL-CR. This tool is based on the theory of organizational readiness proposed by Weiner et al (2008).

Determining a measure for ORL-SD was less straightforward. Drawing on Boy (2022) and Bruno et al. (2020) for descriptors and prior research into organizational system modelling by the lead author (Kennedy et al., 2020), the organizational system descriptors shown in Table 33.1 were derived. For each item, the participant identified to what extent their organization had considered the change

TABLE 33.1
Organizational System Design Descriptors

Item	ORL-SD Descriptor	Supporting Question
SD1	Processes and functions	What are the tasks and activities to be undertaken to achieve the goals?
SD2	Information	What information flows are there, where will information be stored and how will it be retrieved?
SD3	Roles	Who are the stakeholders and what roles will they play? What competencies and training will be needed?
SD4	Infrastructure	What resources are needed?
SD5	Organizational structure	Any changes to the lines of communication or governance required?
SD6	Organizational culture	Any changes to attitudes or values within the organization required?
SD7	Decision-making	What are the key decisions, who makes them and how they influence the organizations?
SD8	Organizational embedding	How will you actually roll the technology out and manage the change in your organization?
SD9	Time alignment	Are the right aspects are available at the right time for the organization and is the timeline is aligned with any technology development?
SD10	Strategy alignment	Is there a strategy for change and is it aligned to the overall organizational strategy?
SD11	Pilot studies	Will you plan to roll out the technology with limited groups first?
SD12	Impact assessment	Have impacts of the effects of the technology adoption been considered?
SD13	Risk	Is there a risk management plan for the adoption of the technology?
SD14	Review	Will the change actually deliver the envisaged benefits to the organization and how will it be monitored?

using the six-point scale: (1) no plans, (2) initial or high-level plans, (3) detailed plans, (4) refined plans, (5) embedded the plan's outcomes, (6) change has been established. A seventh option was added to the instrument to distinguish between (1) no plans and (NA) I do not know, which was removed during analysis.

Step 3: Design a data-gathering exercise to measure the end-user organization (s) using the instruments proposed in Step 2. Ensure that appropriate ethics committee approval has been attained before administering the instruments. Administer the instrument to a cross-section of different types of users in the end-user organizations.

For this case study, the authors administered an online survey to organizational users involved in the IVC procedure with the aims of: (1) ascertaining the extent to which organizations consider the planning, measurement and implementation of organizational change when introducing VR, and (2) identifying if, and how, end-user organizations are involved in the system development of VR medical education systems. The research plan was approved by the Health and Medical HREC of the University of Wollongong (Ethics Protocol 2023/294).

The survey utilized several widely accepted and validated instruments in its design, in addition to the ORIC instrument and the items related to Table 33.1. For conciseness, however, this chapter only discusses the results related to the assessment of ORL-SD and ORL-CR.

Step 4: Analyze the results from Step 3. Aggregate the indices from each instrument to assess the ORL by calculating global measures for ORL-SD and ORL-CR, respectively.

The online survey was administered to stakeholders in the organizations involved in user trials of the IVC activity and who self-identified as members of one of four groups: the learner (47.6%), the learning team (23.8%), organization supporters or enablers (9.5%), and strategic decision makers (19.1%). User trials of the IVC with Learning Assistant were performed with medical students, nursing students, medical staff, and nurses in two locations: a medical school training lab at a university and a hospital simulation center. A total of 28 respondents completed the survey; it is a relatively low participation rate due to the voluntary nature of the survey. A total of 40 participants (learners) experienced the user trial, so the overall survey participation rate is estimated to be approximately 33%.

The ORIC instrument has 12 items spread across two components: Change Commitment and Change Efficiency. Participants respond to each item using a five-point Likert scale ranging from (1) strongly disagree to (5) strongly agree. The results showed a general skew toward the agreement end of the scale for change commitment and efficiency. However, the data also showed that views were inconsistent across the organization, with two change commitment items indicating firmly held views at both ends of the scale. For change efficiency, there are higher percentages of respondents in the *agree* or *strongly agree* bands (between 55% and 77% positive), suggesting that while some respondents do not see a commitment to implementing change, they can see that the organization will be supportive in enabling change to happen.

Using a procedure similar to the calculation HRL_{global}, we propose a $ORL\text{-}CR_{global}$ score aggregating individual item measures. Assuming an even weighting between indices, the average raw scores can be determined for each index and then linearly scaled to match the nine-point scales used by the other readiness levels. Performing this calculation gives a final $ORL\text{-}CR_{global}$ score of **6.44.**

The organizational system questions described in Table 33.1 were administered to participants only in the Supporters/Enablers and the Strategic Decision Makers categories. This decision was taken as it was deemed unlikely that learners or teachers would have sufficient visibility of the organizational change to make informed judgements of ORL. This reduced the number of participants for this section from 28 to 7. This is not unexpected as there are not many decision makers in the organization; however, the findings should be considered as indicative only. Many respondents reported no plans or only initial plans across all indicators. This indicates that the maturity of the organizational system design

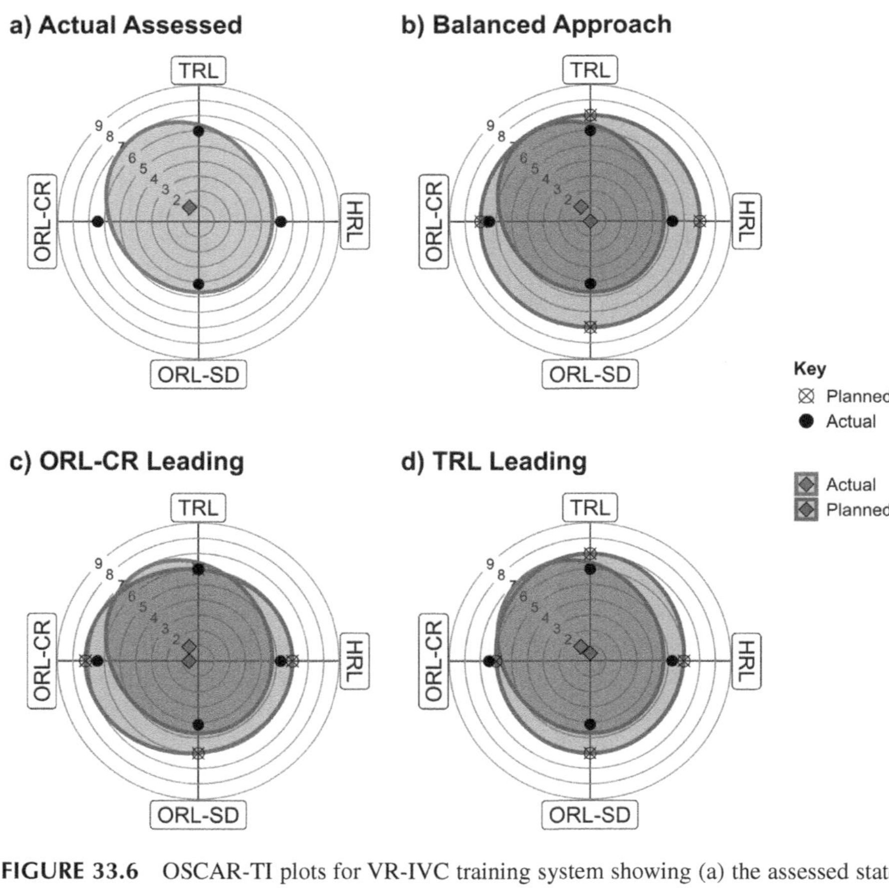

FIGURE 33.6 OSCAR-TI plots for VR-IVC training system showing (a) the assessed state and (b–d) three possible next steps.

is low when considering that the technology is at TRL6. Notably, all seven respondents reported that there were no plans to manage changes to organizational structure due to the introduction of technology. Using the same mathematical approach, a final measure of $ORL\text{-}SD_{global}$ was calculated at **4.13**.

Step 5: Populate the OSCAR-TI framework using the assessed TRL, HRL, ORL-SD, and ORL-CR.

After having gathered the required information, assessed and analyzed the data, global values for all four dimensions of the OSCAR-TI framework are now available to populate the OSCAR-TI plot (Figure 33.6a).

DISCUSSION AND CONCLUSION

For HSI practitioners, this OSCAR-TI enables snapshots of progress to be made and allows for planning of subsequent moves. For example, if we assert that a balanced approach is desired for this case study (readiness level 6 across all measures), then attention should be given to move the centroid closer to the center of Figure 33.6a. It would therefore be desirable to improve the HRL or the ORL-SD. In this case study example, strategies for achieving this can be ascertained by examining individual item responses, HRL_{global} which might be addressed by exploring the factors behind the responses to *managing the politics of implementing change*. Similarly $ORL\text{-}SD_{global}$, delving into the perception that *no planning is done around the organizational structure* item may be explored.

The framework can also be used prospectively to plan the maturity of the technology and organizational system design from a holistic viewpoint. We have assumed that the balanced approach is the "ideal," yet there may be some project and/or organization-specific justifications for not following a balanced approach. For example, a slight lag may be reasonable if the project risk during that development phase is acceptable. Systems engineers need to maintain plans for programmatic planning and justifications for the intended trajectory of the readiness levels (whether balanced or not) throughout the development. (See also Chapter 26, which further discusses using TRL, HRL, and ORLs and alignment with programmatic planning.)

Given the current assessment, what options does the HSI practitioner have for planning the next step in maturity? Figure 33.6b–d shows three hypothetical planning options. In each case, the turquoise ellipse represents the current assessed level, the pink ellipse represents the future planned state to be attained, and the intersection between the two appears purple.

Balanced Approach (Figure 33.6b): Rather than stall project progress, the balanced approach simultaneously pushes forward on all levels but exposes the project to greater risk (the pink shaded area). The HSI practitioner understands that more resources/effort may be required and plans accordingly.

ORL-CR Leading Approach (Figure 33.6c): This approach pushes forward with ORL-CR to level 7.2 but retains the planned TRL, HRL, and ORL-SD at level 6 until the lagging measures "catch up." Assume the organization is reticent to progress formal organizational system design transformation until the leaders are convinced that their organization will be able to implement change in the first place. Therefore, they want the ORL-CR to be at least 7.2. In this case, the project risks remain the same as the current snapshot, but increase on the left-hand side of the plot. Although the organization may become more committed and ready to implement the change, the design has not yet matured sufficiently to roll out the change.

TRL Leading Approach (Figure 33.6d): For development projects where human-centered design or HSI is not explicitly planned or understood, the development may be entirely centered on the progress of the TRL. In this case, the TRL is advanced to level 7, but the HRL and ORLs remain at a goal of 6. The project risk seems at a similar level to the ORL-CR lead approach, and whilst there is still work to "catch up" on the lagging measures, there is now greater risk appearing in the top right-hand side of the plot. This is problematic if the TRL progresses too far ahead, as eventually there will be a significant step change for the HRL and ORLs requiring greater effort. More importantly, the human and organizational considerations may be missing as inputs to the technological design, and issues may not be identified until it is too late in the design process.

We have shown in this chapter that measures of organizational readiness can be integrated with TRL and HRLs using the OSCAR-TI framework to identify mismatched progress during system development. This approach allows an HSI practitioner to better understand the progress of each facet during the system development phases and make informed decisions around planning, resourcing, and the required activities based on project risk. Furthermore, this framework allows system developers to be more aware of the human and organizational aspects of the socio-technical system design, including the issues around planned adoption, and end-user organizations to achieve improved agency in managing their organizational transformation.

REFERENCES

Ainsworth, D., & Feyerherm, A. E. (2016). Higher order change: A transorganizational system diagnostic model. *Journal of Organizational Change Management*, *29*(5), 769–781. https://doi.org/10.1108/JOCM-11-2015-0209

Annosi, M. C., Martini, A., Brunetta, F., & Marchegiani, L. (2020). Learning in an agile setting: A multilevel research study on the evolution of organizational routines. *Journal of Business Research*, *110*, 554–566. https://doi.org/10.1016/j.jbusres.2018.05.011

ANSI/HFES. (2021). *ANSI/HFES 400-2021 Human Readiness Level Scale in the System Development Process*. Human Factors and Ergonomics Society.

Bednar, P. M., & Welch, C. (2020). Socio-technical perspectives on smart working: Creating meaningful and sustainable systems. *Information Systems Frontiers*, *22*(2), 281–298. https://doi.org/10.1007/s10796-019-09921-1

Beer, S. (1984). The viable system model: Its provenance, development, methodology and pathology. *Journal of the Operational Research Society*, *35*(1), 7–25.

Boy, G. A. (2022). Socioergonomics: A few clarifications on the technology-organizations-people tryptic. In *INCOSE International Symposium*, Wiley Online Library.

Bruno, I., Lobo, G., Covino, B. V., Donarelli, A., Marchetti, V., Panni, A. S., & Molinari, F. (2020). Technology readiness revisited: A proposal for extending the scope of impact assessment of European public services. In *Proceedings of the 13th International*

Conference on Theory and Practice of Electronic Governance. Association for Computing Machinery. https://doi.org/10.1145/3428502.3428552

Carmichael, D. G. (2018). Organisations as systems—difficulties in model development and validation. *Civil Engineering and Environmental Systems*, *35*(1–4), 41–56. https://doi.org/10.1080/10286608.2019.1577395

Donaldson, W. (2017). *Simple Complexity: A Management Book for the Rest of Us: A Guide to Systems Thinking*. Morgan James Publishing.

Glegg, S. M. N., & Levac, D. E. (2018). Barriers, facilitators and interventions to support virtual reality implementation in rehabilitation: A scoping review. *PM&R*, *10*(11), 1237–1251, e1231.

Handley, H. A. (2022). A retrospective of the human viewpoint and its applicability to future systems. *Systems Engineering*, *25*(1), 91–103.

Hirshorn, S., & Jefferies, S. (2016). *Final Report of the NASA Technology Readiness Assessment (TRA) Study Team*. Report No. HQ-E-DAA-TN43005, NASA. https://ntrs.nasa.gov/citations/20170005794

Hoverstadt, P. (2020). The viable system model. In M. Reynolds & S. Holwell (Eds.), *Systems approaches to making change: A practical guide* (pp. 89–138). Springer London. https://doi.org/10.1007/978-1-4471-7472-1_3

INCOSE. (2021). *Systems Engineering Body of Knowledge*. www.sebokwiki.org/wiki/Guide_to_the_Systems_Engineering_Body_of_Knowledge_(SEBoK)

International Organization for Standardization [ISO]. (2013). *ISO 16290:2013 Space Systems—Definition of the Technology Readiness Levels (TRLs) and Their Criteria of Assessment*. Geneva, Switzerland: ISO.

Kennedy, G. A., Scott, W. R., Shirvani, F., & Campbell, A. P. (2020). Modeling the evolution of organizational systems for the digital transformation of heavy rail. In Holly Handley & Andreas Tolk (Eds.), *A framework of human systems engineering: Applications and case studies* (pp. 63–96). Wiley.

Kennedy, G. A. L. (2024). A planning and readiness framework for organisational systems change within human systems integration. In *The 3rd INCOSE International Conference on Human Systems Integration Jeju, Republic of Korea*. INCOSE Library.

Mankins, J. C. (1995). Technology readiness levels: A white paper. https://web.archive.org/web/20130626215632/www.hq.nasa.gov/office/codeq/trl/trl.pdf

Pedram, S., & Kennedy, G. (2024). Assessing the validity of VR as a training tool for medical students. *Virtual Reality*, *28*(1), 15.

Robbins, S. P., & Judge, T. A. (2007). *Organizational behavior* (12 ed.). Pearson Prentice Hall.

Schein, E. H. (2010). *Organizational culture and leadership* (Vol. 2). Wiley.

Shea, C. M., Jacobs, S. R., Esserman, D. A., Bruce, K., & Weiner, B. J. (2014). Organizational readiness for implementing change: A psychometric assessment of a new measure. *Implementation Science*, *9*(1), 1–15.

Urban, W. (2017). Amoeba management system transformation in the light of organisational change literature. *Management and Production Engineering Review*, *8*(1), 16–23. https://doi.org/10.1515/mper-2017-0002

Vik, J., Melås, A. M., Stræte, E. P., & Søraa, R. A. (2021). Balanced readiness level assessment (BRLa): A tool for exploring new and emerging technologies. *Technological Forecasting and Social Change*, *169*, 120854.

Weiner, B. J., Amick, H., & Lee, S.-Y. D. (2008). Conceptualization and measurement of organizational readiness for change: a review of the literature in health services research and other fields. *Medical care research and review*, *65*(4), 379-436.

APPENDIX 1

TABLE A1.1
Applications of Organizational Readiness Found in Literature

Organizational Readiness Groupings	Scope (Source)
Factor Affecting Organizational Readiness	• Customer satisfaction (Durant-Tyson, 2022) • Organizational climate (Manik & Ginting, 2019) • Knowledge management (Ezeruigbo, 2023; Rusly et al., 2015)
Change to Paradigm or Process	• Internationalization (Siriphattrasophon & Saiyasopon, 2013) • Circular supply chain implementation (Kayikci et al., 2022) • Lean transformation (Prasad & Vasugi, 2023) • eCommerce practices (Kurnia et al., 2015)
Change to a Technology	• A new technology (Vik et al., 2021) • Open data (Wang & Lo, 2016) • IoT (Ancarani et al., 2020; Viamianni et al., 2023) • Cloud computing (Al Mudawi et al., 2020) • Digital technologies (Bruno et al., 2020)
Organizational Readiness Level (ORL) Scale	• Balanced readiness level assessment (Vik et al., 2021) • Organizational impacts of testing/adopting innovation (Bruno et al., 2020) • Proposed ORL within socioergonomics (Boy, 2022)
Readiness for Change / Instruments for Assessment	• Organizational Readiness for Implementing Change (ORIC) (Shea et al., 2014) • Organizational Readiness to Change Assessment (ORCA) (Crittendon et al., 2020) • Diagnostic tool for assessing organizational readiness for complex change (Blackman et al., 2013)

TABLE A1.2
Mapping of Candidate Organizational Readiness Level Scales

ORL Level	Vik et al. (2021)	Boy (2022)	Bruno et al. (2020)
0	-	About first principles where potential organizational models are explored	-
1	The technology represents a fundamental break with existing work processes or organizing	Goal-oriented research that requires making choices from first principles to practical fully digital organizational setups	Identification of the organizational need (infrastructures, capabilities, skills) and associated organizational readiness aspects

TABLE A1.2 (Continued)
Mapping of Candidate Organizational Readiness Level Scales

ORL Level	Vik et al. (2021)	Boy (2022)	Bruno et al. (2020)
2	Unclear how the technology might be adapted to existing work processes/ organization	Proof of principle development and active R&D is started in a virtual environment	Formulation of proposed solution concept and potential impacts; appraisal of organizational readiness issues; identification of relevant roles, processes, functions, and structures for the solution
3	An idea about integration domestication exists	Virtual agile organizational prototype development and first HITLS (virtual HCD)	Comprehensive description of proposed solution's impacts within the organization in terms of roles, competences and skills, and physical infrastructures required
4	Integration with work processes/ organization is formulated	Proof of organizational concept development using concrete scenario-based design from fully virtual to more tangible environments	Solution validated through simulation of major induced changes to substantiate proposed impacts and organizational readiness: the organization which is developing the solution starts to acquire roles, competences and skills, and physical infrastructures required
5	A concrete plan for integration with existing work processes is formulated	Assessing organization capability in terms of authority sharing (responsibility, accountability, and control), trust, collaboration and coordination, for example	Proposed solution validated through pilot testing in real or realistic organizational environments:theorganization which is developing the solution achieves roles, competences and skills, and physical infrastructures required
6	Large/fundamental organizational changes are needed in order to use the technology	Real-world use-case tests in a wider variety of situations – tangibilizationcontinues	Solution demonstrated in real-world environments and in cooperation with relevant stakeholders to gain feedback in order to improve roles, processes, functions, and infrastructures required

(continued)

TABLE A1.2 (Continued)
Mapping of Candidate Organizational Readiness Level Scales

ORL Level	Vik et al. (2021)	Boy (2022)	Bruno et al. (2020)
7	Small organizational changes are needed in order to use the technology	Practical integration with respect to criteria such as safety, efficiency, and comfort, at various levels of granularity of the organization – tangibilizationcontinues	Refinement of the roles, processes, functions, and infrastructures required and retesting of the solution in relevant organizational environments
8	Technology is adapted to work processes and/or existing technology	Readiness for effective implementation on a real site (fully tangible) based on personnel feedback for deployment approval	Targeted solution, as well as a plan for organizational embedment, complete, and qualified: roles, processes, functions, and infrastructures are available
9	The technology works seamlessly with existing technology	Deployment involving both personnel and real machines	Actual solution proven in relevant organizational environments: roles, processes, functions, and infrastructures are correctly used for the solution on the market

APPENDIX 2

TABLE A2.1
HRL Assessment Questions and Interview Comments

HRL	HRL Question	Finding	Interview Comments
6.1	Have the functioning prototype of the human-system interface and simulations of mission tasks and conditions been fully developed to support assessment of critical human performance issues?	Y	Has been initially tested based on the experience of the development team (including clinical leads who are experienced medical practitioners) rather than explicit identification with the learners. Current user trials seek to verify this now.
6.2	Has the full range of user scenarios and tasks been tested in high-fidelity simulated or actual environments?	Y	Beta testing has been undertaken with the development team, with a video of the VR task; however, this was not efficient and breaks the realism of the experience. The new development is the provision of ghost hands that can be summoned on demand by learners when they need assistance. This has now been trailed with 50 different participants.
6.3	Have task analyses been updated, based on testing in high-fidelity simulated or actual environments?	Y	Initial task lists are prepared by clinical leads in the company. The software developers receive the steps in the list and then decide with the clinical leads which steps are appropriate for the VR simulation, and which require proxy tasks. The current new functionality (ghost hands) has been developed from feedback from learners. Task analyses have been created for design as well as skill/performance assessment.
6.4	Have human–machine teaming strategies and human–machine function allocations been updated, based on testing in high-fidelity simulated or actual environments?	N/A	The function allocation is constrained by the procedural technique required for IVC training.

(continued)

TABLE A2.1 (Continued)
HRL Assessment Questions and Interview Comments

HRL	HRL Question	Finding	Interview Comments
6.5	Has a system to track and resolve human systems issues after fielding been developed and evaluated in high-fidelity simulated or actual environments?	Y	User trials have uncovered issues, in addition to support requests from users.
6.6	Have strategies to mitigate safety implications for human users been updated, based on testing in high-fidelity simulated or actual environments?	N	At present, the developers are not considering the direct safety or implications within the simulation; there is, however, a functionality that shows users metrics on poor performance and hygiene tasks are an inherent part of training. Observational controlled studies have compared the adherence to safety and hygiene practices; however, these are not part of the VR training. *There is an opportunity to also teach safe practices in the manipulation of the simulated equipment (e.g., awareness of stick injury, touching of insertion sites). Similar feedback has been provided in other VR procedures; this may be rolled out for this procedure in future development.*
6.7	Have strategies to accommodate manpower, personnel, and training concerns been updated, based on testing in high-fidelity simulated or actual environments?	In Progress	No, not yet. But in progress.
6.8	Have strategies to address environmental constraints and impacts been updated, based on testing in high-fidelity simulated or actual environments?	N/A	Fully simulated environment

TABLE A2.1 (Continued)
HRL Assessment Questions and Interview Comments

HRL	HRL Question	Finding	Interview Comments
6.9	Have strategies to address implications for other relevant HSI domains been updated, based on testing in high-fidelity simulated or actual environments?	N	There has naturally been a focus on the training domain. Awareness of HSI domains pre TRL6 has been low; therefore, these may not have explicitly been measured/covered. *Further investigation needed to identify which HSI domains are appropriate and then ascertain potential gaps.*
6.10	Have strategies to address human interactions during maintenance and sustainment been updated, based on testing in high-fidelity simulated or actual environments?	In Progress	No, not yet. *Developers are trialing the functionality and are waiting to decide whether to continue with the ghost hands addition or not. If they do, then this will be considered at the next stage.*
6.11	Has testing in high-fidelity simulated or actual environments been used to finalize procedures for human user roles throughout the lifecycle?	Y	In terms of end-users, the procedures have been tested as part of the training simulation.
6.12	Have relevant human performance data been collected and evaluated to determine whether human performance metrics are successfully met, based on testing in high-fidelity simulated or actual environments?	Y	Test features performance metrics collection (automated through the software for the teacher dashboard) and also through post-trial feedback from users. Where issues uncovered, rectification and verification activities are referenced against the design.
6.13	Have strategies to support human use been identified and recommended, based on testing in high-fidelity simulated or actual environments?	Y	This descriptor is difficult to assess for this scenario. The new functionality is the support strategy.

(continued)

TABLE A2.1 (Continued)
HRL Assessment Questions and Interview Comments

HRL	HRL Question	Finding	Interview Comments
6.14	Has conformance of functional prototypes to human performance requirements, design principles, standards, and guidance been verified?	N	While the conformance has been undertaken for the IVC procedure, the ghost hands functionality is experimental. *This additional functionality needs to be verified. User testing highlighted design issues (e.g., ghost hand summoning was too easy to do accidentally, and ghost hands could obscure the view) that were improved in later trials.*

34 Integrating Systems into Socio-Technical Structures through Change Management

Christophe Merlo and Véronique Pilnière

INTRODUCTION

Developing a complex system requires approaches that can deal with this complexity. This complexity is of different kinds depending on the adopted viewpoint. The system itself may be inherently complex due to the many technologies it incorporates or the number of its components, and therefore its architecture. Considering Product Lifecycle Management or Digital Twin concepts, complexity can be seen as the result of a global vision that encompasses different stages of the system life cycle. Model-Based System Engineering (MBSE) approaches (Henderson & Salado, 2021) support the vision of a whole-life-cycle approach when developing a system. Different Socio-Technical Systems (STS), as defined by Sony and Naik (2020), can be identified and considered at each stage of interaction with the system. Each STS can be regarded as a system of systems (Boardman & Sauser, 2006) depending on the nature of its internal relationships and activities or the number of stakeholders and organizations it covers. Aggregating all these STSs may be seen as a super-STS of systems.

In this chapter, we intend to highlight how an STS is transformed by the system under development and/or to be implemented, whether in the design/redesign, use, or operation/maintenance stages. Through the chosen example, we try to demonstrate that the development of a system, even supported by a Human–System Integration (HSI) approach, remains a complex activity, which requires an adaptation of the involved STSs, and that this system is influenced in return by these STSs. Change management is a key issue to consider during the system development process.

CHANGE AND CHANGE MANAGEMENT: A LITERATURE REVIEW

Change is a polymorphous term that needs to be defined: "change is a rupture between an obsolete existing and a future synonymous with progress" (Autissier & Moutot, 2023). More precisely, change is a break in the way a company operates "when

DOI: 10.1201/9781003666431-41

the following elements are transformed: practices (ways of doing things), working conditions (material environment), tools (information technology [IT] and management tools), organization (areas of power and functional boundaries), business (company know-how), strategy (collective goals pursued and envisaged) and culture (value system)" (Autissier & Moutot, 2023). However, "change only exists through the dynamics of the individuals who implement it" (Autissier & Moutot, 2023). For Crozier and Friedberg (1981), "change [...] can only be understood as a process of collective creation through which the members of a given community learn together, that is, invent and fix new ways of playing the social game of cooperation and conflict – in short, a new social praxis – and acquire the corresponding cognitive, relational and organizational capacities." Individual dynamics then become crucial when it comes to change: it is the people who make a change happen. This aspect led to the development of management techniques, known as "change management," in the 1980s and 1990s, during major IT projects. At the time, change management was a matter of training and communication. In this classic approach, people at work were seen in their negative dimension as a source of errors, and as passive in their decision-making. An individual who had to be told what to do, who was, in a way, a victim of his work. Gradually, in the face of the failures of specific major projects, the notions of buy-in, participation, and ownership of change began to emerge in change management.

Several research studies have contributed to this evolution. Drucker (2008), considered to be the father of modern management, Kanter (2001), and the wheel of change, and Kotter and Rathgeber (2008) are just a few of those who have contributed to this evolution. Gradually, these new approaches integrate change management into projects and recognize people's predominant role in change. In the workplace, they are seen as active participants in their work, managing hazards and solving problems in and through their work performance. Considering individuals' reactions, "the acceptance of change passes through this point of passage, which is the appreciation of the risk of losing what exists for an uncertain future" (Autissier & Moutot, 2023). Between risk and uncertainty, fears of change are born. For individuals affected by change, it seems crucial to foster the construction of positive representations, such as a future synonymous with progress and gaining something.

While Piaget (1959) noticed that this construction and evaluation of representation requires learning, Argyris and Schön (1996) considered the concept of "routines" and theories of action. For Levitt and March (1988), the generic term "routine" encompasses the forms, rules, procedures, conventions, strategies, and technologies on which organizations are built.

Argyris and Schön (1996) highlight two levels of learning, depending on the "degree" or "depth" of change in representations. They distinguish between "single-loop" and "double-loop" learning. Single-loop learning involves adapting to perceived changes without questioning the organization's values. Individuals react to deviating results by modifying their action strategies, not the guiding values and paradigms underlying their action theories (see Figure 34.1). Double-loop learning, on the other hand, induces a change in both values and strategies. The double loop refers to the two feedback loops that link the observed effects of action to values and paradigms.

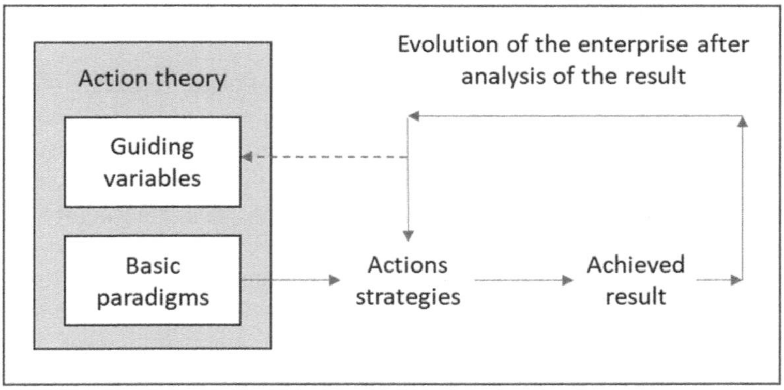

FIGURE 34.1 Single- and double-loop learning.

This kind of learning makes it possible to question the values and norms governing the theories of action at work in organizations. Thus, the more learning occurs in a double loop, the deeper and more deeply rooted the change has a lasting effect on the organization.

Change management creates the right conditions for this double-loop learning process by adopting a methodology that fosters buy-in and ownership, coupled with the change project. Change management then becomes an organizational change management activity.

CASE STUDY: A TECHNOLOGY-CENTRED TRANSFORMATION

Health Service Bayonne (HSB) is a health establishment dedicated to home care services. It coordinates two home care services: home medical care, replacing heavy hospital care (HAD 20%); and nursing home care (SSIAD 80%). It employs more than 250 employees: medical, paramedical, administrative, etc., spread over more than 20 different professions. It manages nearly 500 patients with little or medium loss of autonomy. Despite this, HSB relies on an extensive network of establishments and professionals from its local territory to provide the patient with all necessary services, such as general practitioners, pharmacists, laboratories, and several liberal professions. It therefore offers global support through complex, multiple, geographically distributed business processes and activities.

So, HSB activities are characterized by a collaborative and distributed context, where the patient is at the heart of the concerns of multiple actors. Good coordination between the asynchronous interventions of all actors ensures the quality of care provided. This coordination is facilitated by excellent communication between actors, and one primary shared tool: the patient's "care record,," located at the patient's bedside, in paper format. The challenge of coordination is not only to manage the planning for home visits, but above all to ensure continuity and consistency between the different care activities related to a given patient, despite the asynchronous nature of most interventions, and the random presence of persons performing the necessary functions with the patient. A patient's care record is crucial for tracking and reporting information between stakeholders.

During the past 4 years, HSB has engaged in a project for replacing the patient's care record in paper format, situated at the patient's home, with a digitized patient's care record, accessible using a tablet PC with health care traceability software from every location. This project was initially built to improve care services. It was motivated by different issues Merlo et al. (2023) described: traceability, real-time information exchanges between nurses, and improved decision support. Expected objectives are to reduce information retrieval; to facilitate remote access to information; to improve the readability and completeness of information, and thus decision-making; and finally, to ensure the "Electronic Prescription for the Care Plan." The primary interest of such a project lies in transforming the health organization by introducing a technological system. The issue concerns designing a dedicated system and managing the transformation through change management concerns.

Many stakeholders were involved in this project. First, a management team comprised the doctor–director of the structure, the two health managers, the financial manager, and the research supervisors. Second, an operational team comprised a PhD student, her supervisors, the project's financial manager, and several executive nurses. Other stakeholders were later integrated into the project for specific activities, mostly health professionals, independent from HSB, but integrated into the care process. Finally, an external but very important stakeholder was part of the project: the IT system company, which develops and implements the tool to meet HSB requirements.

APPLIED METHODOLOGY, MODELS, AND RESULTS

A four-step methodology resulting from previous research findings and results has been consolidated and implemented during this project: a *contextualization phase*, a *diagnostic phase*, an *observation phase*, and an extensive and iterative *development phase*, which includes specifications, implementation, evaluations, and guidelines steps.

Contextualization Phase

Method: This phase introduces the project to stakeholders, then interviews with decision-makers, who have control over the project's continuation and finances, and defines the main steps of the project and the involved actors.

Results: The company's demand was discussed with decision-makers. Two 2-hour interviews with each one allowed us to identify HSB's main processes and actors to be involved in the project. By "actor," we mean those impacted by the future system, including users, stakeholders, various HSB employees, and external parties like pharmacists, doctors, or nurses. Initially focused on internal actors, decision-makers gradually became aware of the structure's links with external stakeholders, shaping the scope of change management.

Another important insight from these initial interviews is decision-makers' fears of "resistance" to change from certain socio-occupational groups, who are often reluctant during structural changes. Their concerns led to identifying the "project culture," previously top-down, where management decided, and employees implemented. Late training of employees has generated negative experiences in the past that left decision-makers disillusioned. It was proposed that this project be managed differently by integrating stakeholders throughout the project, integrating change through meetings, outside discussions, and at each stage, including during digital tool development.

Diagnostic Phase

It is carried out to characterize and analyses the organization, its processes, actors' practices, and existing IT tools.

Method: Using Business Process Model and Notation, HSB business processes were modeled globally for an overview and in detail to identify actors' activities.

Results: Key processes are decomposed into a preliminary care process accepting a patient, the care process itself, decomposed into coordination activities and dedicated subprocesses for home care, and a post-care process, alongside several support processes. Professional situations vary greatly and are influenced by patient cases and actor types (HSB employees or external), roles, and tools (paper, tech, IT). Figure 34.2 details pre-care process modelling, involving two HSB employees (secretary and nurse executive), external structures like the patient and their regular doctor. Each activity is described, which allows for identifying the relevant information to be managed by an IT system and the interactions between actors. Considering, e.g., the meeting between the patient, the executive nurse, and the doctor at the patient's home, the aim is to analyze the patient's health situation, to define the possible care plan, and to make the medical prescription for implementing the care plan. This collaborative situation is typical of the activity that generates a significant variability of practices.

Due to this variability, it was found necessary to deepen the diagnosis initiated to promote a better understanding of the structure's activities by highlighting the variability of practices between professionals, for the same activities identified in the previous processes, generating very diverse situations, such as the care of a patient in Home Hospitalization (HAD), which is developed below. That is why, after modelling detailed business processes, it has been necessary to be "in the field," with the actors, for an **observation phase**.

Observation Phase

Context: This phase characterizes more deeply the individual professional practices of the different actors when they achieve selected professional tasks. It focuses on identifying the variability of their functions, as determined in the BPMN processes, when interacting with an IT tool/system, e.g., for making a medical prescription. The concept of practice emphasizes the man in a working situation, while the concept of usage emphasizes the object the man uses (Abi Akle et al., 2016). More generally, when a new IT tool is implemented, it can generate positive or negative impacts on practices. The aim is to facilitate or not facilitate these practices, i.e., fulfilling the

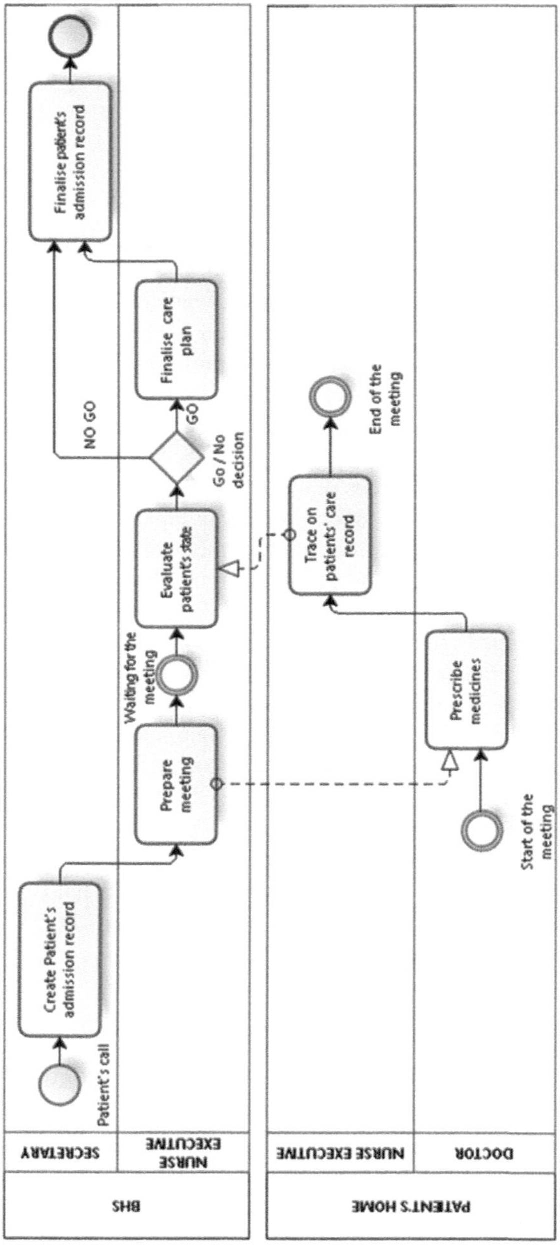

FIGURE 34.2 Collaborative process model: example of pre-care process, detailed view.

work accomplished, considering the change's point of view. Considering usage leads to focusing on the bidirectional link between the tool and the activity it supports. Several dimensions can be regarded for describing usage (Ammenwerth et al., 2006); there is a link between individuals and a group of individuals when using a system (Burton-Jones et al., 2007); and the usages of a tool are built over time (Fidock & Carroll, 2012). The objective of replacing one existing tool with another digital one is to achieve a change "from an old way of doing things to a new way of doing things" (Lorenzi & Riley, 2003) and leads to organizational change. Moreover, as Burton–Jones et al. (2007) point out, different people in the collective may use different functionalities of the (technological) IT system. Different people in the collective can use the system for different tasks. Moreover, more generally, people in the collective can use different features for different tasks.

Method: The variability has been studied by using a dedicated graphical model based on the 5W2H method (illustrated in Figures 34.3 and 34.4), which is often used as a general guide for gathering comprehensive information on a situation (Norman & Crampton, 1998; Nguyen et al., 2020). It is based on seven questions: Who? What? When? Where? Why? How? How much? For our purpose, we focus on four variables only to define a "usage case":

- Who? Individual (the one who acts)
- What? Activity (what is done)
- How? IT tool (object used during the activity)
- Where? Location (of the activity)

Both observations of real life and interviews have led to the description of most practices deduced from professional tasks.

Results: Figure 34.3 illustrates a professional situation (on the left) and the correlated usage case, followed by one possible scenario (on the right).

At the patient's home, one patient, one doctor, and two nurses are involved; they get phones and the patient's care record. In the middle, the "usage case" identifies Dr. Ospital, who prescribes medicines (what) by using the care record (how) at the patient's home (where). Finally, on the right, a specific "usage scenario is defined" including three sub-activities done by Dr. Ospital, with different tools (Borgiel et al., 2015). This model allows for formalizing a working situation (usage case) and then developing other practices (usage scenario). Each scenario may show a variability on one of the variables: the actor may change (who), or the tool they are using (how), or the way the activity is processed (what), or finally, the location (where): sometimes, the patient comes in HSB office or at the doctor's office, as in Figure 34.4.

Such models allow for the characterization and study of the variability of usage. Our observation field demonstrates that from a first usage case, other associated usage cases may exist with different values for one (or more) variable(s), or that several usage scenarios may correspond to the same initial usage case. Therefore, the system (IT tool) to be developed must integrate this variability of practices to facilitate the organization's transformation.

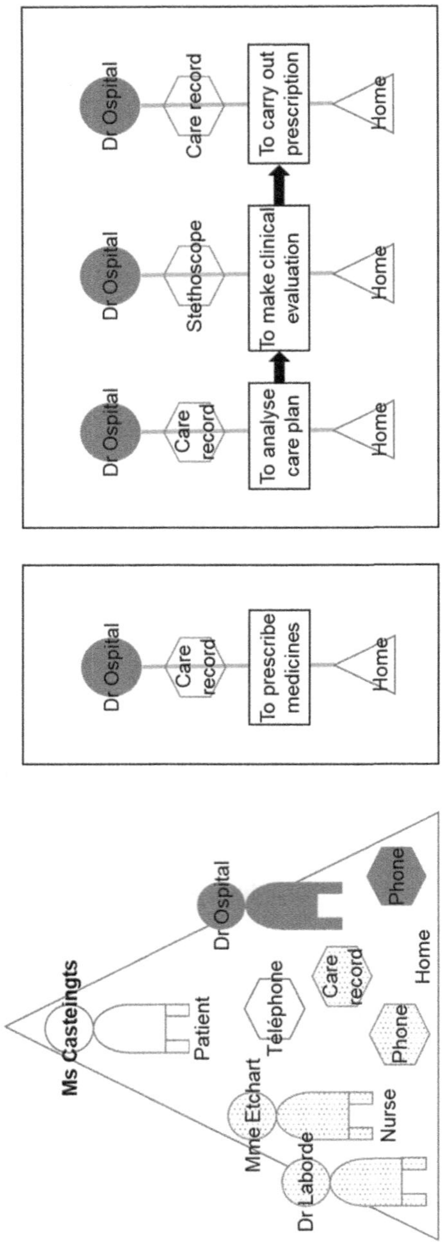

FIGURE 34.3 Usage model: situation, usage case, and usage scenario.

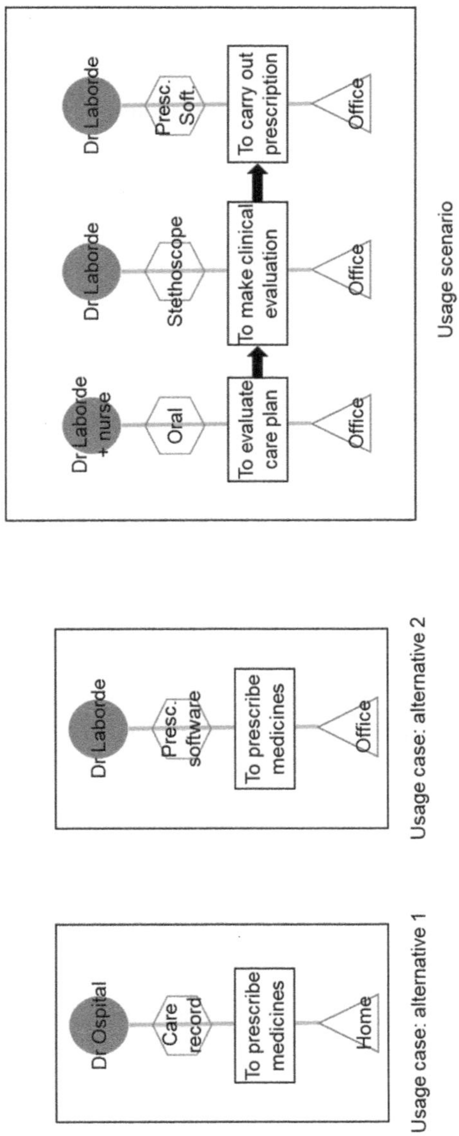

FIGURE 34.4 Usage model: example of variability of usage case "To prescribe medicines."

Development Phase

This phase is iterative after characterizing the available usage cases and their related usage scenarios. During the first analysis step of the design phase, detailed specifications of the future IT system were formalized, integrating relevant variabilities. Then, the IT system was developed using an agile approach. The prototype was tested with internal and external actors at each iteration by applying specific protocols based on the "usage scenarios" models. The test results were used to improve specifications and interactions and, simultaneously, to evaluate future practices and improve them.

The defined usage models were used to manage variability when developing the IT system, supporting the development process and change management simultaneously. Three different iterations were necessary before considering the IT system mature enough to be integrated into HSB's business processes. Developers better understood the required variability: e.g., 50 patients were observed, and nearly 50 ways of achieving the same prescription activity have been identified! Finally, people involved in the project became promoters of the IT tool toward their colleagues, not only for using it but also to explain the new practices they had contributed to characterizing and making the business processes more efficient. The latest IT system has transformed the company through its use.

DISCUSSION: CO-CONSTRUCTION OF CHANGE MANAGEMENT

Several lessons have been drawn from this work, from very detailed and individual concerns to global and organizational aspects. First, focusing on practices and usages simultaneously, usage models were implemented, and several applicative schemes could be identified:

- During the "observation phase," existing usages were formalized based on **existing practices**, with a deep variability analysis.
- When characterizing requirements, the experimentations of implemented prototypes were conceptualized (future) **expected usages** that will foster (future) **expected practices**, keeping, adding, or removing some variability.
- During the experimentations, it has also been possible to identify (future) **emerging usages**, even with variability, that validate or not the expected practices, but also (future) **emerging practices**.
- Finally, after the necessary iterations, the final step of implementation of the IT system led to the implementation of **new effective usages**, corresponding to the implementation of **new effective practices**, with the required variability.

Identifying new usages and characterizing new practices have also allowed the HSB stakeholders to project themselves on their business processes to propose new frames and procedures and prepare themselves and their organization before the "**implementation phase**." The constant back-and-forth observed at each iteration between the level of individual, collective then organizational usages/practices, as described in Figure 34.5, highlights: first, people was able to question the values and

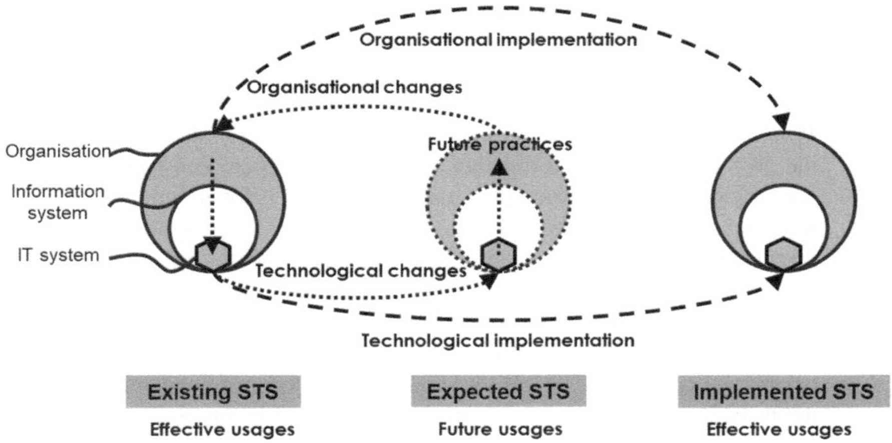

FIGURE 34.5 Intertwined change management and system development methodology.

the paradigms of their work, applying the double loop of Argyris and Schön (1996), and second, change management is deeply intertwined within the development process of the IT system.

Second, being iterative is a key element: to best fit with the requirements, but also, above all, to take a step back from their practices and consider all the impacts on their ways of working and, therefore, their business processes. They can decide which emerging or expected usages/practices to keep. By doing so, their mental representations evolve, and this is the primary purpose of change management: to enable stakeholders to gain empowerment on future changes and to become promoters of the transformation of their organizations.

By doing so, the results of the work carried out within HSB can be generalized to different types of companies and a methodology can be highlighted for managing change in a technological development project. Indeed, this methodology requires prerequisites, mainly to integrate actual and future stakeholders as early as possible in the project. Thus, as Kourilsky (2014) reminds us, a change will be even more effective if we rely on the resources that every human system has to evolve and avoid the use of orders and decrees; this is the basic condition for obtaining the cooperation of the system in its dynamic of evolution.

Nine actions that help stakeholders give sense to the change (Pilnière, 2010) have been established to support the proposed methodology and models and to foster organizational change management:

1. *Consider the Mental Representations of Each Actor:* Mental representations serve as a foundation for collective and collaborative work. The change leader working on these representations to track their evolution helps each actor recognize and find their position in the project by expressing their thoughts and vision.

2. *Putting Everyday Work at the Heart of Exchanges:* It is from the working situation that existing and future functionalities emerge. People can then better understand what they are doing, how they are positioned relative to others, and what others are doing.
3. *Promoting the Problem-Building Approach:* By constructing the objective of the problem to be solved together, exchanges are more constructive; a consensus can be built, and interpersonal conflicts can be overcome.
4. *Use Identifiable Materials:* Photos, videos, and drawings spark constructive discussions and support group dynamics. The mentor must assess their impact, whether fostering or hindering convergence, and adapt as needed with thoughtful adjustments.
5. *Focus on Results and Objectives:* The change leader ensures that the project orientation is the backbone of the group's work, fostering ownership. Indeed, the more people are aware of participating in the construction of results, the more ownership they have and the more meaning they build.
6. *Fostering Confidence-Building among Stakeholders:* This principle fosters psychological security (Diaz-Berrio Döring, 2003), strengthening group cohesion and identity. It encourages public speaking and community building. The accompanying person sets rules of respect and confidentiality, ensuring group approval.
7. *Promote Communication Outside the Group:* The stakeholders involved in the change management activities trust not only in themselves but also in the outside stakeholders, involved later on when implementing at a large scale. Thus, "fostering communication outside the group" allows for the construction of identity within the group and the recognition of group identity outside the group.
8. *Implement Reflexivity:* Reflexivity has three levels: reflecting on practices (Argyris & Shön, 1996) to enhance skills; recognizing learnings and how they occur; and applying them in daily life, for fostering meta-cognitive skills (second loop by Argyris & Shön, 1996). Facilitators prompt participants to review past learnings and behavioral changes at the start of each meeting.
9. *Promoting the Sustainability of the Approach:* Sustainability is built upstream, must be thought upstream, and is never won. The change leader should guide the approach, defining goals and intermediate outcomes, and not hesitate to reinject or challenge them whenever necessary.

CONCLUSION

Integrating a new system into an STS is a complex activity, requiring consideration of design impacts throughout the life cycle. Studying these impacts enables change-support mechanisms, thereby preparing STS for future interactions with the system. Change management activities become part of the HSI approach, where "system" refers to the system to be integrated and "human" refers to the impacted organizational system.

The proposed methodology emphasizes involving stakeholders early and co-constructing tasks and results to facilitate change. This method suits service organizations, especially those with diverse stakeholders varying in technology maturity and collaboration. Such variability in practices must be addressed to ensure smooth transitions.

REFERENCES

Abi Akle, A., Borgiel, K., Merlo, C., Minel, S. (2016). Designing interfaces for seniors in the context of healthcare. In M.M. Soares, F. Rebelo (Eds.), *Ergonomics in Design: Methods and Techniques* (315–333). CRC Press. doi: 10.1201/9781315367668-21

Ammenwerth, E., Iller, C., Mahler, C. (2006). IT-adoption and the interaction of task, technology, and individuals: A fit framework and a case study. *BMC Medical Informatics and Decision Making*, 6(3), 1472–6947. doi: 10.1186/1472-6947-6-3

Argyris, C., Schön, D.A. (1996). *Organizational Learning II: Theory, Method, and Practice.* Reading, MA: Addison-Wesley. ISBN 9780201629835

Autissier, D., Moutot, J. (2023). *Méthode de conduite du changement: Diagnostic, Accompagnement, Performance* (5e édition). Dunod. EAN 9782100848874

Boardman, J., Sauser, B. (2006). System of Systems – the meaning of. In *2006 IEEE/SMC International Conference on System of Systems Engineering, Los Angeles, CA, USA.* doi: 10.1109/SYSOSE.2006.1652284

Borgiel, K., Merlo, C., Minel, S. (2015). A multi-level activity analysis for home healthcare ICT tool redesign. In *Proceedings of the 20th International Conference on Engineering Design (ICED 2015), Vol. 1: Design for Life DS 80-01, 475-484, Published by the Design Society (Glasgow, Scotland),Milan, Italy, 27.-30.07.2015.*

Burton-Jones, A., Gallivan, M.J. (2007). Toward a deeper understanding of system usage in organizations: A multilevel perspective. *MIS Quarterly*, 31(4), 657–579. doi: 10.2307/25148815

Crampton, N.J. (1998). *Preventing Waste at the Source.* CRC Press. ISBN 10: 1566703174

Crozier, M. & Friedberg, E. (1981). *L'Acteur et le système: Les Contraintes de l'action collective.* Edition du Seuil.

Diaz-Berrio Döring, L.A. (2003). *L'influence de la confiance sur les processus d'apprentissage dans une équipe de travail.* Thèse de doctorat, Université du Québec, Montréal.

Drucker, P.-F. (2008). *The Essential Drucker: The Best of Sixty Years of Peter Drucker's Essential Writings on Management.* Collins Business Essentials.

Fidock, J., Carroll, J. (2012). Theorizing about the life cycle of IT use: An appropriation perspective. In D.N. Hart, S.D. Gregor (Eds.), *Information Systems Foundations: Theory Building in Information Systems* (79–112). ANU Press. www.jstor.org/stable/j.ctt24h30p

Henderson, K., Salado, A. (2021). Value and benefits of model-based systems engineering (MBSE): Evidence from the literature. *Systems Engineering*, 24, 51–66. doi: 10.1002/sys.21566

Kanter, R.M. (2001). *Evolve!: Succeeding in the Digital Culture of Tomorrow.* Harvard Business School Press.

Kotter, J.P., Rathgeber, H. (2008). *Alerte sur la banquise! Réussir le changement dans n'importe quelles conditions.* Village Mondial.

Kourilsky, F. (2014). *Du désir au plaisir de changer* (5ème Edition). Dunod.

Levitt, B., March, J. (1988). Organizational learning. *Annual Review of Sociology*, 14(1), 319–338. doi: 10.1146/annurev.so 14.080188.001535

Lorenzi, N., Riley, R.T. (2003). Organizational issues change. *International Journal of Medical Informatics*, 69(2–3), 197–203. doi: 10.1016/s1386-5056(02)00105-3

Merlo, C., Pilnière, V., Borgiel, K. (2023). A Usage Model to Enrich MDSEA Approach. In Archimède, B., Ducq, Y., Young, B., Karray, H. (Eds.), Enterprise Interoperability IX. *I-ESA 2020. Proceedings of the I-ESA Conferences, vol 10. Springer, Cham.* doi: 10.1007/978-3-030-90387-9_18. ISBN 978-3-030-90386-2

Nguyen, V., Nguyen, N., Schumacher, B., Tran, T. (2020). Practical application of Plan–Do–Check–Act cycle for quality improvement of sustainable packaging: A case study. *Applied Sciences*, 10(18), 6332. https://doi.org/10.3390/app10186332

Piaget, J. (1959). *La naissance de l'intelligence chez l'enfant*. Éditions Delachaux et Niestlé.

Pilnière, V. (2010). Pour un management du risque dans lequel l'individu est acteur: l'enjeu de l'accompagnement. In B. Claverie, J.-C. Sallaberry, J.-F. Trinquecoste (Eds.), *Management et Cognition – Pilotage des organisations: questions de représentations* (297–315). L'Harmattan. ISBN 978-2-296-11052-6.

Sony, M., Naik, S. (2020). Industry 4.0 integration with socio-technical systems theory: A systematic review and proposed theoretical model. *Technology in Society*, 61. doi: 10.1016/j.techsoc.2020.101248

35 Dynamic Capabilities for Industrial Supply Chain Resilience Development

Suvi Leinonen, Arto Reiman, Osmo Kauppila, Marko Djapan, and Aleksandar Djordjevic

INTRODUCTION

Different crises, disruptions, and unexpected changes can be considered the new normal in industrial business environments, and ground-breaking challenges will also occur in the future, bringing various and even unpredictable impacts on businesses (Moosavi et al., 2022). To stay competitive in times of uncertainty and respond to changes, companies and supply chains (SCs) must build their resilience by developing organizational capabilities. Developing such organizational capabilities is a social process within a human network (Mitki et al., 2019). The inherent characteristics of SCs, with technological and social interactions between people and companies and their structural embeddedness, indicate SCs are complex sociotechnical systems (Mitki et al., 2019) or simply complex systems (e.g., Boy, 2017). From a system efficiency perspective, it might be problematic if those dominant companies (e.g., buyer companies) focus too strictly on their own businesses and overlook interdependencies with their SC network (Wiedmer & Griffis, 2021).

In this chapter, we approach system resilience in SCs through organizational resilience (OR) and supply chain resilience (SCRes) (Moosavi et al., 2022; Stadtfeld & Gruchmann, 2024). We see that SCRes is strongly interlinked with the OR of the individual organizations in the SC. SCRes are built through resilient organizations, but also require interorganizational capabilities in the SC. Hence, SCRes development improves OR and vice versa. SCRes and OR are built on various capabilities that proactively identify potential problems, threats, and opportunities. Further, they facilitate agile responses, adaptation, and recovery to unpredictable disruptions, and the ability to learn from them (Bento et al., 2021). We study the premises of SCRes and the organizational capabilities needed (Leinonen et al., 2023; Moosavi et al., 2022), including more routine operational capabilities as well as dynamic ones that consist of patterned elements (i.e., processes and routines) rather than spontaneous reactions to crises and disruptions (e.g., Stadtfeld & Gruchmann, 2024). Dynamic capabilities (DCs) are needed in organizations and SCs to reconfigure, integrate, and

DOI: 10.1201/9781003666431-42

transform internal and external resources and competences for adaptation to changes and building SCRes (Kalubanga & Gudegan, 2022; Teece et al., 1997).

We define the premises for the future capabilities needed for SCRes and consider them with potential future threats in the business environment so that the system can be better prepared for meeting future crises and adapt to fast changes. Adding a systems engineering perspective (see Jackson & Ferris, 2013) to our considerations, we examine how capable the organizations are in influencing the development of these capabilities internally in their organization and externally in their SCs.

LITERATURE REVIEW

DCs for SCRes and OR

As a forward-looking and proactive approach, we adopt the DCs view theory (Teece et al., 1997). DCs refer to companies' (and SCs') abilities to identify and forecast future threats and opportunities, seize those considered relevant, and transform or reconfigure their resources and competences to meet changing circumstances. DCs can be viewed on three domain levels. Micro-foundations are routines and processes that drive the organization's capability development for SCRes and OR. Low-order capabilities are immediate outcomes of micro-foundations, the routine activities that keep the organization running. Finally, high-order capabilities are complex and strategic organizational competencies, which involve integrating and leveraging low-order capabilities in innovative ways to adapt to changing environments, identify new opportunities, and respond to challenges (Kalubanga & Gudegan, 2022; Leinonen et al., 2023; Teece, 2012).

Alternatively, SCRes and DCs are concepts that are closely related to DCs. Alternatively, it can be defined from reactive, adaptive, and transformative perspectives, defining organizations' agility, robustness, and capability to resist and respond to unexpected crises, disruptions, and changes in the business environment without losses in business (He et al., 2023). Alternatively, it is based on organizational values and adaptive leadership strategy, promoting organizations' performance and employees' capability to maintain performance and learning during a crisis (He et al., 2023; ISO 22316, 2017; Tvedt et al., 2023). SCRes is the ability to tolerate crises, build preparedness, and recover fast (Bento et al., 2021). SCRes is constructed by combining absorptive, response, and recovery capabilities, as well as reactive and proactive capabilities to mitigate the impacts of crises (Zhao et al., 2023).

Strategic business continuity planning in the short and long term supports proactive planning (ISO/TS 22318, 2021; Moosavi et al., 2022; Raj et al., 2022; Sanchez & De Batista, 2023). As complex sociotechnical systems, SCs must have the capabilities to identify crises proactively, decide on the measures, actions, and follow-up procedures in collaborative strategic teams with their stakeholders (ISO/TS 22330, 2018; Kumar & Sharma, 2021; Moosavi et al., 2022; Raj et al., 2022). Here, systems engineering helps understand the contingencies between the organizations involved in the SC and defines the requirements for the sociotechnical system to be effective (Nuerk & Dařena, 2023). SCRes development needs prioritization and actions by management to build interorganizational responsiveness to prevent the effects of the crises on SC (Moosavi et al., 2022).

In sustainable SC systems, SCRes require trust, commitment, knowledge sharing, continuous innovativeness in and between organizations, and collaboration at the horizontal and vertical levels (Gebhardt et al., 2022; ISO/TS 22330, 2018; Kumar & Sharma, 2021; Sanchez & De Batista, 2023). It is recommended that buyer–supplier partnership and relationship management in SCs are based on open and continuous communication based on defined processes and practices (Han et al., 2022; Kalubanga & Gudegan, 2022). Combining systematic supplier relationship management, collaborative partnership, knowledge sharing, commitment, trust, and commonly shared objectives, assessments, and commonly agreed upon development activities leads to long-term sustainable supplier performance and SCRes (e.g., Gebhardt et al., 2022).

To mitigate the impact of crises, the perspectives of supplier collaboration, risk management, partnerships, business continuity planning, decision-making, and SC visibility should be acknowledged (Chen & Chen, 2019; Han et al., 2022; Moosavi et al., 2022; Raj et al., 2022). Buyer and supplier companies can identify and anticipate crises more efficiently, react agilely, adjust operations, and build SCRes when they monitor and follow-up on their SC and environment in collaborative monitoring teams with stakeholders (Kumar & Sharma, 2021; Gligor et al., 2019). Management needs to create supportive leadership and an inclusive organizational culture while relying on employees to activate them to be innovative and creative and to use their capabilities to advance responsiveness and adaptation (He et al., 2023).

Overarching Capabilities

Complementing the above theoretical reflection on the micro-foundations and low-order capabilities for OR and SCRes, we found two overarching elements: "innovativeness" and the "capability to utilize digitalization." We acknowledge how, in principle, these can be seen as a part of the capabilities discussed above, but to promote their topicality, we propose these as separate entities. Innovativeness and entrepreneurial behavior refer to creativity in finding new and suitable solutions for future crises and disruptions (Golgeci & Ponomarov, 2013; Teece, 2007).

Concerning digitalization, the literature highlights digital strategy creation and implementation; digitalized processes and business models in SC; digital competence development; diverse supplier networks; win–win collaboration with partners; and collaborative risk management as characteristics that can be seen as strengthening sustainable SCRes (Zhao et al., 2023). Here, artificial intelligence (AI) has been highlighted to improve the possibility of evaluating the effectiveness of response strategies, increasing traceability and transparency, and advancing the utilization of data analytics (He et al., 2023; Moosavi et al., 2022). Digital transformation, when facing crises, requires transformative leadership complementing absorptive, response, and recovery capabilities to strengthen SCRes, but digital solutions will not develop employees' capabilities for adaptation (He et al., 2023; Zhao et al., 2023).

MATERIALS AND METHODS

Measures

To gain further insight into the micro-foundations and low-order capabilities needed to operationalize SCRes, we created an online survey to collect data from industry experts. The survey was conducted in early 2023. The survey consisted of pairs of questions related to 35 low-order capabilities divided into six micro-foundations: "pro-active planning," "communication and information sharing," "collaboration," "risk management," "monitoring and follow-up," and "leadership and management," defined based on our earlier research in this area (Leinonen et al., 2023, 2025).

For each low-order capability, we asked (1) the respondent's view on the impact of the stated capability on SCRes and (2) the ability of the respondent's organization to affect the development of said capability in its supply network. The respondents were asked to estimate both viewpoints on a five-point scale ranging from 1 (poor) to 5 (great). Additionally, the survey included two open-ended questions. We asked if important capabilities were not captured and what potential supply chain disruptions the respondent anticipated. The responses were collected and handled anonymously.

Participants

The participants were invited mainly from Finnish industrial collaboration network HSEQ® cluster member companies and their suppliers. The HSEQ® cluster is a network (or system of systems) of 19 primary Finnish process and energy industry companies. It has developed supplier HSEQ management practices over 15 years and HSEQ performance in supplier companies through systematic audit procedures (e.g., Kauppila et al., 2020). In addition, participation possibilities were promoted through the Finnish Production Control and Planning Society, Chambers of Commerce, alumni of Industrial Engineering and Management at the University of Oulu, and Serbian and Italian industrial companies through research project partners.

Procedure

We conducted the study between January and April 2023, invited 237 participants from the above sources, and received 52 (23%) responses. Of the participants, 16 were from procurement, 13 were from top management, and the rest were from other units, such as production, sales and marketing, R&D, customer service, and HSEQ. At the country level, we received 38 responses from Finland and 14 from other countries. Of the 52 respondents, 14 worked in small- and medium-sized companies and 38 in large companies.

Analytical Approach

Both category and individual question levels were considered when analyzing the quantitative responses. To verify that our domain knowledge-based classification of the low-order capabilities and micro-foundations made sense, we calculated

TABLE 35.1
Micro-Foundations' Impact on SCRes and Ability to Influence Their Development

SCRes Micro-foundation	SCRes Impact	Cronbach's Alpha	Ability to Influence Development	Cronbach's Alpha	Difference
Proactive planning	4.2	0.71	3.7	0.77	0.5***
Collaboration	4.1	0.90	3.6	0.86	0.5***
Risk management	4.0	0.86	3.6	0.90	0.5**
Communication and information sharing	3.9	0.74	3.5	0.73	0.5**
Monitoring and follow-up	3.8	0.80	3.6	0.85	0.2
Leadership and HR Management	3.6	0.84	3.2	0.91	0.4**

Note: ***$p \leq 0.001$; **$p \leq 0.01$.

Cronbach's alphas for each micro-foundation and both the "impact" and "influence" questions. Cronbach's alpha values (Table 35.1) ranged from 0.71 to 0.91, within the bounds of what is usually considered an acceptable indication of reliability (Tavakol et al., 2011). No significant differences existed between the answers from different units or between the respondents representing large-, small-, and medium-sized companies. At the country level, the results showed that respondents in Finland experienced a higher possibility of influencing the development of SCRes than foreign companies.

We analyzed the perceived impacts and influencing abilities, first on a micro-foundation and then on a low-order capability level. Next, we investigated the differences between the perceived impact of a low-order capability and its ability to influence its development in the supply chain. To analyze the differences on the micro-foundation level, we considered the micro-foundation mean as a continuous variable. We performed two-sample t-tests to test for statistical significance of the difference between the perceived impact and the ability to influence. The results are shown in Table 35.1, where *** denotes statistical significance at $p \leq 0.001$ and ** denotes statistical significance at $p \leq 0.01$. To analyze the differences on the low-order capability level, we used the nonparametric Kruskal–Wallis test instead of the t-test because we did not consider individual questions fulfilling the requirements for using a parametric test. Minitab 21 software was used for statistical analysis.

The open questions complemented earlier questions and were concerned with the forecasted future threats from the SCRes and OR perspectives. They also required future micro-foundations and low-order capabilities for SCRes and OR to prepare for such threats. The results were analyzed using a summative content analysis approach (Hsieh & Shannon, 2005).

RESULTS

To discuss the future micro-foundations for SCRes and OR, quantitative survey results on the perceived impact are presented next, followed by an analysis of open-ended questions on future SCRes challenges and additional required capabilities. Finally, we analyze how the respondents assessed their ability to influence the development of the surveyed low-order capabilities.

IMPACT ON SCRES

A summary of the micro-foundation-specific mean scores for impact and influencing ability, the differences between these two, the statistical significance of the difference, and Cronbach's alphas are presented in Table 35.1. The standard deviations for the scores ranged between 0.6 and 0.75, and the ability to influence leadership development was the only exception, with a slightly higher standard deviation of 0.84.

As can be seen from Table 35.1, the mean scores were relatively high for all micro-foundations, ranging from 4.2 (proactive planning) to 3.6 (leadership and HR management) for SCRes impact and 3.7 (proactive planning) to 3.2. (leadership and HR management). This was expected as the micro-foundations had been identified as important in earlier research. In this light, it is equally important to observe which micro-foundations and low-order capabilities were not viewed as essential or within their scope of influence by the respondents.

Proactive planning was considered the most critical micro-foundation by a small margin. From a proactive planning perspective, the respondents indicated that all three low-order capabilities, "business continuity planning through the whole SC," "proactive procurement planning," and "material availability planning," were critical capabilities that impact SCRes and OR.

Collaboration was the second most crucial micro-foundation. The respondents indicated several critical low-order capabilities to SCRes from the collaboration perspective: "Identifying key stakeholders and partners," "managing suppliers systematically," "implementing solution-oriented collaboration with suppliers," and "building trust and long-term partnerships." Solution-oriented collaboration with companies in the same industrial field and company networks was considered less important.

Most low-order capabilities in risk management were essential for SCRes. The respondents reported that "capability to identify different kinds of threats" was the most critical low-order capability. Capabilities to "define corrective actions in case of threats," "creating flexibility by resource planning and preparedness," and "creating fast decision-making practices in case of crises" were seen as impacting SCRes and OR. Encouraging creativity in the development of SCRes was less critical.

Concerning communication and information sharing, "open communication and interaction in one's organization" and "between external stakeholders" were considered the most impactful for SCRes and OR, more so than organizing common training and vertical knowledge sharing. Monitoring and follow-up were seen as slightly less impactful on SCRes, with the single most crucial low-order capability being organizing follow-up and monitoring meetings between supplier and buyer companies. Regarding leadership and human resources management, the impact on

SCRes and OR was viewed as the lowest. Developing leadership of remote work, virtual team leadership, and using IT systems in HR management were ranked particularly low.

Identified Challenges and Capabilities

We asked what challenges SCRes the respondents expect to encounter in the next three to five years. A total of 34 challenges were identified, and the most perceived were the following:

- Various crises: wars, pandemics, and economic crises.
- Material and component availability: geopolitical tension, tightening laws and regulations, and related import and export restrictions.
- Challenges in proactive planning related to material and component availability and prices because of an unstable global economy and geopolitical situation, with fast and continuous change.
- Increased logistics and material costs.

In addition, some respondents expressed concern about increasingly complex supply chains related to wider supplier networks, inflexible SCs, traceability, and strategic management and alignments. Some indicated that cybersecurity and AI were future threats.

To complement the low-order capabilities, we asked which other capabilities they would need in their organization or supply chain to meet future challenges. The respondents suggested a total of 22 capabilities, from which two additional micro-foundations were identified:

- Management of digitalization, including the capability to implement digital solutions like AI, business intelligence, and business analytics.
- Managing knowledge availability and information flow related to global market changes and the geopolitical landscape.

Influencing Abilities

For nearly all micro-foundations (Table 35.1), the respondents found their ability to influence the development of low-order capabilities to be lower than their impact on SCRes. For the most impactful micro-foundations, "proactive planning" and "collaboration," the differences were also the most significant (Table 35.1, $p < 0.001$). Only for the "monitoring and follow-up" micro-foundation was there no statistically significant difference between the perceived impact and influencing ability.

By analyzing the differences in the low-order capability level, themes common to the most significant differences can be identified. Issues related to end-to-end predictability, proactivity, risk management, and business continuity were found to be hard to influence. Collaboration, strategic long-term horizontal and vertical collaboration, was viewed as challenging to influence. Additionally, the respondents found human resources and top leadership issues related to agile decision-making and crisis management beyond their immediate reach.

DISCUSSION

Theoretical Reflections

The global situation has been highly challenging to SCs, with fast and unpredictable changes in the business environment; hence, the importance of SCRes and OR capabilities has been highlighted. Adding a systems engineering perspective to the discussion on DCs provides a framework for SCRes development. The present study aimed to define future DCs for SCRes and OR and examine how capable companies considered themselves in developing these capabilities internally and externally in their SCs. As a result of our analysis focusing on these SCs that can be regarded as complex interorganizational systems, we propose that the required low-order capabilities can be classified under eight micro-foundations: proactive planning, collaboration, risk management, communication and information sharing, monitoring and follow-up, leadership and HR management, global knowledge management, and management of digitalization.

We urge industrial SCs to emphasize these capabilities when improving SC system resilience. Our study further highlights the importance of DC thinking, proactive threat identification, responsiveness through flexible and innovative solutions, adaptability, and agility to implement changes to strengthen SCRes and OR. In the turbulent business environment, industrial SCs live; traditional (operational) capabilities are not sufficient to build SCRes and OR, so organizations and SCs must develop their DCs and innovativeness (Golgeci & Ponomarov, 2013; Kähkönen et al., 2023; Stadtfeld & Gruchmann, 2024; Teece, 2007).

Our results highlight the importance of partnerships and supplier relationship management and development. SCRes and OR require close, internal communication with suppliers and SC partners in organizations. This collaboration, driven by systematic strategic management and strong leadership, forms a base for proactive threat identification, responsive actions, and agile change throughout the SC. Our findings complement the literature, considering collaborative strategic teams and communication with suppliers important in crisis management and in buyer–supplier partnership (Han et al., 2022; ISO 22316, 2017; Kalubanga & Gudegan, 2022; Kumar & Sharma, 2021; Moosavi et al., 2022; Raj et al., 2022).

Our respondents did not consider virtual leadership or management of remote work to have an impact on SCRes. At the same time, the respondents highlighted important micro foundations and low-order capabilities for SCRes, e.g., building trust, collaboration, interaction, and knowledge sharing, that are promoted through leadership. This discrepancy can relate to a lack of leadership competencies and allocated resources to invest in leadership. In earlier research, virtual leadership is essential in building OR (Degbey & Einola, 2020). Furthermore, it can be strongly argued that, in the current changing business environment, adaptive leadership is generally needed in virtual leadership and leadership of remote work, both internally and in supplier networks.

Adaptive leadership can promote the learning capability of employees and organizational performance, thus building OR (He et al., 2023; ISO 22316, 2017; Tvedt et al., 2023).

Based on our results, monitoring and follow-up were rated slightly lower than the highest-rated micro-foundations regarding their impact and ability to influence development. Based on this research, we still have not drawn any definite conclusions. Monitoring and follow-up are still relevant for businesses, with previous studies considering traceability and collaborative monitoring teams as promoters for SCRes, promoting risk identification, anticipation, and agile responsiveness in crisis management (Gligor et al., 2019; ISO 22316, 2017; Kumar & Sharma, 2021).

Digitalization and AI are anticipated to provide solutions that build SCRes (Zouari et al., 2021). Supplemented with a systems engineering approach, they can lead to agile value creation, as Nuerk and Dařena (2023) discussed. We must consider situationally and strategically whether AI adds value alongside soft values and DCs. The prerequisites for utilizing AI also need to be understood. These findings are supported by previous studies that considered combining digitalization, collaboration, and leadership to create a digital strategy (He et al., 2023; Zhao et al., 2023). Another important aspect is that digitalization and AI can simultaneously strengthen capability and be a source of fragility. Finally, it remains unclear what benefits AI creates and how reliably it can collect information globally from black swan events.

Our respondents felt that their perceived possibilities of influencing the development of several essential SCRes and OR capabilities were limited, which should be considered when aiming to improve SCRes. The reasons for this may include unclear strategies, practices, or responsibilities in the organization; lack of leadership; centralized decision-making; or lack of empowerment, all of which are perspectives in which systems engineering could contribute as a framework for development by delving deeper into organizational, managerial, and policy levels (Keating et al., 2008). This influence must be strengthened to build SCRes and OR using these capabilities. Our findings complement earlier research stating that supportive, communicative, and collaborative leadership activates employees for innovativeness and to take initiatives for risk responsiveness (He et al., 2023).

Our findings show that technological and social interaction internally and externally with organizations belonging to the SC is an impactful capability to OR and SCRes. Managing SCs as sociotechnical systems improves SCRes when interaction is promoted through collaboration between employees and buyer–supplier partnerships by considering the business perspectives of both parties, as found in previous research (Wiedmer & Griffis, 2021). In addition to social interaction, from a technological standpoint, when more data are available and efficiently utilized, proactive planning, risk management, business continuity management, and monitoring and follow-up of SC performance are more proactive. Organizations need to collaboratively consider not only how to manage and implement digital solutions, but also how to manage the knowledge and information flow globally. These findings are supported by previous studies considering sociotechnical processes and human network connections promoting learning and organizational capability development (Mitki et al., 2019).

Managerial Implications

We found that the capabilities related to business continuity management and interactive buyer–supplier collaboration were viewed as the most impactful capabilities to SCRes and OR. In addition to traditional capabilities, DCs play a significant role in SCRes and OR management. We argue that management must consider their organization and SCs as sociotechnical systems and develop technological and social interaction internally and with suppliers to promote influencing possibilities for SCRes-related capabilities. In addition, utilizing AI and digitalization for data and knowledge management will increase buyer–supplier interaction, resulting in improved threat identification and response capability.

Organizations need to invest in virtual leadership and leadership of remote work in their organizations and SCs, and empower employees to increase influencing possibilities for SCRes and OR-related capabilities development. Even though the professionals in this survey did not see virtual leadership and leadership of remote work as important, we do based on earlier research results, and that managers should reflect on the importance of this micro-foundation.

Organizations need to create systematic supplier relationship management practices and empower employees working with key partners to develop interactions, collaborations, and trust within partnerships. Organizations should consider the value of utilizing AI and digitalization strategies and solutions in their business, and what kinds of capabilities would require internal and external SC. When creating a digitalization strategy, management must consider soft values and DCs because these will be essential in the future.

In conclusion, interactive buyer–supplier collaboration, empowering virtual leadership, leadership of remote work, and utilizing digitalization enable organizations to share knowledge to strengthen business continuity management and DC to proactively identify threats and innovate responsive actions in collaboration with stakeholders that develop SCRes and OR.

CONCLUSIONS

Recent events have shown that turbulence in the business environment has become a new normal. Hence, industrial companies and SCs need the readiness and capabilities to adapt to any threats or changes to ensure business continuity. In the present study, we investigated the experiences of Finnish industrial companies. We aimed to identify the needed SCRes and OR capabilities in the future and whether industrial organizations can influence the development of these capabilities in their organizations and SCs.

We classified future low-order DCs for SCRes and OR into eight micro-foundations: proactive planning, collaboration, risk management, communication and information sharing, monitoring and follow-up, leadership and HR management, global knowledge management, and management of digitalization. We recommend that management invest in sociotechnical system development to increase technological and social interaction with suppliers.

Based on earlier knowledge, we also recommend that organizations investigate the development of these capabilities by investing in virtual leadership and leadership of remote work in their organizations and SC to empower employees to take responsibility and build interactive collaboration internally and with SC partners. Our respondents did not find this impactful, but a lack of awareness could cause this.

Defined responsibilities, systematic supplier relationship management practices, and shared decision-making responsibilities increase the influencing possibilities for capability development. Using AI and digitalization advances global data availability and knowledge management, thereby influencing the possibilities of developing risk management, business continuity management, monitoring, and follow-up.

The findings highlight the importance of interactive buyer–supplier collaboration, empowerment internally in one's organization and with key partners in SC, systematic supplier relationship management, virtual leadership, and utilizing digitalization in SCRes and OR management in the future. These capabilities create a foundation and strengthen the development of communication and information sharing, knowledge management, proactive planning, and business continuity management.

REFERENCES

Bento, F., Garotti, L. & Mercano, M. P. (2021). Organizational resilience in the oil and gas industry: A scoping view. *Safety Science,* 105036. https://doi.org/10.1016/j.ssci.2020.105036

Boy, G. A. (2017). Human-centered design of complex systems: An experience-based approach. *Design Science, 3*, Article e8. https://doi.org/10.1017/dsj.2017.8

Chen, Y., Chen, I. J. (2019). Mediated power and sustainable supplier management (SSM): linking power use, justice, and supplier performance. *International Journal of Physical Distribution and Logistics Management, 49*(8), 861–878. https://doi.org/10.1108/IJPDLM-12-2018-0393

Degbey, K. & Einola, K. (2020). Resilience in virtual teams: Developing the capacity to bounce back. *Applied Psychology, 69*(4), 1115–1572. https://doi.org/10.1111/apps.12220

Gebhardt, M., Spieske, A., Kopyto, M. & Birkel, H. (2022). Increasing global supply chains' resilience after the COVID-19 pandemic: Empirical results from a Delphi study. *Journal of Business Research, 150*, 59–72. https://doi.org/10.1016/j.jbusres.2022.06.008

Gligor, D., Gligor, N., Holcomb, M. & Bozkurt, S. (2019). Distinguishing between the concepts of supply chain agility and resilience: A multidisciplinary literature review. *International Journal of Logistics Management, 30*(2), 467–487. https://doi.org/10.1108/IJLM-10-2017-0259

Golgeci, I. & Ponomarov, S. Y. (2013). Does firm innovativeness enable effective responses to supply chain disruptions? An empirical study. *Supply Chain Management, 18*(6), 604–617. https://doi.org/10.1108/SCM-10-2012-0331

Han, Z., Handfield, R. B., Huo, B. & Tian, Y. (2022). Effects of power use in buyer–supplier relationships: The moderating role of communication. *Industrial Marketing Management, 102*, 45–57. https://doi.org/10.1016/j.indmarman.2022.01.001

He, Z., Huang, H., Choi, H. & Bilgihan, A. (2023). Building organizational resilience with digital transformation. *Journal of Service Management, 34*(1), 147–171. https://doi.org/10.1108/JOSM-06-2021-0216

Hsieh, H. F. & Shannon, S. E. (2005). Three approaches to qualitative content analysis. *Qualitative Health Research, 5*(9), 1277–1288. https://doi.org/10.1177/1049732305276687

ISO 22316 (2017). *Security and resilience – organizational resilience – principles and attributes*. International Standardization Organisation.

ISO/TS 22318 (2021). *Security and resilience – business continuity management systems – guidelines for supply chain continuity management.* International Standardization Organisation.

ISO/TS 22330 (2018). *Security and resilience – business continuity management systems – guidelines for people aspects of business continuity.* International Standardization Organisation.

Jackson, S. & Ferris, T. L. J. (2013). Resilience principles for engineered systems. *Systems Engineering, 16*(2), 152–164. https://doi.org/10.1002/sys.21228

Kalubanga, M. & Gudegan, S. (2022). The impact of dynamic capabilities in disrupted supply chains – the role of turbulence and dependence. *Industrial Marketing Management, 103*, 154–169. https://doi.org/10.1016/j.indmarman.2022.03.005

Kauppila, O., Jounila, H. & Reiman, A. (2020). Networking for supplier development; evidence from a Finnish industrial cluster. *International Journal of Occupational and Environmental Safety, 4*(2), 22–36. https://doi.org/10.24840/2184-0954_004.002_0003

Kähkönen, A. K., Evangelista, P., Hallikas, J., Immonen, M., & Lintukangas, K. (2023). COVID-19 as a trigger for dynamic capability development and supply chain resilience improvement. *International Journal of Production Research, 61*(8), 2696–2715. https://doi.org/10.1080/00207543.2021.2009588

Keating, C. B., Padilla, J. J. & Adams, K. (2008). System of systems engineering requirements: Challenges and guidelines. *Engineering Management Journal, 20*(4), 24–31. https://doi.org/10.1080/10429247.2008.11431785

Kumar, B. & Sharma, A. (2021). Managing the supply chain during disruptions: Developing a framework for decision-making. *Industrial Marketing Management, 97*, 159–172. https://doi.org/10.1016/j.indmarman.2021.07.007

Leinonen, S., Reiman, A. & Kauppila, O. (2023). Dynamic capabilities in supply chain resilience management in the Finnish energy industry. *International Journal for Quality Research, 17*(4), 1061–1084. https://doi.org/10.24874/IJQR17.04-06

Leinonen, S., Reiman, A., Kauppila, O., Djapan, M., Djordjevic, A. & Demichela, M. (2025). Capabilities for organizational and supply chain resilience in industrial customer–supplier partnerships. *International Journal of Value Chain Management, 15*(2), 119–141. https://doi.org/10.1504/IJVCM.2024.143656

Mitki, Y., Shani, A. B. & Greenbaum, B. E. (2019). Developing new capabilities: A longitudinal study of sociotechnical system redesign. *Journal of Change Management, 19*(3), 167–182. https://doi.org/10.1080/14697017.2018.1490337

Moosavi, J., Fathollahi-Fard, A. M. & Dulebenets, M. A. (2022). Supply chain disruption during the COVID-19 pandemic: Recognizing potential disruption management strategies. *International Journal of Disaster Risk Reduction, 75*, Article 102983. https://doi.org/10.1016/j.ijdrr.2022.102983

Nuerk, J. & Dařena, F. (2023). Activating supply chain business models' value potentials through Systems Engineering. *Systems Engineering, 26*(5), 660– 674. https://doi.org/10.1002/sys.21676

Raj, A., Mukherjee, A. A., Jabbour, A. B. L. S. & Srivastava, S. K. (2022). Supply chain management during and post-COVID-19 pandemic: Mitigation strategies and practical lessons learned. *Journal of Business Research, 142*, 1125–1139. https://doi.org/10.1016/j.jbusres.2022.01.037

Sánchez, M. A. & De Batista, M. (2023). Business continuity for times of vulnerability: Empirical evidence. *Journal of Contingencies and Crisis Management, 31*(3), 431–440. https://doi.org/10.1111/1468-5973.12449

Stadtfeld, G. M. & Gruchmann, T. (2024). Dynamic capabilities for supply chain resilience: A meta-review. *International Journal of Logistics Management. 35*(2), 623–648. https://doi.org/10.1108/IJLM-09-2022-0373

Tavakol, S., Dennick, R. & Tavakol, M. (2011). Making sense of Cronbach's alpha. *International Journal of Medical Education, 11*(54), 53–55. https://doi.org/10.5116/ijme.4dfb.8dfd

Teece, D. J. (2007). Explicating dynamic capabilities: The nature and microfoundations of (sustainable) enterprise performance. *Strategic Management Journal, 28*(13), 1319–1350. https://doi.org/10.1002/smj.640

Teece, D. J. (2012). Dynamic capabilities: Routines versus entrepreneurial action. *Journal of Management Studies, 49*(8), 1395–1401. https://doi.org/10.1111/j.1467-6486.2012.01080.x

Teece, D. J., Pisano, G. & Shuen, A. (1997). Dynamic capabilities and strategic management. *Strategic Management Journal, 18*(7), 509–533. https://doi.org/10.1002/(SICI)1097-0266(199708)18:7%3C509::AID-SMJ882%3E3.0.CO;2-Z

Tvedt, I. M., Tommelein, I. D., Klakegg, O. J. & Wong, J.-M. (2023). Organizational values in support of leadership styles fostering organizational resilience: A process perspective. *International Journal of Managing Projects in Business, 16*(2), 258–278. https://doi.org/10.1108/IJMPB-05-2022-0121

Wiedmer, R. & Griffis, S. E. (2021). Structural characteristics of complex supply chain networks. *Journal of Business Logistics, 42*(2), 264–290. https://doi.org/10.1111/jbl.12283

Zhao, N., Hong, J. & Lau, H. K. (2023). Impact of digitalization on supply chain resilience and performance: A multi-mediation model. *International Journal of Production Economics, 259,* Article 108817. https://doi.org/10.1016/j.ijpe.2023.108817

Zouari, D., Ruel, S. & Viale, L. (2021). Does digitalizing the supply chain contribute to its resilience? *International Journal of Physical Distribution & Logistics Management, 51*(2), 149–180. https://doi.org/10.1108/IJPDLM-01-2020-0038

Section VIII

Human–AI Collaboration and Autonomy

36 Human–Systems Cooperation and Symbiosis in Times of Increasing Autonomous Abilities of Technology

Frank Ole Flemisch, Marie-Pierre Pacaux-Lemoine, Yuichi Saito, and Marcel Usai

FROM COOPERATION TO AUTOMATION TO COOPERATIVE AND SYMBIOTIC SYSTEMS: A HISTORIC OVERVIEW

Introduction: Tools, Short-Term Cooperation Partners, and Long-Term Symbiotic Partnerships

To understand the challenges of automation, artificial intelligence (AI), cooperation, and symbiosis of today and the future, looking back into history can make some sense. Artifacts like the spears of Schoeningen, which were found in 1994 in an open-cast coal mine in the North of Germany, and with an age of three hundred thousand years, were identified as the oldest wooden artifacts found so far, showing that *Homo sapiens* were creating artifacts and tools already at that time. What makes these tools so special is their use in a cooperative manner, which made them effective against a much more agile prey. For us, these spears are one of the earliest examples of human factors as an early adaptation of the tool to the user.
Human Systems Integration denoted integrating humans, technology, organization, and the environment into a meaningful whole that serves our purpose.

Many thousands of years after mastering spears, *Homo sapiens* started to use more intelligent "tools" like dogs (e.g., Hancock, 1996) or horses (e.g., Chamberlin, 2010), began to team up with them, and even got into a strong symbiosis with them, as described by Harari (2014).

DOI: 10.1201/9781003666431-44

Early Examples and a Still-Relevant Concept: Automation

In contrast to *Homo sapiens'* cooperation and symbiosis with animals, the development of automation started much simpler. The classic aim of automation is to wholly or partially "replace" functions that humans have performed with machines (e.g., Parasuraman et al., 2000). Function allocation, which determines what a human or machine should do to achieve a goal, is a key issue in automation. "What" should be automated? The first strategy was to automate all tasks that can be automated (Rouse, 1991). Systems engineers challenge that goal to extend the boundary of what is technically feasible. One motive for replacing human operators with automation is to save them from hazardous work environments. This first strategy for function allocation is in terms of technological feasibility. The second strategy is automating certain functions to minimize the cost of system development and operation (Rouse, 1991). The functions of humans and machines are allocated in terms of economic rationality. For example, automation improves productivity. These function's allocation strategies do not consider the characteristics of human functions. Tasks that "cannot" be automated by machines, or that are more difficult, remain with humans, who struggle to fulfill their roles, which is one of the ironies of automation (Bainbridge, 1983).

In contrast, the third strategy compares the capabilities of humans and machines, and the functions in which machines are superior to humans are automated (Rouse, 1991; Hancock et al., 1996; Hoc, 2000). This strategy considers the characteristics of human functions. Humans and machines are different. The list of things humans and machines are better at is called the MABA-MABA list (Fitts, 1951).

One significant disadvantage of this function allocation strategy is that if machines get better and better, nothing might be left for humans. This might be an irony of automation on its own. As humans also like to perform functions, work can give meaning to life. A future where machines do everything might not be desirable or even stable. What if we could design a strategy of function allocation where machines can take over undesirable tasks while humans retain meaningful and valuable work and functions if they want? Such a strategy would use the advantages of machines while maintaining human values.

The classic allocation of functions is determined from the perspective of "who should do what." This strategy is static (Hancock et al., 1996; Inagaki, 2003). The agent to which the function is allocated is responsible for fulfilling its role to achieve the goal "at all times." The question may arise regarding whether this is "always" appropriate for function allocation based on comparing human and machine capabilities. The answer to this is no. Human functions can be unreliable and unpredictable. The same can be true for functions assigned to the machine. Their performance varies dynamically depending on environmental factors (e.g., time pressure and uncertainty) and human factors (e.g., attention allocation, situation awareness, and effort) (Endsley et al., 2004). Even if a human or a machine is assigned to a task they are better at, they may not always be better at it. This explains the need for "situation-adaptive function allocation" that can dynamically change the functions of humans and machines in response to changes in the environment and human performance (Hancock et al., 1996; Rouse, 1998; Scerbo, 1996; Inagaki, 2003). In addition to this, an essential perspective from Fitts' (1951) MABA-MABA list is that humans and machines can be "complementary," rather than seeing human and machine functions in "opposition," and cooperative functions of humans and machines contribute to identifying and applying such a complementarity, so-called "Know-How-to-Cooperate" (Pacaux-Lemoine, 2020). Another perspective beyond functions is to look at abilities, authority, control, and responsibilities (Flemisch et al., 2011; 2024a, 2024b, 2024c).

Only humans (and organizations) can take responsibility for legal and moral reasons. Humans are responsible for ensuring the safety of systems where man and machine work together (Rouse, 1991). In the field of aircraft and automobiles, many regulations (e.g., Billings, 1996) stipulate that humans must be in command. The function of a human or a machine can be structured in four stages: (1) information acquisition, (2) situation diagnosis, (3) action selection, and (4) action implementation (Parasuraman et al., 2000).

When and how can human functions complement each other to achieve the common goals of human–machine systems? Sheridan (1978, 1992) proposed levels of automation. This concept describes the interaction by which humans and machines respond to each other. The term "autonomy" is used for accomplishing tasks independently without human involvement (Calhoun, 2022). Higher levels of automation provide machines with more autonomy. The use of intermediate levels of automation is proposed in systems positioned as teammates with shared goals (Inagaki, 2006, 2008). There are the concepts of trading and sharing authority (Inagaki, 2003) or trading or sharing of control (e.g., Griffiths & Gillespie, 2005; Abbink et al., 2012; Abbink et al., 2018; Flemisch et al., 2019). Who makes the decisions about the trading or sharing of authority? The strategy for invoking trading or sharing authority depends on the locus of decision authority, which can be distinguished into two types (Sauer et al., 2012). The first type is "adaptive automation," in which machines retain the authority to make decisions (Scerbo, 1996). The other is "adaptable automation," in which humans retain the authority to make decisions (Li et al., 2013). Adaptable automation strategies increase the predictability of the system, i.e., the ability to know what the automation will do and when (Miller et al., 2007). However, it also has the negative effect that the human workload is higher in managing the automation tasks, which can lead to failure to invoke the automation. The control of cooperation, by humans, by machines, or by both, may help to mitigate these negative effects. Levels of cooperation have been defined to reach this objective (Pacaux-Lemoine et al., 2023; Pacaux-Lemoine & Vanderhaegen, 2013).

First Glimpse at Control, Cooperation, and Symbiosis

To fully understand the steps from automation to cooperative and symbiotic systems, one needs to look back at one of the roots of automation and take a closer look at cybernetics. Norbert Wiener is considered the pioneer of cybernetics with his eponymous book (Wiener, 2019), first published in 1948. Recognizing the effects of feedback in control systems not only in technology but also in disciplines such as biology, Wiener heavily influenced control theory, which opened the possibilities for artificial systems to gain access to abilities like autonomy, learning, or cooperating with other systems. This laid the fundamentals for today's cooperative and artificial intelligence systems.

Licklider (1960) used the approach to formulate an early version of human–machine cooperation, which he called Man–Computer Symbiosis, on the systemic level, providing us with ideas on the core principles of how to form new systems of humans joined by machines to control complex situations using the strengths of all involved. In the concept of Licklider (1960), humans in a symbiotic system can step back from routine tasks and focus on the supervision and decision making rather than execution itself, by, as stated in modern words, moving from being *in* the loop to being *on* the loop (Flemisch et al., 2024a, 2024b, 2024c).

In alignment with these ideas and the general framework of task allocation, Rasmussen (1983) introduced the three levels of performance of skilled human operators: skill-based, rule-based, and knowledge-based behavior, which are still widely used and adapted today to explain human behavior as well as roles in human–machine systems. Following Licklider's (1960) ideas on symbiosis, a human operator could substitute their skill-based behavior with a machine by forming a symbiotic or joint system.

Rasmussen's model on skill-, rule-, and knowledge-based behavior is applied to automation by Onken (1998) (see also Onken & Schulte, 2010). While conventional automation is more on the skill-based level and partially on the rule-based level, cognitive automation is also capable of the knowledge-based level.

Cognitive automation (Onken & Schulte, 2010) is an essential prerequisite of Licklider's (1960) ideas on joint systems working on intellectual tasks, which was taken to a new level by Hollnagel & Woods (2005) and Woods & Hollnagel (2006) by proposing to integrate humans and machines into joint cognitive systems. With this step, cooperation between human and machine is no longer limited to a distribution along a certain task classification. Instead, it involves joint decisions and actions following different distributions of authority, responsibility, and control within the joint system, as later described by (Flemisch et al., 2012).

A Closer Handle on Human–Machine Cooperation

Based on the ideas of Licklider (1960) and Rasmussen (1983), Hoc (2000) proposed a shift from the, at that time dominant, paradigm of (human–computer) interaction, meaning a relatively static interfacing of machines to give humans access to controls, to a paradigm of cooperation to allow systems to be designed with dynamic shifts of control (function allocation) between human and machine in mind. Onken (2002) then described the cooperative integration of automations into work systems, foreshadowing the ideas of Hollnagel and Woods (2005) of a joint cognitive system. As with many, these ideas were first prepared by Hoc and Lemoine (1998), subsequently implemented in the aviation domain (e.g., Schulte 2009; Onken & Schulte, 2010; Flemisch & Onken, 1998) and spread to other domains, e.g., manufacturing (Pacaux-Lemoine et al., 2017) and on handicap (Pacaux-Lemoine et al., 2023).

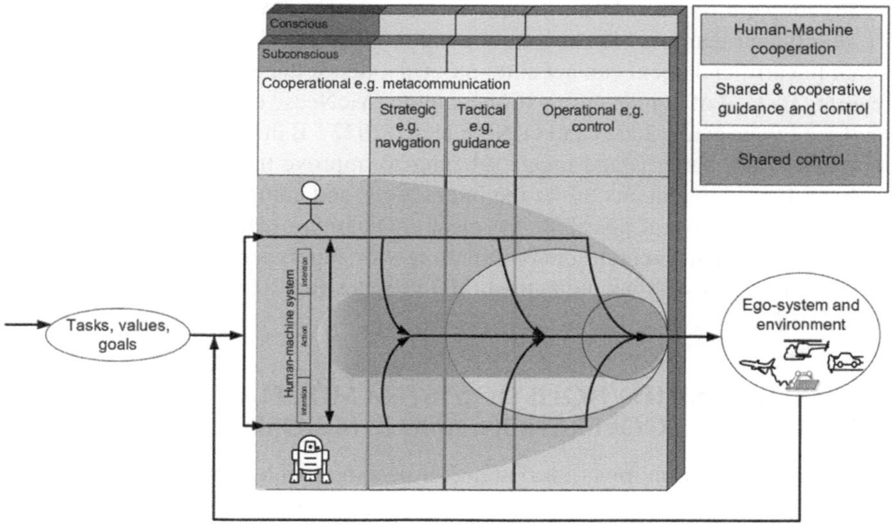

FIGURE 36.1 Human–machine cooperation framework to show the relationship between shared control, shared and cooperative guidance and control, and human–machine cooperation (Usai et al. 2020, based on Pacaux-Lemoine et al. 2017; Flemisch et al., 2019). It also shows that cooperation may happen in a conscious as well as subconscious manner.

With the spread to multiple domains and the adaptation of concepts to an application in reality, today's taxonomy of human–machine systems is not (yet) a common one. Especially in the broad field of robotics, there are still quite the differences (Onnasch & Roesler, 2021), with proposals of categorizations (Yanco & Drury, 2004) dating back to 1990 (Granda et al., 1990).

Pacaux-Lemoine and Flemisch (2016) and Flemisch et al. (2019) introduced a human–machine cooperation framework to include the concepts of traded, supervisory, shared, and cooperative control in a single model (Figure 36.1). It structures a human–machine system from both actors' perceptions and intentions towards their actions, merging into a joint action on the different layers. In addition to the three traditional layers of strategic, tactical, and operational behavior, an additional "cooperational" layer deals mainly with the meta-communication in the sociotechnical system, e.g., between humans and machines.

Human Autonomy Teaming

Another term to describe human–machine cooperation is Human–Autonomy Teaming (HAT). HAT is extending the framework for human–machine collaboration to a higher autonomy of the single agents and possibly multiple artificial agents (automations), which follow a common goal. Stemming from Rasmussen's

(1986) cooperation idea, Miller (2005) formulated the playbook metaphor to allow for an efficient teaming "for keeping operators in charge." A crucial aspect of that is that the artificial agent can act autonomously, hence the term Human–Autonomy Teaming instead of human automation teaming (McNeese et al., 2018), for an overview see Lyons et al. (2021) and O'Neill et al. (2022). Baltzer et al. (2019) show an approach to use patterns and image schemas to improve the teaming aspect, especially with artificial agents, to let the interactions and communication between the agents be as intuitive as possible. A typical application of HAT is the (cooperative) control of a primary vehicle in combination with supervisory control by a human over multiple autonomous unmanned (air) vehicles (e.g., Draper, 2008, or Draper et al., 2018).

TOWARD A UNIFIED HOLISTIC SYSTEM MODEL OF INTERACTION, COOPERATION, AND SYMBIOSIS

Analyzing the literature mentioned above, it becomes evident that concepts like coexistence, interaction, cooperation, collaboration, teaming, and symbiosis are often used differently in different domains, and in a fuzzy way. This is not bad, per se, as long as everyone involved in one project or socio-technical system consistently understands it. It becomes critical if knowledge and know-how are intended to be exchanged between domains. To increase the compatibility between the different domains, we find the following description helpful. It is important to note that this is not an exclusive Aristotelian way of definition,[1] but more in terms of Wittgenstein's family resemblances theory, which allows an inclusive description of concepts (Wenneberg, 1967).

Interaction is "when two or more people or things communicate with or react to each other" (*Cambridge Dictionary*, Interaction). Applied to Human Systems Integration, we define (Human Systems) Interaction as an occasion where two or more socio-technical subsystems communicate with or react to each other.

Collaboration is "working with others or organizations to create or achieve something" (*Cambridge Dictionary*, Collaboration). Commonly observed in robotics, collaboration describes an alliance wherein humans and robots operate in a shared workspace. Tasks often overlap in real time, requiring synchronous engagement to reach a predefined goal. Although it involves active partnership, collaboration does not inherently demand a deep fusion of capabilities (Nissoul et al., 2025).

Cooperation is "working with another company, organization, or country to achieve something" (*Cambridge Dictionary*, Cooperation). Applied to Human Systems Integration, we define Human Systems Cooperation as "the process of two or more subsystems of a socio-technical system working together to achieve something, making the work easier for each other." "Something" could be a common goal, an achievable situation, etc.

Symbiosis is "a close connection between different types of organisms in which they live together and benefit from each other" (*Cambridge Dictionary*, Symbiosis). Applied to Human Systems Integration, we define (Human Systems) Symbiosis as a "close connection between different subsystems of a sociotechnical system, in which they exist together and profit from one another."

It is important to note here that in contrast to the common understanding that in a symbiotic relationship all partners profit from another, biology has a more refined understanding of symbiosis, where parasitism is regarded as a special type of symbiosis as well, where a kind of dependence is imposed, dependence that can lead to risky situations if human is not able to control the situation anymore (Pacaux-Lemoine & Trentesaux, 2019).

What do interaction, cooperation, and symbiosis have in common? What is the difference? A strong common point is that *two or more subsystems that do something together are somehow connected*. While interaction could also be positive, neutral, or hostile, cooperation is usually positive, directed toward a common goal, and requires more cognitive abilities. Symbiosis and cooperation are traditionally more long term than just interaction. We understand interaction, cooperation, and symbiosis as nested family concepts: Symbiosis is generally formed by cooperation, and cooperation and symbiosis usually need interaction.

Applied to Human Systems Integration, a common central aspect of interaction, cooperation, and symbiosis is that human and technical subsystems, like computers, do something together, and are somehow connected. Interaction can be positive, neutral, or negative, and does not necessarily need cooperative abilities. Human-machine cooperation (or collaboration, or teaming) is usually positive and requires a certain level of cognition. Human-machine symbiosis needs cooperation and interaction and is generally understood as a more long-term relationship where both partners profit from each other.

For Human–Machine Cooperation, a key idea is to use the models derived from human cognition and apply them to the cooperation of humans and machines. An example is the Rasmussen "decision ladder" model (Rasmussen et al., 1990) shown in Figure 36.2 (left), which explains cognition and action as a stepwise process between activation and execution, stepping up a cognitive ladder by observing, identifying, interpreting, evaluating, defining tasks, and finally executing tasks.

Figure 36.2 (right) shows how Rasmussen's decision ladder can be used to describe human–machine cooperation. Here, two or more cognitive ladders of humans and machines are combined to define an air defense task. Each step of the ladder can be shared or traded to stabilize the cognitive process overall and enhance the capability to act jointly.

Figure 36.3 shows another example of combining models originally formulated for human decision-making, now applied to human-machine cooperation and symbiosis. The starting point is John Boyd's model of military decision-making, which structures the process into the steps Observe, Orient, Decide, and Act (Boyd, 1996). To structure cooperative decision-making processes, e.g., for human AI teaming, Flemisch et al. (2024a, 2024b, 2024c) couple two or more OODA loops and describe how decisions under uncertainties about action or non-action are made jointly, and which dilemmas might arise.

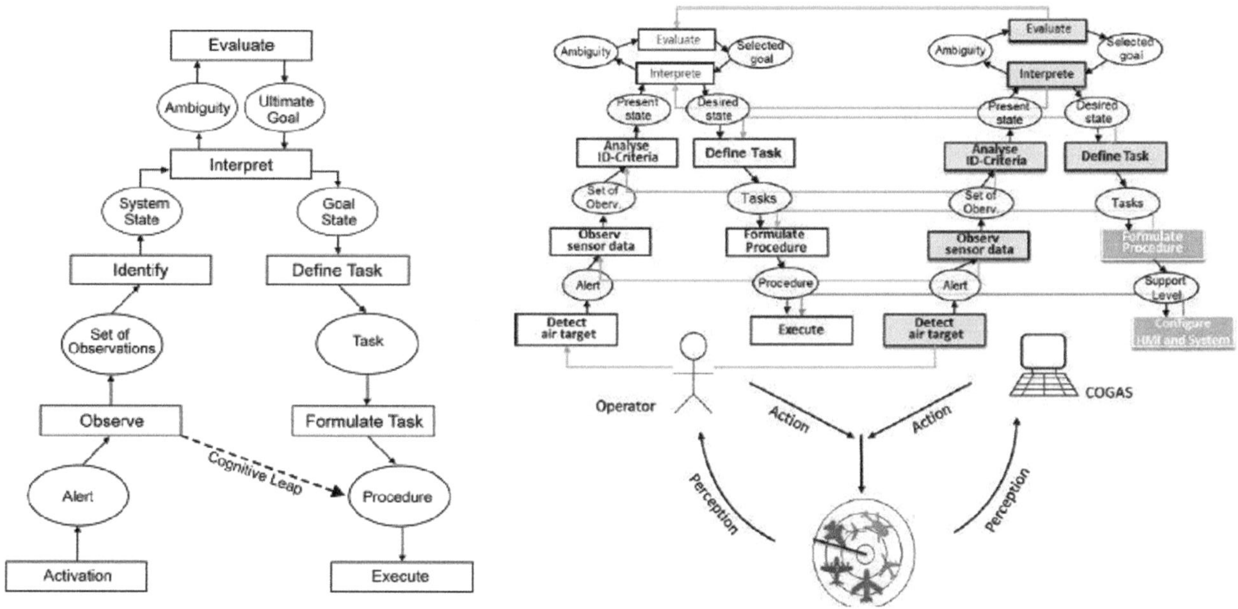

FIGURE 36.2 Left: Rasmussen's decision ladder (Rasmussen et al., 1990) and Human. Right: machine cooperation ladder applied to a cooperative air defense task (Özyurt et al., 2015 based on Rasmussen, 1990 and Flemisch et al., 2008).

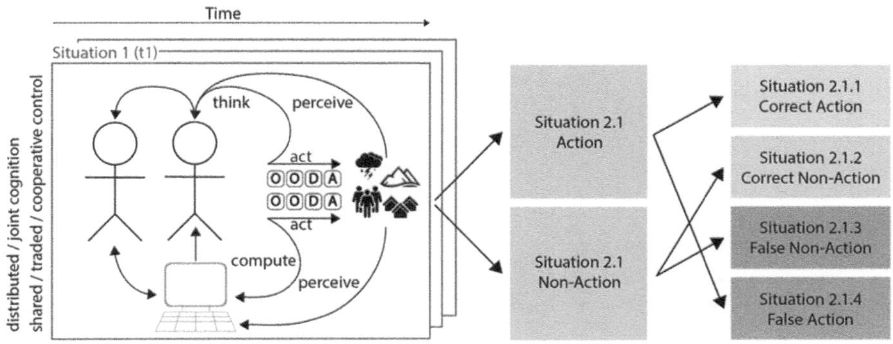

FIGURE 36.3 Dilemma model of human–systems cooperation (Flemisch et al., 2024, based on Boyd, 1996).

CONCLUSION AND PERSPECTIVE: TOWARD AND BEYOND THE HORIZONS OF HUMAN-MACHINE COOPERATION AND SYMBIOSIS

In 2024, we are starting to understand the implications of cooperative and/or symbiotic relationships with automation and AI. Many questions are still open, e.g., about the multimodal mixture of interaction and cooperation patterns. To design and evaluate these complex cooperative systems, the interplay of development and environment, e.g., in an increasing level of realism of the models and AI task, could play a decisive role in mastering the complexity, as Figure 36.4 shows.

Mastering agile processes like Figure 36.4 is just one step of a longer journey with more challenges. It is especially unclear how this interplay between humans and machines develops over time, when cooperation leads to more long-term symbiosis.

In 2025, we will have much machine thinking power in many different domains, but humans are still strongly involved. On the other hand, increasing abilities of machines, combined with increasing time pressure, might lead to a dynamic of pushing the humans out of the loop, as described for societies, e.g., by Rosa (2013), for weapon systems by Scharre (2018), and in industry (Pacaux-Lemoine & Trentesaux, 2019). In combination, this might lead to a non-reversible symbiosis of humans and computers, in contrast to a more desirable reversible symbiosis (Flemisch et al., 2022), that can be controlled by cooperation when humans are losing competencies and skills, being too dependent on machines (Nissoul et al., 2025). Norman (1994) calls for building systems "that make us smart" is not only a niceness, but also a necessity to stay, or become even more "*Homo sapiens*."

Talking about survival: A majority of the discussion in the scientific community has so far focused on algorithms, or single human–machine systems. This is a good starting point, but not sufficient to understand and solve more complex problems like

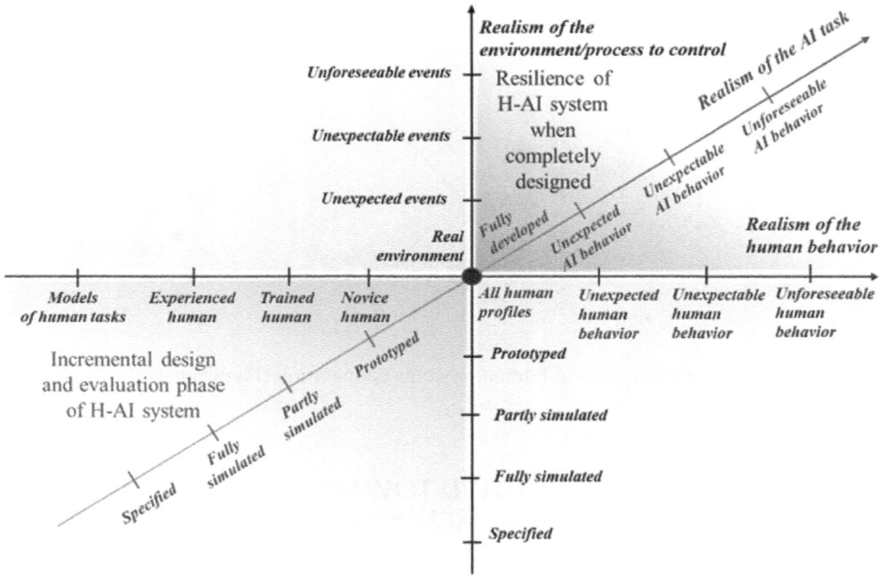

FIGURE 36.4 Three dimensions to design and evaluate human–AI cooperation (Pacaux-Lemoine et al., 2023).

global warming or global arms races. Like on a small scale, problem-solving on a bigger scale also starts with cognition and cooperation.

Figure 36.5 shows a holistic bowtie model as an example of such holistic thinking. The holistic bowtie model connects a local system in the middle, here a human–machine system in a dilemma situation, with levels of surrounding systems in an increasing level of abstraction, from the system-of-systems, the organizations, to societies, and finally planet Earth. Further, the model shows how cooperation is formed over different layers of physical, cognitive, legal, and ethical interrelationships, building on top of each other. The model also integrates the concept of a shared workspace of Hoc (2000) and Pacaux-Lemoine & Debernard (2007).

Thinking and acting not only about human–machine cooperation and symbiosis on a local scale, but also on a global scale, and not only about its positive sides, but also about its opposites, like war or disconnection with nature, might look highly ambitious to our limited brains. However, what would be the alternative in the age of global warming and global arms races? Thinking *and* acting, locally *and* globally, focused *and* holistically is the only chance we have to survive, and to earn the name that our ancestors gave to our species, not to describe a fact, but to encourage and inspire us for a long-term program: *Homo sapiens*, in cooperation, and maybe reversible symbiosis with artifacts and machines including AI, and in symbiosis with each other, our co-creatures and our home planet Earth.

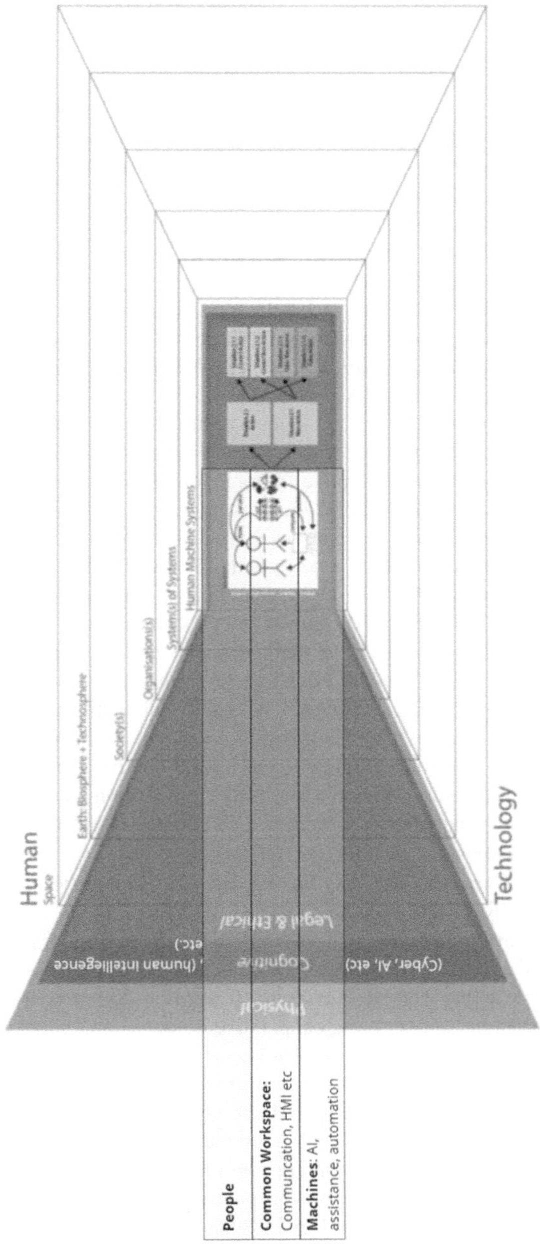

FIGURE 36.5 Holistic bowtie model of human–systems cooperation and symbiosis.

NOTE

1 See www.uni-kassel.de/philosophie/Heinemann/Materialien%20zur%20antiken%20Philosophie/6.5.%20Aristoteles%201030%20Met.%20VII%2005.pdf, for a German annotated version of Aristoteles.

REFERENCES

Abbink, D. A., Carlson, T., Mulder, M., De Winter, J. C., Aminravan, F., Gibo, T. L., & Boer, E. R. (2018). A topology of shared control systems – finding common ground in diversity. *IEEE Transactions on Human–Machine Systems*, 48(5), 509–525. IEEE.

Abbink, D. A., Mulder, M., & Boer, E. R. (2012). Haptic shared control: smoothly shifting control authority? *Cognition, Technology & Work*, 14, 19–28.

Bainbridge, L. (1983). Ironies of automation. In *Analysis, Design and Evaluation of Man–Machine Systems* (pp. 129–135). Pergamon.

Baltzer, M. C., López, D., & Flemisch, F. (2019). Human autonomy teaming using cooperative automation, tnteraction patterns, and image schemes.

Billings, C. E. (1996). *Aviation Automation: The Search For A Human-Centered Approach.* Hillsdale, NJ: Lawrence Erlbaum Associates.

Boyd, J. R. (1996). The essence of winning and losing. https://ooda.de/media/john_boyd_the_essence_of_winning_and_losing.pdf (accessed 9 November 2023).

Calhoun, G. (2022). Adaptable (not adaptive) automation: Forefront of human–automation teaming. *Human Factors*, 64(2), 269–277. https://doi.org/10.1177/00187208211037457

Cambridge Dictionary. Interaction, Cooperation, Symbiosis https://dictionary.cambridge.org/dictionary/english/interaction (Last accessed on: 25 July 2024).

Chamberlin, J. E. (2010). *Horse: How the Horse Has Shaped Civilizations*. Vintage Canada.

Draper, M. H. (2008, September). Supervisory control of multiple unmanned vehicles: International perspectives on methodologies and enabling interface technologies. In *Proceedings of the Human Factors and Ergonomics Society Annual Meeting* (Vol. 52, No. 1, pp. 59–63). Los Angeles, CA: SAGE Publications.

Draper, M. H., Rowe, A., Douglass, S., Calhoun, G., Spriggs, S., Kingston, D., ... & Frost, E. (2018). Realizing autonomy via intelligent hybrid control: Adaptable autonomy for achieving UxV RSTA team decision superiority (also known as intelligent multi-UxV planner with adaptive collaborative/control technologies (IMPACT)). *DTIC Scientific Technology*. https://apps.dtic.mil/sti/pdfs/AD1048921.pdf

Endsley, M. R. (2004). *Designing for Situation Awareness: An Approach to User-Centered Design* (2nd ed.). CRC Press. https://doi.org/10.1201/b11371

Fitts, P. M. (1951). Human engineering for an effective air navigation and traffic-control system.

Flemisch, F., Abbink, D. A., Itoh, M., Pacaux-Lemoine, M. P., & Wessel, G. (2019). Joining the blunt and the pointy end of the spear: Towards a common framework of joint action, human–machine cooperation, cooperative guidance and control, shared, traded, and supervisory control. *Cognition, Technology & Work*, 21, 555–568. https://doi.org/10.1007/s10111-019-00576-1

Flemisch, F., & Baltzer, M. (2022). Are Rider-Horse or Centaurs intelligent human systems integration? First sketch of reversible and non-reversible human technology/machine/AI symbiosis. In *iHSI International Conference on Intelligent Human Systems Integration*. Venice, Italy: AHFE International. https://openaccess.cms-conferences.org/publications/book/978-1-7923-8988-7/article/978-1-7923-8988-7_107

Flemisch, F., Baltzer, M., Abbink, D., Siebert, L., Diggelen, J., Draper, M., Boardman, M., & Pacaux-Lemoine, M. (2024a). Towards a dynamic balance between humans and AI-based systems: System of system perspective on ability, responsibility, authority, autonomy, meaningful and effective control, and accountability. In L. Siebert & D. Abbink (Eds.), *Handbook on Meaningful Human Control*. Edward Elgar Research Handbook.

Flemisch, F., Heesen, M., Hesse, T., Kelsch, J., Schieben, A., & Beller, J. (2012). Towards a dynamic balance between humans and automation: Authority, ability, responsibility, and control in shared and cooperative control situations. *Cognition, Technology & Work*, 14, 3–18.

Flemisch, F., Herzberger, N., Usai, M., & Wasser, J. (2024b). Beyond levels of automation: Human system modes, levels, functions and patterns for an intelligent human systems integration of humans and co-systems. In T. Ahram, W. Karwowski, D. Russo, & G. Di Bucchianico (Eds.), *Intelligent Human Systems Integration (IHSI 2024): Integrating People and Intelligent Systems. AHFE (2024) International Conference. AHFE Open Access* (Vol. 119). USA: AHFE International. http://doi.org/10.54941/ahfe1004516

Flemisch, F., Preutenborbeck, M., Burlage, L., & Ripkens, A. (2024c). Dilemma model of cooperative guidance and control to solve the speed-certainty-dilemma in human–autonomy teaming: First Sketch. In M. Hou, T. Falk, A. Mohammadi, A. Guerrieri & D. Kaber (Eds.), *IC-HMS Conference*. Proceedings of the 2024 IEEE International Conference on Human-Machine Systems. Toronto, Canada.

Flemisch, F., Schieben, A., Kelsch, J., & Löper, C. (2008). Automation spectrum, inner/outer compatibility and other potentially useful human factors concepts for assistance and automation. In D. Waard, F. Flemisch, B. Lorenz, H. Oberheid, & K. Brookhuis (Eds.), *Human Factors for Assistance and Automation*. Maastricht: Shaker.

Flemisch, F. O., & Onken, R. (1998). The Cognitive Assistant System and its contribution to effective man/machine interaction. In *NATO RTO MP-3*. Monterey, USA.

Granda, T. M., Kirkpatrick, M., Julien, T. D., & Peterson, L. A. (1990). The evolutionary role of humans in the human-robot system. In *Proceedings of the Human Factors Society Annual Meeting* (Vol. 34, No. 9, pp. 664–668). Los Angeles, CA: SAGE Publications. https://doi.org/10.1177/154193129003400919

Griffiths, P. G., & Gillespie, R. B. (2005). Sharing control between humans and automation using haptic interface: Primary and secondary task performance benefits. *Human Factors*, 47(3), 574–590.

Hancock, P. A. (1996). Teleology for technology. In R. Parasuraman & M. Mouloua (Eds.), *Human Factors in Transportation. Automation and Human Performance: Theory and Applications* (pp. 461–497). CRC Press.

Hancock, P. A., & Scallen, S. F. (1996). The future of function allocation. *Ergonomics in Design*, 4(4), 24–29. https://doi.org/10.1177/106480469600400406

Harari, Y. N. (2014). *Sapiens: A Brief History of Humankind*. Random House.

Hoc, J. M. (2000). From human–machine interaction to human–machine cooperation. *Ergonomics*, 43(7), 833–843.

Hoc, J. M., & Lemoine, M. P. (1998). Cognitive evaluation of human–human and human–machine cooperation modes in air traffic control. *The International Journal of Aviation Psychology*, 8(1), 1–32.

Hollnagel, E., & Woods, D. D. (2005). *Joint Cognitive Systems: Foundations of Cognitive Systems Engineering*. CRC Press.

Inagaki, T. (2003). Adaptive automation: Sharing and trading of control. In *Handbook of Cognitive Task Design* (pp. 147–170). Taylor & France, Boca Raton, FL, USA: CRC Press.

Inagaki, T. (2006). Design human–machine interactions in light of the domain-dependence of human-centered automation. *Cognition, Technology & Work*, 8, 161–167. https://doi.org/10.1007/s10111-006-0034-z

Inagaki, T. (2008). Innovative collaboration between humans and machines is based on mutual understanding. *Annual Reviews in Control*, 32(2), 253–261. https://doi.org/10.1016/j.arcontrol.2008.07.003

Li, H., Sarter, N., Wickens, C., & Sebok, A. (2013). Supporting human–automation collaboration through dynamic function allocation: The case of space teleoperation. *Proceedings of the Human Factors and Ergonomics Society Annual Meeting*, 57, 359–363. https://doi.org/10.1177/1541931213571078

Licklider, J. C. (1960). Man–computer symbiosis. In *IRE Transactions on Human Factors in Electronics* (Vol. 1, pp. 4–11). IEEE.

Lyons, J. B., Sycara, K., Lewis, M., & Capiola, A. (2021). Human–autonomy teaming: Definitions, debates, and directions. *Frontiers in Psychology*, 12, 589585.

McNeese, N. J., Demir, M., Cooke, N. J., & Myers, C. (2018). Teaming with a synthetic teammate: Insights into human-autonomy teaming. *Human Factors*, 60(2), 262–273.

Miller, C. A. (2005). Delegation architectures: Playbooks and policy for keeping operators in charge. Submitted for inclusion in Proceedings of the Workshop on Mixed-Initiative Planning And Scheduling. Workshop conducted at The 15th International Conference on Automated Planning and Scheduling (ICAPS), Monterey, California. Workshop on Mixed-Initiative Planning And Scheduling at ICAPS05.

Miller, C. A., & Parasuraman, R. (2007). Designing for flexible interaction between humans and automation: Delegation interfaces for supervisory control. *Human Factors*, 49, 57–75. https://doi.org/10.1518/001872007779598037

Nissoul, S., Pacaux-Lemoine, M-P., & Chaabane, S. (2025). Cooperation grid framework: Analyzing and improving Human-CPS agents' cooperation in multi-level decision-making. In *11th IFAC Conference on Manufacturing Modelling, Management and Control*, 30 June–3 July. Trondheim, Norway.

Norman, D.A. (1994). Things That Make Us Smart: Defending Human Attributes In *The Age Of The Machine.* Greenwood Village, CO, USA: Basic Books. ISBN: 978-0201626957.

O'Neill, T., McNeese, N., Barron, A., & Schelble, B. (2022). Human–autonomy teaming: A review and analysis of the empirical literature. *Human Factors*, 64(5), 904–938.

Onken, R. (1998). Die zwei Gesichter der Automation. In *3rd Workshop Verbundvorhaben Cockpit, University of German Armed Forces, Munich.* Neubiberg, Germany: Institute for System Dynamics and Flight Mechanics.

Onken, R. (2002, October). Cognitive cooperation for the sake of the human–machine team effectiveness. In *Proceedings of RTO Human Factors and Medicine Panel (HFM) Symposium.* NATO. https://apps.dtic.mil/sti/tr/pdf/ADA422247.pdf

Onken, R., & Schulte, A. (2010). *System-ergonomic Design of Cognitive Automation: Dual-mode Cognitive Design of Vehicle Guidance and Control Work Systems* (Vol. 235). Springer.

Onnasch, L., & Roesler, E. (2021). A taxonomy to structure and analyze human–robot interaction. *International Journal of Social Robotics*, 13(4), 833–849.

Özyurt, E., Flemisch, F., & Döring, B. (2015). A cooperative automation concept for user-oriented support of air target identification. In *6th International Conference on Applied Human Factors and Ergonomics* (Vol. 3). Las Vegas, USA: AHFE 2015.

Pacaux-Lemoine, M. P., & Debernard, S. (2007). Shared workspace or how to support cooperative activities between human operators: Application to fighter aircraft. In *Engineering Psychology and Cognitive Ergonomics: 7th International Conference, EPCE 2007, Held as Part of HCI International 2007, Beijing, China, July 22-27, 2007. Proceedings 7* (pp. 796–805). Springer Berlin Heidelberg.

Pacaux-Lemoine, M. P., & Flemisch, F. (2016). Layers of shared and cooperative control, assistance, and automation. *IFAC-PapersOnLine*, *49*(19), 159–164.

Pacaux-Lemoine, M. P. (2020). Human–machine Cooperation: Adaptability of shared functions between Humans and machines- Design and evaluation aspects. Dissertation, Université Polytechnique Hauts-de-France. https://hal.science/tel-02959402

Pacaux-Lemoine, M. P., Flemisch, F., & Chaabane, S. (2023). Calibration and resilience of humans-AI systems cooperation in industry. In *Proceedings of SOHOMA International Conference.* Annecy, France.

Pacaux-Lemoine, M. P., Habib, L., & Carlson, T. (2023). Levels of cooperation in human–machine systems: A Human-BCI-Robot example. In G. Fortino, D. Kaber, A. Nurnberger, & D. Mendoca (Eds.), *Handbook of Human-Machine Systems*. Wiley-IEEE Press, ISBN 1119863635.

Pacaux-Lemoine, M. P., Trentesaux, D., Rey, G. Z., & Millot, P. (2017). Designing intelligent manufacturing systems through Human–Machine Cooperation principles: A human-centered approach. *Computers & Industrial Engineering*, 111, 581–595.

Pacaux-Lemoine, M. P., & Trentesaux, D. (2019). Ethical risks of human-machine symbiosis in industry 4.0: Insights from the human–machine cooperation approach. In *IFAC-PapersOnLine, 14th IFAC Symposium on Analysis, Design and Evaluation of Human Machine Systems (IFAC HMS'19)* (Vol. 52, No. 19, pp. 19–24). Tallinn, Estonia: Elsevier.

Pacaux-Lemoine M.-P., & Vanderhaegen F. (2013, October). *Towards Levels of Cooperation.* anchester, UK: IEEE SMC Conference.

Parasuraman, R., Sheridan, T., & Wickens, C. (2000). A model for types and levels of human interaction with automation. *IEEE Transactions on Systems, Man, and Cybernetics – Part A: Systems and Humans*, 30, 286–297. https://doi.org/10.1109/3468.844354

Rasmussen, J. (1983). Skills, rules, knowledge, signals, signs, symbols, and other distinctions in human performance models. *IEEE Transactions on Systems, Man, and Cybernetics*, 3, 257–266.

Rasmussen, J. (1986). The role of hierarchical knowledge representation in decision-making and system management. In *IEEE Transactions on Systems, Man and Cybernetics.* https://doi.org/10.1109/TSMC.1985.6313353

Rasmussen, J., Pejtersen, A. M., & Schmidt, K. (1990). Taxonomy for cognitive work analysis. Riso, Denmark: Risø National Laboratory.

Rosa, H. (2013). *Social Acceleration: A New Theory of Modernity*. Columbia University Press.

Rouse, W. B. (1988). Adaptive aiding for human/computer control. *Human Factors*, 30(4), 431–443.

Rouse, W. B. (1991). *Design for Success: A Human-Centered Approach to Designing Successful Products and Systems*. New York: Wiley.

Sauer, J., Kao, C.-S., & Wastell, D. (2012). A comparison of adaptive and adaptable automation under different levels of environmental stress. *Ergonomics*, 55, 840–853. https://doi.org/10.1080/00140139.2012.676673

Scerbo, M. W. (1996). Theoretical perspectives on adaptive automation. In R. Parasuraman & M. Mouloua (Eds.), *Automation and Human Performance: Theory and Applications* (pp. 37–63). Lawrence Erlbaum Associates, Inc.

Scharre, P. (2018). *Army of None: Autonomous Weapons and the Future of War*. WW Norton & Company.

Schulte, A. (2009). Enhancing human–machine system performance by introducing artificial cognition in vehicle guidance work systems. In *NATO RTO HFM-181 Symposium on Human Performance Enhancement for NATO Military Operations*.

Sheridan, T. B. (1978). Supervisory control of remote manipulators for undersea applications. In *Proceedings: International Conference on Cybernetics and Society* (p. 237). Institute of Electrical and Electronics Engineers. https://apps.dtic.mil/sti/tr/pdf/ADA116236.pdf

Sheridan, T. B. (1992). *Telerobotics, Automation, and Human Supervisory Control*. MIT Press.

Usai, M., Meyer, R., Nagahara, H., Takeda, Y., & Flemisch, F. (2020). Towards a truly cooperative guidance and control: Generic architecture for intuitive human–machine cooperation. In *Advances in Human Aspects of Transportation: Proceedings of the AHFE 2020 Virtual Conference on Human Aspects of Transportation*, July 16–20, 2020 (pp. 92–98). USA: Springer International Publishing.

Wennerberg, H. (1967). The concept of family resemblance in Wittgenstein's later philosophy. *Theoria*, 33(2), 107–132. https://doi.org/10.1111/j.1755-2567.1967.tb00614.x

Wiener, N. (2019). *Cybernetics or Control and Communication in the Animal and the Machine*. MIT Press.

Woods, D. D., & Hollnagel, E. (2006). *Joint Cognitive Systems: Patterns in Cognitive Systems Engineering*. CRC Press.

Yanco, H. A., & Drury, J. (2004). Classifying human–robot interaction: An updated taxonomy. In *2004 IEEE International Conference on Systems, Man and Cybernetics (IEEE Cat. No. 04CH37583)* (Vol. 3, pp. 2841–2846). The Hague, Netherlands. https://doi.org/10.1109/ICSMC.2004.1400763

37 Human–System Integration and Cognitive Consonance for Artificial Intelligence

Gopal R. Jani and Parag Sanghani

INTRODUCTION: ARTIFICIAL INTELLIGENCE MEETS HUMAN COGNITION

The convergence of human cognition and artificial intelligence (AI) is transforming the digital landscape. Public digital services must cater to large-scale, demographically diverse populations, especially in developing nations like India. Let us discuss the PM-WANI Framework and its implementation by Gavedu—an AI-powered, human-centric platform delivering free public Wi-Fi under the Department of Telecommunications, India. Through seamless authentication, real-time network optimization, and fraud detection, the system demonstrates the application of human–system integration (HSI) principles in large-scale digital deployments.

THEORETICAL FOUNDATIONS: HSI AND COGNITIVE SCIENCE

Understanding Human–System Integration

HSI optimizes system performance by integrating human capabilities, limitations, and behavior into technology design. It includes physiological, cognitive, and contextual factors. In Gavedu, HSI principles are applied in SD-WAN-based authentication systems that adapt to user diversity while ensuring scalability and efficiency (Boy, 2023).

HSI is a multidisciplinary framework that emphasizes aligning technological systems with human capabilities, limitations, and behavior to optimize performance, safety, and user satisfaction. At its core, HSI addresses the physiological (e.g., sensory–motor functions, physical ergonomics), cognitive (e.g., memory, attention, decision-making), and contextual (e.g., cultural, linguistic, environmental) dimensions of human interaction with complex systems (Boy, 2023; Madni & Sievers, 2013).

In the case of Gavedu's PM-WANI Framework, these HSI principles are operationalized by implementing Software-Defined Wide Area Network (SD-WAN)-based authentication and access mechanisms. The system dynamically adapts authentication methods and bandwidth allocation to suit its users' cognitive and behavioral

DOI: 10.1201/9781003666431-45

profiles, whether urban professionals, rural users with low digital literacy, or first-time internet users. This alignment with HSI ensures that system performance is not merely a function of network efficiency but also human usability and acceptance.

Correlation with HSI Frameworks and Models

Boy's HSI Framework (2023)

Gavedu reflects Boy's layered model of HSI, which advocates for integrating human-centered design from the conceptual phase to the operational deployment of technology. Specifically:

Functional Layer: Gavedu's SD-WAN adapts services like login authentication and content delivery based on user behavior.
Cognitive Layer: The User Interface (UI)/User Experience (UX) accommodates users with varying literacy levels through audio prompts, iconography, and intuitive flows.
Organizational Layer: The design respects contextual norms of bandwidth use, rural device access, and regional ad preferences.

The HSI Technical Framework by INCOSE

The International Council on Systems Engineering (INCOSE) HSI model includes seven domains: workforce, personnel, training, human factors engineering, safety, habitability, and survivability. In the Gavedu system:

Training: Minimal training is needed due to intuitive onboarding interfaces.
Human Factors Engineering: Interfaces are tested for cognitive load minimization and accessibility.
Safety and Survivability: Adaptive fraud detection supports system robustness and user protection.

Cognitive Load Theory Integration

Through user-centric interface design and progressive disclosure, Gavedu reduces extraneous cognitive load (Sweller, 1988). By presenting only the most relevant information at each interaction step and providing real-time feedback during transactions, the system improves learnability and cognitive consonance—a condition where user mental models align with system responses.

Hollnagel's Contextual Control Model

Gavedu's adaptive behavior aligns with Contextual Control Model (COCOM) by shifting system behavior from strategic (predefined flows) to opportunistic control (dynamic ad delivery, bandwidth scaling) depending on real-time context and user interaction.

The Role of Cognitive Science

Cognitive science contributes decision-making, memory, perception, and attention models to system design (Sweller, 1988; Mayer & Moreno, 2003). Gavedu applies these principles in ad-sponsored Internet delivery, minimizing cognitive load and building trust via intuitive user interfaces and feedback mechanisms.

HUMAN–MACHINE SYNERGY: COGNITIVE CONSONANCE

Bridging AI and Cognition

AI technologies like deep learning, Natural Language Processing (NLP), and adaptive algorithms personalize interactions and optimize backend operations. Gavedu incorporates dynamic ad delivery, behavior-based threat detection, and bandwidth allocation to meet user needs in real-time (Calikli & Bener, 2014; Hanham, 2023).

AI-powered adaptivity in Gavedu reflects **closed-loop HSI architectures** where the system monitors human input, adjusts responses, and feeds information back to the user.

- *Feedback Loops:* Real-time content adjustments and fraud alerts create bidirectional communication.
- *Adaptive Cognition:* Interfaces adjust in complexity and presentation based on observed behavior (e.g., more visuals for novice users, detailed stats for power users).

Norman's Model of Interaction (Seven Stages of Action)

Gavedu ensures that AI-enhanced features support all seven stages—from **goal formation** to **execution and feedback**.

Example: A user intending to watch an ad for internet credits is guided through selection (NLP-driven), interaction (ad engine), and real-time feedback (confirmation prompts), minimizing the gulf of execution and evaluation.

Cognitive Engineering Principles

AI facilitates **situation awareness** by predicting and highlighting relevant data. For instance:

- Threat alerts are not just triggered but **contextualized**—e.g., "Suspicious login from new location. Was this you?"
- Bandwidth usage stats are simplified into cognitive-friendly dashboards, reducing decision fatigue.

Calikli and Bener's Framework on Confirmation Bias in Intelligent Systems

Gavedu counters confirmation bias by dynamically surfacing **disconfirming evidence**—e.g., showing alternative ad paths or login suggestions when a user repeats suboptimal behavior. AI ensures the system does not reinforce unhelpful habits.

Trust Calibration Models

Behavior-based threat detection and adaptive prompts support **trust calibration** (i.e., the user's trust matches the system's reliability). Clear, explainable AI outputs prevent over-trust or disuse of features—both common pitfalls in HSI design (Drummond & Fischhoff, 2017).

Defining Cognitive Consonance

Cognitive consonance is the alignment between human mental models and AI system behavior, minimizing friction and maximizing trust (Picard, 1997). Gavedu achieves this via voice prompts, biometric login, and progressive disclosures—interfaces that speak the language of cognition.

ARCHITECTURE OF THE GAVEDU PM-WANI FRAMEWORK

Beyond Security: Scalable AI-Driven Systems

Gavedu provides voice feedback, multilingual UI, and visual cues, tailored to users with low digital literacy. Embedding cognitive load theory into design enables equitable access across user demographics (Kalyuga et al., 2003).

User-Centric Design

The BharatNet project catalyzes cognitive empowerment by providing high-speed internet to rural India. It enables access to education, digital services, and economic opportunities, aligning HSI with social cognition and equity.

The BharatNet project, a flagship initiative by the Government of India, is designed to provide **affordable high-speed broadband access to over 250,000 Gram Panchayats** (village-level governance bodies), bridging the rural–urban digital divide. More than a technological rollout, BharatNet represents a **cognitive infrastructure**—an enabler of learning, digital awareness, and economic participation for populations historically excluded from the digital economy.

This access catalyzes cognitive empowerment in multiple ways:

- *Education:* Rural students gain exposure to online content, virtual classrooms, and MOOCs, promoting memory retention, conceptual learning, and digital literacy.
- *Services:* E-governance portals, health platforms, and online banking reduce cognitive and logistical load for tasks previously requiring physical travel or third-party mediation.
- *Opportunities:* Small businesses and entrepreneurs use digital platforms to access markets, information, and financial tools, empowering decision-making and self-efficacy.

By providing the foundational access layer, BharatNet aligns with HSI by recognizing that technological systems must first account for access equity, usability, and contextual relevance to be effective.

BharatNet fits into the **macro-HSI layer,** which includes infrastructure, policy, and sociotechnical systems. It does not just affect individual users but reshapes the **ecosystem of access, inclusion, and human development**. Connectivity becomes the entry point to **system-wide human–system integration**.

Social Cognition and Collective Efficacy (Bandura, 1997), Access to information platforms boosts self-efficacy in individuals and collective efficacy in communities. For instance, farmers using BharatNet for weather updates or market prices act with greater confidence and collaborative potential, which are critical cognitive traits for rural resilience.

Cognitive Justice Framework (Visvanathan, 2009), BharatNet promotes cognitive justice by recognizing the right of all communities to both access and produce knowledge in their own cultural and linguistic contexts. Interfaces in regional languages, visual navigation, and voice-based portals align with inclusive HSI principles.

USE CASES AND APPLICATIONS

Digital Payments and Cognitive Trust

Gavedu streamlines ad-based payments and validates user sessions via blockchain and AI anomaly detection. This fosters cognitive trust by reinforcing safety and reducing ambiguity (Drummond & Fischhoff, 2017).

Fraud Prevention

AI models trained on behavioral anomalies can flag unusual user or advertiser behavior. Paired with human-centric alerts, this reduces cognitive overload and increases system transparency (Bureau & Praxis, 2023).

Adaptive Authentication

The system supports biometric and pattern-based login that adapts based on user history, reducing entry barriers for semi-literate users. Feedback loops promote learnability and confidence (Meske & Amojo, 2020).

COGNITIVE CONSONANCE AS A DESIGN PHILOSOPHY

Cognitive consonance shifts the design approach from human-in-the-loop to human-at-the-core. It promotes inclusivity, predictability, and emotional resonance in AI systems (Mohanani et al., 2017).

Behavioral Indicators

Metrics like time-to-connect, reauthentication frequency, and click-through rates are used to fine-tune the interface and content delivery (Nanda, 2023).

System Resilience

Human–machine synergy enables real-time network recovery and adaptive load balancing—essential for India's rural and semi-urban areas (Gupta & Balasubramaniam, 2024).

FUTURE DIRECTIONS

Future applications include:

- IoT-based cognitive mesh networks
- Neuro-symbolic AI for adaptive learning
- Quantum-secure blockchain systems

These systems must embody cognitive consonance to be accepted at the population scale (Kalyuga, 2007).

CONCLUSION

Gavedu PM-WANI exemplifies how HSI and cognitive science principles can create scalable, secure, and human-centered digital systems when fused with AI. Through cognitive consonance, it redefines public infrastructure to be both technologically advanced and emotionally and cognitively intuitive. This model sets a benchmark for future initiatives that strive for a balance between automation and human values.

REFERENCES

Bandura, A. (1997). Self efficacy: The exercise of control. New York: W. H. Freeman and Company.

Boy, G. A. (2023). Model-based human systems integration. In A. M. Madni & N. Augustine (Eds.), *Handbook of model-based systems engineering* (pp 471–499). Springer. https://doi.org/10.1007/978-3-030-97835-2_14

Bureau & Praxis. (2023). *The anatomy of a fraud report*. Bureau. www.bureau.id

Calikli, G., & Bener, A. (2014). Empirical analysis of factors affecting confirmation bias levels of software engineers. *Software Quality Journal*, 22(3), 387–405. https://doi.org/10.1007/s11219-014-9250-6

Drummond, C., & Fischhoff, B. (2017). Development and validation of the scientific reasoning scale. *Journal of Behavioral Decision Making*, 30(1), 26–38. https://doi.org/10.1002/bdm.1906

Gupta, H., & Balasubramaniam, V. S. (2024). Behavioral economics in UI/UX: Reducing cognitive load. *International Journal of Modern Engineering & Emerging Tech*, 12(12), 1–10. https://ijrmeet.org

Hanham, J. (2023). Integrating cognitive load theory with other theories. *British Journal of Educational Psychology*, 93(1), 239–250. https://doi.org/10.1111/bjep.12612

Kalyuga, S., Ayres, P., Chandler, P., & Sweller, J. (2003). The expertise reversal effect. *Educational Psychologist*, 38(1), 23–31. https://doi.org/10.1207/S15326985EP3801_4

Kalyuga, S. (2007). Expertise reversal effect in learner-tailored instruction. *Educational Psychology Review*, 19(4), 509–539. https://doi.org/10.1007/s10648-007-9054-3

Madni, A.M. & Sievers, M. (2013). Systems Integration: Key Perspectives, Experiences, and Challenges. Systems Engineering Journal, Vol. 17, Issues 1, pp. 37–51. https://doi.org/10.1002/sys.21249.

Mayer, R. E., & Moreno, R. (2003). Nine ways to reduce cognitive load in multimedia learning. *Educational Psychologist*, 38(1), 43–52. https://doi.org/10.1207/S15326985EP3801_6

Meske, C., & Amojo, I. (2020). Ethical guidelines for digital nudges. *arXiv preprint arXiv:2003.05249*. https://arxiv.org/abs/2003.05249

Mohanani, R., Salman, I., Turhan, B., Rodriguez, P., & Ralph, P. (2017). Cognitive biases in software engineering: A mapping study. *arXiv preprint arXiv:1707.03869*. https://arxiv.org/abs/1707.03869

Nanda, N. (2023). Cognitive load and screen design. *International Journal of Human-Computer Interaction and Data Mining*, 6(2), 1–10.

Picard, R. W. (1997). *Affective computing*. MIT Press.

Sweller, J. (1988). Cognitive load during problem solving: Effects on learning. *Cognitive Science*, 12(2), 257–285. https://doi.org/10.1207/s15516709cog1202_4

38 Dynamic Task Allocation in Human–Autonomy Teaming

Kahina Amokrane-Ferka

INTRODUCTION

Until now, systems have been designed to address users' needs, which is reflected in common systems engineering standards and handbooks (Haskins et al., 2007; Kapurch, 2010). In these systems, humans are considered an external component or just actors who act as a last line of defense against the consequences of events that exceed the design scope by interacting with the system through human–machine interfaces. Today, with continuous advancements in artificial intelligence (AI) and the emergence of autonomous systems, needs have evolved toward human–system collaboration and cooperation to achieve a common goal. Thus, the autonomous AI-based system must coexist with humans, carefully selecting the right type and appropriate level of automation. Therefore, some human tasks can be delegated to the AI-based system (Hoc, 2000; Boy, 2017). This system can perform more dynamic functions/ tasks in teamwork and taskwork than ever. AI-based systems are starting to be viewed as teammates rather than tools. For instance, they can increasingly participate in teamwork activities involving coordination, task reallocation, and continuous interaction with humans and other autonomous agents (Chen, 2016; O'Neill, 2023. This cannot be done by considering automation alone. Still, it must include a simultaneous assessment of changes like the operators' activity, their role, and the structure of their activity (Hollnagel, 2002). In this chapter, we will present some principles of human–autonomy teaming, the role an AI-based system can have in this team, and the evolution of interaction to cover the new objectives related to this form of collaboration.

HUMAN–AUTONOMY TEAMING: SOME PRINCIPLES

The human–system collaboration paradigm is known under several designations, such as human–autonomy teaming (HAT), man–machine teaming (MMT), man–machine partnerships, etc. These concepts describe humans and intelligent or autonomous agents working interdependently toward a common goal (Chen, 2016; Lyons, 2021). HAT has been described as at least one human working cooperatively with at least one autonomous agent, where the autonomous agent is a computer entity with

DOI: 10.1201/9781003666431-46

a partial or high degree of self-governance concerning decision-making, adaptation, and communication (O'Neill, 2020).

Lyons et al. (2021) clearly distinguish between automation and autonomy. Automation is doing something in place of a human that the human no longer needs to do. Indeed, the use of automation does not eliminate the need for a human; instead, it changes the role of a human, often to one of a supervisor of the automated system. However, several research studies have shown that automation usually introduces additional complexity that humans must manage, monitor, and control, even when their numbers are reduced. Autonomy allows machines to make decisions without humans in the control loop. Autonomy focuses on decision-making, adapting to changing demands, and improving performance over time with a certain level of self-governance.

Collaboration is the process by which agents work together to achieve a common goal by planning, assigning responsibilities, and coordinating. There are two main approaches related to how humans and computers can collaborate (Terveen, 1995):

- *Human Emulation Approach:* which requires giving the computer capabilities like those of a human to make human–computer collaboration as similar as possible to human–human collaboration.
- *Complementary Human Approach:* This approach aims to give the computer complementary capabilities, assign different roles and responsibilities to each agent, and use interaction techniques to facilitate communication between man and computer.

The principal question revolves around the role of AI within the human–AI team. Firstly, the changing role of humans is the result of automation or the integration of AI into their activities. The idea that automation replaces and/or offloads the human being, making their role less critical. Some studies have demonstrated the opposite. For example, Lee (2008), in his review of Parasuraman and Riley's (1997) paper on human–machine interaction, pointed out that, counterintuitively, the required learning level of humans increases when automation is introduced. Furthermore, they highlighted that the design of automation interfaces becomes more challenging and gains increased significance in such a context.

Furthermore, automation does not generally replace humans, but instead changes the nature of their work (Parasuraman and Manzey, 2010). For example, with the increasing cockpit automation, commercial pilots have abandoned their traditional role of direct aircraft management for a more critical role of automation supervision. Pohl's (1997) research identified areas of assistance to the human–computer partnership in decision support systems and concluded with design guidelines centered on assisting rather than replacing the user.

According to Dubey et al. (2020), four roles have been identified:

- *A personal assistant* is a task-oriented assistant that helps a human team member and can also provide them with personalized help for their task.
- *An associate (or teammate)*: an assistant capable of performing cognitive tasks like a human teammate.

- *A team-working facilitator* is a coordination-oriented assistant who supports communication, facilitates human coordination, and focuses on attention.
- *A collective moderator:* an assistant capable of moderating the ad hoc collectives comprising traditional teammates, and temporary new team members acquired through social media and crowdsourcing platforms.

The nature of tasks an AI can accomplish in collaboration with humans is also fundamental in this emerging field. In Fabri et al. (2023), the authors distinguished five types of thinking, called cognitive functions (see next section). In Dellermann (2021), the authors highlighted four task types:

- *Action:* Action tasks consist of operational tasks (physical or cognitive) that must be done by an agent (human or machine) to reach a goal. It usually changes something in the world and takes time to complete. It can be simple (elementary action) or complex (decomposable into subtasks or actions).
- *Prediction:* These tasks aim to predict future events based on previous data, such as stock prices or market dynamics.
- *Reasoning:* These tasks focus on understanding data by inductively building (mental) models of a particular phenomenon, making solving complex problems with small data possible.
- *Recognition:* defines tasks that recognize objects, images, or natural language.

It is worth noting that the type of task or mission of the AI depends on its level of autonomy (Madni & Madni, 2018), where the authors distinguished three levels:

- *Fully autonomous systems* are characterized as "human-out-of-loop" systems, designed to execute a limited set of tasks. In other words, humans cannot intervene in this real-time system. For example, order-picking systems.
- *Supervised autonomous systems,* also known as "human-on-the-loop" systems, are systems in which humans can intervene in real time in the system's operations in progress. Such systems are in demand in both the defense and commercial sectors. Robot vacuum cleaners, for example.
- *Semi-autonomous systems* are also known as "human-in-the-loop" systems. In this system, the machine waits for human intervention before taking control. Examples include automated or semi-automated vehicles.

Several researchers have delved into taxonomy in this field. The taxonomy proposed by Dellermann (2021) can be mentioned. It describes various features related to interactions between a human and an AI-based system, the mechanisms through which they learn, and the classification of task characteristics. The second noteworthy taxonomy is that of Dubey et al. (2020), in which they focus particularly on task characteristics, learning paradigms, and team properties, then present a proof-of-concept implementation of their Human–AI Collaboration (HACO) solution architecture. In the same context, the framework proposed by Caldwell et al. (2022) covers the areas of bidirectional situational awareness, human–AI interaction, intelligent decision-making, human–AI operations, and the transferability of safe development environments to real-world contexts.

ROLE ATTRIBUTION BETWEEN HUMAN AND AI-BASED SYSTEMS IN HAT

The level of automation has fundamental implications in changing the nature of the operator's activity and human–machine interaction in human autonomy teaming. It is, therefore, necessary to have practical principles and methods that can be used to allocate functions between humans and the AI-based system. This must be done so that the entire human and machine system behaves optimally. This requires a set of design concepts and methods that integrate the best from human factors, cognitive systems engineering, and control theory (Bye, 1999; Rauffet, 2022).

Today, the machine is not just a tool; it is expected to take on the role of teammate. This notion of team goes beyond the classic human–machine combination, which emphasizes the tool and the action; we move to an abstraction associating humans and autonomous systems, which emphasizes the pursued goal and the prescribed task. From a human-centered perspective, the machine (or system) takes its share of the work and becomes an individual subject. It is no longer there to serve as a substitute for human agents; replacements always remain unsatisfactory, because they are never complete (Zhang et al., 2020).

Some researchers believe we must move toward a complementary relationship between "human judgment" and AI algorithms by reversing this partnership. Rather than seeing machines as providers of data to be used by humans to make decisions or judgments about a situation, they propose that human judgment and "intuitions" should serve as data for AI algorithms. Then, they consider that all tasks that can be automated should be delegated to machines, even if it means taking humans "out of the loop." To justify this division of roles between human and machine, which they see as a "partnership," in addition to improving the performance of computing machines, the authors also point out the limits of human cognition (McAfee & Brynjolfsson, 2017; Zouinar, 2020).

Another approach is distributing tasks between humans and machines according to their strengths and weaknesses (McAfee & Brynjolfsson, 2017). This approach is based on the following principle: if a machine outperforms a human in a specific function, the function must be automated; if this is not possible, automation makes no sense (Winter & Dodou, 2014). For example, it states that, on one side, humans can detect tiny variations in visual or acoustic stimuli, improvise or use flexible procedures, or reason deductively. Conversely, machines can respond quickly to signals, perform repetitive tasks, reason deductively, and process complex operations. This approach has been widely questioned. One of the problematic points in this approach concerns the role of humans in decision-making. As underlined in the paper, it is not necessary, and is even preferable, to put the human "out of the decision loop" when the system is more efficient.

For Shively et al. (2018), three principles should guide the design of the human–machine teaming: bidirectional communication, transparency, and dynamic delegation.

Joint task execution is made possible by the first two principles; maintaining the human operator's situational awareness derives from the third principle. This third principle must also consider the workload and time pressure to propose a more flexible work allocation. The autonomous system will then be entrusted with tasks within its area of competence.

The concept of hybrid intelligence aims to combine the strengths of human intelligence and AI to achieve superior performance compared to either one considered separately (Dellermann, 2019). This approach seeks to enhance human thinking rather than replace it or reinforce human biases. The hybrid intelligence developed by Akata et al. (2020) identified four research challenges for implementing effective human–machine interaction:

- collaborating synergistically with humans,
- learning and adapting to human needs and the environment,
- behaving responsibly and ethically,
- moreover, explaining awareness, goals, and strategies.

Recently, Fabri et al (2023) have been interested in interworking human agents and AI-enabled systems that characterize human–AI hybrids, when human agents and AI-enabled systems closely collaborate. The authors propose a taxonomy of human–AI hybrids in this work, including sociomaterial entities, dimensions, and characteristics. It comprises three distinct sociomaterial entities: the human (human agency), the AI (material agency), and sociomaterial practices, and nine dimensions: human cognitive functions, interaction human to AI, human focus, AI cognitive functions, interaction AI to human, AI focus, form of interworking, and learning. The key cognitive functions of human agents are perceiving, reasoning, predicting, planning, decision-making, explaining, interacting, creating, and empathizing. Moreover, perceiving, reasoning, predicting, planning, decision-making, interacting, and creating are relevant (current) cognitive functions of AI-enabled systems.

SOME PRINCIPLES FOR SUCCESSFUL TASK ALLOCATION IN HAT

For a successful collaboration, humans must understand the decisions of the system. On the other hand, the system must understand humans (their behavior and characteristics), and this system must be designed to reason or to behave like a human. For example, in the context of automated driving, the car needs to understand how humans behave during the handover and to act so that the human can understand its subsequent actions. The human and the system must have the same reference model describing the tasks performed (Amokrane-Ferka, 2022). This reference model is known as the task model in ergonomics. It is based on ergonomist activity analysis languages (Amokrane, 2010). In Amokrane-Ferka (2022), the authors present how they extended the SysML language to consider task description language concepts. First, they proposed a meta-model with a principal component of task description language based on a particular representation of human–system cooperation. Then they added new concepts such as the distinction between human agent and system agent, and they distinguished three types of tasks: collaborative, cooperative, and individual. They finally proposed a design-specific language by adding several stereotypes. For example, a task stereotype from SysML can inherit from the concepts of activity because each task and activity describes the behavior of an agent or a block. Constructor stereotype: We use three existing activity diagram concepts to describe temporal and logical relations between subtasks of a complex task. For the sequential (serial) (SEQ), we use the rational organization of the activity diagram.

From the same perspective, one specific challenge in implementing any human–automation interaction system lies in creating an underlying representation for the user and the automation to communicate tasks, resources, and intent (Miller, 2005). The authors are interested in how the elements of the activity model can be shared between humans and machines and how humans can control this sharing. To do so, they identified a set of everyday tasks, grouping them into plays, and parameterizing elements such as time and location, so that a set of play templates can be created. Play is to be executed, so the operator can select a play template and bind the parameter values as appropriate to his/her needs. The operator and the automation have a similar model of the sequence of tasks.

Human–automation interaction has long been recognized as a critical component of automation design, and guidelines have been developed for specific domains, e.g., aerospace (Billings, 1996), robots, and unmanned systems (Chen, 2014), human–AI interaction (Amershi et al., 2019 , Li et al., 2023). In the last one, the authors separate four levels. The interaction level contains the following guidelines:

- Time services are based on context, which is related to determining the appropriate moment to act or interrupt by considering the user's current task and environment.
- Show contextually relevant information (display information pertinent to the user's current task and environment).
- Match relevant social norms (ensure the experience is delivered in a way that users would expect, given their social and cultural context).
- Match relevant social norms (ensure the experience is delivered in a way that users would expect, given their social and cultural context).

Other research focuses on the factors influencing co-comprehension between man and machine. In other words, one approach concerns the possibility of giving machines access to eminently human elements to be processed to enrich human–machine interaction (Metge et al., 2022; Le Blanc & Kaltenbach, 2016). This is notably the case for emotions, which we find in many works on HMIs (Xiao et al., 2020), which propose a very formal modeling of human emotional signals. Another approach focuses on comparisons between machines when faced with a specific usage activity carried out by people. Strayer et al. (2017), e.g., compare the influence of different smartphones' voice control on users' mental load in a driving task context. These comparisons between solutions or proposed machines have led to new concepts for evaluating machine performance. For machine intelligence quotient, MIQ would enable humans to know what they are dealing with (Miller, 2019).

Today, the scientific literature is exploring new interaction modalities and information sharing between human and AI systems. In Flathmann et al.s (2020), the authors proposed a framework for human–agent interaction. This framework details interactions in HAT, which future groupware developers and researchers can utilize to ensure groupware systems fully support HATs. This framework has split team functionality between human, agent, and support functionality by considering functional roles and AI limitations. This framework consists of three layers. *Human Teammate layer* that illustrates problems and solutions studied in the human-to-human communication research area. For example, many different modalities and contexts can convey information. These teams move toward shared cognitive constructs through communication, which may occur in physical and virtual settings using verbal or nonverbal methods. Nonverbal communication includes gestures, smiling or frowning, widening eyes, and many more. The level of complexity in human communication and successful teamwork makes integrating the two in HAT a considerable challenge. *Support Technologies Layer* that details the interactions that groupware needs to facilitate at the current stage of human–agent teaming. This layer is designed based on detected compartmentalization that allows humans and agents to work together seamlessly and allows the implementation of various features offered by the groupware. Data-oriented communication created by human and artificial teammates is leveraged to support the team through monitoring and coordination systems. Both these systems serve as supporting technology and enhance the interaction between human and artificial teammates. *Autonomous Agent Layer* shifts toward explainable systems that may use multiple agents rather than a single, black-box system. For instance, regarding a team that manages a drone system, individual worker agents may oversee monitoring a single quadrant of a map, and the large agent would interpret the four quadrant worker agents to complete their goal of analyzing the entire map. Thus, it is also not enough for groupware to allow this clustering; the software must also allow for the dynamic nature of these clusters as their composition, and therefore interactions, will change based on the required workload. This layer will allow agents to learn from other agents and humans, making them better team members throughout their team's existence. Human interaction could also contribute to transparency, a critical factor in building trust, and is vital to efficient HAT. Without these agent designs, trust in artificial teammates could be lost, and, as a result, overall team performance could see significant drops.

The following are some concrete examples of HAT applications focusing on human–AI task allocation and corresponding interactions.

Miller et al. (2005) developed an architecture for machine-based delegation systems based on a sports team's "playbook" metaphor. This approach streamlines delegation by offering a compiled set of plans, or "plays," with short, easily commanded labels that can be further modified. Wherein the operator can elicit these plays to improve performance during high workload mission segments, they contain predefined behavior patterns understood by all team participants (including the automation). In terms of HMI, this playbook UI comprises five primary regions that appear after the user has interacted with it to develop a fully executable airfield denial mission. The first is a mission composition space showing the present plan composition. The second one is an available resource space, in their use case, presenting the set of aircraft available for use. The third one contains an interactive context zone (terrain map) of the area of interest, facilitating interactions with significant geographic content. The space between these windows (empty at the beginning) is a resource in use space; once resources (UAVs, munitions, etc.) are selected, they will be moved to this workspace, where they can be interacted with in more detail. Finally, the lower set of control buttons is present for interaction. This includes options such as "finish plan" for handing the partial plan off to the APC for completion and "show schedule" for obtaining a Gantt chart timeline of the activities planned. The Playbook paradigm has been used successfully to simulate UAV flights and has been used in simulation and field testing of UAV automation for the U.S. Army. In a recent experiment, Miller et al. (2007) examined the effects of nonoptimal Playbook solutions on operator performance. The advantage of adjustable systems is that the machine never usurps the decision authority; the disadvantage is that in time-constrained and high workload environments, management of automation allocation decisions may be too time-consuming or too complex to be practical.

Cummings et al. (2012) advised exercising caution in using visualization tools for mixed-initiative multivariate path-planning tasks, particularly when dealing with cost functions that exhibit significant variability in their sensitivity to changes in plan variables. Operators' decision-making process may be susceptible to errors, especially when managing more sensitive functions, as determining the accurate costs of various plan options becomes challenging. An experiment was conducted to assess the impact on operator performance empirically and involved a prototype path planner under diverse conditions. Participants were tasked with creating and optimizing paths, considering increasingly complex path cost functions. Various map visualizations, including a novel representation based on a numerical potential field algorithm, were employed during the experiment.

Additionally, participants were required to plan paths under conditions of degraded automation. A prototype path planning interface, Planetary Aid for Traversing Humans (PATH), was developed in this context. The PATH allows users to create, adjust, and submit perceived least-cost paths for geospatial path planning on the Moon. Participants are provided with a lunar terrain map featuring designated start and goal locations, marking the beginning and end of a path, along with a critical way area. This vital area, akin to a lunar exploration site, signifies an area of particular interest for exploration. The total path cost, determined by a predefined cost function, is computed upon constructing a comprehensive path. The results are then presented in a tabular and graphic format. This graphical representation includes a path cost profile, illustrating the cost along the path, and a path elevation profile, detailing the elevations encountered throughout the path. Path elevation becomes crucial considering the significance of minimizing ascents while walking on the Moon. Participants can refine path solutions by adjusting the position of waypoints, adding new ones, or removing existing ones. The PATH program diligently records participant interactions by storing every user mouse-click along with corresponding time stamps and path costs, thereby tracking the evolution of the path planning process (Cummings et al., 2012).

Similarly, the Cockpit Assistant Bidirectional (CAB) project at SystemX aims to define and evaluate an intelligent cockpit integrating a bidirectional assistant that will augment operators' capabilities in real-time in complex and/or atypical situations. This research project aims to allow a human operator to collaborate with an AI-based virtual assistant in the decision-making process and the learning and knowledge formalization process, while integrating human factors. The human–machine interaction was studied in this project, and a generic IHM framework was proposed (Zouinar, 2023). This framework has four zones: *notification center*, through this zone, the CAB assistant gathers and communicates all the events coming from the simulators of the real work environment (like, e.g., Grid2Op (Marot et al., 2022) to simulate electric grid, or XPlane 11 to simulate aviation cockpit with its environment), and/or tasks to be processed by the operator. The context zone allows the visualization of spatial contextual information through a map or graph that localizes the event in space. Timeline: allows the temporal visualization of contextual information. It will enable a chronological display of events and some recommendations-related information. Assistance and exchange zone: the CAB assistant communicates and displays through this zone, particularly the proposed recommendations and the associated choice criteria. The operator can request, answer, correct, or share information with the CAB assistant through this zone.

CONCLUSION

With this state-of-the-art technology that depicts the progress made in the field of HAT, it is undeniable that there is still much work to be done. The identified challenges and gaps underscore the need for continuous research and development to advance in this domain. Additional efforts are required to address unresolved issues, explore new avenues, and enhance existing solutions, especially in collaboration/cooperation, sharing tasks/functions, or control between human and AI-based systems. Considering the opportunities and challenges that arise, long-term commitment and collaboration among researchers in different domains (HMI, IA, human factors, system engineering, HSI) remain essential to make meaningful strides in advancing this field.

REFERENCES

Akata, Z., Balliet, D., de Rijke, M., Dignum, F., Dignum, V., Eiben, G., Fokkens, A., Grossi, D., Hindriks, K., Hoos, H., Hung, H., Jonker, C., Monz, C., Neerincx, M., Oliehoek, F., Prakken, H., Schlobach, S., van der Gaag, L., van Harmelen, F., … Welling, M. (2020). A research agenda for hybrid intelligence: Augmenting human intellect with collaborative, adaptive, responsible, and explainable artificial intelligence. *Computer*, 53(8), 18-28. https://doi.org/10.1109/MC.2020.2996587

Amershi, S., Weld, D., Vorvoreanu, M., Fourney, A., Nushi, B., Collisson, P., Suh, J., Iqbal, S., Bennett, P. N., Inkpen, K., et al. (2019). Guidelines for human–AI interaction. In *Proceedings of the 2019 CHI Conference on Human Factors in Computing Systems* (pp. 1–13).

Amershi, S., Weld, D., Vorvoreanu, M., Fourney, A., Nushi, B., Collisson, P., & Horvitz, E. (2019). Guidelines for human-AI interaction. In *Proceedings of the 2019 chi conference on human factors in computing systems* (pp. 1–13).

Amokrane, K. (2010). Suivi de l'apprenant en environnement virtuel pour la formation à la prévention des risques sur des sites Seveso. octoral dissertation, Compiègne.

Amokrane-Ferka, K., & Hein, A. M. (2022, February). A meta-model for integrating human factors into system engineering. In *INCOSE International Symposium* (Vol. 32, pp. 118–130).

Billings, C. E. (1996). *Human-centered aviation automation: Principles and guidelines* (No. NASA-TM-110381).

Boy, G. A. (2017). *The Handbook of Human–Machine Interaction: A Human-Centered Design Approach*. CRC Press. eBook ISBN: 9781315557380.

Bye, A., Hollnagel, E., & Brendeford, T. S. (1999). Human–machine function allocation: A functional modelling approach. *Reliability Engineering & System Safety*, 64(2), 291–300. https://doi.org/10.1016/S0951-8320(98)00069-6

Caldwell, S., Sweetser, P., O'Donnell, N., Knight, M. J., Aitchison, M., Gedeon, T., ... & Conroy, D. (2022). An agile new research framework for hybrid Human–AI teaming: Trust, transparency, and transferability. *ACM Transactions on Interactive Intelligent Systems (TiiS)*, 12(3), 1–36.

Chen, J. Y., Procci, K., Boyce, M., Wright, J. L., Garcia, A., & Barnes, M. (2014). Situation awareness-based agent transparency.

Chen, Y., Mao, X., Hou, F., Wang, Q., & Yang, S. (2016). Combining re-allocating and re-scheduling for dynamic multi-robot task allocation. In *2016 IEEE International Conference on Systems, Man, and Cybernetics (SMC)* (pp. 000395-000400). IEEE.

Cummings, M. L., Marquez, J., & Roy, N. (2012). Human-automated path planning optimization and decision support. *International Journal of Human–Computer Studies*, 70, 116–128.

Dellermann, D., Calma, A., Lipusch, N., Weber, T., Weigel, S., & Ebel, P. (2021). The future of human–AI collaboration: A taxonomy of design knowledge for hybrid intelligence systems. *arXiv preprint arXiv:2105.03354*.

Dubey, A., Abhinav, K., Jain, S., Arora, V., & Puttaveerana, A. (2020, February). HACO: A framework for developing human–AI teaming. In *Proceedings of the 13th Innovations in Software Engineering Conference on Formerly known as India Software Engineering Conference* (pp. 1–9).

Fabri, L., Häckel, B., Oberländer, A. M., Rieg, M., & Stohr, A. (2023). Disentangling Human-AI hybrids: Conceptualizing the interworking of humans and AI-enabled systems. *Business & Information Systems Engineering*, 65(6), 623–641.

Flathmann, C., Schelble, B., Tubre, B., McNeese, N., & Rodeghero, P. (2020, November). Invoking principles of groupware to develop and evaluate present and future human–agent teams. In *Proceedings of the 8th International Conference on Human–Agent Interaction* (pp. 15–24).

Haskins, C., Forsberg, K., & Krueger, M. (2007). *INCOSE Systems Engineering Handbook*. International Council on Systems Engineering INCOSE.

Hoc, J. M. (2000). From human–machine interaction to human–machine cooperation. *Ergonomics*, 43(7), 833–843. https://doi.org/10.1080/001401300409044

Hollnagel, E. (2002). Time and time again. *Theoretical Issues in Ergonomics Science*, *3*(2), 143-158.

Kapurch, S. (2010). *NASA Systems Engineering Handbook*. NASA.

Le Blanc, B., & Kaltenbach, M. (2016). Knowledge "Utilisateur" and knowledge "Usager": The need for an evolving assistance. In *European Conference on Intangibles and Intellectual Capital* (p. 159–163). Academic Conferences International Limited.

Lee, J. D. (2008). Review of a pivotal human factors article: "Humans and automation: Use, misuse, disuse, abuse." *Human Factors*, 50(3), 404–410.

Li, T., Vorvoreanu, M., DeBellis, D., & Amershi, S. (2023). Assessing human–AI interaction early through factorial surveys: A study on the guidelines for human-AI interaction. *ACM Transactions on Computer–Human Interaction*, 30(5), 1–45.

Lyons, J. B., Sycara, K., Lewis, M., & Capiola, A. (2021). Human–autonomy teaming: Definitions, debates, and directions. *Frontiers in Psychology*, 12, 589585.

Madni, A. M., & Madni, C. C. (2018). Architectural framework for exploring adaptive human–machine teaming options in simulated dynamic environments. *Systems*, 6(4), 44.

Marot, A., Donnot, B., Chaouache, K., Kelly, A., Huang, Q., Hossain, R.-R., & Cremer, J. L. (2022). Learning to run a power network with trust. *Electric Power Systems Research*, 212, 108487.

McAfee, A., & Brynjolfsson, E. (2017). *Machine Platform Crowd: Harnessing our Digital Future*. New York: W.W. Norton and Company.

Metge, A., Maille, N., & Le Blanc, B. (2022, June). Décision assistée par un système intelligent: de l'utilisation de l'IA à l'intention de l'opérateur dans la replanification. In *Conférence Nationale en Intelligence Artificielle 2022 (CNIA 2022)*.

Miller, C., Funk, H., Wu, P., Goldman, R., Meisner, J., & Chapman, M. (2005, September). The Playbook™ approach to adaptive automation. In *Proceedings of the human factors and ergonomics society annual meeting* (Vol. 49, No. 1, pp. 15-19). Sage CA: Los Angeles, CA: Sage Publications.

Miller, C. A., & Parasuraman, R. (2007). Designing for flexible interaction between humans and automation: Delegation interfaces for supervisory control. *Human factors*, *49*(1), 57-75.

Miller, A. (2019). The intrinsically linked future for human and Artificial Intelligence interaction. *Journal of Big Data*, 6(1), 38.

O'Neill, T. A., Flathmann, C., McNeese, N. J., & Salas, E. (2023). 21st Century teaming and beyond: Advances in human–autonomy teamwork. *Computers in Human Behavior*, 147, 107865.

Parasuraman, R., & Manzey, D. H. (2010). Complacency and bias in human use of automation: An attentional integration. *Human Factors*, 52(3), 381–410.

Parasuraman, R., & Riley, V. (1997). Humans and automation: Use, misuse, disuse, abuse. *Human Factors*, 39(2), 230–253.

Pohl, J. (1997, January). Human-Computer Partnership in Decision-Support Systems: Some Design Guidelines. In *Proceedings of Intersymp 1997 the 9th International Conference on Systems Research Informatics and Cybernetics Baden Baden Germany*.

Rauffet, P. (2022). Tools and methods for Human–Autonomy Teaming: Contributions to cognitive state monitoring and system adaptation. *arXiv preprint arXiv:2212.01435*.

Shively, R. J., Lachter, J., Brandt, S. L., Matessa, M., Batste, V., & Johnson, W. W. (2018). Why human–autonomy teaming? In *Advances in Neuroergonomics and Cognitive Engineering: Proceedings of the AHFE 2017 International Conference on Neuroergonomics and Cognitive Engineering, July 17–21, 2017, The Westin Bonaventure Hotel, Los Angeles, California, USA 8* (pp. 3–11). Springer International Publishing.

Strayer, D. L., Cooper, J. M., Turrill, J., Coleman, J. R., & Hopman, R. J. (2017). The smartphone and the driver's cognitive workload: A comparison of Apple, Google, and Microsoft's intelligent personal assistants. *Canadian Journal of Expérimental Psychology/Revue canadienne de psychologie expérimentale*, 71(2), 93.

Terveen, L. G. (1995). Overview of human–computer collaboration. *Knowledge-Based Systems*, 8(2–3), 67–81.

Winter, J.C.F. (de), & Dodou, D. (2014). Why the Fitts list has persisted throughout the history of function allocation. *Cognition, Technology & Work*, 16(1), 1–11.

Xiao, G., Ma, Y., Liu, C., & Jiang, D. (2020). A machine emotion transfer model for intelligent human–machine interaction based on group division. *Mechanical Systems and Signal Processing*, 142, 106736.

Zhang, R., McNeese, N. J., Freeman, G., & Musick, G. (2020). "An ideal human" expectations of AI teammates in human–AI teaming. *Proceedings of the ACM on Human–Computer Interaction*, 4(CSCW3), 1–25.

Zouinar, M. (2020). Évolutions de l'Intelligence Artificielle: quels enjeux pour l'activité humaine et la relation Humain-Machine au travail?. *Activités*, (17-1).

Zouinar, M., Amokrane-Ferka, K., Rousseaux, V., & Vallet, F. (2023, October). Conception d'un assistant bidirectionnel générique à base d'IA pour la supervision de réseaux et le pilotage d'aéronefs. In *Ergo'IA 2023*.

39 Designing Engineering Copilots for Industrial Systems and Process Engineering

Peter Denno, Carolyn Psenka, Nicolas Lelouche, Rachael Sexton, Jaemun Sim, and Kyoung-Yun Kim

INTRODUCTION

Large portions of industry are being left out of the digital revolution because the barriers to learning to use complex tools are too high. The consequences of this are myriad. Small manufacturers, for example, are significantly less capitalized than large manufacturers on a per capita basis (Armstrong, Berger & Bonvillian, 2021). One reason for this may be that, for small manufacturers, the risks of investment are too high. It is particularly debilitating for small manufacturers to invest in the use of a complex tool only to find it unfit for the purpose they intended, or that becoming skilled in its use takes too long. Though manufacturers typically have deep expertise in their production processes, they may be unfamiliar with *supporting technologies* that have become increasingly possible with digitalization, such as digital twins and decision support systems. Learning to use the complex tools of supporting technologies can be challenging.

In this chapter, we consider the design of *DSL-based engineering copilots* to reduce barriers to the use of supporting technologies. Domain-specific languages (DSLs) are formal languages designed to facilitate a specific task in a particular domain.[1] Copilots are artificial intelligence (AI)-based systems that assist users within compatible applications using large language models (LLMs) and conversational interfaces (NIST, 2024a). Engineering copilots are copilots that employ the machine-mentors-human teaming model in a relationship that spans the lifetime of an engineering endeavor. The principal goal of an engineering copilot is to enable teaming on complex engineering tasks by bridging the gap between persons inexperienced in a discipline and the tools and practices of that discipline, and to do this in ways that are trustworthy and empowering. DSL-based engineering copilots use DSLs to model concepts in the users' world and capture user requirements in long-running relationships with the users' goal problem.[2] A quintessential example of long-running relationships is the optimization of manufacturing processes and operations. Manufacturing is a knowledge-intensive activity where process and operations

DOI: 10.1201/9781003666431-47

optimizations often translate into profitability and therefore also to the survival of the enterprise. Manufacturing processes and operations are typically improved little by little over months or years. Measurement, experimentation, and theorizing about causal relationships is commonplace in the effort to optimize.

Work with DSL-based copilots could begin with an introduction to the given supporting technology using AI-mediated conversation to elicit from users' expressions of their needs and perspectives. From this start, a human/AI collaborative learning environment emerges. Because the collaboration is captured in a shareable, DSL-based work product, the learning can be viewed as constructionist (Papert & Harel, 1991) (i.e., "learning by making").

The challenges discussed in the chapter can be divided into two broad areas (1) system design for trustworthy use, and (2) DSL design for constructionist learning and collaboration. The trustworthiness challenge concerns how expertise is harnessed to ensure that the DSL-based product of the collaboration is fit for purpose.[3] Teaming presents challenges to verification and validation (V&V) of the solution of the goal problem. V&V is the essential measurement science of systems engineering. The problem being solved in engineering copilots is typically a difficult one; it is unlikely to be something solvable by a programming copilot. For example, solutions to the problems of scheduling manufacturing production or integrating a digital twin into a production facility typically depend on analyses of complex requirements. The systems engineering of complex systems requires input from multiple engineering disciplines and the use of viewpoints specific to those disciplines. Accordingly, the chapter's discussion of trustworthy engineering copilot design emphasizes the verification of *corroborating models* generated in the analysis of requirements. The discussion of this challenge leans heavily on the systems engineering body of knowledge.

In contrast, the challenge of DSL design for constructionist learning and collaboration is widely transdisciplinary. The principal goal here is to provide an environment where users feel empowered, where they can gain control over the goal problem.[4] The assumption in using engineering copilots, however, is that the users may not be familiar with supporting technology and the discipline central to their goal problem. Broadly speaking, the challenge is one of establishing common ground at the intersection of the users' world, the technology's capabilities, and the consensus best practice in use of the DSL-based tool, which may be established by a community of practice.

There is much more to designing an engineering copilot than just designing the DSL.[5] However, as the section "Design of DSLs for Engineering Copilots" argues, characteristics of the DSL determine, to a large degree, how easily users are able to express their problem in formal terms (in the DSL) and navigate the relationships between the DSL and the body of knowledge that inspired it (see Figure 39.2). The DSL is what makes constructionist learning tractable. Incidentally, we have not created a new DSL in our exploratory implementation; we illustrate points with an existing DSL, MiniZinc (Stuckey, Marriott & Tack, 2018), that has many of the beneficial attributes discussed in the section "Design of DSLs for Engineering Copilots."

We consider two instances of the machine–mentors–human teaming model: In *adaptive integration mentoring*, a supporting technology is being integrated into operations. For example, suppose a new machine tool and its digital twin are being integrated into a manufacturer's production system. The manufacturer of the machine tool may provide a basic digital twin, but for best results, idiosyncrasies of the operating environment (e.g., process details, workpiece material, and supervisory system interface) must be considered. Adaptive integration mentoring in this setting is intended to help bridge this gap from the basic digital twin to a more effective one adapted to its actual operating environment.

In *formulation mentoring*, the supporting technology is analytical software capable of solving a particular class of problems relevant to the user. Formulation mentoring is joint work towards the conception and refinement of an analytical (e.g., physics-based) model. Here, mentoring involves (1) demonstrating how the goal problem posed by users (analysts) can be solved by the software and (2) providing users with the conceptual foundation to solve a range of similar problems. With this foundation, the AI mentoring agent and users work together on the users' problem, for example, the development of a production scheduling system for their manufacturing operations. A goal of the formulation copilot is that, through the experience of collaborating with the machine agent, the analyst becomes sufficiently familiar with the subject technology that they might see opportunities for its use elsewhere. It is also valuable that they see the shortcomings of the technology in particular situations. An example of formulation mentoring used throughout this chapter is the formulation of production scheduling system solutions using the MiniZinc DSL and solvers for combinatorial optimization problems.

Formulation mentoring is distinct from adaptive integration mentoring in that formulation mentoring does not involve an established artifact of interest (such as a digital twin), but rather may refer to entirely unanticipated system components in bespoke formulation. For example, a formulation copilot might be used to model production scheduling of beer brewing without having prior models of fermentation tanks; these models would be conceived in conversation with users.

Both mentoring types support the formulation and maintenance of a solution objectified in the language of copilot's DSL. For example, where the problem being addressed is production scheduling, the solution may be objectified in MiniZinc. From the perspective of social interactions, the DSL-based solution serves to anchor discussion and highlight facets of the approach applied. From the perspective of achieving the technical goal, the DSL-based object serves as formal requirements to meeting users' goals with the supporting technology. Further, the history of modifications to the DSL-based object may serve to describe the evolution of thought about the goal problem.

The remainder of the chapter is organized into four sections. The section "Agency, Verification, and Validation" discusses the trustworthiness challenge and reviews work in agency and the use of corroborating models for V&V. The section "Chatbots, Constructionism, and DSLs" discusses DSL design and reviews work in educational chatbots, communities of practice (CoPs), and learning. This section makes the argument for the use of DSLs in a constructionist learning environment. The section "Engineering Copilot Design Example" discusses an engineering copilot design example informed in part by our experience implementing a MiniZinc scheduling copilot and in part by the precepts developed in the sections "Agency, Verification, and Validation" and "Chatbots, Constructionism, and DSLs". The last section concludes the chapter.

AGENCY, VERIFICATION, AND VALIDATION

Trustworthy use of engineering copilots requires mentoring reliably and producing work products fit for purpose. This section focuses on methods to ensure work products are fit to purpose, specifically, V&V. There are two principal means by which the design of engineering copilots can safeguard against shortcomings of state-of-the-art generative AI in trustworthiness. The first is simply not to rely on generative AI for risk-sensitive design tasks. For example, in our scheduling copilot, the MiniZinc solution is predominantly realized from parametric templates of solutions that were designed by a human expert. In this arrangement, the role of the copilot is to elicit requirements so as to classify the users' problem and choose the correct template to instantiate.

The second means toward trustability is to generate models intended to corroborate expressed requirements. Corroborating models, which can be checked by humans, might include visual models such as the systems modeling language SysML, or a business process modeling language. They might also be executable simulations of some sort. This section of the chapter discusses corroborating models and means of their construction.

Engineering uses many methods of verification, and its processes are generally mature; what is not at all mature is the verification of results from LLM-based agents for use in engineering processes. Key tasks in the engineering of complex systems include eliciting requirements, representing them, allocating them to components, designing and testing the components, and validating the resulting system. We assume that all of these tasks could be part of the work performed with an engineering copilot. Repeatable and maintainable methods, the mainstay of complex system verification, include documented design processes, validation of analytical tools used, and formal methods of specification. Associated with each of these methods are means to dissect and inspect the decision process before using it in practice. It is the absence of this ability to dissect and inspect before use (we will call it *previewability*) that is the greatest contrast with how LLM-based results are typically used.

The absence of previewability in the use of LLMs is mostly owing to LLM's emergent attributes of agency, attributes that are also the greatest strength of LLM-based problem solving. Psychological and philosophical conceptions of agency entail conscious intention and adaptive means to achieve goals (Schlosser, 2019). The emerging capabilities of AI accentuate the notion that non-human components of sociotechnical systems may have agency. Flexibility in assigning agency is emerging as a powerful organizing paradigm in LLM-based architectures (Suzgun, & Kalai, 2024). LLM-based agents use *system instructions* that allows developers to describe, for subsequent chat interaction, the role and nature of the agency of the artificial agent. Defining artificial agents by means of a system instruction is a useful tool for tasks such as testing, focused problem solving, and agent orchestration. Both human and artificial agents are adaptive problem solvers, interacting with other system components by being delegated to serve some purpose or perform some task. The agent's ability to adapt to unforeseen circumstances is what makes delegation feasible. LLM-based dialog, for example, would not be possible were things otherwise. However, delegation is the antithesis of previewability. The need for more rigorous verification that the intent of supervising elements is being carried out is especially needed with LLM-based agents, owing to LLM's tendency to confabulate (NIST, 2024b) and to assert confidently even when wrong.

As a means to verify agent actions, seeking explanations from LLM-based agents is problematic. Ostensibly, explainable AI (XAI), the study of AI's ability to provide rationale, would provide methods with this ability. However, explanations from XAI techniques tend to explain in terms of the training set and algorithmic mechanisms that produced the result (Zhao et al., 2023). This is entirely understandable from the standpoint of the goals of analysts training the AI system; you might wonder how or why things would be otherwise. After all, AI system developers use explanations of bad behavior to guide refinement of their training set and algorithmic mechanisms. However, users of copilots are more likely to value explanations that highlight notions they expressed and the plausibility of alternatives. Compounding the challenge is the possibility that the users hold assumptions that have not been expressed in dialog.

We provide example usage of V&V with corroborating models in the section "Engineering Copilot Design Example" after a discussion of the role of DSLs to support the mentoring relationship with users below.

CHATBOTS, CONSTRUCTIONISM, AND DSLS

The challenges discussed in this section concern the problem-solving and social dynamics considerations of using DSLs with engineering copilots. These include attributes of the DSL that tend to empower users to learn more about the subject discipline and technology. This is a goal consistent with human-centered AI (HCAI), and the oft-cited concern that AI might diminish human control over decision processes. Constructionist learning (Papert & Harel, 1991) and soft-system methodology (Checkland, 2000) are defining characteristics of engineering copilots addressing these concerns.

Conversational interfaces help address some requirements of engineering copilots, but many requirements can be addressed with computational notebooks, which typically do not have conversational interfaces. Our first foray into the problem space of mentoring agents (Demo, 2020) predates the emergence of robust LLM-based chatbots. The research prototype implemented in that work used a Jupyter-notebook-based tool to formulate production scheduling solutions using the MiniZinc DSL. Principal aims of the work were to demonstrate the value of (1) plan recognition (Raghavan, Singla & Mooney, 2014) to facilitate joint human/machine work, and (2) constructionist learning (Papert & Harel, 1991) in production scheduling with the MiniZinc DSL. A field study of the system involved two analysts at a medium-sized manufacturing firm using it to schedule production. The analysts learned MiniZinc by working with notebooks and human tutoring that gradually removed educational scaffolding. An example MiniZinc solution similar to ones used in the field study is shown in Figure 39.1.

The aim of the system developed in the field study was to investigate basic ideas of joint work with DSLs, not to create a usable mentoring system. Nonetheless, combined with some expert human supervision, computational notebooks proved useful toward scaffolding and capturing (by means of program comments) the intent

```
int: nProducts = 5;
set of int: Product 1..nProducts; Objects in the domain: beer products.
set of int: Day 36..365; Objects in the domain: days where beer is being made.
enum Task (mill, ferment, bright); % Objects in the domain: tasks to make brew the beer.
array [Product, Task] of int: taskDuration [ | 1, 20, 20 | Processing times for each task.
                                             1, 20, 20 | The three tasks and 5 beers.
                                             1, 30, 11 |
                                             1, 21, 11 |
                                             1, 18, 19 |];
% The following are data concerning the objects and their interrelation
array [Product] of Day: neededDate [50, 80, 130, 140, 150];

% We assume that a task uses one major resource; resources are synonymous with tasks.
array [Product, Task] of var Day: taskStarts; 'var' means it is a decision variable.
array [Product, Task] of var Day: taskEnds; % vars will be determined by the solver.

& We'll use this below to ensure a resource can only be used by one product at a time.
array [Task, Day] of var Product: busyWith;

% The first two Product are already in the fermentation tank and bright tank respectively.
constraint taskStarts [1,mill] 35; % Product 1 is therefore in the bright tank....
constraint taskStarts [2, mill] = -5; %...and out before this one moves there.

% The next task must start right after previous task ends. No place to put WIP!
constraint forall (p in Product) (taskEnds [p, mill ]+1 == taskStarts [p, ferment]);
constraint forall (p in Product) (taskEnds [p, ferment] + 1 taskStarts [p, bright]);

% A task ends taskDuration days after it starts.
constraint forall (p in Product, t in Task) (taskEnds [p, t] taskDuration [p, t]); taskStarts [p, t] +

% A resource (task) can only be used with one product at a time.
constraint forall (t in Task, pin Product, din Day where d > taskStarts [p,t] /
  \ d <= taskEnds [p,t])
(busyWith [t,d] == p);

% Aim to have each beer completed when it is needed, an no sooner.
solve minimize sum (pin Product) (abs (neededDate [p] taskEnds [p,bright]));
```

FIGURE 39.1 A simple MiniZinc formulation for scheduling beer production.

of the formulated MiniZinc model. Much more capability to elicit requirements from users would have been necessary to produce a usable system. The ability to capture unstructured requirements with LLM-based chatbots appears to provide this missing capability. Starting with discussion of chatbots and ordinary copilots, we illustrate how the additional requirements of long-running relationships with technology can be achieved with DSL-based engineering copilots.

One could also extrapolate to usages more ambitious than DSL-based engineering copilots. Broadly speaking, teaming systems supporting long-running relationships could serve as the enterprise's principal repository of acquired knowledge, providing a competitive advantage. It could be argued that by recording conversation of requirements, decision processes, and tradeoffs made, and by encoding requirements in the supported DSL language, engineering copilots are well suited to this goal.

Ordinary Copilots

Ordinary (non-educational) copilots built on LLMs are proving to be useful in many areas. Copilots can provide programming snippets from text descriptions of requirements, rewrite paragraphs for better readability, and provide background discussion on many topics. All of these scenarios concern short, independent interactions with the tool. However, even in a future where technology designs *and explains* solutions consistent with a user's requirements, there are significant challenges in scaling such capabilities, for example, in scaling from providing program snippets to providing a comprehensive generative capability for engineering design. The problems introduced by ill-considered use of copilots include many of the things systems engineering best practice aims to avoid:

Ambiguous Aim: When users do not have sufficient background in the discipline, they are less likely to be able to give expression to design requirements.

Unclear Roles in Composition: When individual system components are conceived without a shared understanding of allocated roles, understanding system behavior may become difficult.

Inconsistent Iinterface in Composition: When individual system components are conceived without reference to a common technology and protocols for interfacing, system complexity may increase needlessly.

Knowledge Attrition: When the experience of designing the solution does not leave a lasting impression, the import of what was learned may be lost. This includes lost opportunities to express principles, policies, and practices that might have been used to improve enterprise operations.

Increased Technical Debt: When solutions are provided to isolated requirements without knowledge of the bigger picture, the solution might miss opportunities and be nearsighted with respect to the system's maintenance and future evolution.

It is because engineering copilots aim to support users in areas unfamiliar to them, and in an integrated, long-running relationship that the system's nature differs. Programming copilots, for example, are being used successfully because the users are programmers skilled in the discipline and are aware of the issues cited above.

Educational Chatbots

Educational chatbots are chatbots designed to serve pedagogical goals. They have been found to reduce the load on teachers and allow students to learn at their own pace (Katchapakirin, Anutariya & Supnithi, 2022). A systematic review of 36 papers about educational chatbots by Kuhail et al. (Kuhail, Alturki & Alramlawi, 2023) noted that more than half of the chatbots acted as teaching agents, and more than a third acted at peer agents. Engineering copilots, however, do not fit either of these categories; the engineering copilot is typically the *more knowledgeable other* with respect to the DSL and the discipline of the joint work, but only the users know their problem domain intimately. For our purposes, however, the salient issue to discuss is that educational chatbots typically do not employ constructionist learning with DSLs. This point is taken up directly after a short discussion of constructionist learning.

Seymour Papert developed constructionism (Papert & Harel, 1991) from Piaget's theory of constructivism in child development. Learning in constructionism occurs in the process of producing an artifact in an environment that affords rich experimentation. The product made is sometimes referred to as a *public object* owing to it being possible to share it with others. We assume that the elements from which a learner constructs a public object comprise a DSL, possessing rules for composition and an operational semantics. Conceiving of and refining the public object involves problem solving with the DSL.

The idea that a DSL has a role in engineering copilots stems from the observation that even children can learn to perform a very technical task (programming) with an appropriately conceived DSL and supporting environment. The very successful Scratch programming environment, a direct descendant of Papert's work, is an example of this (though not a copilot nor chatbot). In the Scratch programming environment and similar earlier works, the elements of the DSL are visual puzzle pieces. The puzzle pieces only fit together in syntactically correct ways. Further, the operational semantics of the piece types are easily demonstrated in small programs. Together, these attributes work to invite experimentation and reduce uncertainty about how things in Scratch work. Many children learn Scratch without guidance from a teacher or chatbot.

For the sole purpose of learning the DSL, it is possible that engineering copilots would not need a chat capability either. The chatbot capabilities of engineering copilots, however, serve an additional purpose of guiding formulation and providing a record of decision-making that can be curated in a long-running relationship with the technology. These issues are taken up in the section "Design of DSLs for Engineering Copilots."

Compared to educational systems, engineering copilots are less likely to be focused on the development of the individual than on the development of enterprise knowledge made concrete in recorded conversation and formalized in DSL. Owing to the DSL, engineering copilots also encourage a deep relation with a consensus body of knowledge and community of practice in the supported industrial practice; this is uncommon in educational chatbots.

Design of DSLs for Engineering Copilots

We next consider the advantages of DSLs for empowering users and supporting a trustworthy process. In bridging the gap in understanding, we should not presume that users are programmers and systems engineers. But further, learning supporting technology may entail integrating knowledge of some academic field (e.g., operations research or physics-based modeling) with the semantics of an unfamiliar analytical tool. It may be that none of this is familiar to the users. For these reasons, in our work, as described in the section "Engineering Copilot Design Example," in interacting with users, we demonstrate the basic features of the DSL early in the requirements gathering conversation using basic notions from their world. In this conversation, we incrementally refine a DSL-based solution, where the intermediate versions of the solution may fall short of meeting system requirements, some of which may not yet have been expressed. Demonstrating features of the DSL in this way becomes an entry point into learning the tool and the academic field.

DSLs are designed for simplicity and tailored to the needs of domain practitioners. They, and their supporting environment, are effective by virtue of what they omit perhaps more than by what they include. Scratch, for example, includes very few affordances for program control, and the ones provided can be demonstrated easily in programs that contain just a few elements and provide immediate feedback upon execution. For example, a control block may test whether a moving object is touching some other object of a certain color and change the behavior of the moving object based on the test. By using the idea of one color touching another, the designers are leveraging attunement, specifically that this concept is familiar to children and is easily seen as useful to designing objects that move on the screen. Similarly, the MiniZinc DSL defines only four principal constructs, one each to describe (1) objects in the problem domain, (2) properties of the objects and their interrelation, (3) constraints on interrelations of objects, and (4) a goal expression scoring solutions (feasible properties and interrelations) of the objects. See Figure 39.1.

The DSL should be conceived to foster peer interaction, an integral part of the educational philosophies of Piaget and Papert. A *community of practice* is a social learning system arising out of its history of learning itself. One motivation for using a DSL is the possibility that a community of practice might form (or may already exist) around the DSL. Strong online communities already exist for some DSLs, including MiniZinc and Modelica, for example. Cloud-based team communication platforms are numerous, easy to create, and flexible to the various requirements that might emerge for DSL-based engineering copilots. A factor that encourages intramural peer interaction is simply that system requirements might involve many functions in the enterprise. Speaking with the people supporting those functions, expressing project objectives, and highlighting how their work fits in can work to foster their participation.

There are three principal universes of discourse (and lexicons) operating in engineering copilots: the users' domain lexicon. This is shown in Figure 39.2.

There is benefit in establishing common ground that goes beyond the isolated task of formulating a solution. Users who learn the lexicon of an established body of knowledge can more easily and more confidently interact with experts of that body of knowledge. This is especially beneficial in that it encourages a virtuous cycle in engagement with the corresponding community of practice.

Finally, the use of DSLs can also ease the tasks of V&V. This is especially the case where (1) the DSL is declarative (describing what is required rather than how it is achieved) rather than procedural (describing how it is achieved), and (2) the DSL

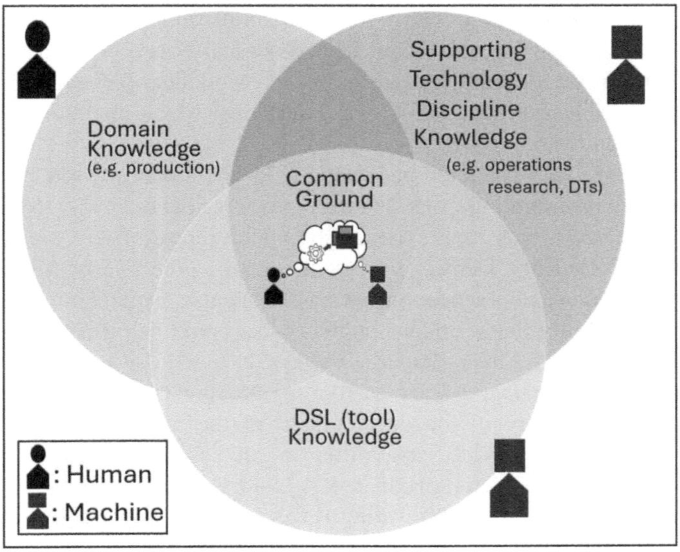

FIGURE 39.2 Establishing common ground with users engaged in a teaming activity involves assimilation and accommodation among the agents.

content is compiled to an intermediate form that drives a separate solver. Roughly speaking, by organizing the implementation this way, the design of the DSL can serve as a contract or type system, either of which can specify what the solver should be able to compute. In summary, we enumerate several DSL design criteria below.

Unitary Public Object: Public object refers to the constructionist learning notion of a designed artifact that can be shared and comprehensible to others familiar with the technology in which it was conceived. For example, in teaching programming to children with the Scratch programming language, the child's Scratch program is the public object. A *unitary* public object provides a single comprehensive viewpoint by which the necessary cognitive work of design can be performed. Where there is not a single comprehensive viewpoint, such as in multi-physics simulation, it is presumed that there is a greater cognitive burden on users, owing to the need to juggle multiple abstract viewpoints.

Declarative Metamodel: Metamodel, as used here, refers to a model of a modeling language detailed enough to serve as a storage form for models. Declarative modeling paradigms abstract away messy implementation details. Further, declarative statement simplifies plan recognition (Sukthankar et al., 2014). MiniZinc, used in constraint-based optimization, is a declarative language that translates to forms that are processable by combinatorial optimization solvers. Modelica, used in physics-based modeling projects, combines declarative constructs with object-oriented classification.

Compositional Metamodel: Compositionality in modeling has significantly advanced the state of the art. To accommodate compositionality in this context is to allow units of functionality (such as simulations) to interoperate despite differences in their internal logic. Compositionality can be seen as a positive viewpoint on the non-unitary public object. Modelica, for example, owing to its object orientation and use of the Functional Mock-up Interface, supports compositionality.

Community-Forming: The interconnectedness of the modern world has facilitated forming and proliferating CoPs. For example, it is not hard today to find communities dedicated to tiny topics – through GitHub repositories. Quality programs have had CoPs since Deming's work in Japan in the 1950s. MiniZinc has a strong community of practice; online classes are also available. Intramural CoPs and the possibility that a unitary public object could help form them, however remote, is a powerful idea (Papert, 1993).

Linkage to Requirements: As far as the engineering copilot is responsible for recording the history of the development of the model, it would be useful were the DSL to provide capabilities to link each DSL expression to the requirements from which it originated. A history describing how the model has evolved and the rationale motivating it can be useful towards (a) on-boarding new participants and (b) not repeating past errors. Strongly declarative DSLs sometimes have constructs that clearly correspond to specific requirements. For example, MiniZinc constraints do this. Such capabilities are rare, however, and the best place to make this linkage might not always be in the DSL itself; other development environment features could support it. Ideally, established common ground terminology would accommodate and assimilate concepts cited in requirements without loss of the original sense of the requirements.

ENGINEERING COPILOT DESIGN EXAMPLE

In this section, we illustrate concepts discussed earlier using SchedulingTBD [17], an engineering copilot we are implementing to explore design possibilities. SchedulingTBD is intended to mentor users in formulating and refining scheduling systems using MiniZinc. Currently, SchedulingTBD can generate complete scheduling systems through interaction with surrogates of human experts, LLM-based agents instructed to behave as humans skilled in manufacturing a given product. Figure 39.3 depicts the instruction given to an LLM assistant (agent) to produce a surrogate expert.

This feature was conceived for use in testing; we intend to test the system on the products featured in each of the 1,600 segments of the Science Channel's *How It's Made*, an educational television series that ran from 2001 to 2019. Figure 39.4 depicts the system.

In our implementation, conversation is factored into three interrelated topics: (1) the process conversation focuses on constructing models of the production process, including the production tasks to be performed, typical task durations, and opportunities for concurrency; (2) the resource conversation focuses on constructing models of actual resources of the production facility, their number and their capabilities with

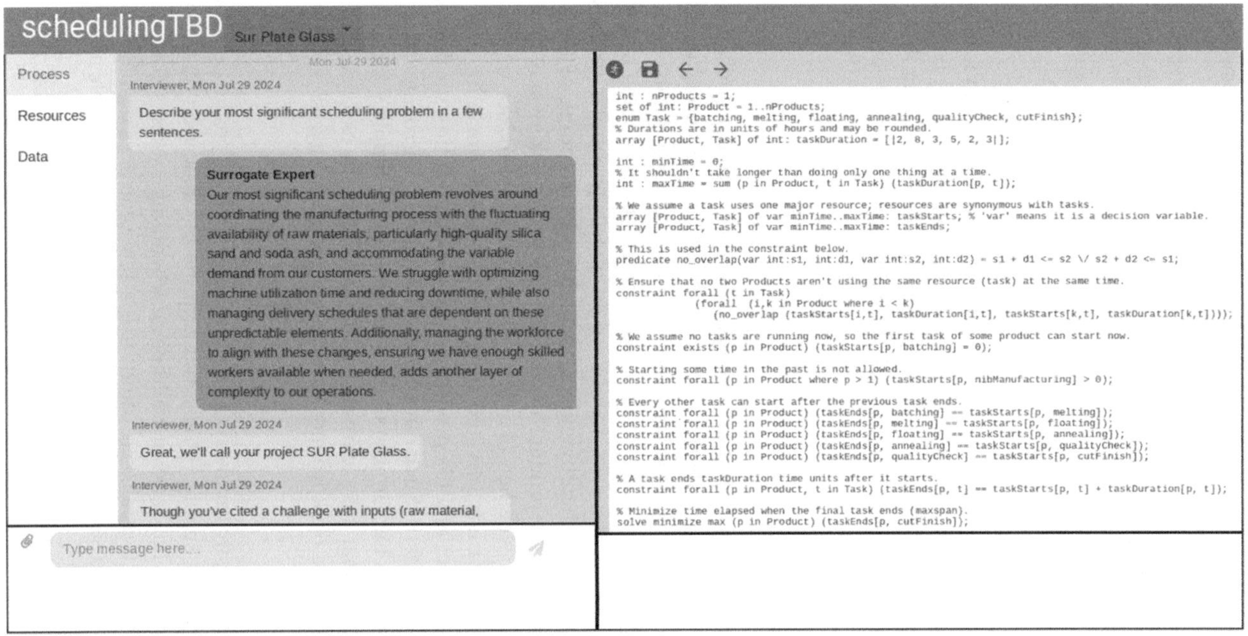

FIGURE 39.3 Scheduling TBD while interviewing a surrogate expert about plate glass production. The leftmost area of the user interface allows switching between three interrelated conversations (Process, Resources, and Data). The conversation shown is between the interviewer agent (light gray) and surrogate expert (dark blue). Except for choosing the topic (plate glass), there is no human interaction in such testing with surrogates. The upper right region depicts a MiniZinc implementation at some intermediate level of refinement.

"You manage a company that makes $X. You are an expert in production and manage your company's supply chains. You help me by answering questions that will allow us to collaborate in building a scheduling systems for your company. Your answers typically are short, just a few sentences each. If you don't have information to answer my questions, you provide a plausible answer nonetheless."

FIGURE 39.4 Instruction provided to a surrogate agent. A product type, for example, plate glass, is substituted for the variable $X in the creation of the agent. Such an instruction could be used to instantiate a surrogate expert for testing, or a parallel surrogate for gaining default background information.

respect to processes; and (3) the data conversation discusses with the user the intended interpretation of data from user-uploaded spreadsheets. Interpretation of uploaded spreadsheets is particularly practical in that it obviates the need to collect these data by other means. Small businesses, for example, often maintain spreadsheets of customer orders.

Parallel Surrogate Experts

In development it became apparent that surrogate experts, initially conceived for testing, could serve two other important roles in the system. A surrogate expert can be initiated as soon as it is clear for what sort of production the users expect to make a scheduling system. Such a *parallel surrogate expert*, running as a background process invisible to the user, can be asked questions similar to those to be asked of human uses, and thereby gain a "default" understanding of the kinds of production typical for that product class. The questions asked may include, for example, determining whether (1) the system is for a product or service; (2) production (if product) is motivated by a make-to-stock, make-to-order, or engineer-to-order incentive; and (3) production (of product) follows a flow-shop or job-shop production pattern. Such questions to the parallel surrogate can either eliminate some discussion human users may find tedious or simplify the conversation to only confirm what was learned by the surrogate. In essence, parallel surrogates provide default background knowledge about the users' scheduling domain.

You manage a company that makes $X. You are an expert in production and manage your company's supply chains. You help me by answering questions that will allow us to collaborate in building a scheduling systems for your company. Your answers typically are short, just a few sentences each. If you don't have information to answer my questions, you provide a plausible answer nonetheless.

The second purpose of a parallel surrogate is closely related; an interview with the parallel expert can be used to construct corroborating models of default knowledge.

The models generated this way might be quite similar to the those generated through interviews with users, as the questions put to both agents can be similar. Differences in the models can be used to check understanding of the human interview, and to note how the actual process differs from the default knowledge.

Leading an Interview

In our implementation, knowledge of how things are made, the processes that are subject to scheduling, is not encoded anywhere in the system. Instead, this knowledge is drawn out from the users through interviews. Each of the three interview topics (process, resources, and data) requires distinct expertise. Specifically, (1) the process interview should reflect mechanistic knowledge of the transformation of raw materials into product, (2) the resources interview should reflect knowledge of machines and production facility organization, and (3) the data interview should reflect knowledge of data representation schemes and business concepts such as orders, customers, quantities, and due dates. It is assumed that, associated with each area of expertise, there may be particular ways of navigating the body of knowledge that experts view as being most productive. For example, in understanding the operation of a production facility, an expert might start with questions that determine whether the production is discrete, batch, and continuous. Discrete production might be further distinguished as synchronous or asynchronous, and so on.

CONCLUSION

The chapter considers how human/AI teaming systems can be designed to bridge the gap between users inexperienced in a discipline and the tools and practices of that discipline. Human/AI teaming systems, such as engineering copilots, are sociotechnical systems where some agents are machine-implemented. DSL-based engineering copilots differ from educational chatbots and computational notebooks by mentoring the user over the entire lifetime of an engineering endeavor. The principal design challenges faced concern trustworthy operation and empowering DSLs.

Regarding the trustworthy aspect, engineering copilots lean heavily on LLM-based conversation, but the challenge of trustworthy operation is attributed to shortcomings of explainability in LLM-based agents. Rather than seeking explanation from AI agents, we explore processes to produce corroborating models by guiding conversation to express requirements relevant to constructing relevant models. Incremental refinement of these models to executable or visually meaningful forms that provide a conventional form of verification.

Attributes of DSL beneficial to the goals of engineering copilots were discussed. DSLs can be designed to foster constructionist learning of a supporting technology, system integration, and linkage to requirements. The DSL can serve as an on-ramp to deeper investigation of the associated scholarly subject.

In the ongoing study, use of the scheduling copilot system has been limited to AI surrogate users. With these, the system is able to generate correct, rudimentary scheduling systems for some kinds of scheduling problems. Without providing much direction to the AI surrogates, however, the surrogates do not express a wide variety of scheduling scenarios. In the future, we will introduce humans to the system to examine the human-AI relationship. Success of the copilot teaming relationship will be gauged by its ability to support humans in learning the DSL to formulate effective scheduling system solutions.

NOTES

1 Example of DSLs includes Hypertext Markup Language (HTML) for web pages, Structured Query Language (SQL) for database management, MiniZinc for constraint-based optimization, and Modelica for multi-domain physical modeling.
2 The users' goal problem is the problem being pursued through formulation and refinement of a DSL-based model. For example, the problem might be efficient scheduling of production, and the model formulated in MiniZinc.
3 Note that we are talking about design from two perspectives: the perspective of designers of DSL-based engineering copilot, and the perspective of the users of such systems. Users of copilot systems express their design in the DSL language. Just as engineers using a Modelica multi-physics modeling environment (not an AI-based engineering copilot) express their design as components in Modelica's object-oriented DSL.
4 One perspective on the chapter is that it presents a new technical approach to the notion of *learning systems* as described in the soft system methodology [**?**]
5 Another important design problem for copilots, for example, is that of knowing "what questions to ask when" in the effort to understand the users' world.

REFERENCES

Ben Armstrong, Suzanne Berger, and Bill Bonvillian. *Advanced Technology, Advanced Training: A New Policy Agenda For U.S. Manufacturing*. MIT Initiative for Knowledge and Innovation in Manufacturing, May 2021.

Gita Sukthankar, Robert P. Goldman, Christopher Geib, V. Pynadath David, and Hung H. Bui. *Plan, Activity and Intent Recognition*. Morgan Kaufmann, Amsterdam, 2014.

Kantinee Katchapakirin, Chutiporn Anutariya, and Thepchai Supnithi. ScratchThAI: A Conversation-Based Learning Support Framework for Computational Thinking Development. *Education and Information Technologies*, 27(6):8533–8560, July 2022.

Markus Schlosser. *Stanford Encyclopedia of Philosophy Entry on Agency*. Stanford Encyclopedia of Philosophy, 2019.

Mirac Suzgun and Adam Tauman Kalai. Meta-Prompting: Enhancing Language Models with Task-Agnostic Scaffolding, January 2024.

Mohammad Amin Kuhail, Nazik Alturki, Salwa Alramlawi, and Kholood Alhejori. Interacting with Educational Chatbots: A Systematic Review. *Education and Information Technologies*, 28(1):973–1018, January 2023.

NIST. *Artificial Intelligence Risk Management Framework: Generative Artificial Intelligence Profile*. Technical Report NIST AI NIST AI 600-1, National Institute of Standards and Technology, Gaithersburg, MD, 2024.

NIST. *The Language of Trustworthy AI: An In-Depth Glossary of Terms* (updated August 4, 2024). Technical Report, 2024.

Peter Checkland. Soft Systems Methodology: A Thirty Year Retrospective. *Systems Research and Behavioral Science*, 17:11–58, 2000.

Peter Denno. *Production Scheduling as Joint Cognitive Work*. PhD thesis, Loughborough University, 2020.

Peter Denno. SchedulingTBD: Exploratory code for human/AI teaming in MiniZinc model formulation. https://github.com/pdenno/schedulingTBD, 2024.

Peter J. Stuckey, Kim Marriott, and Guido Tack. *MiniZinc Handbook Release 2.2.1*. Technical Report, Creative Commons, 2018.

Seymour Papert and Idit Harel. Situating Constructionism. In *Constructionism*. Ablex Publishing Corporation, 1991.

Seymour Papert. *Mindstorms: Children, Computers, and Powerful Ideas*. New York, NY, HarperCollins Press, 1993.

Sindhu Raghavan, Parag Singla, Raymond J. Mooney. Plan Recognition using Statistical Relational Models. In Gita Sukthanker, Robert P. Goldman, Christopher Geib, V. Pynadath, David, and Hung H. Bui, editors, *Plan, Activity, and Intent Recognition*, pp. 57–85. Morgan Kaufmann, Amsterdam, 2014.

Wayne Xin Zhao, Kun Zhou, Junyi Li, Tianyi Tang, Xiaolei Wang, Yupeng Hou, Yingqian Min, Beichen Zhang, Junjie Zhang, Zican Dong, Yifan Du, Chen Yang, Yushuo Chen, Zhipeng Chen, Jinhao Jiang, Ruiyang Ren, Yifan Li, Xinyu Tang, Zikang Liu, Peiyu Liu, Jian-Yun Nie, and Ji-Rong Wen. A Survey of Large Language Models, June 2023.

40 AI Compass

A Design Framework for Human–AI Integration

Gudela Grote, Sharon K. Parker, and Kevin Crowston

INTRODUCTION

Human–system integration has never been easy, with decades of human factors research bearing testimony to difficulties arising from interdisciplinary collaboration, different speeds of technological innovation and organizational change, and most fundamentally differences in professionals' worldviews on what constitutes good human–system integration (Davis et al., 2014; Orlikowski & Scott, 2008; Parker & Grote, 2022). In the tradition of socio-technical systems theory, a system's capacity to self-regulate and thereby cope with variances and disturbances in work processes at their source, high-quality jobs for all human actors, and human control over technology have been proposed as key indicators of successful socio-technical and human–systems integration (Clegg, 2000; Grote et al., 2000). However, technology development and implementation still follow a techno-centric approach, with economic efficiency gains being the main driver and indicator of success (Boy, 2013; Dul et al., 2012; Emmenger & Norman, 2019). Moreover, the unprecedented speed and complexity of innovation in the field of artificial intelligence (AI) have led many to call into question whether human–system integration can still follow the same principles as in more traditional technology development and implementation (Berente et al., 2021; Salmon et al., 2023; Slota et al., 2023).

In the following, we first outline existing design principles, focusing on the KOMPASS method as one example of how these principles can be put into practice (see section "Human–System Integration for Traditional Technologies: The Example of KOMPASS"). Subsequently, we discuss challenges arising in the context of AI development, implementation, and use, arguing that the fundamentals of socio-technical system integration still hold, but that some new design principles and processes are needed to capture the complexities of AI (see section "Challenges in the Design of AI") fully. We then present the design framework AI Compass, which supports developers and users in safely and effectively employing these emerging technologies (see section "AI Compass: Principles, Processes, and Criteria for the

 DOI: 10.1201/9781003666431-48

Design of AI Systems"). In the final section "Conclusion," we draw some overarching conclusions.

HUMAN–SYSTEM INTEGRATION FOR TRADITIONAL TECHNOLOGIES: THE EXAMPLE OF KOMPASS

Core to human–system integration is the allocation of functions between human operators and technical systems, which determines the level and type of automation and the role(s) human operators are to fulfil in "joint cognitive systems" (Hollnagel & Woods, 2005; Parasuraman et al., 2000). Over many decades, an understanding of function allocation has evolved, emphasizing the complementarities of humans and technical systems rather than focusing on the competitive advantage of either (Jordan, 1963; Bailey, 1989). This thinking has been captured in the design principle of human-in-the-loop, requiring that human operators stay in control even in largely automated systems (Billings, 1997; Hollnagel & Woods, 2005). Such control concerns not only options for influencing the functioning and outcomes of technical systems, but also transparency and predictability of technical processes that enable situation awareness for the human operators and thereby provide the prerequisites for adequately intervening in these processes if needed (Endsley, 2017; Norman, 1986; Parasuraman & Riley, 1997).

Function allocation follows levels of automation, which determine the specific distribution of tasks between humans and technology, from light technological support for task performance by humans, such as warning signals when certain performance limits are reached, to full automation, with, under normal circumstances, no human involvement required. When functions are allocated in a way that humans are still involved in task completion, receive support from the technology, and can override decisions by the technology, this has been termed augmentation based on the assumption that in these cases human capabilities are enhanced in a complementary fashion (Parasuraman et al., 2000; Parker & Grote, 2022).

Beyond complementary function allocation, human–system integration also benefits from a broader systems perspective that helps to embed the interaction between individual human operators and technical systems into larger socio-technical systems. By designing work processes in line with the basic tenet of socio-technical systems theory—enabling work systems to control variances at their source (Cherns, 1976)—individual empowerment and collective self-regulation are fostered, which are crucial requisites for human operators to employ their control over technical systems effectively. Both knowledge and motivation are built and maintained for use, but not to over-rely on technology, and to intervene in automated processes when needed. Thereby, a capacity for adaptive performance is created, which allows workers to "manage the unexpected" (Bienefeld & Grote, in press; Boy, 2021; Parker & Boeing, 2023; Parker & Grote, 2022).

TABLE 40.1
KOMPASS Design Heuristic

1 Expert analysis of existing work systems
Evaluation of the current system design concerning complementarity and control of variances and disturbances

2 Development of shared design philosophy
2.1 Definition of primary task and functions of the new work system
2.2 Collection of potential variances and disturbances in the new work system
2.3 Definition of goals to be fulfilled for a successful work system
2.4 Identification of contributions to a successful work system by human operators, technology, and organization
2.5 Specification of the working conditions required for human operators to make their specific contribution

3 Derivation of design requirements
3.1 Definition of demands on system design according to KOMPASS criteria
3.2 Clustering of design requirements into work packages for realization

Source: Adapted from Grote et al. (2000).

KOMPASS is one of many methods to support technology developers in following this path to socio-technical and human–system integration (Grote et al., 2000; Wäfler et al., 2003). It comprises a design process and a set of design criteria. The design process spells out specific steps that design teams can follow to create a shared design mindset that fosters more holistic design considerations (Boos et al., 2013; Wäfler et al., 2003) (see Table 40.1). The design criteria specify how human-in-the-loop, high-quality jobs and self-regulating teams can be ascertained while developing and implementing a technical system (see Table 40.2). It should be noted that the criteria across the levels of human–technology interaction, individuals' jobs, and team collaboration are interlinked. For instance, expecting individuals and teams to make decisions in response to non-routine situations requires that the technology they interact with be transparent and controllable in support of effective decision-making.

KOMPASS has been employed in a variety of technology development projects and, over the years, has also been adapted to cover new kinds of technologies, such as Internet of Things applications (Boos et al., 2013). KOMPASS and similar methods have been found to raise awareness and consideration of technology users' needs and concerns.

Nevertheless, properly including work and organizational design in technology development remains a significant challenge (Kahlert & Grote, 2024; Parker & Grote, 2022). The fundamental requirement that human technology users can only be held accountable for system functioning and outcomes if they can sufficiently understand and influence the technology has always been challenging to implement. The "ironies of automation" (Bainbridge, 1983) point to the impossibilities for humans to control technical systems created to surpass human capabilities. AI-based systems exacerbate these ironies as users and developers have insufficient influence over these systems, which we expand on in the following section.

TABLE 40.2
KOMPASS Design Criteria

Work System: Self-Regulating Teams

Task Completeness: Processing depth, functional integration, complexity of production

Independence of Work System: Buffers, flexibility of workflow, local quality control

Fit between Regulation Requirements and Opportunities: Type and extent of system variances and disturbances, task interdependence, and cooperation forms.

Polyvalence: Proportion of tasks for which operators have complete skills

Autonomy: Joint decision-making regarding internal and external coordination, internal personnel management, and continuous improvement within the work system

Boundary Regulation: Ratio between superiors' internal and external coordination tasks

Human Work Tasks: High Job Quality

Task Completeness: Preparation, planning, execution, controlling and maintenance/repair

Planning/Decision-Making Requirements: Choices for work content/outcomes, equipment, work flow

Communication Requirements: Amount and kinds of common planning and decision-making

Opportunities for Learning/Personal Development: Using existing qualifications, acquisition of new skills

Variety: Dealing with different materials, procedures, tools, products, persons

Transparency of Workflow: Knowledge of how one's own work tasks fit into the overall workflow

Influence over Working Conditions: Controlling orders, work schedule, task distribution, production goals

Temporal Flexibility: Planning horizon, temporal coupling in workflow, time pressure

Human-Technology System: Human-in-the-Loop

Process Transparency: Forming and maintaining mental models of technological processes and required interventions

Dynamic Coupling: Technically provided options regarding time, place, work procedures, and required cognitive effort in task execution

Decision Authority: Roles of human operator and technical system regarding information access and process control

Flexibility: Variability of function allocation between the human operator and the technical system

Source: Adapted from Grote et al. (2000).

CHALLENGES IN THE DESIGN OF AI

Many applications of AI raise similar questions to those of previous technologies. Basic decisions on whether humans should be replaced through automation or augmented by technology to improve their performance and on how work processes among humans should be changed to best take advantage of new technological opportunities remain very relevant (Choudhary et al., 2023). As does the fundamental challenge to align the control humans (should) have over the technology with the accountability for system functioning and outcomes they (should) have (Boos et al., 2013; McLean et al., 2023; Salmon et al., 2021).

However, design challenges are amplified by several interconnected factors (Berente et al., 2021; Castelvecchi, 2016; Slota et al., 2023). The capacity of AI to act autonomously has grown and will further grow dramatically. Organizations increasingly rely on fully autonomous decision-making by AI, shifting preferred forms of function allocation between humans and technology towards automation rather than augmentation. Algorithmic management is among the hotly debated examples of this shift, for instance, in the context of platforms such as Uber, where price setting, ride assignment, and even discontinuing drivers happen fully automatically based on algorithms only known to Uber (Cameron & Rahman, 2022; Rahman, 2021). The availability of "big data" has facilitated new forms of machine learning, such as deep neural networks. Self-learning systems can handle increasingly complex decision-making in classification and prediction tasks and autonomously generate new content, such as text, images, and music. The range of tasks that AI-based technologies can carry out, autonomously or by augmenting human performance, is substantially broadened.

Furthermore, as such systems continuously change their functioning, they lose transparency and predictability and become more human-like, requiring new forms of interaction with their developers and users (Glassman et al., 2024). The growing complexity of AI-based systems and their ability to learn and act autonomously render them opaque for anyone interacting with them, which has fueled discussions about who is to be held to account for the functioning and outcomes of AI. At the same time, the many efforts to make AI systems more explainable or interpretable are hoped to at least partially restore human understanding and control (Glassman et al., 2024; Kim & Doshi-Velez, 2021; Rudin, 2019).

These characteristics of emerging AI technologies require new approaches to achieving human–system integration (Grote, 2023). Even when leaving the macroeconomic questions of job loss through automation aside (Arntz et al., 2016; Frey & Osborne, 2017), the potential of AI to automate increasingly complex tasks challenges current attempts to keep humans in the loop. The necessity for human involvement is reduced, as is the ability of humans to interact with self-learning systems due to their inherent opacity (Castelvecchi, 2016). As Hancock (2019, p. 483) has stated, "the slow, steady and apparently inexorable extinction of any human contribution may be the natural sequella of automation myopia." At the same time, it remains humans who are held to account when such systems malfunction or produce undesired outcomes, such as biased decision-making.

Autonomously learning AI systems are the first technology continuously changing through its use (Faraj et al., 2018; Slota et al., 2023). These systems become opaque and unpredictable even for their developers, which raises the question of where accountability should be situated if neither users nor developers have sufficient control over these systems. Additionally, it creates task interdependencies between system development and use that are not present in other kinds of technologies. A whole AI supply chain must be considered, including data providers, companies selling access to large AI models, and knowledge brokers translating system outputs for AI users (Cobbe et al., 2023). These issues need to be addressed in methods aimed at supporting human–system integration within larger socio-technical systems or "sociotechnical landscapes" (Slota et al., 2023, p. 1287), which include multiple stakeholders besides those directly involved in the development of AI systems and the ultimate users of those systems. Importantly, regulators must consider who plays a significant role in effective AI governance (European Union, 2023; NIST, 2023).

Moreover, there is a general apprehension that power imbalances among stakeholders involved in technology development and use will worsen for AI-based systems, with Uber and Tesla frequently cited examples (Endsley, 2023; Möhlmann et al., 2021). Design methods need to address such power imbalances not only through participatory design approaches (e.g., Mumford, 2000) but also through more equalitarian stakeholder involvement (Bridoux & Stoelhorst, 2022), which has been captured in Slota et al.'s (2023) concept of networks of accountability. Such approaches can help to manage goal conflicts in pursuit of the overarching goal to mitigate risks and improve effectiveness (Grote et al., in press; Waterson, 2014).

AI COMPASS: PRINCIPLES, PROCESSES, AND CRITERIA FOR THE DESIGN OF AI SYSTEMS

Following the above considerations, a design framework, AI Compass, is proposed to address the challenges arising from the socio-technical integration of AI-based technologies. We first derive design principles in response to these challenges and then describe requirements for the design process and criteria for assessing successful socio-technical integration (see Table 40.3).

TABLE 40.3
Design Framework of AI Compass

	Design Process	Criteria for Successful Implementation of Design Principle
Design principle 1. Establish control–accountability as overarching design objective	The overarching goal of socio-technical integration, that is alignment of control and accountability to permit effective control of variances at their source, is established from the onset of AI development by project initiators.	• Agreement on control-accountability alignment as core organizational principle for socio-technical integration
Design principle 2. Establish network of accountability	All relevant stakeholders are identified by project initiators and engaged in reconsidering and possibly adapting the purpose of the new AI system, its desired impact on work processes, and how emerging challenges throughout the AI life cycle are to be handled.	• Accountability as a shared norm • Acceptance of rationalist and constructivist perspectives

(*continued*)

TABLE 40.3 (Continued)
Design Framework of AI Compass

	Design Process	Criteria for Successful Implementation of Design Principle
Design principle 3. Empower AI developers and users to jointly manage task interdependencies	The stakeholders agree on an operational envelope within which AI developers and users together can act autonomously and in line with the task interdependencies to be managed by them; stakeholders support collaboration between AI developers and users by creating exchange platforms and feedback mechanisms as needed.	• Joint decision-making power on optimal level of AI capabilities for AI developers and users • Capacity for self-regulation among AI developers and users • High job quality for AI developers and users

Design Principles

Design principle 1: Establish control-accountability alignment as an overarching design objective. A core requirement for socio-technical systems to function effectively is handling variances and disturbances in work processes as close to their source. This control is achieved through socio-technical integration, which renders humans capable of controlling these processes at the level of the immediate human-technology interaction, their individual work tasks, and the collaboration within and across work units. If this control can be achieved, accountability can be assigned to the respective actors. In the case of AI, control may not be achievable for users anymore, but only for developers, and even for them, often not entirely, due to the black box nature of advanced AI systems.

AI creates new opportunities for automation that reach into ever more organizational processes, while also rendering control for developers and users more difficult. For instance, algorithmic management allows for monitoring, controlling, and incentivizing task performance without any direct involvement of managerial decision-makers, which raises issues of uncontrolled biases introduced through training data and proliferated in the fully automated application of decision algorithms (Kellogg et al., 2020; Parent-Rocheleau & Parker, 2022). AI is frequently discussed as a "new team member," stressing the fact that not only the interaction between single users and the technology needs to be considered in system design, but also the interaction within teams of human users and AI systems, rendering the unequivocal allocation of control and accountability more difficult (Choudhary et al., 2023). Adequately handling these emerging issues requires establishing the fundamental principle of control-accountability alignment as a quintessential requirement of socio-technical integration at the start of any new system development, thereby promoting early and in-depth discussions of how this alignment can be achieved for all relevant actors.[1]

Importantly, these discussions need to be especially prudent regarding the human-in-the-loop or "human oversight" (European Union, 2023) requirement, as control (and accountability) for end users may be an impossibility in advanced AI systems. Accordingly, accountability may have to be shifted to developers or jointly to developers and users, as spelled out in more detail in design principle 3.

Design principle 2: Establish a network of accountability. For AI-based systems, the boundaries of the socio-technical system must be considered to reach well beyond the organization within which the technology is implemented and used. Besides organizations that want to use AI, AI supply chains include organizations that develop and test training data, machine learning algorithms, and system interfaces to enhance explainability, as well as organizations that monitor system development and use. Thus, the relevant socio-technical system can be considered more of a socio-technical landscape in which a wide range of stakeholders must collaborate and be held accountable for system functioning and outcomes, as captured in the term "network of accountability" (Slota et al., 2023).

This broader set of relevant actors with potentially large power differentials among them needs to be governed in ways that help to establish a shared sense of accountability while also assigning accountability for specific decisions and actions to particular actors, along with establishing the actors' control over these decisions and actions. More egalitarian forms of governance based on shared norms and the willingness to accept different perspectives of what would be considered desirable system characteristics and outcomes are likely to be conducive to successful negotiations about control–accountability alignment among stakeholders (Grote et al., 2000, in press; Hardy et al., 2020). Collaboration across boundaries of stakeholder organizations is required, and a view toward achieving joint goals rather than maximizing individual gains (Bridoux & Stoelhorst, 2022).

Design principle 3: Empower AI developers and users to manage task interdependencies jointly. In the case of AI, system development and use are much more intertwined than in traditional systems because machine learning algorithms are continuously or periodically updated based on new data that the users themselves may have created. As the end users, such as physicians making diagnostic decisions based on an AI system, and the developers of such systems in their various roles along the AI supply chain, potentially have insufficient control over AI system performance and outcomes, allocating accountability becomes a complicated problem. In the attempt to open the black box of advanced AI algorithms and work toward control–accountability alignment, AI developers and users[2] require joint autonomy to make decisions about the capabilities of AI systems and function allocation between AI and human users early on, but also need to revisit these decisions continuously as AI systems evolve and new work routines emerge.

The need to collaboratively handle variances and disturbances during AI development and use creates high task interdependencies between AI developers and users. It demands joint control over the relevant work processes. Users get involved already in the early stages of development and as equal partners of developers rather than only in user experience exercises during system implementation. Developers and users must collaborate in rigorous testing, controlled experimentation, and deliberately challenge the new systems through practices such as red teaming. As experience with a new system increases, attention may shift from system requirements for routine work processes to handling exceptions (Baruwal Chhetri et al., 2024). To enable AI developers and users to manage the resulting interdependencies of their tasks collaboratively, it is crucial that stakeholders jointly define the decision latitude AI developers and users have in making design choices and ensure working conditions that allow them to make the best use of that decision latitude, for instance, by creating exchange platforms and feedback mechanisms. This requirement explicitly considers the design of AI developers' and users' jobs (Bienefeld & Grote, in press).

Comparing the foundational principles of KOMPASS and AI Compass, the most important difference is a more differentiated approach to the human-in-the-loop postulate by arguing that with the increasing complexity of AI systems, end users cannot be kept in the loop any longer, but control (and accountability!) for system outcomes needs to be transferred to developers. A second difference is the expanded view on participation, which argues the importance of involving more stakeholders in accountability networks to permit control-accountability alignment along the AI supply chain. A third difference is the emphasis on working conditions for users and developers that permit joint decision-making in the highly interdependent tasks required for steering system development and use in line with requirements for control-accountability alignment, a prerequisite for risk mitigation and effective performance.

Design Process and Criteria

Design process for implementing design principle 1. Any project to develop and use a new AI system will begin with a core group of decision-makers who create a first vision of the system's purpose, functionality, and specific use cases. This group may be larger or smaller depending on the organizational setting within which the system is to be developed and for which. To name a few such settings, an organization may build their own AI systems based on own data; an organization may create their systems, but using models developed by another organization, as currently happens with tailored applications of Large Language Models (LLMs); an organization may buy an existing AI system from a supplier or commission a supplier to develop a bespoke system.

In initial meetings among the project initiators, the first design principle of establishing control–accountability alignment as an overarching design objective must be discussed and agreed upon. A process similar to the second step outlined in the KOMPASS design heuristic (see Table 40.3) can be helpful for such discussions. By reflecting on the goal to be achieved with the new AI system and the expectations regarding new work processes and the role of humans in those processes, differences in perspectives can be identified and eventually integrated into a shared design philosophy with the alignment of control and accountability at its core, which also constitutes the criterion for successfully implementing the first design principle. As Grote et al. (in press) examined in detail, a driver for constructive dialogue should be the common interest to mitigate AI risks. By ensuring that actors are in control of the decisions and work processes for which they are held accountable and vice versa that

TABLE 40.4
Considerations for Choices among Levels of Autonomous Adaptivity of AI Systems

	Low	Medium	High
AI Capabilities for Autonomous Adaptivity	**(e.g., Programmed System; Trained System with Few Parameters)**	**(e.g., Trained System with Many Parameters, but Frozen)**	**(e.g., Continuously Learning Deep Neural Networks)**
Control and accountability regarding system outcomes	AI users are in control and held accountable if the system is explainable and leaves final decision-making to them.	AI users are at best partially in control if the system is explainable and still leaves certain decisions to them. AI developers are in control and held accountable as they define and maintain boundary conditions for system use.	AI users have no control and fully rely on the system as part of their work tasks. AI developers are partially in control by intense testing and monitoring of system outcomes within an operating envelope. AI developers are accountable in as much as senior management empowers them to decide on system capabilities.
Control and Accountability regarding system functioning	AI developers are in control and held accountable if they have decision power over ML techniques in line with the chosen system functionality.	AI developers are in control if they have decision power over ML techniques in line with the chosen system functionality.	AI developers are partially in control by intense testing and monitoring of system functioning within an operating envelope. AI developers are accountable in as much as senior management empowers them to decide on system capabilities.

Source: Abridged from Grote et al. (in press).

actors who are in control of decisions and work processes accept the accountability for the outcomes of those decisions and processes, actors can be enabled (through control) and motivated (through accountability) to achieve desired and avoid undesired outcomes of AI development and use.

Design process for implementing design principle 2. As the project becomes more established, further stakeholders should be integrated, and an accountability network should be formed. This network includes AI developers and users and the senior management of the respective organizations as key stakeholders, but possibly also representatives of organizations that supply training data or computing capacity, employee representatives, venture capitalists, brokers who translate requirements between developers and users, and, very importantly, representatives of regulatory agencies or other oversight bodies. By reconsidering and possibly adapting the purpose of the AI system, its desired impact on work processes and outcomes along the AI life cycle, and the necessary steps of risk mitigation, an awareness of the importance of accountability as a shared norm is established. Such a shared norm does not imply that the accountability for all specific decisions and actions is shared, but that all stakeholders agree that explicitly allocating accountability for all decisions and actions is crucial for safe and effective system development and use, and that they are willing to accept the accountability assigned to them. Fulfilling the criterion of a shared commitment to accountability will be easier if stakeholder governance is decentralized, permitting equalitarian interactions among all stakeholders in the network.

Once stakeholders form a network of accountability, considering different perspectives on AI development and use is facilitated. In particular, rationalist and constructivist perspectives need to be bridged for a more complete understanding of how the impact of the new AI system on the organization of work can and should be shaped (Grote et al., in press; Orlikowski & Scott, 2008). A rationalist perspective focuses on scientific knowledge, objectivity, and quantification, and stresses deterministic influences of technology on organizational and work processes. Constructivist views emphasize subjective meaning-making in social discourse and highlight the entanglement of technology and social reality as new work practices and routines emerge. Specific desired effects of technology, such as unbiased decision-making or increased efficiency of work processes, that the rationalist perspective would propagate, are questioned through discussions of how people may understand and use the technology from a constructivist perspective. Accepting these different perspectives provides the grounds for an equitable allocation of control and accountability in the newly emerging work system. It is therefore a second important criterion for successfully implementing the second design principle.

Design process for implementing design principle 3. By implementing design principles 1 and 2, preconditions are established to empower AI developers and users to jointly manage task interdependencies as postulated by the third and final design principle. As emphasized earlier, a significant new challenge arising from AI is the closer collaboration between developers and users to decide how limited control of both can be compensated for in line with their accountabilities. Together, they should be able to choose the adequate level of autonomous adaptivity of the newly developed AI system, including options in which AI systems are kept simple to provide still control over system outcomes for users and control over system functioning for developers (see Table 40.4 for relevant considerations in these decisions). Such decision latitude is the first criterion for successfully implementing the third design principle. Especially for very advanced AI systems, Asatiani et al. (2021) have proposed the use of operating envelopes to help AI developers keep some control over system functioning and outcomes. For instance, system use can be restricted only to cases that are sufficiently close to cases encountered in training data (e.g., quality of photos), or specific undesirable outcomes can be excluded as decision options for the system (e.g., driving above the speed limit for autonomous vehicles). However, the definition of operating envelopes becomes more difficult as systems and the data they learn from become more complex.

The new task interdependencies between AI developers and users require intense collaboration to decide on system functionalities and monitor and correct system outcomes. Therefore, AI developers and users need a capacity for self-regulation, rather than alone, constituting a second criterion for design principle 3. Compared to traditional systems, where users may be asked at specific points during the development to give feedback on system requirements or to test user interfaces, the risks encountered in developing and using AI systems can best be mitigated by establishing mechanisms for continuous learning by AI developers and users. Detecting and correcting faulty or biased decision-making, uncontrolled use of private data, or unvetted reliance on outcomes produced by generative AI is a joint task for developers and users who must be jointly empowered.

Not only do changes in job quality for users of new technologies require attention, but so does the design of developers' work, as stated in the third criterion for design principle 3.

Hagtvedt et al. (2025) found that some developers are much more open to acknowledging their limited control and increasing accountability. Identifying working conditions that make such an openness more likely and encourage developers to find new ways of collaborating with users is an important question for future research. Existing job design criteria included in methods such as KOMPASS (Grote et al., 2000) or SMART work design (Parker & Boeing, 2023) can inform such research and provide a generic framework for high-quality jobs.

CONCLUSION

Many methods have been developed to foster human-system integration within a sociotechnical systems theory framework. We have discussed some new demands on such methods that arise from specific characteristics of AI-based technologies. By outlining three design principles and how they may be implemented during AI development and use, we hope to have guided how these new demands could be met. We have also identified avenues for further research, especially concerning the required intense collaboration between AI developers and users, and the consequences for job design of these key stakeholder groups. Moreover, the principles are currently very generic, rendering them applicable to the many organizational settings within which AI development and use may happen. Experience with employing the principles will allow us to detail them further and improve their value for practitioners.

NOTES

1 At this stage, we only consider human actors. However, it is conceivable that with further development of autonomous AI agents, they become actors who are to be held to account for their actions and accordingly need to be in control of those actions. How these conditions are to be established is a new field of inquiry which is beyond the scope of this chapter.

2 Users can be the end users of AI systems who are normally farthest removed from the development process, but users could also be individuals who use AI for system development. In AI Compass we conceive of users mostly as end users as they are the most vulnerable group concerning a mismatch between control and accountability.

REFERENCES

Arntz, M., Gregory, T., & Zierahn, U. (2016). The risk of automation for jobs in OECD countries: A comparative analysis. OECD Social, Employment and Migration Working Papers, No. 189. OECD, Paris.

Asatiani, A., Malo, P., Nagbol, P. R., Penttinen, E., Rinta-Kahila, T., & Salovaara, A. (2021). Sociotechnical envelopment of artificial intelligence: An approach to organizational deployment of inscrutable artificial intelligence systems. *Journal of the Association for Information Systems,* 22, 325–352.

Bailey, R. W. (1989). *Human performance engineering* (2nd edn). London: Prentice-Hall International.

Bainbridge, L. (1983). Ironies of automation. *Automatica,* 19, 775–779.

Baruwal Chhetri, M., Tariq, S., Singh, R., Jalalvand, F., Paris, C., & Nepal, S. (2024). Towards human–AI teaming to mitigate alert fatigue in security operations centres. *ACM Transactions on Internet Technology*, 24/3, Article 12.

Berente, N., Gu, B., Recker, J., & Santhanam, R. (2021). Managing artificial intelligence. *MIS Quarterly*, 45, 1433–1450.

Bienefeld, N., & Grote, G. (in press). Artificial intelligence: How it changes the landscape for work design research and practice. In S. Parker, F. Klonek, C. Knight, & F. Zhang (Eds.), *Transformative work design: Synthesis and new directions*. SIOP Organizational Frontiers Series, Oxford University Press.

Billings, C. E. (1997). *Aviation automation: The search for a human-centered approach.* Mahwah, NJ: Lawrence Erlbaum.

Boos, D., Grote, G., & Guenter, H. (2013). A toolbox for managing organisational issues in the early stage of the development of a ubiquitous computing application. *Personal and Ubiquitous Computing*, 17, 1261–1279.

Boy, G. (2013). *Orchestrating human-centred design.* London: Springer.

Boy, G. (2021). *Design for flexibility: A human systems integration approach.* Cham, Switzerland: Springer Nature.

Bridoux, F., & Stoelhorst, J. W. (2022). Stakeholder governance: Solving the collective action problems in joint value creation. *Academy of Management Review*, 47, 214– 236.

Cameron, L. D., & Rahman, H. (2022). Expanding the locus of resistance: Understanding the co-constitution of control and resistance in the gig economy. *Organization Science*, 33, 38–58.

Castelvecchi, D. (2016). Can we open the black box of AI? *Nature*, 538(7623), 20–23.

Cherns, A. (1976). The principles of sociotechnical design. *Human Relations*, 29, 783–792.

Choudhary, V., Marchetti, A., Shresta, Y. R., & Purunam, P. (2023). Human–AI ensembles: When can they work? *Journal of Management*, 51, 536–569.

Clegg, C. W. (2000). Sociotechnical principles for system design. *Applied Ergonomics*, 31, 463–477.

Cobbe, J., Veale, M., & Singh, J. (2023). Understanding accountability in algorithmic supply chains. In *Proceedings of FAccT '23, June 12–15, 2023, Chicago, IL, USA*, pp. 1186–1197.

Davis, M. C., Challenger, R., Jayewardene, D. N., & Clegg, C. W. (2014). Advancing socio-technical systems thinking: A call for bravery. *Applied Ergonomics*, 45, 171–180.

Dul, J., Bruder, R., Buckle, P., Carayon, P., Falzon, P., Marras, W. S., Wilson, J. R., & van der Doelen, B. (2012). A strategy for human factors/ergonomics: Developing the discipline and profession. *Ergonomics*, 55, 377–395.

Emmenegger, C., & Norman, D. (2019). The challenges of automation in the automobile industry. *Ergonomics*, 62, 512–513.

Endsley, M. R. (2017). From here to autonomy: Lessons learned from human–automation research. *Human Factors*, 59, 5–27.

Endsley, M. R. (2023). Supporting human–AI teams: Transparency, explainability, and situation awareness. *Computers in Human Behavior*, 140, 107574.

European Union (2023). Artificial Intelligence Act. Retrieved from www.europarl.europa.eu/RegData/etudes/BRIE/2021/698792/EPRS_BRI(2021)698792_EN.pdf

Faraj, S., Pachidi, S., & Sayegh, K. (2018). Working and organizing in the age of the learning algorithm. *Information and Organization*, 28, 62–70.

Frey, C. B., & Osborne, M. A. (2017). The future of employment: How susceptible are jobs to computerization? *Technological Forecasting and Social Change*, 114, 254–280.

Glassman, E. L., Gu, Z., & Kummerfeld, J. K. (2024). AI-resilient interfaces. *arXiv:2405.08447v1*.

Grote, G. (2023). Shaping the development and use of artificial intelligence: How human factors and ergonomics expertise can become more pertinent. *Ergonomics*, 66, 1702–1710.

Grote, G., Parker, S. K., & Crowston, K. (in press). Taming artificial intelligence: A theory of control-accountability alignment among AI developers and users. *Academy of Management Review*.

Grote, G., Ryser, C., Waefler, T., Windischer, A., & Weik, S. (2000). KOMPASS: A method for complementary function allocation in automated work systems. *International Journal of Human–Computer Studies*, 52(2), 267–287.

Hagtvedt, L. P., Harvey, S., Demir-Caliskan, O., & Hagtvedt, H. (2025). Bright and dark imagining: How creators navigate moral consequences of developing ideas for artificial intelligence. *Academy of Management Journal,* 68(1), 19–49.

Hancock, P. A. (2019). Some pitfalls in the promises of automated and autonomous vehicles. *Ergonomics*, 62, 479–495.

Hardy, C., Maguire, S., Power, M., & Tsoukas, H. (2020). Organizing risk: Organization and management theory for the risk society. *Academy of Management Annals*, 14, 1032–1066.

Hollnagel, E., & Woods, D. D. (2005). *Joint cognitive systems: Foundation of cognitive systems engineering*. New York: Taylor and Francis.

Jordan, N. (1963). Allocation of functions between man and machines in automated systems. *Journal of Applied Psychology*, 47, 161–165.

Kahlert, A., & Grote, G. (2024). "Why should I care?" Understanding technology developers' design mindsets in relation to prospective work design. *European Journal of Work and Organizational Psychology*, 33, 230–244.

Kellogg, K. C., Valentine, M., & Christin, A. (2020). Algorithms at work: The new contested terrain of control. *Academy of Management Annals*, 14, 366–410.

Kim, B., & Doshi-Velez, F. (2021). Machine learning techniques for accountability. *AI Magazine*, 42, 47–52.

McLean, S., Read, G. J. M., Thompson, J., Baber, C., Stanton, N. A., & Salmon, P. M. (2023). The risks associated with artificial general intelligence: A systematic review. *Journal of Experimental & Theoretical Artificial Intelligence,* 35(5), 649–663.

Möhlmann, M., Zalmanson, L., Henfridsson, O., & Gregory, R. W. (2021). Algorithmic management of work on online labor platforms: When matching meets control. *MIS Quarterly,* 45, 1999–2022.

Mumford, E. (2000). A socio-technical approach to systems design. *Requirements Engineering*, 5, 125–133.

NIST (2023). Artificial Intelligence risk management framework (AI RMF 1.0). Retrieved from https://doi.org/10.6028/NIST.AI.100-1

Norman, D. A. (1986). New views of information processing: implications for intelligent decision support systems. In E. Hollnagel, G. Mancini, & D. D. Woods (Eds.), *Intelligent decision support in process environments* (pp. 123–136). Berlin: Springer.

Orlikowski, W. J., & Scott, S. V. (2008). Sociomateriality: Challenging the separation of technology, work, and organization. *Academy of Management Annals*, 2, 433–474.

Parasuraman, R., & Riley, V. (1997). Humans and automation: Use, misuse, disuse, abuse. *Human Factors*, 39, 230–253.

Parasuraman, R., Sheridan, T. B., & Wickens, C. D. (2000). A model for types and levels of human interaction with automation. *IEEE Transactions on Systems, Man, and Cybernetics–Part A: Systems and Humans*, 30, 286–297.

Parent-Rocheleau, X., & Parker, S. K. (2022). Algorithms as work designers: How algorithmic management influences the design of jobs. *Human Resource Management Review*, 32, 100838.

Parker, S. K., & Grote, G. (2022). Automation, algorithms, and beyond: Why work design matters more than ever in a digital world. *Applied Psychology: An International Review*, 71, 1171–1204.

Parker, S. K., & Boeing, A. A. (2023). Workplace innovation in the digital era: A role for SMART work design. In P. Oeij, S. Dhondt, & A. McMurray (Eds.), *A research agenda for workplace innovation* (pp. 91–112). Northampton, MA: Edward Elgar Publishing.

Rahman, H. A. (2021). The invisible cage: Workers' reactivity to opaque algorithmic evaluations. *Administrative Science Quarterly*, 66, 945–988.

Rudin, C. (2019). Stop explaining black box machine learning models for high-stakes decisions and use interpretable models instead. *Nature Machine Intelligence,* 1, 206– 215.

Salmon, P. M., Baber, C., Burns, C., Carden, T., Cooke, N., Cummings, M., Hancock, P., McLean, S., Read, G. J. M., & Stanton, N. A. (2023). Managing the risks of artificial general intelligence: A human factors and ergonomics perspective. *Human Factors and Ergonomics in Manufacturing & Service Industries*, 33, 366–378.

Salmon, P. M., Carden, T., & Hancock, P. (2021). Putting the humanity into inhuman systems: How human factors and ergonomics can be used to manage the risks associated with artificial general intelligence. *Human Factors and Ergonomics in Manufacturing & Service Industries*, 31, 223–236.

Slota, S. C., Fleischmann, K . R., Greenberg, S., Verma, N., Cummings, B., Li, L., & Shenefiel, C. (2023). Many hands make many fingers to point: Challenges in creating accountable AI. *AI & Society*, 38, 1287–1299.

Waterson, P. (2014). Health information technology and sociotechnical systems: A progress report on recent developments within the UK National Health Service (NHS). *Applied Ergonomics*, 45, 150–161.

Wäfler, T., Grote, G., Windischer, A., & Ryser, C. (2003). KOMPASS: A method for complementary system design. In E. Hollnagel (Ed.), *Handbook of cognitive task design* (pp. 477–502). Mahwah, NJ: Lawrence Erlbaum.

Section IX

Future Directions and Applications

41 Engineering, Society, and Environment

Interaction and Integration

Avigdor Zonnenshain and Arnon Bentur

INTRODUCTION

Engineering has traditionally played a vital role in the development of civilization, integrated well within all human deeds. It has been going through continuous changes over centuries, some of which are categorized as paradigm shifts. Nevertheless, the engineering profession is currently facing serious challenges due to rapidly changing technological advancements, globalization, sustainability concerns, and increasing societal expectations. The concept of engineering is also changing rapidly. Engineers no longer work in isolation; they must collaborate with other professionals from various disciplines and engage with society. These challenges call for a critical reflection on engineering's role in society and its relationship with the environment. It is crucial to redefine this concept to meet current and future needs, ensuring that engineering solutions are not only technically sound but also environmentally sustainable and socially responsible. This paper is based on our experience, expertise, and the attached references.

ENGINEERING AS A SOCIETAL RESOURCE

Engineering should be considered a societal resource, not simply a technical endeavor. This perspective emphasizes that the primary purpose of engineering is to serve society by addressing its needs and improving the quality of life. This requires engineers to move beyond a narrow technical focus and consider the broader societal impacts of their work. Engineering activities should contribute to the advancement of society's well-being and progress. This encompasses not only economic development, but also social equity, environmental protection, and cultural preservation. By viewing engineering as a societal resource, it enables engineers to contribute effectively to sustainable development and address global challenges such as climate change, poverty, and inequality.

SOCIETAL GOALS

The overarching goal of society is to improve the overall quality of life for its members. This is a complex and multifaceted goal that can be achieved through various means,

DOI: 10.1201/9781003666431-50

including economic growth, social progress, environmental sustainability, and technological advancement. These objectives should be pursued in a balanced and integrated manner. Economic growth should not come at the expense of environmental degradation or social inequality. Technological advancements should be harnessed to address societal challenges and improve the well-being of all members of society. Engineers play a critical role in achieving these societal goals. Their work directly impacts the economic, social, and environmental aspects of society. Therefore, it is essential for engineers to understand and align their activities with these goals.

MODES OF ACTION AND ENGINEERS TO MEET SOCIETAL GOALS

To meet these goals, engineers must adopt a more holistic and integrated approach to their work. This involves:

- *Understanding Societal Needs:* Engineers need to actively engage with society to understand its needs and aspirations. This requires effective communication and collaboration with various stakeholders, including the public, policymakers, and other professionals.
- *Considering Social and Environmental Impacts:* Engineers must carefully assess the potential social and environmental impacts of their projects. This includes conducting thorough impact assessments and incorporating sustainability principles into their designs.
- *Developing Sustainable Solutions:* Engineers should strive to develop solutions that are not only technically feasible but also environmentally sustainable and socially responsible. This requires innovation, creativity, and a willingness to challenge traditional engineering practices.
- *Promoting Ethical Practice:* Engineers must adhere to the highest ethical standards in their professional practice. This includes acting with integrity, honesty, and transparency, and prioritizing the public good over personal gain.
- *Adopting a system thinking approach:* By taking on the complexity and interdependency of our social-technical systems that consist of several components and processes of interactions to fulfil a purpose or goal, engineers need to consider the complexity of the systems they interact with, to consider the whole life cycle, and to anticipate unintended consequences of their work.
- *Embracing Lifelong Learning:* Engineering is a constantly evolving field. Engineers must commit to lifelong learning to stay abreast of new technologies, methods, and societal challenges.

THE ENGINEERING PROFESSION: PROFESSIONALISM, ETHICS, AND HOLISTIC APPROACH

The engineering profession is characterized by its commitment to professionalism, ethical conduct, and a holistic approach to problem-solving. Professionalism in engineering entails technical competence, integrity, accountability, and a dedication to serving the public good. It involves adhering to codes of ethics, upholding standards of practice, and prioritizing the safety, health, and welfare of the public.

Ethical considerations are central to engineering practice. Engineers are expected to make decisions that are not only technically sound, but also morally defensible. This requires considering the potential impacts of their work on individuals, communities, and the environment. Codes of ethics provide guidance on ethical dilemmas and promote responsible decision-making.

A holistic approach in engineering emphasizes the interconnectedness of technical, social, and environmental factors. It involves considering the entire life cycle of a project, from design and construction to operation and disposal. Engineers need to collaborate with other professionals from various disciplines to address complex challenges that transcend traditional engineering boundaries. This holistic perspective ensures that engineering solutions are comprehensive, sustainable, and aligned with societal goals.

QUANTIFYING SOCIAL RESPONSIBILITY

Quantifying social responsibility is a challenging task, but it is essential for measuring and improving the societal impact of engineering activities. While traditional engineering metrics often focus on technical performance and economic efficiency, social responsibility requires considering broader social and environmental outcomes.

One approach to quantifying social responsibility is through the development of social indicators. These indicators can measure various aspects of societal impact, such as employment creation, community development, environmental protection, and social equity. Life Cycle Assessment (LCA) is also used to evaluate the environmental impact of a product. In the future, the use of LCA may be enhanced by the inclusion of Social Life Cycle Assessment (SLCA) indicators and data. By tracking these indicators over time, it is possible to assess the progress of engineering activities toward achieving social and environmental goals.

Another approach is to integrate social and environmental considerations into project evaluation frameworks. This involves incorporating social and environmental costs and benefits into cost–benefit analyses and other decision-making tools. This ensures that social and environmental factors are weighed in engineering decisions.

ENGINEERING–SOCIETY–ENVIRONMENTAL INTERACTIONS

Engineering, society, and the environment are interconnected and interdependent. Engineering activities significantly impact society and the environment, while societal needs and environmental conditions shape the direction of engineering development.

Engineering to Society: Engineering provides essential infrastructure, goods, and services that improve the quality of life. However, engineering projects can also have negative social impacts, such as displacement, inequality, and cultural disruption.

Engineering on the Environment: Engineering activities can have positive and negative environmental impacts. Positive impacts include the development of sustainable technologies, renewable energy sources, and pollution control systems. Negative impacts include resource depletion, pollution, habitat destruction, and climate change.

Society to Engineering: Societal needs, values, and priorities influence the direction of engineering research, development, and application. Public policies, regulations, and social movements can also shape engineering practices.

Environment to Engineering: Environmental conditions and constraints can limit or influence engineering activities. For example, climate change is driving the development of more resilient and sustainable infrastructure.

Understanding these complex interactions is crucial for promoting sustainable development. Engineering decisions must consider the potential social and environmental consequences, while societal policies and practices must support engineering solutions that are both technically feasible and environmentally responsible.

QUANTIFICATION OF ENGINEERING–SOCIETY–ENVIRONMENT INTERACTIONS

Quantifying the complex interactions between engineering, society, and the environment is essential for effective decision-making and sustainable development. This requires developing appropriate metrics, tools, and methodologies to measure and assess the impacts of engineering activities on social and environmental systems.

Environmental Impact Assessment (EIA) is a well-established tool for evaluating the potential environmental consequences of engineering projects. EIA involves identifying, predicting, and assessing the environmental effects of a proposed project or development. It also includes measures to mitigate adverse impacts and enhance positive ones.

LCA is another important methodology for quantifying the environmental impacts of products and services throughout their entire life cycle, from raw material extraction to disposal. LCA can help engineers identify opportunities to reduce ecological footprints and promote sustainable consumption.

Social Impact Assessment (SIA) assesses the potential social effects of planned interventions or developments. SIA involves analyzing, monitoring, and managing the intended and unintended social consequences of policies, programs, and projects.

Input–output analysis can also be used to quantify the economic, social, and environmental impacts of different sectors of the economy, including engineering. This method can help to understand the interdependencies between various industries and assess the overall effects of engineering activities on the economy, society, and the environment.

SOCIETAL CONSIDERATIONS IN ENGINEERING DESIGN

Societal considerations should be integrated into every stage of the engineering design process. This requires shifting from a purely technical approach to a more holistic and human-centered design philosophy.

- *Needs Assessment:* Engineers must start by understanding the needs and aspirations of the people affected by their designs. This involves conducting thorough research, engaging with stakeholders, and considering diverse perspectives.
- *Value-Sensitive Design:* Engineers should strive to design systems that are functional and align with society's values and ethical principles. This includes considering fairness, equity, privacy, and security issues.
- *Participatory Design:* Involving users and other stakeholders in the design process can lead to more acceptable and effective solutions. Participatory design empowers communities and ensures that their voices are heard.
- *Universal Design:* Engineering designs should be accessible and usable by people of all ages, abilities, and backgrounds. Universal design principles promote inclusivity and reduce barriers to participation.
- *Sustainability:* Sustainability should be a central consideration in engineering design. This involves minimizing environmental impacts, conserving resources, and promoting long-term resilience.

By integrating societal considerations into engineering design, engineers can create solutions that are not only technically sound but also socially responsible, environmentally sustainable, and aligned with the needs and values of society.

CONCLUDING REMARKS

Integrating engineering, society, and the environment is a critical challenge and opportunity for the engineering profession. Engineers can play a pivotal role in creating a sustainable and equitable future by redefining engineering as a societal resource, adopting a holistic and ethical approach, and actively engaging with society and the environment.

Quantifying the complex interactions between engineering, society, and the environment is essential for effective decision-making and sustainable development. This requires developing and applying appropriate metrics, tools, and methodologies to measure and assess engineering activities' social and environmental impacts.

Societal considerations must be embedded in every stage of the engineering design process, from needs assessment to implementation. This shift toward a human-centered and values-driven design philosophy will enable engineers to create solutions that are not only technically sound but also socially responsible, environmentally sustainable, and aligned with the diverse needs and aspirations of society. By embracing these principles, engineers can contribute meaningfully to advancing society's well-being and preserving our planet for future generations.

BIBLIOGRAPHY

ABET (2021). *Criteria for Accrediting Engineering Programs 2022–2023*, Engineering Accreditation Commission.

Bjørn, A., Owsianiak, M., Molin, C. & Hauschild, M.Z. (2018). LCA history. Chapter 3 in Hauschild, M.Z., Rosenbaum, R.K., & Olsen, S.I. (Eds.), *Life Cycle Assessment: Theory and Practice*. Springer, pp. 17–30.

Boy, G.A. (2020). *Human Systems Integration: From Virtual to Tangible*. CRC Press – Taylor & Francis Group, USA.

Christensen, S.H., Delahousse, B., Didier, C., Meganck, M. & Murphy, M. (2019). *The Engineering-Business Nexus*. Springer.

Conlon, E. (2019). Prisoners of the capitalist machine: Captivity and the corporate engineer. In Christensen, S.H. et al. (Eds.), *The Engineering-Business Nexus*. Springer.

Davis, M. (2019). Engineering and business management: The odd couple. Chapter 2 in Christensen, S.H., Delahousse, B., Didier, C., Meganck, M., & Murphy, M., (Eds.), *The Engineering-Business Nexus*. Springer.

Deane, P. (1978). *The First Industrial Revolution*. Cambridge University Press.

Ferguson, E.S. (1993). *Engineering and the Mind's Eye*. The MIT Press.

Fledderman, C.B. (2012). *Engineering Ethics*, Fourth Edition. Prentice Hall.

Garrison, E.G. (1991). *A History of Engineering and Technology: Artful Methods*. CRC Press.

Gill, J. & Zonnenshain, A. (2021). *Opportunities and Challenges of Sociotechnical Systems Engineering, IS 2021*, INCOSE International Symposium.

Grayson, L.P. (1993). *The Making of an Engineer: An Illustrated History of Engineering Education in the United States and Canada*. Wiley.

Hajek, P., Ctislav, F., Birit Buhr, J. & Ivana, L. (2013). Integrated life cycle assessment of concrete structures. *FIB Bulletin*, 71.

Harris, C.E. (2008). The good engineer: Giving virtue its due in engineering ethics. *Science and Engineering Ethics*, 14, Article 153.

INCOSE (2023). *The INCOSE SE Handbook*, 5th Edition.

John, T. (2016). *Introduction to Systems-Theoretic Process Analysis (STPA)*. MIT Press.

Karp, B., Zonneshain, A. & Bentur, A. (2023). *Engineering, Society and Environment, Relations and Interaction (in Hebrew)*, Research Report, Neaman Institute for National Policy Research, Technion.

Kenett, R., Swarz, R.S. & Zonnenshain, A. (2020). *Systems Engineering in the Fourth Industrial Revolution, Big Data, Novel Technologies, and Modern Systems Engineering*. Wiley.

Leveson, N. (2012). *Engineering a Safer World. Systems Thinking Applied to Safety*. MIT Press.

McNeil, I. (Ed.) (1990). *An Encyclopaedia of the History of Technology*. Routledge.

Meiksins, P. (1988). The "Revolt of the Engineers" reconsidered. *Technology and Culture*, 29(2), 219–246.

Mitcham, C. & Briggle, A. (2009). The Interaction of Ethics and Technology in Historical Perspective. In Meijers A. (Ed.), *Philosophy of Technology and Engineering Sciences*. Elsevier, pp. 1147–1191.

OECD (2019). Future of education and skills 2030: OECD learning compass 2030, OECD 2019.

Parrillo, V.N. (Ed.) (2008). *Encyclopedia of Social Problems*. SAGE Publications, Inc.

Royal Academy of Engineering (2011). *Engineering Ethics in Practice: A Guide for Engineers*. The Royal Academy of Engineering.

Samuel, A. & Weir, J. (1999). *Introduction to Engineering Design*. Elsevier Science & Technology Books.

Sarfraz, Z., Sarfraz, A., Iftikar, H.M. & Akhund, R. (2021). Is COVID-19 pushing us to the Fifth Industrial Revolution (Society 5.0)? *Pakistan Journal of Medical Sciences*, 37(2), 591–594. https://doi.org/10.12669/pjms.37.2.3387

Savitz, A.W. (2006). *The Triple Bottom Line*. Wiley.

Snooks, G. (1994). *Was the Industrial Revolution Necessary*? Palgrave.

UN, Filho, W.L. (Ed.) (2021). *Encyclopedia of the UN Sustainable Development Goals.* Springer. https://link.springer.com/content/pdf/bfm:978-3-319-95873-6/1.pdf

UNEP (2020). *Guidelines for Social Life Cycle Assessment of Products and Organizations 2020.* United Nations Environment Programme (UNEP).

Yang, S., Maa, K., Liua, Z., Renb, J. & Manc, Y. (2020). Development and applicability of life methodologies. Chapter 5 in *Life Cycle Sustainability Assessment for Decision-Making*, pp. 95–124. Elsevier.

Zonnenshain, A. (2006). *The Corporate Social Responsibility (CSR), The System Perspective and The System Engineering Role.* INCOSE IS 2008 – The International Symposium of INCOSE.

Zonnenshain, A. & Sheps, I. (2008). *Design for Sustainability: The Challenge for Systems Engineering.* INCOSE IS 2008 – The International Symposium of INCOSE.

42 Pragmatic Tools for Human–System Integration

Ondrej Doule

INTRODUCTION

Human–Systems Interaction (HSI) surrounds us everywhere and anywhere in our time-space, in a perceivable envelope. HSI starts as any other project with a broad description of needs and goals to inform about the project or problem envelope. However, it appears to be wide and complex due to the emphasis on the human element. This dynamic, ever-evolving actor affects the established goals, mission, project, or program. To gain the required knowledge and to verify the needs that are relevant to the project scope, a principal analysis of the human behavior scope is essential to avoid so-called scope creep, to mitigate the correct set of risks to prevent "over" or "under design" problems, and to reach results of robust and well-functioning solutions.

A structured process is as important as freedom of creative and ad hoc development, as each level of order also affords different tools and environments to achieve what a specific HSI project requires. HSI, unlike other domains, can be part of a program or project from the very beginning, an integrative part of the project (A), assuring ideal human performance by design, or HSI can serve as a supporting domain (B) to optimization of an existing phase of the project/product during its lifecycle development or optimization of an existing project/product while reverse engineering its operational, organizational context, forms, and functions.

In this chapter, we will look at the main strategic elements of a successful project from the perspective of a dynamic and complex system, which we can consider as tools and their contribution in different contexts, from broad organizational to product-specific. The concept or principle of the "*tool*" blends with *method* or *approach* in this chapter and attempts to emphasize that traditional "toolbox" of businessmen as well as strategists or system designers—and architects is moving in the environment of mental modeling, comprehension, decision-making, abstraction, and conceptualization rather than the utilization of specific software or instruments as a software tool. The "tools" in our sense, thus, are becoming cognitive guides that help us to succeed in our businesses. Why this shift from a specific procedure to emphasis on comprehension? There is a simple reason: the increasing omnipresence of automation. We will shape our world around us by the correct comprehension and capacity to abstract, conceptualize, and understand dynamic and complex systems, rather than

DOI: 10.1201/9781003666431-51

the capability to engineer a system. In contrast, that capability will be increasingly more automated. The capacity of value abstraction, conceptualization, and determination of the right options without tangibilizing the solutions will enable less costly and more fitting integration of human and environmental needs.

ABSTRACT MODELING

A detailed, abstract record of a project (client's environment or system context) is not always the standard way to start a project. While originating in aerospace and the building industry, the methods of detailed graphical description of existing environments in the project's scope, or the product development scope, are becoming more common in many industries. Diligent focus on human needs in this early phase of a project is critical for HSI. Nowadays, the software industry provides various tools that are also available due to free common modeling standards, e.g., by the Object Modeling Group (OMG).

Proven by practice and originating from older modeling languages, the SysML/UML languages became well known and distributed through many domains. They became a standard for abstractly modeling various system elements before the physical design and detailed engineering.

Graphical object modeling languages and related methods and applications that may be perceived as unintuitive to creative product developers are expanding the world of human-system integration significantly, as they allow describing the multidimensional world of relationships between humans and their existing or needed interactions. They require immersion into and detailing an abstract world that precedes the creative phase of physical concept or solution initiation and the description of many intangible components, starting from the project requirements parameters based on the contextual models.

Based on Unified Modeling Language and originating from other systems and environment modeling efforts, such as Object Process Modeling language and method, Rationale Unified Process (RUP), a Systems Modeling Language was created as well structured and optimized for abstract systems modeling.

Systems engineering has developed a detailed practice of abstract modeling of product context by breaking the design problem into four categories: the client's requirements, requested product functions, logical flows, and physical modeling of the solutions. While the client's or product requirements belong to the so-called "Human and environment" problem or request domain, the other three are solution-driven. While the first three are abstract (intangible), the physical category represents a non-abstract, tangible, product-oriented solution. Physical modeling is the easiest for human imagination and a source for developing abstract modeling practices and skills.

While the physical model can be considered a direct representation of the solution that represents the required functional and logical project needs, the requirements model could be viewed as an indirect solution model or metamodel of the solution.

The power of the abstract models (Figure 42.1) lies in their simplicity and speed of deployment, simulation, and testing with minimum resources. While physical

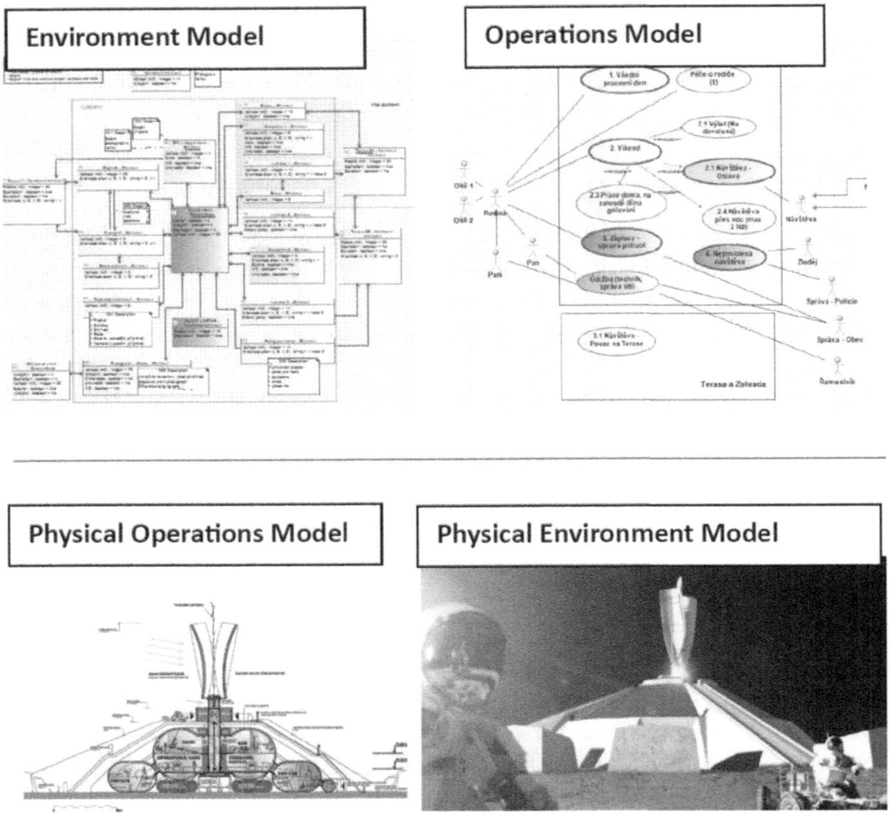

FIGURE 42.1 Abstract (top) versus physical models (bottom) of one product/project, considering the physical model as the lowest level of abstraction. Bottom models show the Lunar Base concept LB10 by Ondrej Doule.

models require more time, processing power, and workforce, and their modification can be costly, abstract models allow rapid changes and account for human and system behavior modeling capability compared to physical models. On the other hand, the drawbacks of abstract models include requirements on the mental ability of abstract thinking, concept thinking, and experience with actual potential solutions.

Human abstraction and mental capability are required, for example, during environment modeling (here the "Environment" is thought of as a context (surrounding and impacting/impacted elements) of the client's product or problem that poses a continuous challenge due to its enormous scope, complexity, and dynamic or distributed character). A well-comprehended environment and its abstraction are the basis for a quality integration of the human in the system and the system in the surrounding context. HSI can primarily benefit from the methods, frameworks, and languages developed for the most complex systems humans have ever created in aerospace, automotive, and building industries, designed to decompose, master, and control the complexity through a structured modeling approach. Therefore, abstract modeling will soon penetrate all sectors, research, and academic domains and become essential to human business and design thinking processes.

ENVIRONMENT OR CONTEXT MODELING

While OMG provides us with comprehensive languages, there are numerous metamodels, methods, and frameworks to support the usage of such languages. Depending on the needed application, specific frameworks are chosen to simplify, structure, and speed up the abstract modeling process. Predefined abstract ontologies, taxonomy, and domain typologies are always useful as guides when proven by industry application or research data.

Based on the focus and scope of the project, we should be able to select the proper framework for our abstract models. Therefore, every project requires correct classification and definition of precedents that informed the creation of a framework.

- Value models are business or philosophical abstract models that interpret the basic ideological framework of business, organization, society, or an individual. This area has been addressed from many perspectives. Business projects often work with value models to explain the monetary values, while societal value models may be closer to human-centered humanistic topics.
- Enterprise modeling frameworks support the enterprise environment. For example, pragmatic enterprise operation use is supported by the EA3 Cube framework, while the Unified Architecture Frameworks support enterprise acquisition needs. Organizational structure and various elements are expressed in detail, including relationships and their classes, programs, and types of allocation. These are just examples, while many others are available for enterprise-level human–system integration.
- Technical abstract metamodels support the actual definition of artificial and natural systems at the part level, as well as a description of their elements and behavior, etc.

Each framework includes elements based on the focus, domain, and human involvement. Understanding the complexity and variability of human systems, Prof. Guy Andre Boy developed a practical tool for environmental modeling for human–system integration without any domain architecture level bias, applicable to any problem. The Artifact User Technology Organization Scenario (AUTOS) tool affords interpretation and abstraction of highly complex systems with humans in the loop while preserving and explicating the fundamental relationships that define the existence of the environment or context.

AUTOS (Boy, 2020) is a multidimensional metamodel of environment, as each of the AUTOS elements exhibits three relationships (potentially dimensions) that further allow for detailed description using tools such as situational awareness model, functional analysis model, etc. The AUTOS tool, combined with another system engineering tool called N-square, enables a two-dimensional expression of a context and a detailed description of the environment in which human and product co-exist (considering that the product is the focus of our study). Such a metamodel (tool) can be used for any problem society or human exhibits. The only challenge is properly comprehending the

AUTOS elements and their relationships, their subsequent interpretation of the studied subject, and the following steps to develop the abstract or physical solution models.

AUTOS is one of the most comprehensive abstraction supporting tools for context modeling. This tool is a highly integrated metamodel of the human environment (meaning it represents multiple relationships within itself, interconnecting its five essential meta-elements), which affords decomposition of the problem and its environment as well as meta-solution, i.e., requirement's decomposition and definition of the metamodel relationships. Utilizing this tool requires practice and abstraction skills, the capability of context perception and decomposition, and abstract problem-solving (definition of requirements). AUTOS also requires an interdisciplinary background that enables an architect-level perspective on the human in the loop of the designed system usage in dedicated environments.

DESIGN AND MODELING PROCESS

Frameworks also exist as aids or tools to guide the development of standard project problem solving. Such frameworks differ based on the domain or field of interest to which the project belongs.

One of the comprehensive and industry-proven frameworks is Dassault Systèmes "MagicGrid," developed for the execution of the design process that also considers the human role and respects standard SysML syntax and diagram topology (Structural, Behavioral, Requirements, and Parametrics). This framework is a derivative of the standard OMG SysML/UML language typology and traditional system engineering practices reflected by the ISO 15288 standard (Figure 42.2). This framework can be perceived as a universal project lifecycle structure that guides the project from its inception, starting with the client's requirements, to a physical solution to be delivered to the client as a final product. Standards such as Human Spaceflight ANSI-*AIAA S-153-2020 emphasize* Human–System integration, explicitly stating the need for Human–System testing and simulation.

These standards represent abstract frameworks, in the form of a product lifecycle metamodel that can be used to its full capacity or partially, based on the required development phase. Specifically, the modeling-oriented OMG-based MagicGrid applies to tangible or non-tangible product solutions.

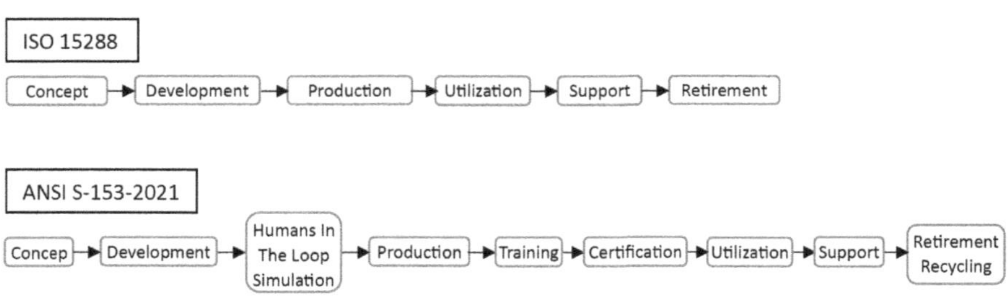

FIGURE 42.2 Standard frameworks for the product lifecycle. Left System Engineering ISO 15288 right ANSI-AIAA S-153-2021 emphasizes HITLS and training phases for any business, stemming from the human spaceflight domain (Refs. 1, 2).

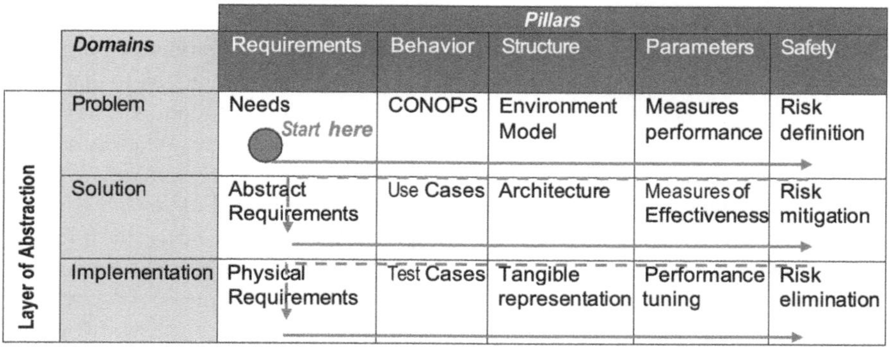

FIGURE 42.3 Design and modeling process framework by Dassault Systèmes, "MagicGrid" structure. Note that the content of the matrix may change based on the project, while the domains and pillars stay the same. Blue arrows show a linear approach to the design or modeling process, sequentially addressing the Problem, Solution, and Implementation Domains (Refs. 3–5).

MagicGrid considers three domains of information and data: problem, solution, and implementation. The term "problem" selected by this framework is perceived as a "design challenge" rather than a domain with a negative connotation. Figure 42.3 represents one of many interpretations of MagicGrid for a specific human safety-focused project. Elements such as Use Cases can further follow the taxonomy Use Case-Scenario explained in the next chapter. Similarly, the element Environment model can be represented by the tools introduced in the previous chapter, "Engineering, Society, and Environment: Interaction and Integration."

Such a framework is similar to or identical to traditional system engineering, such as "RFLP," which consists of functional, logical, and physical requirements. A linear approach is the most intuitive, starting with defining the Problem (Challenge) domain, continuing from top left to bottom right.

ORGANIZATIONAL OR HUMAN BEHAVIORAL MODELING

The essence of human activities modeling goes far back to before the OMG, but the typology or taxonomy of behavioral components is far from unified until now. Many disciplines are prone to developing their approaches based on immediate needs. For HSI, we attempt to create a complete scope and integrate various approaches, so they do not conflict. Therefore, the definitions of the main objects are fundamental, starting with Use Case, Scenario, and Concept of Operations, and ending with scenarios classification to nominal, off-nominal, or emergency, while understanding the difference regarding Process or Business Process, and attempting to follow industry standards such as American Productivity & Quality Center (APQC). First, let us look at simplified definitions to create a uniform mental framework for ourselves. Then, the relationship between these behavioral meta-elements is depicted in the metamodel.

Definition of Use Case: Interaction of systems, humans, or other actors within the relevant scope and environment, usually with a designed product or organization, or by behaving toward a common goal to generate value. Note that a Use Case is not a behavior describing when something goes wrong. A Use Case is always a behavior description with a positive value outcome. When something goes wrong, we define such behavior as Off-Nominal or Emergency Scenario when executing a specific Use Case that can be composed of several Business Processes, for example, and we can say that the Use Case has the following scenarios and is executed according to the following Business Processes.

Definition of Scenario: Scenario is one type or style of execution of a Use Case, business process(s), etc. It can have a few forms, whether successful or not. It is the actual behavior of the actor (human or system). Scenarios are often distinguished as nominal, off-nominal, and emergency. The three types are usually grouped into nominal and off-nominal, with the latter containing emergency scenarios. The difference between the "Standard Off-Nominal" scenario and the "Emergency Off-Nominal" scenario is in the time component. Emergency scenarios require rapid action or response to unwanted systems or actors' behavior. At the same time, standard off-nominal scenarios can be addressed without urgency. Scenarios are sometimes also discovered during testing or iterative prototyping. We call such scenarios "Emergent." Such scenarios must be considered in the Use Cases, mission, business process, or product design. They may require transforming initial requirements, business goals, or a design change. Emergent scenarios that are not discovered may cause a system failure or increase the risk of operations. Often, emergent scenarios result from unpredictable human behavior or influences of complex systems such as weather or interactions with society. It is therefore essential to execute thorough systems behavioral modeling and Human-in-the-Loop (HITL) testing since, without it,

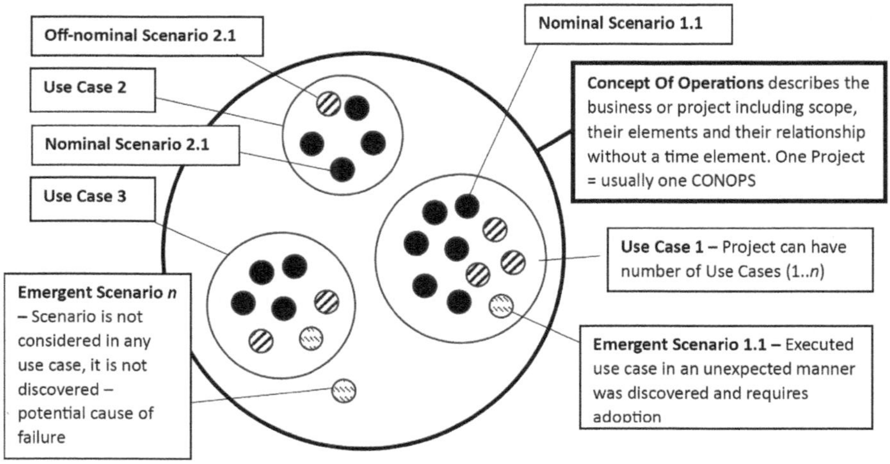

FIGURE 42.4 Graphical explanation of behavior model types based on Use Case, Scenario, and CONOPS classification.

the emergent scenarios will most certainly unexpectedly affect the system performance (see Figure 42.4).

Definition of Concept of Operations: Often called Concept of Operations (CONOPS), it is usually a depiction or written description of business or project elements and their mutual interaction within the relevant scope and environment. This timeless depiction shows all crucial actors indicating behavior or functional or logical exchanges, such as data communication, movement, or transformation that could happen in time. On the other hand, a "**Mission**" describes the usage of the CONOPS over time as a sequential/chronological action with a concrete start and end.

Definition of (Business) Process: A Process is usually a prescribed set of actions or activities to be executed according to some rules. Someone thus completes the process, and when it is being executed, the description of the execution is in the **Scenario.**

HUMANS IN THE LOOP TESTING AND SIMULATIONS

HITL in this chapter represents the simulation used for system development, design, verification, and validation. The HITL in this chapter is not considered a type of training or use of an artificial agent that supplements the human element.

To perform developmental and testing HITLS as early as possible in the product or value lifecycle (development) has always been challenging for all businesses. The costliest type of testing—testing with real human users—has been a "privilege" of the systems with the highest criticality or risk to human safety. Such testing can often be omitted in non-life-critical systems development due to high cost or complexity and is replaced by component or subsystem testing, with some prototype usability studies only.

Nonetheless, nowadays we understand that the sooner we involve the user in the product development loop, the cheaper and more effective the delivered solution will be. Therefore, it is reasonable to consider other than physical types of

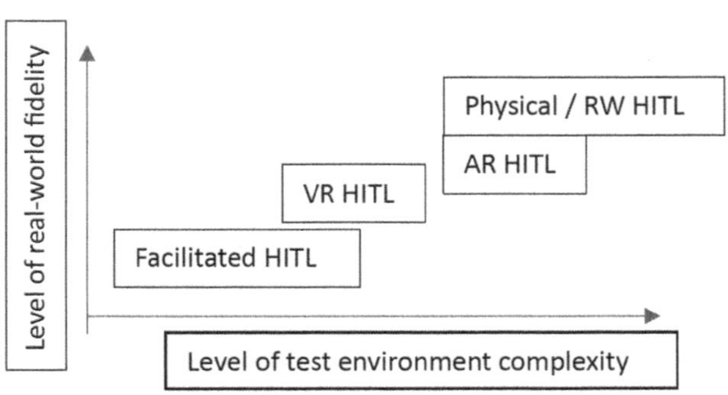

FIGURE 42.5 Level of HITL fidelity relative to test complexity.

humans in the loop tests that are less complex, less costly, and less time-consuming (Figure 42.5).

We can distinguish four types of HITLS that utilize and benefit from the most recent technology and allow for very early human–system tests:

- *Facilitated Virtual Tests and Simulations:* Tests executed with users remotely by guiding users through tested environments, either by a human or a system guide, are a very effective way of conducting any system test. Such testing takes different forms in different industries, but can be executed effectively in real-time or offline. Such a test requires a kind of offline autonomous system preparation that is either auto-executed based on the activation of the test by the user or is conducted by a remote operator based on a scheduled virtual meeting. The test scope is entirely under the facilitator's control and can contain static, dynamic virtual elements, or operations in real time. The user is an observer who requests or comments on operations in a virtual environment observed and shared by the facilitator. The user and the facilitator may be in different places (continents, time zones). This test provides an affordable way to test even the most complex systems without needing powerful IT infrastructure. It enables guided walk-throughs with ad-hoc investigations based on users' behavior or prescribed, pre-recorded, or process-based testing with standard questionnaire-type feedback.
- *Virtual HITLS*: Test users with access to the minimum required technical infrastructure for testing execution can participate in virtual HITLS. This testing type is a step up regarding information and data sharing with the test users while enabling higher autonomy and potentially keeping the geographical location enabled by the Facilitate virtual tests and simulations. Virtual HITLS may serve as a complete supplement to physical HITL when in-house testing infrastructure does not allow for physical testing due to cost to manufacturing, environmental access, or technical complexity. Virtual HITLS enables full autonomy and high fidelity for test users, although it is only tangible in virtual settings. Similarly, environment fidelity is limited due to human—real world physics, scenario, interactivity, technical or social complexity, and emulation capacity.

AR HITLS: Similar to VR HITLS, with the difference of integration of a real-world environment or real-world physical components. In certain situations, AR HITL is the most suitable due to emulation of complex parameters or systems that would otherwise be very costly; in other situations, AR HITL may be more expensive and more complex than physical HTIL due to its IT component and implementation requirements. AR HITLS are used in special situations where, for example, access to specific components is restricted, or where physics or environmental context is, for some reason, unavailable, or conversely, when the product is required to be tested in extreme or complex environmental contexts and its manufacturing is too costly. AR HITLS connects the best of both worlds but has to be carefully evaluated for its cost and real-world emulation efficiency.

Real World (RW)/Physical HITLs—DITL—Considering that a physical product or environment is required for the product or business testing, it represents the highest fidelity. Often defined by alpha and beta prototypes in the product or software development world, RW/physical HITLs are the most rewarding, highest fidelity simulations that may or may not be the costliest and may or may not carry the highest return on investment. Similarly, these HITLS carry the safety risk associated with the human operations and may be combined with VR testing or focused only on specific business components. The business's division into test components carries another risk that may result in emergent risks stemming from the components' implementation or integration and the complete business system operations. Therefore, a Day In The Life (DITL) scenario-based HITLS is often executed, requiring a complete product (business) to be in the loop and the actual nominal scenario for which the product/business is being executed.

NASA HUMAN SYSTEM INTEGRATION HANDBOOK

NASA has often brought scientific innovation that the general industrial or services landscape could have adopted. NASA Spin-Offs publications have many published and commercialized examples. The NASA Human–System Integration Guide is adept at becoming another publicly accessible innovation spin-off that was initially developed for human spaceflight systems and that provides tangible tools for industry, research, and academic branches to benefit from a framework that has proven to deliver cost savings, as well as product efficiency in the space sector.

This sub-chapter introduces and summarizes the core elements of the NASA HSI guide and emphasizes areas that are transferable to other industries, such as NASA-specific HSI methods, goals, NASA HSI domains that develop and compose the HSI discipline, and HSI economic effectiveness from a product development perspective according to the *NASA HSI Handbook and Guide*. The HSI implementation process is introduced from an aerospace domain perspective.

Why Has NASA Developed a Detailed HSI Guide for the Aerospace Industry?

NASA HSI has been developed for the most extreme situations outside of Earth. Systems and businesses developed for extraterrestrial environments are incredibly costly and time-consuming prototypes. NASA has identified that one of the significant roles related to system failures is the human role. Therefore, NASA introduced the HSI guide to save on project costs, increase system and human safety, save development and operation time, and improve business or product resiliency and durability. Most importantly, the NASA HSI methodology supports the minimization of risks. All these improvement factors or HSI Key Performance Indicators (KPIs) are supported by specific case studies originating in NASA projects or missions. Generally, the NASA HSI methods and principles can be well utilized in the Earth environment to benefit HSI on business, architectural, or ergonomics levels. Although HSI should be part of every project from its inception and should be scoped according to the industry and project topics, HSI is also applied to existing, running projects or businesses that seek to enhance their performance according to the above-noted KPIs.

THE NASA HSI method starts with the specification of the HSI organization. The HSI team is composed along with the HSI Plan in a standard or agile project management manner where the team has a hierarchical structure led by an HSI Lead, supported by HSI Practitioners, or also sometimes called Domain leads (the HSI Domains are introduced later in this sub-chapter), and external HSI practitioners, for example, associated with project contractors. The role of the HSI Lead is similar to the role of an architect in an engineering project, which focuses on HSI planning, management, and coordination of the domain tasks executed by the practitioners, facilitation of the interdisciplinary collaboration, decision-making in terms of priority and trade-offs between the HSI domains, and monitoring of risks and opportunities to address the HSI KPIs.

The HSI domains that determine the required specialization of the HSI team roles for accomplishing the project's HSI include human factors engineering, operations, maintainability, habitability, safety, and training domains. The HSI lead evaluates each domain and determines whether the project requires a specialist for the domain. The size of the business or project also determines whether and which HSI roles are represented.

Economically, HSI Guide emphasizes that large projects can improve their performance several times the total value of the expected project, considering a minor investment of 1–6% of the project development cost. Budgeting can be performed per HSI requirement or by % of system engineering effort. Standard project management parameters such as time, resources, scope, complexity, experts' availability, number of standards required for implementation, or other infrastructure requirements help determine the total cost of the HSI effort. HSI budgeting can also be supported by similar, industry-successful projects that allow for building costing parameters based on empirical data.

NASA HSI Guides also emphasize that its implementation and execution have the most significant impact in the early development phases of the business/project. An effective HSI plan supports minimizing training costs and reducing maintenance time and life-cycle costs. Nonetheless, it is noted that HSI must be actively advocated for throughout the business phases to successfully impact its performance, as it is often omitted by inexperienced project management teams.

REFERENCES

1. https://arc.aiaa.org/doi/book/10.2514/4.106286
2. www.iso.org/standard/63711.html
3. www.linkedin.com/pulse/whats-coming-2nd-edition-magicgrid-bok-aiste-aleksandraviciene/
4. www.3ds.com/assets/invest/2023-10/catia-cyber-magicgrid-methodology-flyer.pdf
5. https://discover.3ds.com/magicgrid-book-of-knowledge

43 HSI Design of Digital Twin in the Oil and Gas Domain

Stélian Camara Dit Pinto

INTRODUCTION

Studying human system integration provides critical guidelines for understanding how systems should be designed by and for humans. As expressed in this chapter, human-in-the-loop (HITL) modeling and simulation are powerful tools for designing HSI-oriented systems. HITL modeling and simulation enable a system's design team to test human interaction with a system at a limited cost. This becomes even more important while working with complex systems, as their behavior is, by definition, difficult to predict (Mitchell & Newman, 2001). This increases the complexity of the design process and the difficulty of predicting the impact of humans on the systems and vice versa. HITL modeling and simulation will therefore be the key to understanding those interactions and, through the help of studies and prototypes, will enable us to design an HSI-oriented system (Boy, 2013).

This chapter presents how HITL modeling and simulation were integrated in a design methodology and applied to design Digital Twins (DTs) in the oil and gas domain. This will start by introducing the DT concept and its close relation to the HITL modeling and simulation practices. Then the principles of the methodology and their impact on the designed solution are illustrated. Finally, the possibility of applying a similar methodology to create other complex systems is explored.

UNDERSTANDING DTS

DTs are not trivial engineering solutions that show a virtual image of a system. They are complex systems that use data and information to virtually mimic a real-life complex system and make it manageable by humans. In a more complete definition, they are "*dynamic representations of a physical system and its environment using interconnected data, models, and processes to enable access to knowledge of past, present, and future states to manage action on that system*" (Camara Dit Pinto et al., 2021).

The concept originates from the Product Lean Management (PLM) concept that aims to follow a system state through its life cycle. Multiple definitions have been proposed to either try to encapsulate the concept and its holistic abilities or to define it based on one of its specific roles (Brenner & Hummel, 2017; Grieves,

DOI: 10.1201/9781003666431-52

2014; Kritzinger et al., 2018; Rosen et al., 2015). However, it is essential to highlight the difference between DTs, digital shadows, and digital models (Kritzinger et al., 2018). Those three concepts are close, but they explain the difference between a virtual representation of a system and a fully capable DT (Figure 43.1).

Digital models are virtual models of a system (existing or not). They can be updated manually to reflect the following changes on a real system or used to change the real system manually. Those models are the most common types of digital systems in industry and literature. Digital shadows are virtual models of existing systems that are updated automatically based on the system state to represent its current state as closely as needed. However, if changes need to occur on the real system, they will have to be conducted manually as the digital shadow does not have automated actions on the systems. Finally, DTs are virtual models that will fully automatically adapt their state based on the real system and conduct changes on the real system. Those digital systems open the possibility of automatically managing the real system using the digital one. They are more complex, and their applications in industry or literature are still scarce. However, they are a goal to aim for when implementing a digital system because of their holistic approach and ability to enable humans to interact more with the managed systems. They are, therefore, a great tool to facilitate integration between humans and complex systems.

Another important distinction to make when dealing with DTs is modeling and simulation from an engineering and HSI point of view. As part of the capabilities and techniques used by DT, modeling and simulation are among the most important ones. Modeling is commonly defined as implementing a simple system description to make calculations or predict what might happen. In engineering, dealing with complex systems consists of decomposing a complex system into multiple simpler components that we can describe mathematically (Sillitto et al., 2019). For example, in the case of an oil-and-gas platform, the platform will be modeled into different sub-sections depending on the processes taking place in each. The final model will be used to run simulations and mathematically predict the various outputs of the system. For example, chemical separation of the compounds in the mixture will happen in a separation tank, and a specific pressure will build up in the tank. To manage the mixture input, the user must be careful that the pressure stays below a particular threshold. Simulation will enable you to predict the pressure built up and input the best amount of mixture.

HITL modeling and simulation in the context of HSI is an adaptation of the techniques used in engineering to model human functions and simulate human activities on a prototype to ensure the best system design. This is important as modeling is, by nature, not going to be holistic, as it is a simple representation of an infinitely complex real system. Therefore, it is essential to identify the elements and behavior of the system that are important for the model to be used. That is where we can see that modeling a system needs to be performed with the humans using the model in mind. For example, suppose we want to use a model to simulate the output pressure of the equipment. In that case, its three-dimensional (3D) geometry is not as important as modeling the chemical reaction happening in the equipment. The same occurs if a simulation of the reaction gives the user the percentage of compound contained in

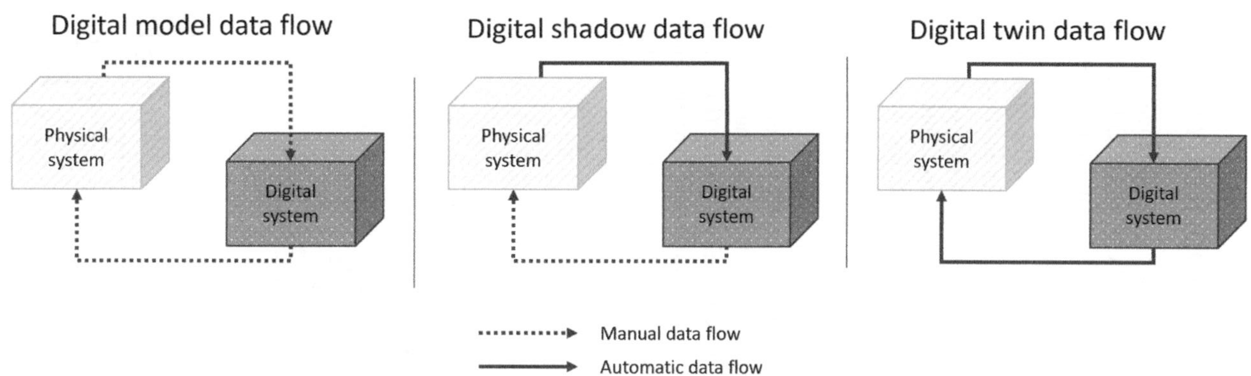

FIGURE 43.1 Data flow in a digital model, digital shadow, and DT.

the chemical reaction mixture, but not the needed output pressure. Therefore, if a DT uses engineering modeling and simulation as its core component, HITL modeling and simulation design techniques must ensure that the DT supports its users.

THE REALITY ANCHOR METHODOLOGY FOR DT DESIGN

The oil-and-gas domain understands the importance of having tools designed with humans in mind, as their safety can be at stake. Oil-and-gas sites are managed and under surveillance from control rooms to ensure human safety. Control room operators manage the process and the platform's safety from those rooms using different tools that provide them with data from the system. To improve safety on their site, the domain's major companies are turning to DTs in those control rooms. They aim to use more precise and complete up-to-date models to ensure control room operators better understand the on-site situation. In this context, a HSI design methodology using human-in-the-loop modeling and simulation was defined to ensure that the DT could bridge the gap between the users and the situation and support their decision-making process.

Representing reality through a virtual model (partial image of the reality) raises a challenge, as the elements of reality naturally perceived by humans differ from those available from sensors. To tackle this challenge, the elements of reality that the user needs to connect to reality, called reality anchors (Figure 43.2), must be identified and implemented in the model. An example of a reality anchor in the case of an oil-and-gas platform is the pressure in a storage tank. Knowing the pressure in a tank is necessary to ensure the safety of the platform and its inhabitants, and this value would

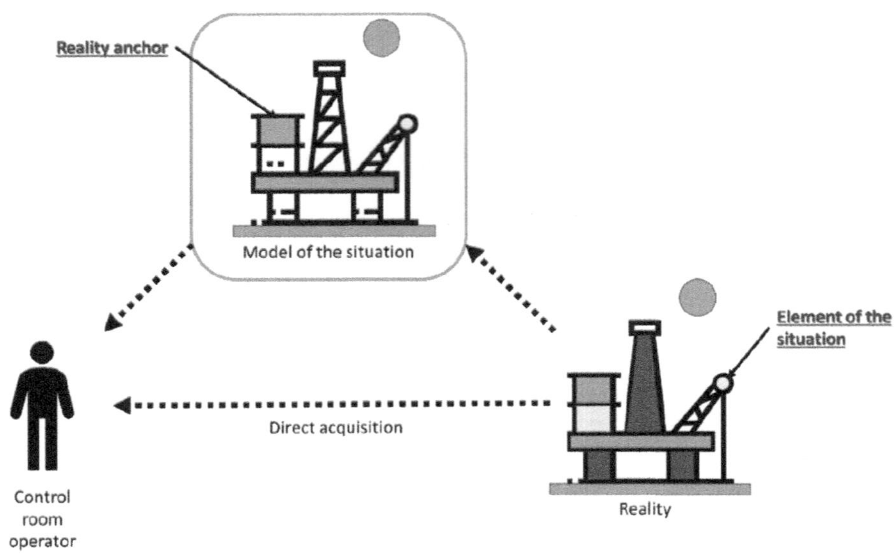

FIGURE 43.2 Elements of the situation perception using situation models.

be available using a gauge on the equipment. Therefore, providing this value in the DT is necessary to ensure that the user in the control room can keep the link to reality.

A specific methodology was developed to identify and implement reality anchors based on human feedback from control room operators: the Reality Anchor Methodology (RAM) (Camara Dit Pinto et al., 2021). The methodology is organized into three parts (Figure 43.3). The first part focuses on identifying the reality anchors through a cognitive function analysis. Performing interviews with experts in the field enables us to identify the tasks to be performed by the user and the elements of reality necessary to perform those tasks. The tasks are then decomposed into activities, and the required reality anchors are elicited. The second part focuses on implementing a prototype that enables HITL simulation. To facilitate the implementation of an otherwise complex and costly DT, the "Wizard of Oz" concept is implemented, and the identified reality anchors are implemented in the prototype. The third part evaluates the results through HITL simulation and iterates to design the best DT possible. This evaluation includes multiple aspects such as the user's situation awareness, experience

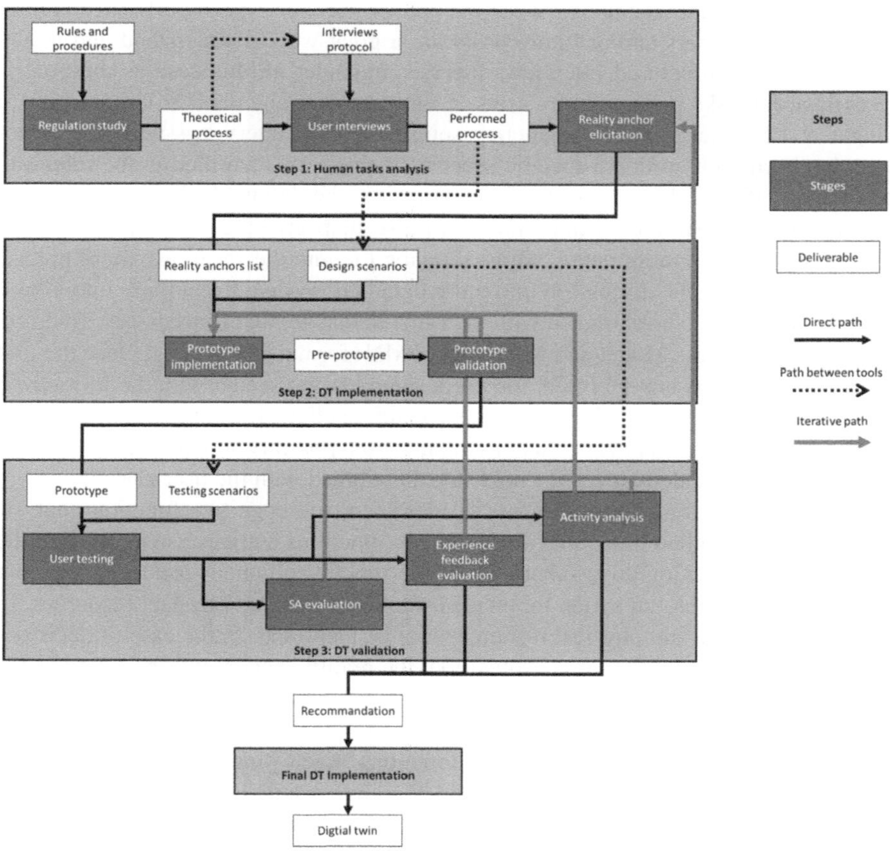

FIGURE 43.3 Reality anchor-based methodology representation.

feedback, and the activity performed during the tests. Those three aspects give one a complete picture of the reality anchors needed or to be removed, as well as their priority. At the end of this process, the design team is left with clear guidelines for implementing a DT that enables users to access reality anchors and have the best link to reality possible.

REALITY ANCHOR-BASED HUMAN-IN-THE-LOOP MODELING

The first step of every HSI-inspired methodology should focus on modeling the human interactions with the system. This enables one to understand human needs better and direct the rest of the work to consider the integration of humans and systems. In the presented use case, this means modeling the tasks to be performed by users and their impact on other humans and the systems. If the system to be designed is to replace or upgrade a current process and system, rules and regulations are a good first step to understand the tasks to be performed, and they give a guideline to understand the emerging tasks that the user might be performing. Rules and regulations, if rarely defined to support humans directly, express the constraints and objectives of the process. To go into more details, or when the process has never been performed before, interviews with users and/or future users are a great way to understand and model the tasks. Semi-structured interviews are recommended in this case as the goal is to extract the tasks performed by users in finer detail (Lallemand & Gronier, 2015; Unger & Chandler, 2009). Conducting semi-structured interviews will help structure the discussion around a specific process (the process identified in the rules and regulations, if available), as a set of questions is to be defined before the interview. Conducting semi-structured interviews will also help give the user the freedom to explain the tasks in more detail, while allowing the interviewer to ask more precise questions to direct the discussion into those details if needed. From these interviews, a list of tasks should be extracted with their link to one another if necessary. To do so, the Business Process Model and Notation (BPMN) format is engaging. Once the first round of interviews is performed, the users should be presented with the formalized process and precision, or additional tasks should be elicited. Looping with the users to discuss the extracted process enables them to both possibly add missing details and validate that the guidelines for the next step are aligned with the user's needs.

Once the decision-making process is identified, the cognitive function analysis method can be applied to identify the cognitive functions humans use to perform the tasks (Boy, 1998). Cognitive functions are the functions humans use to transform a task (theoretical) into an action (applied) using cognitive and physical resources. In addition, cognitive and physical resources will be identified. In the case of decision-making and situation awareness acquisition, the reality anchors will be found in the physical resources.

Applied to the detection and management of gas leaks, four major tasks were decomposed into a set of 29 cognitive functions. Each function was decomposed into its role and the associated physical and cognitive resources. Cognitive resources, being cognitive functions, the decomposition was performed down to primary cognitive functions such as "See a text" or "Hear a sound." At that point, a set of 29 reality anchors was identified in the physical resources. Those 29 reality anchors were then implemented in the first iteration of the prototype to be validated.

REALITY ANCHOR-BASED HUMAN-IN-THE-LOOP PROTOTYPE IMPLEMENTATION

The second step in applying HSI to the design of DT is the implementation of a prototype for HITL simulations. This step is crucial when dealing with complex systems because of the full implementation cost of time and money. Therefore, testing on a prototype with future users should be performed. Creating a good prototype is not cost-free either; however, using the "Wizard of Oz" technique is a good practice to enable simulation flexibility at a limited cost (Dahlbäck et al., 1993). The concept of the "Wizard of Oz" technique works as the prototype; another human, unseen by the users, simulates complex behaviors. This human, called the wizard, should be able to control the prototype's behavior and hence create scenarios more efficiently than if the prototype were to be reimplemented every time. This technique requires an understanding of the user's possible actions. However, following this methodology, we acquired the necessary knowledge from the decision-making process and the tasks and actions taken by the user during the first modeling step of the methodology. In parallel with defining the structure of the prototype, testing scenarios should be defined. These scenarios aim to immerse the user in a situation that the DT will support the user in solving the problem. In the prototype case, this means having the reality anchors take values that transcribe an imaginary situation that the user would have to perceive, understand, project, and decide upon. Once the structure of the prototype is defined, the interface is implemented, and the scenario is determined, the prototype and the scenario are to be reviewed by experts in the field iteratively. This HITL validation aims to ensure the prototype's realism and covers the main functions. One or more experts from the domain are asked to come and try out the prototype and the defined scenarios. During the tests, their comments regarding the interface and their opinion regarding the reaction of the reality anchors to a described scenario are considered to iterate on the prototype and scenarios. Those validation tests are performed semi-structured with the presentation of the scenarios, then the presentation of the interface, and finally the run of the scenarios in the prototype. At the end of the tests, the prototype and the scenarios are updated to reflect the comments from the tests. This process is repeated until the experts agree on the validity of the scenario and the prototype form. After this validation, the tests with different experts will be conducted to ascertain the validity of the reality anchor analysis and its role in the DT.

Applied to the detection and management of gas leaks, the prototype was defined as a two-player game where one "player" would control the values of the reality anchors and the second 'player' would gather information, take actions, and make decisions (see Figure 43.4). The game interface was defined to resemble current oil-and-gas control room tools. As every site has a different interface but the same general structure, the interface was determined by considering guidelines from one specific site. The reality anchors identified during the analysis that were not already part of the classic interface were implemented similarly to the other ones, and the interface

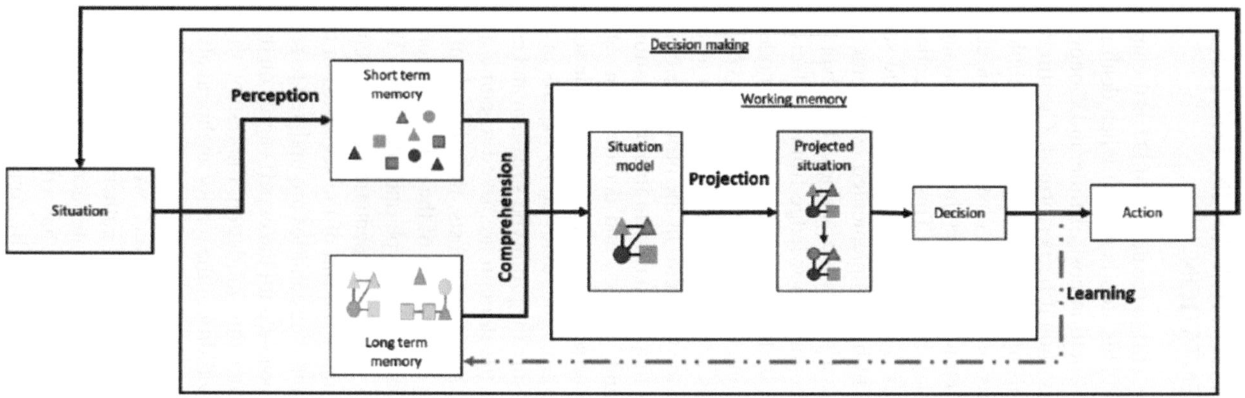

FIGURE 43.4 Decision-making-oriented situation awareness process. (Adapted from Endsley, 1995.)

was reviewed by an expert from the field, currently working as a control room operator trainer. The five defined scenarios were validated immediately, and after three rounds of testing and updating, the prototype was considered realistic and ready for simulations. Changes to the prototype interface focused on the alarm system sound and visual effects, the shape and color of some map icons, and the order in which some data were displayed. Those types of comments are the ones to aim for in trying to create a realistic environment for future simulations.

REALITY ANCHOR-BASED HUMAN-IN-THE-LOOP SIMULATIONS

The final step is to perform human-in-the-loop simulations to validate/evaluate the use of the identified reality anchors and possible emerging human functions. The Reality Anchor methodology proposes combining three approaches to evaluate the prototype. A test protocol must be defined to ensure the repeatability of the tests. This is even more important as using the "Wizard of Oz" technique implies human intervention and timing of the prototype evolution. Applied to the detection and management of gas leaks, a test protocol was defined for the user to (1) train with the prototype, (2) perform the scenarios, and (3) fill out evaluation forms. A "Wizard of Oz" protocol was also defined for each scenario to ensure the scenario's timing was repeated for each user. Five scenarios were defined for the tests and presented in a different order to each of the five expert users who took the test. The users were asked to fill in the evaluation forms at the end of each scenario.

Regarding the evaluation, the first form was an experience feedback report. As it is common in the industry to fill out an experience feedback report at the end of managing abnormal situations, the methodology takes advantage of this habit. It turns it into an evaluation feature for the reality anchors evaluation. Three different types of experience feedback were requested from the user in various scenarios. The first type of experience feedback is a free experience feedback form. This type of experience feedback form gives complete freedom to the user to express their perception, comprehension, and projection of the situation using different reality anchors. The second type is directed experience feedback. This type of experience feedback form asks the user to (1) describe the situation, (2) analyze the situation, and (3) describe their actions and justify them and their impact. This ensures the user expresses his perception, comprehension, and projection in more detail. Finally, a query-based experience feedback form asks him to use the tool to provide the value of the reality anchors. This type of experience feedback ensures the user can find the reality anchors in the tool. Applied to the detection and management of gas leaks, the user provided different types of feedback depending on the form type. The free experience feedback forms were shorter, with a specific expression of some precise reality anchor directly linked to the technical aspects of the gas leak.

On the other hand, the directed experience feedback forms were more descriptive of the context of the situation. They used different reality anchors, not going into the technical details of the leak. This gave us a hierarchy of reality anchors, enabling us to identify the more important ones that should be put upfront in the interface. This analysis helped us identify five reality anchors that never appeared in the experience feedback forms. Coupling this analysis with the two other evaluation forms will provide more detail explaining the reason for this event. This will also determine if those reality anchors will be kept in the next iteration of the prototype and the final DT.

The second evaluation form focuses on situation awareness evaluation and the use of situation awareness methods such as the Situation Awareness Rating Technique (SART) (Taylor, 1990) or the Situation Awareness Global Assessment Technique (SAGAT) (Endsley, 2000) and their derivatives. Situation awareness is defined as a process that takes place at the beginning of the decision-making process and is determined following three different levels: (1) the perception, (2) the comprehension, and (3) the projection (Endsley, 1995) (see Figure 43.4).

The most widely used technique to evaluate situation awareness is the SAGAT (Endsley, 2000). This assessment technique uses queries that appear on the frozen screen of the prototype. It reflects the three levels of situation awareness based on a previously performed analysis. Other derived methodologies like the Direct Questioning Technique (DQT) (Stanners & French, 2005) or the Situation Awareness Control Room Inventory (SACRI) (Hogg et al., 1995) work on the same principle of frozen screen queries to evaluate different, more detailed aspects of the user's situation awareness. However, if those techniques have proven good at evaluating situation awareness, they require a prototype adapted to the screen freeze technique.

On the other hand, SART is a technique that asks the user to self-evaluate in light of nine general constructs at the end of a task. This method was developed for pilot evaluation, and the analysis of the construct evaluation enables the identification of the user's situation awareness. As the evaluation happens at the end of a task, it does not require changing the prototype; therefore, it helps make the prototype cheaper. Applied to detecting and managing gas leaks, a SART form was presented to each user at the end of each scenario. The user then evaluated their situation awareness for the nine constructs on a scale from one to seven. The results from these forms revealed that the user considered having a constant share of attention across the information and having the necessary amount of information without being over-saturated mentally. This validated reality anchors and interfaces with the structure in a general situation awareness.

Finally, the third evaluation is based on the analysis of the user activity on the prototype. As defined in the cognitive analysis section, cognitive function and reality anchors are necessary for humans to transform tasks into activity. If the first analysis of the reality anchors was based on the tasks to be performed by the user, the prototype enables analysis of the coincident activity. Once again, the user activity can be recorded in multiple forms, from recording every action directly in the prototype to using eye tracking technology to identify the user's attention pattern. Applied to the detection and management of gas leaks, we recorded the user's screen and recorded the user speaking as he was encouraged to use the thinking-aloud technique. This technique encourages users to explain and justify their actions while performing the tests. Combining both these techniques enabled us to generate a number representing the percentile of time that the reality anchors' information was visible on the screen for each reality. In addition, we were able to track the user information consultation patterns. Using the percentile of availability value enabled us to correlate the non-use of some reality anchors, particularly the five identified during the experience feedback analysis. The decision was then made to remove those reality anchors from the next iteration of the prototype. Analyzing the reality anchor consultation pattern, the most essential reality anchors were identified, and the decision was made to make them constantly available on the interface when the situation was detected in the next iteration of the prototype.

CONCLUSION AND APPLICATION TO OTHER DOMAINS

The definition and application of this methodology to the oil-and-gas domain highlighted the strength of the HITL modeling and simulation techniques. Those techniques ensured user needs were integrated into the DT design process. HITL modeling focuses on creating a model of the tasks of humans in the decision-making process. This enabled a designer to identify the reality anchors, resources humans use to anchor themselves, necessary for the user to acquire their situational awareness and decide. Looping back with the users through interviews enabled us to refine the decision-making process and validate the tasks model. A prototype was defined based on the defined model to enable human-in-the-loop simulations. Looping back with the users through those simulations enabled us to study their use of the tool and identify the actions taken. This anchored the integration of humans and the system in the process and ensured that the system agreed with human needs.

If this process were performed explicitly in the case of DT development in the oil-and-gas domain, it is clear that the techniques used can and should be applied to design other systems in other domains. Modeling the human tasks and goals is essential to ensure that the developed system can encompass those goals and provide the desired support to the user. Using interviews to communicate directly with the users and loop back with them ensures the validity of the work performed by the design team. It facilitates the acceptance of the new systems, as the users are already part of the design process. Using a prototype to run HITL simulations can be costly but it can be reduced through techniques like the "Wizard of Oz" or creating partial prototypes. In addition, the benefits of using simulations and looping back with the user are greater as you ensure your system will correspond to the user's needs. This iterative process is also a great way to integrate future users in the design process and ensure their support upon exporting the system to new users.

In conclusion, integrating humans in the design process, looping back with them during the different stages, is a great HSI practice that optimizes complex systems design processes and ensures the integration of the user needs. This practice is already extensively used in the aeronautical domain and is slowly spreading to other domains such as the oil and gas industry. In addition, multiple techniques are already available and can be implemented to support this practice, regardless of the domain and system tackled. There is therefore no reason to pass on the application of such practice in your goal to integrate humans and systems.

REFERENCES

Boy, G. A. (1998). Cognitive function analysis for human-centered automation of safety–critical systems. In *Proceedings of the SIGCHI Conference on Human Factors in Computing Systems—CHI'98*, ACM Press/Addison-Wesley Publishing Co., USA, 265–272. https://doi.org/10.1145/274644.274682

Boy, G. A. (2013). *Orchestrating Human-Centered Design*. Springer. https://doi.org/10.1007/978-1-4471-4339-0

Brenner, B. & Hummel, V. (2017). Digital twin as enabler for an innovative digital shopfloor management system in the ESB Logistics Learning Factory at Reutlingen University. *Procedia Manufacturing*, *9*, 198–205. https://doi.org/10.1016/j.promfg.2017.04.039

Camara Dit Pinto, S., Masson, D., Villeneuve, E., Boy, G. A. & Urfels, L. (2021). From requirements to prototyping: Application of human systems integration methodology to digital twin design. *Proceedings of the Design Society*, *1*, 1617–1626. https://doi.org/10.1017/pds.2021.423

Camara Dit Pinto, S., Villeneuve, E., Masson, D., Boy, G. A., Baron, T. & Urfels, L. (2021). Digital twin design requirements in downgraded situations management. *IFAC—PapersOnLine*, *54*(1), 869–873. https://doi.org/10.1016/j.ifacol.2021.08.102

Dahlbäck, N., Jönsson, A. & Ahrenberg, L. (1993). Wizard of Oz studies—why and how. *Knowledge-Based Systems*, *6*(4), 258–266. https://doi.org/10.1016/0950-7051(93)90017-N

Endsley, M. R. (1995). Toward a theory of situation awareness in dynamic systems. *Human Factors: The Journal of the Human Factors and Ergonomics Society*, *37*(1), 32–64. https://doi.org/10.1518/001872095779049543

Endsley, M. R. (2000). Direct measurement of situation awareness: Validity and use of SAGAT. In In M. R. Endsley & D. J. Garland (Eds.), *Situation Awareness Analysis and Measurement* (pp. 1–21). Lawrence Erlbaum Associates Publishers.

Grieves, M. (2014). *Digital Twin: Manufacturing Excellence Through Virtual Factory Replication [white paper]* (pp. 1–7). Dassault Systèmes, the 3DEXPERIENCE Company https://doi.org/10.5281/zenodo.1493930

Hogg, D. N., Follesø, K., Strand-Volden, F. & Torralba, B. (1995). Development of a situation awareness measure to evaluate advanced alarm systems in nuclear power plant control rooms. *Ergonomics*, *38*(11), 2394–2413. https://doi.org/10.1080/00140139508925275

Kritzinger, W., Karner, M., Traar, G., Henjes, J. & Sihn, W. (2018). Digital twin in manufacturing: A categorical literature review and classification. *2018, IFAC (International Federation of Automatic Control)*, *51*(11), 1016–1022. https://doi.org/10.1016/j.ifacol.2018.08.474

Lallemand, C. & Gronier, G. (2015). Exploration, Entretien. In *Methode de design UX* (pp. 109–131). Eyrolles.

Mitchell, M. & Newman, M. (2001). Complex systems theory and evolution. *Encyclopedia of Evolution*, *1*, 1–5.

Rosen, R., Von Wichert, G., Lo, G. & Bettenhausen, K. D. (2015). About the importance of autonomy and digital twins for the future of manufacturing. *IFAC-Papers Online*, *48*(3), 567–572. https://doi.org/10.1016/j.ifacol.2015.06.141

Sillitto, H., Martin, J., McKinney, D., Griego, R., Dori, D., Krob, D., Godfrey, P., Arnold, E. & Jackson, S. (2019). *Systems Engineering and System Definitions*. INCOSE.

Stanners, M. & French, H. T. (2005). *An Empirical Study of the Relationship between Situation Awareness and Decision Making*. PsycEXTRA Dataset. DSTO Systems Sciences Laboratory https://doi.org/10.1037/e424442005-001

Taylor, R. M. (2011). Situational awareness rating technique (SART): The development of a tool for aircrew systems design. In *Situational Awareness (1st ed.)* (pp. 2013–2015). Routledge. https://doi.org/10.4324/9781315087924-8

Unger, R. & Chandler, C. (2009). *A Project Guide to Ux Design*. New Riders Pub.

44 Automation and Sustainable Working

Kine Reegård and Prosper Ameh Kwei-Narh

INTRODUCTION

In the fast-changing work landscape, automation stands at the forefront of shaping the future of labor. Integrating robotics and increasingly intelligent algorithms is transforming work across various sectors. Current discussions on automation center on how it can influence jobs, redistribute tasks, and modify workplace social fabric (Parker & Grote, 2020). These changes significantly impact our lives as our jobs provide us with economic sustenance and shape our identity and purpose, influencing our well-being and contributions to organizations and society (Allan et al., 2020; Blustein et al., 2023). Therefore, the concepts of "decent" and "meaningful" work become imperative for sustainable work experiences as the dynamic between human workers and technological systems evolves.

Decent work "reflects basic workplace conditions that everyone is entitled to that ensure the principal values for freedom, equity, security and human dignity" (ILO, 1999, p. 3). Facilitating such conditions enhances career advancement opportunities and nurtures meaningful careers (Duffy et al., 2016). Recognizing work as meaningful, that is, purposeful and significant (Pratt & Ashforth, 2003), is vital for individuals' well-being and is consistently linked with work engagement, job satisfaction, organizational commitment, and employee retention (Allan, Batz-Barbarich, et al., 2019). Hence, experiences of work fulfillment contribute to individuals' thriving at work and, in turn, to organizational sustainability by fostering positive attitudes toward work among employees.

However, jobs vary in their conditions to promote decent and meaningful work and their susceptibility to automation, which can further influence these conditions. Jobs primarily involve repetitive manual tasks and are particularly vulnerable to technological displacement (Frey & Osborne, 2017). In contrast, jobs emphasizing cognitive and social skills are expected to evolve qualitatively, demanding new skill sets and adaptability (Johnson et al., 2020). This disparity suggests an emerging polarization between jobs requiring social and cognitive skills that offer higher wages, and jobs dependent on physical and sensory skills that provide lower wages (Gravina & Foster-McGregor, 2024).

DOI: 10.1201/9781003666431-53

Such disparities underscore the importance of considering contextual factors when assessing automation's impact on work experiences. Failure to address these factors leaves us ill-equipped to propose effective solutions that foster enriching work and thus successful human–system integration. Despite this, much of the existing literature isolates the effects of automation, offering fragmented insights into broader implications for workers and organizational sustainability.

To advance this dialog, this chapter addresses the relationship between automation and sustainable working within workers' experiences of decent and meaningful work. We ground our discussion in the Psychology of Working Theory (PWT), which posits that decent work fulfills individuals' basic needs for survival while fostering purpose and significance through self-determination, social connectedness, and social contributions (Allan et al., 2020; Blustein et al., 2023; Duffy et al., 2016). Decent and meaningful work is thus fundamental to the sustainability of employee well-being and organizational efficacy, affecting the organization's capacity to maintain a skilled workforce and mitigate the risk of losing vital competencies through employee attrition or sickness absences. We will explore how these work experiences are shaped by automation, as illustrated in Figure 44.1.

We begin by examining PWT's core elements before elaborating on these regarding the impact of automation. Furthermore, we outline principles for integrating automation within the organizational context, considering PWT.

EXPERIENCES OF DECENT AND MEANINGFUL WORK: THE PWT

The PWT proposes that positive occupational experiences of fulfillment, such as meaningfulness, are only possible if basic psychological needs are met through decent work (Allan et al., 2020; Blustein et al., 2023; Duffy et al., 2016).

Experiences of Decent Work

Experiences of decent work are shaped by the contextual factors of economic constraints and marginalization, as well as the psychological aspects of career adaptability and work volition (Duffy et al., 2016). Economic constraints reflect limited economic resources, a critical barrier to securing decent work (Duffy et al., 2016). Lack of economic resources creates a situation where individuals perceive their career options to be limited and are therefore willing to accept work that may be hazardous or lack the conditions for decency. Economic constraints have both a direct effect on experiences of decent work and an indirect impact through career adaptability and worker volition. Career adaptability (Savickas, 2002, p. 156) denotes "an individual's readiness and resources for coping with current and anticipated vocational development tasks." Economic constraints limit the sense of career adaptability because workers might need to work overtime or have multiple jobs to stay financially afloat. As immediate financial needs dominate their daily concerns, individuals are less likely to pursue relevant education and skill acquisition needed to secure better work. Furthermore, economic constraints can diminish worker volition when workers feel they have less freedom to exercise choice, as they must prioritize short-term acquisition of financial

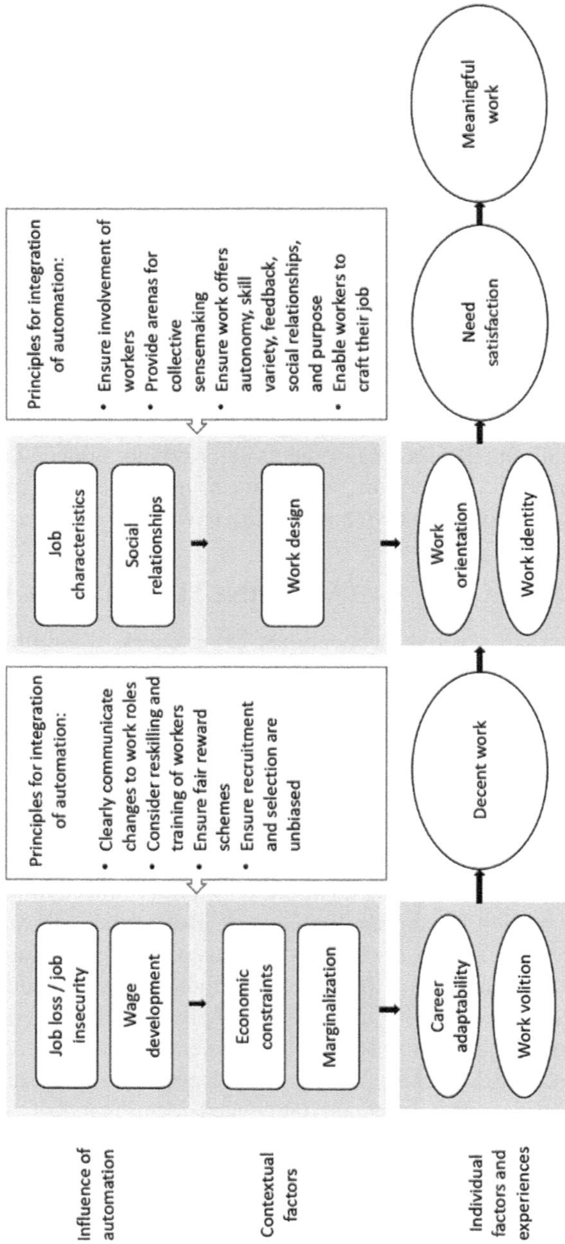

FIGURE 44.1 Framework for understanding the impact of automation on decent and meaningful work.

resources over demands for better worker representation or contracts. For example, workers on revolving short-term contracts may not be members of labor unions that can support them in making significant changes to their work conditions.

Membership in marginalized social groups also impacts the experience of decent work. For example, sexual identity and race impact access to work opportunities, making certain positions inaccessible even in ostensibly progressive societies (Allan, Tebbe, et al., 2019; Duffy et al., 2018). Consequently, individuals from these marginalized groups may lack the agency to shape their work context or initiate necessary career changes. Other marginalized social groups may include immigrant workers and workers with physical and mental disabilities.

While economic constraints and marginalization serve as the basic conditions influencing experiences of decency at work, the psychological factors of career adaptability and worker volition help explain why some individuals experience decency while others do not, even within similar job situations or marginalized groups. Of the two psychological factors, work volition seems to be the stronger predictor of experiences of decent work (Duffy et al., 2020).

Decent work encompasses the fundamental conditions that satisfy human survival needs, referring to "access to resources such as food, shelter, and social capital" (Duffy et al., 2016, p. 138). It does so in three ways: (1) through the provision of basic work conditions with a reliable income, (2) creating paths to developing human connections within the workplace, and (3) through experiences of connecting to one's community by contributing to its functioning (Allan et al., 2020). By satisfying these survival needs, decent work enables workers to also experience work as meaningful, which is central to their well-being and sustained work motivation (Allan et al., 2020; Blustein et al., 2023; Duffy et al., 2016).

Experiences of Meaningful Work

Experiences of meaningful work refer to an individual's subjective perception that their job is "[…] at a minimum, purposeful and significant" (Pratt & Ashforth, 2003). This attribution of significance is closely tied to individuals' identities and is interpreted through their work or social relationships (Pratt et al., 2013). Hence, there are three influential sources of meaningful work: the worker's self-concept, design of work, and social relationships at work.

First, individuals' self-concept—the subjective perception of who they are, including their beliefs and motivations—is central to how they interpret meaningful work. The more work aligns with an individual's self-concept, the more meaningful it is experienced to be. Workers hold subjective evaluations of what makes work worthwhile that act as internalized standards against which they evaluate their work, referred to as work orientations (Pratt et al., 2013). Work can be considered a job, a career, or a calling (Wrzesniewski et al., 1997). Individuals with job orientations evaluate meaningfulness primarily regarding financial security, viewing work as a means to obtain resources needed for activities in other life domains. Career-oriented individuals find meaningful work in jobs that provide opportunities for social promotion and thus personal success and achievements. Those with calling orientations, in contrast, perceive their work activities as valuable in themselves through which they foster social bonds with colleagues and contribute to a greater good (Pratt et al., 2013; Wrzesniewski et al., 1997).

Second, jobs that tend to be experienced as more meaningful are those that grant workers decision latitude in how they carry out their work, opportunities for utilizing different skills, and involve completing whole tasks that impact beneficiaries and through which they receive important feedback (Hackman & Oldham, 1976). Social aspects of a job have also been shown to be important (Humphrey et al., 2007). Employees may also proactively change their work roles to align the job with their needs and preferences and thus enhance meaningfulness (Wrzesniewski & Dutton, 2001).

Third, social interactions and relationships shape individuals' perceptions of the meaning of their work (Allan et al., 2020). Workers observe and engage in conversations with others in the workplace, which informs their interpretations of the meaning and value of their work (Wrzesniewski, 2003). Additionally, leaders can influence employees' sense of meaningfulness through their work framing (Rosso et al., 2010). Individuals' identification with social groups also affects their perception of meaningful work because they derive a sense of who they are and what they value from their group membership (Pratt & Ashforth, 2003).

In summary, by providing decent work conditions, work satisfies basic human needs of survival, social contributions, and self-determination (Blustein et al., 2023; Duffy et al., 2016), enabling meaningful experiences. In turn, workers with such internal motivation exhibit favorable attitudes and behaviors, including improved performance, job satisfaction, and organizational commitment (Allan, Batz-Barbarich, et al., 2019; Humphrey et al., 2007). A global push is necessary toward providing decent work for all because it benefits individual workers and organizations. However, automation is anticipated to transform work that impacts decent, meaningful, and sustainable work. It is therefore crucial to consider how the advent of automation may reshape these experiences, which we turn to next.

HOW CAN AUTOMATION IMPACT EXPERIENCES OF DECENT AND MEANINGFUL WORK?

Automation has three core effects on work: a productivity effect, a displacement effect, and a work role effect (Acemoglu & Restrepo, 2018). The productivity effect is how much automation can increase the overall work output. The displacement effect describes the extent to which automation displaces the worker. The work role effect refers to the experience of work itself and how the worker's connection with identifiable and meaningful aspects of work can be enhanced or reduced. These effects can impact workers' experiences of decency and meaningfulness through structural changes in employment opportunities and wage development, and through changes in work design, that is the tasks, responsibilities, and relationships at work.

Automation's Influence on Opportunities for Decent Work

Automation delivers productivity benefits in that more work can be completed in less time than human labor (Autor & Salomons, 2018). Robots are increasingly being deployed to undertake menial tasks previously performed by humans, rooted in the belief that robots can provide services to customers with greater reliability (e.g., Bear Robotics, 2020). For instance, in the tourism industry, trial robots such as "Penny" can reliably deliver multiple drinks simultaneously, run food between the kitchen and tables, and retrieve dishes. Deployment of automation thus has the potential to reduce the need for human workers, thereby decreasing job opportunities within technologically evolving sectors and prompting a shift in labor demand to other sectors, particularly services (Autor & Salomons, 2018). However, automation commonly targets routine, low-skilled jobs that cater to economically challenged or marginalized groups, such as immigrants who rely on such employment to gain a foothold in the economy. Indeed, Schwabe and Castellacci (2020) showed that automation disproportionately leads to job loss among less skilled workers who are more likely to face economic hardships. This may further reduce the perceived agency of those who remain in these jobs to improve their work conditions when lower workforce numbers diminish their ability to form unions. Automation may thus reduce access to decent work for many, especially those already in unfavorable work conditions.

Automation is also anticipated to impact decent work by influencing wage developments. While automation has been estimated to reduce wages (Acemoglu & Restrepo, 2018), its influence on wages may vary according to workers' skill levels. Gravina and Foster-McGregor (2024) observed that the advent of intangible technologies (i.e., software) tends to result in wage premiums predominantly for individuals with higher education, and that an increase of 1% in robot adoption seems to affect the wage dispersion between high-skilled workers and their medium-skilled and low-skilled counterparts.

However, despite this complex and challenging backdrop, automation also has the potential to level the playing field by creating more inclusive employment opportunities for workers with varied social and ethnic backgrounds. When controlled for historical human biases in its programming, automation may enable more objective evaluations in the hiring process, assisting decision-makers to render more unbiased selection decisions (Raisch & Krakowski, 2021).

Parallel to these shifts, new work models have also sprung from technological advancements, with gig work gaining momentum in the low-skilled labor market. Gig work through online platforms allows workers to perform work flexibly and enables organizations to use algorithmic management for their staff. These technologies provide workers a high degree of control while offering opportunities for task complexity and variety as they choose activities on a task-by-task basis (Myhill et al., 2021; Wood et al., 2019). However, the use of algorithmic control and digital rating systems often results in significant pressures on gig workers due to intense competition in environments where clients hold the power (Myhill et al., 2021; Wood et al., 2019), thus raising new concerns regarding decent work in the digital age.

While the implications of automation on opportunities for decent work are critical, it is equally important to examine how these technological changes can impact meaningful work through their influence on the design of work and workers' identities.

Automation's Influence on Work Design and Work Identity

Integrating automation into workplaces is poised to reshape the tasks we perform and the fulfillment we derive from our work. Work is central to how we view ourselves as individual beings. As automation changes the content and social interactions inherent in our work, it impacts our work identities. For example, case handlers may perceive their role as "seeing the individual applicant" and experience automatic case processing as diminishing this interaction. Conversely, other case handlers may see their role as a gatekeeper and view automation as a means to facilitate the efficient processing of cases. The alignment between the changes brought by automation and workers' identities thus shapes their experiences of work meaningfulness (Pratt & Ashforth, 2003).

Work (re-)design is vital to addressing this alignment challenge and is poised to become more frequent as automation is introduced quickly (Johnson et al., 2020). Work design, "the content and organization of one's work tasks, activities, relationships, and responsibilities" (Parker, 2014, p. 662), can be done to preserve and improve meaningfulness for workers. For instance, it enables workers to spend more time on creative and challenging tasks and allocate mundane, repetitive tasks to automation. Automation can thus be a means to enrich work through designs that facilitate autonomy, feedback, skill variety, and social connections, aligning with workers' self-definitions. However, automation may also harm workers if these critical dimensions are overlooked (Parker & Grote, 2020). There are a variety of ways in which automation can influence these aspects. Here, we only briefly touch upon some of these influences.

To gain the full benefits of automation, front-line employees must be empowered to make decisions (Fountaine et al., 2019). However, algorithmic management is a striking example of how workers' autonomy may be neglected, resulting in work detriment. For example, Duggan et al. (2023) found food delivery and other gig workers to be excluded from decision loops regarding their work assignments, causing experiences of limited volition and autonomy in their tasks.

Automation can also influence feedback by giving workers precise feedback on task performance or enabling direct evaluation from and interaction with customers. This can be a potent motivator by providing workers with grounds for seeking improvements and demonstrating their impact and contribution.

Skill variety is another widely recognized impact of automation on work design. As automation takes over many routine tasks, the share of workers' tasks requiring high-level skills in terms of cognition and social interaction increases (Johnson et al., 2020; Parker & Grote, 2020). This shift occurs by repurposing work roles, allocating tasks between employees, and automating. For example, many services are increasingly implementing self-service systems, allowing customers to access services independently without the immediate aid of front-line workers.

Consequently, the tasks assigned to these workers may transition from routine customer reception to more demanding tasks, such as problem solving, when customers encounter issues. While these new tasks may be more rewarding by utilizing workers' problem-solving capacities, they may also expose them to more emotionally demanding customer interactions when customers are already frustrated (Johnson et al., 2020).

Automation may also cause new tasks to emerge, creating opportunities for worker skills development. For example, Grønsund and Aanestad (2020) observed that as professionals' manual tasks were replaced with automation, they were assigned new responsibilities such as monitoring, evaluating, training, and adjusting the algorithms' performance. In these roles, the professionals' expertise was recognized as indispensable, demonstrating a repurposing effect where workers shift from service delivery to more technology-centered functions. These new roles can be framed to enhance workers' perceptions of their contributions to the organization by highlighting their importance in facilitating efficient automation and improving service delivery. However, such repurposing can also potentially decrease the meaning in work by distancing workers from the outcome or result of the service activities (Parker & Grote, 2020). It may make workers feel less responsible for outcomes (Parasuraman & Manzey, 2010).

The opposite of skill development, skill erosion, may also occur due to automation. A potential side effect of automation is that employees may become less engaged in automated tasks, sometimes mistakenly believing that these tasks are being carried out correctly. This is especially likely to occur in work environments with high workloads because employees must prioritize their attention between demanding tasks (Parasuraman & Manzey, 2010). Over time, this can result in skill erosion and increased reliance on automation, as employees lose competence in performing the automated tasks and assessing their performance quality (Rinta-Kahila et al., 2023). The potential for skill erosion is especially relevant in safety-critical industries, where human control is critical. There is a risk that such control becomes more elusive because not only the complexity of automation may exceed human control capabilities (e.g., Parker & Grote, 2020) but also human skills for these tasks are not maintained over time.

Social dynamics and relationships among employees are certainly also impacted by automation. On the one hand, if automation is applied in ways that relegate workers to roles of supervisors or monitors of automated processes, their opportunities for social interactions are reduced, which may erode the foundation for building strong work relationships through community and mutual efforts. Moreover, automation can lead to a reduction in staffing levels, which increases the work responsibilities and demands on individual workers. On the other hand, automation also has the potential to foster increased social interaction for workers. As noted earlier, when automation takes over routine and straightforward tasks, workers can devote more time to address complex, ambiguous tasks that often require human creativity and problem-solving skills, for which human collaboration is invaluable, or to engage with customers. To illustrate, Staaby et al. (2021) observed that consultants whose repetitive tasks were automated had more time to engage with their clients, thus building a deeper understanding of the client's problems, and organized into a task force to verify the automated activities, prompting novel ways of collaborating.

The impact of automation on work identity depends on how significant workers find the tasks and relationships that are altered. Automation that enables workers to focus on complex and challenging work can enhance their work identities (e.g., Chen et al., 2021). Conversely, when automation encroaches on tasks that employees view as essential to their roles, it can undermine their work identities. For example, Strich et al. (2021) found that introducing AI in core tasks for loan consultants in a bank threatened their professional identity by shifting their role from decision-makers with significant impact on customers, to servants of the AI system, executing scripted exchanges with loan applicants. Furthermore, automation can challenge workers' self-esteem by creating new performance standards against which workers measure their worth. Moulaï et al. (2022) observed that cashiers working alongside self-service checkout machines began comparing their performance to the machines, leading to feelings of inadequacy. Moreover, after introducing these machines, cashiers felt that the emphasis shifted from providing pleasant checkout experiences to prioritizing efficiency, further devaluing their social skills.

These potential effects on opportunities for decent and meaningful work demonstrate that there is no uniform way in which automation will impact work, but that it depends on the decisions made regarding its integration. Understanding how automation may influence work fulfillment provides essential insights into how organizations can promote high-quality jobs and thus sustainable working amidst these changes.

Promoting Work Fulfillment: Principles for Integrating Automation

From the foregoing, organizations must ensure that automation results in decent and meaningful work for their workers, fostering high performance, motivation, and commitment. This is key to the individual worker and the organization's sustainability (Allan et al., 2020). Achieving this objective requires concerted actions from individuals and organizations (Demerouti, 2020; Parker & Grote, 2020; Pratt et al., 2013). In the following, we outline principles for how organizations can promote work fulfillment when automation is integrated into the workplace.

Principles for Organizational Practices

Introducing new technologies leads to qualitative shifts in how work is done and the human roles in the work system, signifying the need to consider how automation affects job roles and identity (Selenko et al., 2022). As automation involves reallocation of tasks and responsibilities, employee involvement becomes crucial. By involving employees in automating processes, their insights into how work should be performed and the motivating aspects of their jobs can be factored into the design phase. Such involvement may include worker representation through unions, which can play a central role in defining work changes that are advantageous for both the quality of work and organizational efficiency (Lloyd & Payne, 2023).

Relatedly, organizational communication is essential regardless of whether the automation substitutes for or augments workers. It reduces uncertainty regarding the change and enables workers to navigate transitions by clarifying how automation will impact their role (Johnson et al., 2020; Raisch & Krakowski, 2021). Such communication shapes employees' perceptions of meaningfulness by signaling values and practices important to the organization (Pratt & Ashforth, 2003) beyond the initial planning and implementation phases. Leaders can also facilitate workers' appraisals of automation by creating arenas for collective sense-making as employees' task conditions evolve.

Recruiting and selecting individuals likely to fit the job and the organization is a key practice promoting work fulfillment (Pratt & Ashforth, 2003). Conscious recruitment and selection practices are essential because automation may inadvertently lead to job displacement or exacerbate existing inequalities. However, there have already been reports that automated recruitment and selection tools may favor the already privileged group due to biases in the underlying data. The workers who possess relevant talents to perform the job may not necessarily come from the privileged backgrounds that many organizations draw from as their resource pool (Carey & Smith, 2016), and whose data may have served as training input for the algorithms employed in these tools. To address these challenges, organizations should actively seek to mitigate biases, ensuring that recruitment and selection decisions are based on relevant skills and competencies.

Organizations undertaking automation should also pay careful attention to their reward policies. A core condition for experiencing decency is that the wages are fair. Workers can become agitated if profits are disproportionately directed toward executive compensation rather than shared profit schemes. Consequently, organizations should ensure workers notice a collective progressive remuneration when implementing automation in their processes, viewing it as rewarding effort rather than justifying wage suppression.

Finally, organizations that employ workers at high risk of job loss due to automation should closely consider opportunities to reskill and retain employees, even as they increase their use of automation in production. When automation fails and manual control is needed, employees who understand the manual routines can quickly intervene and resolve problems. Organizations retain essential human capital by training and upskilling these employees while remaining adaptive to potential automation failures. Such upskilling of employees can facilitate the organization's ability to effectively integrate automation across time and tasks, gradually enhancing human skills and automation capabilities (Raisch & Krakowski, 2021).

Principles for the Design of Work

Automation can serve as a resource that supports work goals and stimulates personal growth rather than creating unnecessary strain (Demerouti, 2020). To this end, automation should be used for tasks where the technology has an advantage over workers' abilities or where workers are put at physical risk (Raisch & Krakowski,

2021). Workers should prioritize decision-making and leverage their core abilities, with involvement in defining those abilities central to success.

Generally, work that grants employees control over how they work (e.g., which methods they employ, scheduling, and order of tasks) incorporates the use and development of skills in tasks that have an impact on others, and offers constructive feedback, fulfills individuals' psychological needs, and fosters experiences of meaningfulness (Hackman & Oldham, 1976). These principles should also guide automation integration (Parker & Grote, 2020).

Moreover, organizations should pay attention to workers' opportunities for direct interaction with and impact on the beneficiaries of their jobs. This relational architecture helps workers recognize the significance of their work, contributing to their prosocial identities (Grant, 2007). Relational aspects of work are essential for individuals who derive meaning primarily from their perceived effect on others (Pratt et al., 2013), such as in nursing or teaching. However, relational aspects may become increasingly important as automation reduces human interaction, potentially leading to feelings of isolation (Tang et al., 2023). Organizations must know this potential dilemma and find strategies to maintain social relationships.

Because the effects of automation on work are emergent, many may not be anticipated during work design. Enabling employees to craft their jobs to better align with what is important to them can be a complementary strategy to promote meaningfulness in work (Demerouti, 2020; Parker & Grote, 2020). Organizations can facilitate this crafting by providing flexibility in task and relational boundaries or training (Demerouti, 2020). As advances in automation lead to more frequent redesigns of work (Johnson et al., 2020), such training can become increasingly important.

CONCLUSION

A workforce that finds its work meaningful and fulfilling is more engaged, productive, and loyal to the organization, contributing to the long-term viability and success of their working lives and the organization. Thus, workers' aspirations for decent work foster meaningful work and sustainability, thereby benefiting individuals and organizations. The accelerating pace of automation challenges these fundamental goals of decent and meaningful work, especially for economically constrained or marginalized workers. Therefore, it is crucial to consider the impacts of automation when deciding on and designing its integration into work.

We identify organizational practices in involvement, communication, recruitment and selection, compensation and reward systems, training, and work design as key mechanisms to ensure that work remains enriching in the ongoing shift toward increased automation. Technology can enable high-quality jobs by recognizing the human element at the heart of work and proactively shaping workplace practices to support this. As we continue to harness the power of automation, the overarching aim should be to provide every individual with the opportunity to engage in work that is both productive and meaningful, securing not only economic prosperity but also the collective well-being of our society.

REFERENCES

Acemoglu, D., & Restrepo, P. (2018). Artificial intelligence, automation and work. *Economics of Artificial Intelligence*, Issue. Available at SSRN: https://ssrn.com/abstract=3101994

Allan, B. A., Autin, K. L., Duffy, R. D., & Sterling, H. M. (2020). Decent and meaningful work: A longitudinal study. *Journal of Counseling Psychology*, 67(6), 669–679. https://doi.org/10.1037/cou0000432

Allan, B. A., Batz-Barbarich, C., Sterling, H. M., & Tay, L. (2019). Outcomes of meaningful work: A meta-analysis. *Journal of Management Studies*, 56(3), 500–528. https://doi.org/10.1111/joms.12406

Allan, B. A., Tebbe, E. A., Bouchard, L. M., & Duffy, R. D. (2019). Access to decent and meaningful work in a sexual minority population. *Journal of Career Assessment*, 27(3), 408–421.

Autor, D., & Salomons, A. (2018). Is automation labor-displacing? Productivity growth, employment, and the labor share. *Brookings Papers on Economic Activity*, 1–63.

Bear Robotics (2020). *Elevate your restaurant operations with server robots*. www.bearrobotics.ai/restaurants.

Blustein, D. L., Lysova, E. I., & Duffy, R. D. (2023). Understanding decent work and meaningful work. *Annual Review of Organizational Psychology and Organizational Behavior*, 10(1), 289–314. https://doi.org/10.1146/annurev-orgpsych-031921-024847

Carey, D., & Smith, M. (2016). How companies are using simulations, competitions, and analytics to hire. *Harvard Business Review*. Available from https://hbr.org/2016/04/how-companies-are-using-simulations-competitions-and-analytics-to-hire

Chen, Y., Stavropoulou, C., Narasinkan, R., Baker, A., & Scarbrough, H. (2021). Professionals' responses to the introduction of AI innovations in radiology and their implications for future adoption: A qualitative study. *BMC Health Services Research*, 21(1), 813. https://doi.org/10.1186/s12913-021-06861-y

Demerouti, E. (2020). Turn digitalization and automation to a job resource. *Applied Psychology*, 71(4), 1205–1209. https://doi.org/10.1111/apps.12270

Duffy, R., Blustein, D. L., Diemer, M. A., & Autin, K. L. (2016). The psychology of working theory. *Journal of Counseling Psychology*, 63(2), 127–148. https://doi.org/10.1037/cou0000140

Duffy, R. D., Kim, H. J., Allan, B. A., & Prieto, C. G. (2020). Predictors of decent work across time: Testing propositions from Psychology of Working Theory. *Journal of Vocational Behavior*, 123, 103507. https://doi.org/10.1016/j.jvb.2020.103507

Duffy, R. D., Velez, B. L., England, J. W., Autin, K. L., Douglass, R. P., Allan, B. A., & Blustein, D. L. (2018). An examination of the Psychology of Working Theory with racially and ethnically diverse employed adults. *Journal of Counseling Psychology*, 65(3), 280.

Duggan, J., Carbery, R., McDonnell, A., & Sherman, U. (2023). Algorithmic HRM control in the gig economy: The app-worker perspective. *Human Resource Management*, 62(6), 883–899. https://doi.org/https://doi.org/10.1002/hrm.22168

Fountaine, T., McCarthy, B., & Saleh, T. (2019). Building the AI-powered organization: Technology isn't the biggest challenge. Culture is. *Harvard Business Review*, 97(4), 62.

Frey, C. B., & Osborne, M. A. (2017). The future of employment: How susceptible are jobs to computerisation? *Technological Forecasting and Social Change*, 114(Supplement C), 254–280. https://doi.org/https://doi.org/10.1016/j.techfore.2016.08.019

Grant, A. M. (2007). Relational job design and the motivation to make a prosocial difference. *The Academy of Management Review*, 32(2), 393–417. https://doi.org/10.2307/20159308

Gravina, A. F., & Foster-McGregor, N. (2024). Unraveling wage inequality: Tangible and intangible assets, globalization and labor market regulations. *Empirical Economics*, 67(4), 1375–1420. https://doi.org/10.1007/s00181-024-02587-y

Grønsund, T., & Aanestad, M. (2020). Augmenting the algorithm: Emerging human-in-the-loop work configurations. *The Journal of Strategic Information Systems*, 29(2), 101614. https://doi.org/10.1016/j.jsis.2020.101614

Hackman, J. R., & Oldham, G. R. (1976). Motivation through the design of work: Test of a theory. *Organizational Behavior and Human Performance*, 16(2), 250–279. https://doi.org/10.1016/0030-5073(76)90016-7

Humphrey, S. E., Nahrgang, J. D., & Morgeson, F. P. (2007). Integrating motivational, social, and contextual work design features: A meta-analytic summary and theoretical extension of the work design literature. *Journal of Applied Psychology*, 92(5), 1332–1356. https://doi.org/10.1037/0021-9010.92.5.1332

International Labor Organization (1999). *Report of the Director-General: Decent work*. International Labour Conference, 87th Session, Geneva. Available from www.ilo.org/public/english/standards/relm/ilc/ilc87/rep-i.htm

Johnson, A., Dey, S., Nguyen, H., Groth, M., Joyce, S., Tan, L., Glozier, N., & Harvey, S. B. (2020). A review and agenda for examining how technology-driven changes at work will impact workplace mental health and employee well-being. *Australian Journal of Management*, 45(3), 402–424. https://doi.org/10.1177/0312896220922292

Lloyd, C., & Payne, J. (2023). Food for thought: Robots, jobs and skills in food and drink processing in Norway and the UK. *New Technology, Work and Employment*, 38(2), 272–290. https://doi.org/10.1111/ntwe.12194

Moulaï, K., Islam, G., Manning, S., & Terlinden, L. (2022). "All too human" or the emergence of a techno-induced feeling of being less-able: Identity work, ableism, and new service technologies. *The International Journal of Human Resource Management*, 33(22), 4499–4531. https://doi.org/10.1080/09585192.2022.2066982

Myhill, K., Richards, J., & Sang, K. (2021). Job quality, fair work, and gig work: The lived experience of gig workers. *International Journal of Human Resource Management*, 32(19), 4110–4135. https://doi.org/10.1080/09585192.2020.1867612

Parasuraman, R., & Manzey, D. H. (2010). Complacency and bias in human use of automation: An attentional integration. *Human Factors*, 52(3), 381–410. https://doi.org/10.1177/0018720810376055

Parker, S. K. (2014). Beyond motivation: Job and work design for development, health, ambidexterity, and more. *Annual Review of Psychology*, 65(1), 661–691. https://doi.org/10.1146/annurev-psych-010213-115208

Parker, S. K., & Grote, G. (2020). Automation, algorithms, and beyond: Why work design matters more than ever in a digital world. *Applied Psychology*, 71(4), 1171–1204. https://doi.org/10.1111/apps.12241

Pratt, M. G., & Ashforth, B. E. (2003). Fostering meaningfulness in working and at work. In K. S. Cameron, J. E. Dutton, & R. E. Quinn (Eds.), *Positive Organizational Scholarship*. Berrett-Koehler Publishers Inc.

Pratt, M. G., Pradies, C., & Lepisto, D. A. (2013). Doing well, doing good, and doing with organizational practices for effectively cultivating meaningful work. In B. J. Dik, Z. S. Byrne, & M. F. Steger (Eds.), *Purpose and Meaning in the Workplace* (pp. 173–196). American Psychological Association. www.jstor.org/stable/j.ctv1chs448.13

Raisch, S., & Krakowski, S. (2021). Artificial intelligence and management: The automation–augmentation paradox. *Academy of Management Review*, 46(1), 192–210. https://doi.org/10.5465/amr.2018.0072

Rinta-Kahila, T., Penttinen, E., Salovaara, A., Soliman, W., & Ruissalo, J. (2023). The vicious circles of skill erosion: A case study of cognitive automation. *Journal of the Association for Information Systems*, 24(5), 1378–1412. https://doi.org/10.17705/1jais.00829

Rosso, B. D., Dekas, K. H., & Wrzesniewski, A. (2010). On the meaning of work: A theoretical integration and review. *Research in Organizational Behavior*, 30, 91–127. https://doi.org/10.1016/j.riob.2010.09.001

Savickas, M. L. (2002). Career construction: A developmental theory of vocational behavior. In D. Brown (Ed.), Career choice and development (pp. 149–205) (4th ed). San Francisco: Jossey-Bass.

Schwabe, H., & Castellacci, F. (2020). Automation, workers' skills, and job satisfaction. *PLoS One*, 15(11), e0242929.

Selenko, E., Bankins, S., Shoss, M., Warburton, J., & Restubog, S. L. D. (2022). Artificial intelligence and the future of work: A functional-identity perspective. *Current Directions in Psychological Science*, 31(3), 272–279. https://doi.org/10.1177/09637214221091823

Staaby, A., Hansen, K. S., & Grønli, T.-M. (2021). Automation of routine work: A case study of employees' experiences of work meaningfulness. 54th Annual Hawaii International Conference on System Sciences, HICSS 2021.

Strich, F., Mayer, A.-S., & Fiedler, M. (2021). What do I do in a world of artificial intelligence? Investigating the impact of substitutive decision-making AI systems on employees' professional role identity. *Journal of the Association for Information Systems*, 22(2), 304–324. https://doi.org/10.17705/1jais.00663

Tang, P. M., Koopman, J., Mai, K. M., De Cremer, D., Zhang, J. H., Reynders, P., Ng, C. T. S., & Chen, I. H. (2023). No person is an island: Unpacking the work and after-work consequences of interacting with artificial intelligence. *Journal of Applied Psychology* 108(11): 1766–1789.

Wood, A. J., Graham, M., Lehdonvirta, V., & Hjorth, I. (2019). Good gig, bad gig: Autonomy and algorithmic control in the global gig economy. *Work Employment and Society*, 33(1), 56–75. https://doi.org/10.1177/0950017018785616

Wrzesniewski, A. (2003). Finding positive meaning in work. In K. S. Cameron, J. E. Dutton & R. E. Quinn (Eds.), *Positive Organizational Scholarship*. Berrett-Koehler Publishers Inc.

Wrzesniewski, A., & Dutton, J. E. (2001). Crafting a job: Revisioning employees as active crafters of their work. *The Academy of Management Review*, 26(2), 179–201. https://doi.org/10.5465/AMR.2001.4378011

Wrzesniewski, A., McCauley, C., Rozin, P., & Schwartz, B. (1997). Jobs, careers, and callings: People's relations to their work. *Journal of Research in Personality*, 31(1), 21–33. https://doi.org/10.1006/jrpe.1997.2162

45 Conclusion and Perspectives

Guy André Boy

INTRODUCTION

Sociotechnical systems (STSs) cannot be thought of in silo any longer. They are systems organized from various purposeful perspectives, where decision-makers must simultaneously grasp the social, economic, and environmental challenges and the links and impacts they can have on each other. The systemic approaches presented in this handbook demonstrate the need to manage the diversity of the various elements and the relationships between them.

Our sociotechnical world has become so complex that we must cope with ever-increasing delays. The paradox lies in the fact that we are developing techniques and tools that improve local efficiency, speed of execution, safety, and comfort, but which considerably delay the search for a concrete solution for satisfactory human–systems integration (HSI) on a global scale. We observe chaotic processes and behaviors leading us toward dangerous outcomes. It is time to understand and master the science of complexity, systems science, engineering, and above all, an HSI better.

HSI has become a solid field at the intersection of several contributing disciplines: systems engineering (SE), human factors and ergonomics, information technology, and various industrial and public sectors. HSI is investigated and implemented according to several perspectives: safety, competencies and professionalism, sustainability, habitability, occupational health, socio-cultural and organizational factors, training, HSI planning, human factors engineering, workforce planning, integrated logistics support, and maintenance. HSI is a triptych's heart: technology, organizations, and people. This triptych is always immersed in natural and artificial environments (Figure 45.1).

We have seen in this handbook that HSI is a systemic extension of Human–Computer Interaction (HCI) and, more generally, Human Factors and Ergonomics (HFE). Over the years, the field expanded from when the focus was on mechanical engineering and biomedical issues at work to cognitive engineering when microcomputers disseminated worldwide at the end of the past century, and today to HSI, where social problems have become prominent with the rise of our highly connected digital society. Our technology-infused world has shifted human factors issues from the physical to the cognitive and then to the social realm, while maintaining the interdependence between these issues (Figure 45.2).

 DOI: 10.1201/9781003666431-54

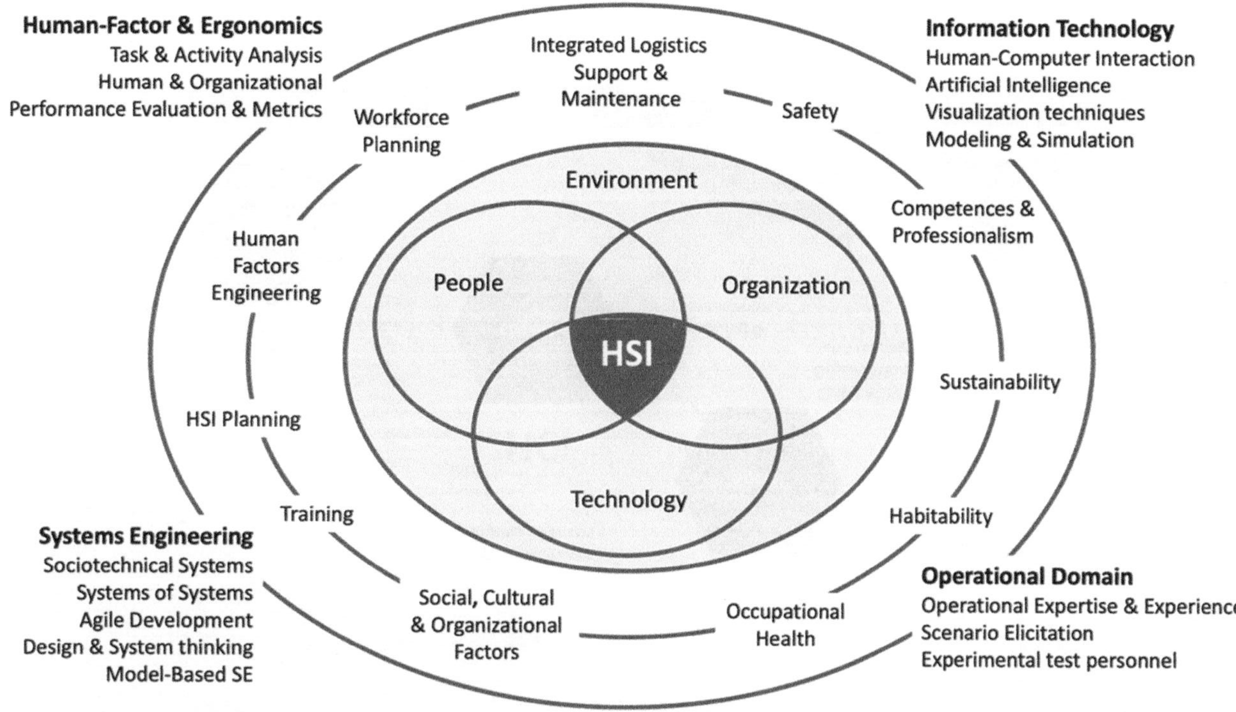

FIGURE 45.1 HSI technology, organization, and people within an environment (Boy, 2019, further used in the *INCOSE Systems Engineering Handbook*, 5th Edition).

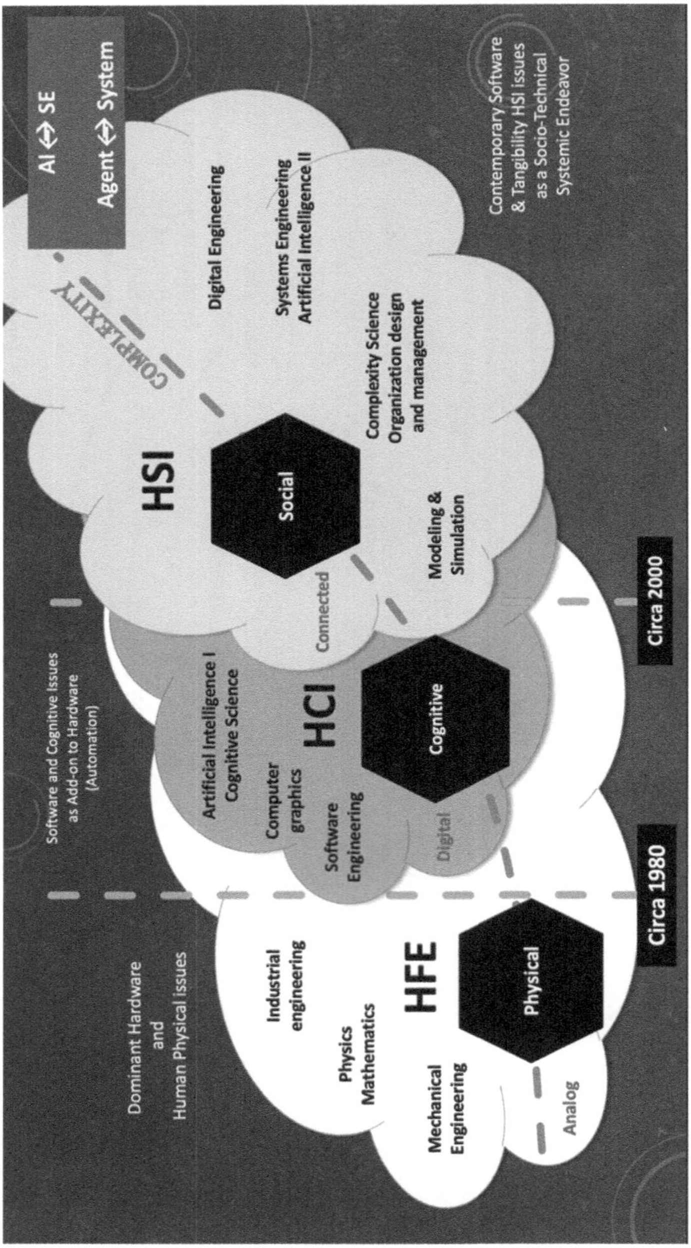

FIGURE 45.2 From HFE to HCI to HSI evolution.

HSI could be considered a new branding for HFE associated with industrial engineering. However, for many reasons, HSI redefines several concepts in human-centered engineering design. HSI emphasizes the crucial concept of integration from the start and throughout the life cycle of an STS. This involves the integration of various kinds of human and machine systems. The concept of "system" is now clearly defined as a representation of complex entities, whether natural or artificial. HSI has become a mandatory discipline with the growth of digitalization and artificial intelligence as it is possible to develop effective activity models, not only prescribed task models, throughout the life cycle of a system. We are incrementally moving toward digital societies where the interrelated roles of technology, organizations, and people still must be discovered, rationalized, and mastered. This handbook provides a contemporary contribution in this direction.

More specifically, social issues have become central to the evolution of STS's HSI, which can be called socio-ergonomics. Interconnectivity among people is now mediated through computer networks, making communication more artificial (i.e., we must learn how to interact using different languages). Communication is possible anytime, anywhere, remotely, asynchronously, written or spoken, etc. We have more flexibility than ever before, but we often miss face-to-face interaction that provides direct emotional information. If everything is rationalized, which could be suitable for work purposes, we often miss appropriate contextualized information automatically provided in face-to-face dialogs. Context has been a research topic for a long time, especially in HCI (Winograd, 2001). Well-done automation of people (e.g., when they follow procedures and checklists) and machines within mastered operational contexts increases safety, efficiency, and comfort. However, outside of these validity contexts, it can become dangerous and, in some cases, catastrophic. These validity contexts define rigid concepts of operations (CONOPS). Indeed, when operating in different contexts, people in charge need flexibility and competencies to solve problems. They need appropriate technology, called "FlexTech," as well as skills and knowledge that can be individual and collective. Based on the TOP (technology, organizations, and people) model, the aim of HSI must also be to increase flexibility throughout the life cycle of a sociotechnical system (Boy, 2021).

Digital engineering has become a well-established practice. It drastically changed the way engineering is done. During the 20th century, everything started with hardware development, and mechanical engineering was a prominent discipline. In the last decades of the 20th century, electronics and software have invaded mechanical systems to the point where software problems hide hardware problems. Within three decades or so, we have shifted from hardware to software. For example, only 50 years ago, we could repair a car engine problem by hand with a minimum of mechanical engineering knowledge and skills. Today, not only can you not do it yourself, but you also cannot do it without going to an automotive specialist who will connect your car to a machine that will tell you the cost of the repair, and if you do not ask, no one will tell you what it is. All of this is about software. Everything is fine as long as it takes place within the framework of the operations programmed into the computers. However, emerging digital engineering is shifting us into a new world where software is no longer an add-on but is the whole thing.

Digital engineering has inverted the way we used to do engineering. Everything is designed on a computer, using PowerPoint, modeling and simulation tools, video games, and today, digital twins. Everything can be tested in virtual worlds. Human-in-the-loop simulations (HITLS) can be developed early during design. Therefore, human–systems integration can be tested at design time. The reason is that CONOPS is not only a matter of task analyses (i.e., what is prescribed) but can be based on activity analyses (i.e., what is effectively done). Virtual prototypes can be used to observe such activity so that activity analysis results can be considered and integrated into design concepts and implementations. In addition, everything that is hardware-based starts in a virtual world and can be three-dimensional (3D)-printed to become real hardware. We go from software to hardware, which is the opposite of what we used to do. The complex challenge is now tangibility. Can we grab, physically and figuratively (i.e., understand what to do with it and the consequences of using it), this device safely, efficiently, and comfortably? Can we know how it works and how it can be used? This is a matter of figurative tangibility. In this Industry 4.0 framework, physical and figurative tangibility must be articulated appropriately to grasp STSs concretely.

Moving from technology-centered engineering to human-centered design (HCD) is not straightforward for several reasons. First, the engineering sciences have long been based on mathematics and physics, disciplines that have nothing to do with the humanities and social sciences. We often talk about hard and soft sciences. These connotations are generally not flattering for the so-called "soft" sciences. Engineering was a strong and highly valued discipline in the 20th century, so much so that the humanities and social sciences fell by the wayside in many of the sociotechnical fields under construction at the time. People had to adapt to technology more than the other way around. People invented tools to support their survival, but not necessarily to depend on them. Today, can we live without a smartphone? Not in our industrial and digital sociotechnical world. We have created dependencies that are difficult to avoid.

As digital engineering and, more generally, our digital society cannot operate without software, let us use it to design new tools that support a better life from safety, efficiency, and comfort perspectives. New methods and tools are being developed to support HITLS. As already said, HITLS enables activity analysis at an early stage of a sociotechnical system, where people and machines must adapt to benefit people in general. HITLS supports HCD in increasing physical and figurative tangibility and, therefore, the acceptability of new kinds of life. Without tangible demonstrations, nobody can decide correctly. This is why HITLS is essential. HITLS cannot be implemented without appropriate scenario-based design (SBD) techniques.

SBD consists of developing stories that could be anticipated as possible futures. The problem here is not a prediction but a test of anticipated possibilities. Predictions are based on projections from what exists today, based on causal descriptions of the past. A prediction is a derivative, a tangent to a historical curve. If we go too far on the tangent, we will make a big mistake in the prediction. This is the reason why prediction is always short-term and event-driven. When we deal with STSs, such as an aircraft and the infrastructure that goes with it, it is advised to anticipate possible futures that go beyond 30 or 40 years. This is a longer term! This kind of anticipation is goal-driven. This means that the people involved must plan for success. This planning deals with competencies, flexibility, coordination, and resilience. All these factors characterize the complexity of the STS being designed, refined, and maintained.

The complexity of STSs is undoubtedly a significant issue. It involves a deeper analysis of what can be separated from the system being investigated and what cannot. Once this separability analysis is done, the analyst must become familiar with the behavior and properties of the interconnected parts. Becoming familiar with a complex system involves discovering emerging behaviors and phenomena. The more familiar one is, the more STS maturity increases. STS maturity is not only a matter of technology readiness levels but also related to the maturity of practice and societal and/or organizational maturity (Boy, 2022b).

For example, we are producing highly complex cars that we want to use in the most straightforward manner possible. For more than four decades, HCI techniques and tools have been developed to increase the usability of user interfaces, but have we attacked the core business of HSI? That departs from the surface of single-agent interaction to deeper integration of acceptable multi-agent real-life scenarios and organizational configurations. Developing user interfaces is a process that belongs to an *inside-out* philosophy (i.e., going from the kernel of technology means to potential users' purposes) that tends to adapt people to technology. Current HSI approaches are devoted to taking the *outside-in* philosophy (i.e., going from people's purposes to means) that tends to adapt technology and organizations to people. In addition, the incremental nature of technology design and development tends to develop numerous layers of artifacts that complexify resulting STSs. A better understanding of the complexity of STS should lead to *simplexity*. "Amid order, there is chaos; but amid chaos, there is order" (Gribbin, 2005). Some contributors discuss *simplexity* (Berthoz, 2009; Pina e Cunha & Rego, 2010; Berthoz & Petit, 2014). This is why it is urgent to continue developing and mastering systemic methods and techniques toward HSI of different STSs.

By documenting the design process and its solution, HSI processes constantly improve the integration of emergent phenomena and properties into the STS at stake. Model-based systems engineering (MBSE) has been developed to support complex systems design, development, and manufacturing. However, it did not consider enough people and organizations in its development. This is why the primary purpose of HSI is to integrate technology, organizations, and people during the whole life cycle of an STS. We then discuss model-based human systems integration (MB-HSI) (Boy, 2022a). The underlying modeling requires an appropriate ontology to articulate the concepts and design processes. From a systemic perspective, this ontology is based on developing the concept of a system. We have seen that a system is a conceptual representation of any natural or artificial entity. Along these lines, HSI is a systemic approach to analyzing, designing, and evaluating STSs.

Context is still problematic to articulate, understand, and master. All systems have a contextualized purpose in a teleological sense. Sociotechnical systems should be thought of and analyzed in context. "Context thinking" gives operational meaning to systems built (in a constructivist sense), studied, evaluated, and used to support technical design and SE for HSI. Context may take several facets, including time, space, normal and abnormal situations, emergencies, exceptions, etc. Context may be goal-driven or event-driven. Without meaningful context, a system would be a dead description of a supposedly real-world entity. The context description is always linked to the purpose of the problem we are trying to identify, understand, rationalize, and finally attempt to solve.

Automation evolution toward human–AI teaming (HAT) leads to considering HSI phenomena and metrics such as trust, collaboration, and operational performance in various contexts, particularly in supervision, mediation, and cooperation between highly computerized and interconnected human machines. Given the growing complexity of our connected world, managing unforeseen situations throughout the life cycle of an STS (i.e., during design, development, industrialization, validation, operations, maintenance, and dismantling) is a significant challenge. Dealing with the unexpected requires anticipation at design time through HITLS, expert problem-solving at operations time, and experience maintenance always. We see that new activities (specifically jobs) must be better understood and considered at both the individual and organizational levels.

Finally, while SE and human factors and ergonomics (HFE) are different fields and disciplines, HSI forms a new bridge between the two. In the same way, no symphony can be played without music theory; HSI requires an ontology to have SE and HFE players work together toward making sustainable sociotechnical systems. This handbook shows the need to integrate technology, organizations, and people, and consider and respect our environment. This need is crucial today, as digitization has become more critical than physical mechanization. This brings the perspective of the history of industrial revolutions. The fourth one, Industry 4.0, is being replaced by Industry 5.0, where people, environment, and organizations must be central. Consequently, we should talk about Society 5.0.

REFERENCES

Berthoz, A. (2009). *La Simplexité*. Odile Jakob, Paris, France.

Berthoz, A. & Petit, J.L. (2014). *Complexité-Simplexité*. Editions du Collège de France, Paris, France. ISBN: 78-2-7226-0330-1.

Boy, G.A. (2021). *Design for Flexibility – A Human Systems Integration Approach*. Springer Nature, Switzerland. ISBN: 978-3-030-76391-6.

Boy, G.A. (2022a). Model-Based Human Systems Integration. In: Madni, A.M., Augustine, N., Sievers, M. (eds) *Handbook of Model-Based Systems Engineering*. Springer, Cham. https://doi.org/10.1007/978-3-030-27486-3_28-1

Boy, G.A. (2022b). *Socioergonomics: A Few Clarifications on the Technology–Organizations–People Triptych*. Presented at the INCOSE Human Systems Integration Conference, November 16–19, 2021. Wiley Online Library, vol. 32, no. S1, pp. 29–37. https://doi.org/10.1002/iis2.12869.

Gribbin, J. (2005). Profound *Simplicity: Bringing Order to Chaos and Complexity*. Random House, USA. ISBN-13: 978 1400062560.

INCOSE Systems Engineering Handbook, 5th Edition (2023). INCOSE-TP-2003 002 05. Wiley. https://dokumen.pub/qdownload/incose-systems-engineering-handbook-5nbsped-2023022915-9781119814290-9781119814306-9781119814313.html

Pina e Cunha, M. & Rego, A. (2010). Complexity, Simplicity, Ssimplexity. *European Management Journal*, vol. 28, no. 2, pp. 85–94. ISSN 0263-2373. https://doi.org/10.1016/j.emj.2009.04.006

Winograd, T. (2001). Architectures for Context. *Human–Computer Interaction (HCI) Journal*, vol. 16, no. 2, pp. 401–419.

Index of Authors

D

E

F

G

N

O

T

U

V

W

X

Y

Z

Index of Relevent Terms

S

T

U

V

W

For Product Safety Concerns and Information please contact our EU representative GPSR@taylorandfrancis.com Taylor & Francis Verlag GmbH, Kaufingerstraße 24, 80331 München, Germany